ISBN 978-1-332-31352-5
PIBN 10312747

This book is a reproduction of an important historical work. Forgotten Books uses
state-of-the-art technology to digitally reconstruct the work, preserving the original format
whilst repairing imperfections present in the aged copy. In rare cases, an imperfection in
the original, such as a blemish or missing page, may be replicated in our edition. We do,
however, repair the vast majority of imperfections successfully; any imperfections that
remain are intentionally left to preserve the state of such historical works.

1 MONTH OF
FREE
READING

at
www.ForgottenBooks.com

By purchasing this book you are eligible for one month membership to ForgottenBooks.com, giving you unlimited access to our entire collection of over 700,000 titles via our web site and mobile apps.

To claim your free month visit:
www.forgottenbooks.com/free312747

CLIMATE

AND

HEALTH RESORTS.

BY

J. BURNEY YEO, M.D.

"Peu de maladies guérissent dans les circonstances et les lieux où elles naissent, et qui les ont faites. Elles tiennent à certaines habitudes que ses lieux perpétuent et rendent invincibles. Nulle réforme (physique ou morale) pour qui reste obstinément dans son péché original."—*Michelet.*

NEW EDITION.

LONDON: CHAPMAN AND HALL,

LIMITED.

1885.

CHARLES DICKENS AND EVANS,
CRYSTAL PALACE PRESS.

TO

DR. PHILIP FRANK,

OF ÇANNES.

TO WHOSE FRIENDLY COUNSEL THESE STUDIES OWE MUCH,

THEY ARE GRATEFULLY DEDICATED.

59

PREFACE.

THE first edition of this book, which was published in the autumn of 1882, has been some time exhausted. The delay in the production of the present edition has been mainly due to the desire to make it more worthy of the favourable reception which was accorded to its predecessor, and, in part, to the many interruptions, in carrying out this intention, inseparable from an active professional life. When it is remembered that the first edition of this work consisted of only eight chapters, and when it is seen that this volume contains seventeen, and that some of the former chapters have been largely added to, it will be recognised that the production of this new edition has necessarily involved a considerable expenditure of time and labour.

I have endeavoured in the following pages to convey some information, reliable and accurate so far

as it goes, of every Health Resort of any importance at all likely to be visited by invalids from
this country. I have also attempted to give, as
occasion offered, a rational account of the action of
the various agencies, climatic and medicinal, which
come into operation in the different classes of resorts
treated of.

In a few instances, when it has not been
practicable to give an account of a Health Resort
amongst the groups dealt with in the several chapters
of which this volume is composed, a brief but
sufficient account of it has been introduced in the
Index.

I have only to express a hope that this edition
will be found to merit, in some measure, the very
kind and favourable opinions which were expressed
of the first.

My acknowledgments are due to the valuable
writings of Dr. Hermann Weber, Dr. Henry Bennet,
Dr. Sparks, Dr. Lombard, Dr. Jourdanet, Dr. Leudet,
Professor Beneke, Mr. Robert H. Scott, Professor
Geikie, Mr. Buchan, Dr. Cazalis, Dr. Brachet, Dr.
Brandt, Dr. Braun, Dr. Macpherson, and to many
others.

London, July, 1885.

PREFACE TO FIRST EDITION.

THE following studies of various Health Resorts have afforded me a congenial occupation during several of my annual holidays, and have most of them appeared, from time to time, in the pages of *The Fortnightly Review* or in the columns of *The Times* newspaper.

It has often been suggested to me by those, the correctness of whose judgment I could not doubt, that they would prove of some value and usefulness if they were republished in a collected form; in adopting this suggestion I have been careful to revise all these studies thoroughly, and to rewrite parts of most of them. The chapters on the Engadine are founded on a little volume I published in 1870, entitled "A Season at St. Moritz." This has been revised throughout, and to a great extent rewritten. It will no longer appear in a separate form.

CONTENTS.

CHAPTER VIII.

CHAPTER XVI.

CHAPTER XVII.

CLIMATE AND HEALTH RESORTS.

CHAPTER I.

CLIMATE AND CLIMATES.

WHAT do we mean by climate? What is the kind of information we require when we ask the question —what kind of climate has any particular place? We generally mean by a question of this kind that we desire to know how the natural conditions and surroundings of a particular place affect the life and well-being of its human inhabitants. That, practically, is what we want to know when we make inquiries about climate; for we are not concerned, in this place, with the geographical distribution of plants and animals, nor with any other considerations connected with climate, only in so far as they affect the well-being of mankind.

It is evident that the climate of a place must depend, first, on the conditions and character of its atmosphere—its aërial cloak—which may be hot or cold, moist or dry, still or agitated, pure or foul, uniform or variable, etc.; and, second, on the nature of its surface, which may be land or water, low or elevated, level or broken, barren or cultivated, sand, clay, rock, marsh, wood, meadow, country, or town, etc.

The characters and quality of the atmosphere differ considerably in different places on the surface of the globe. The atmosphere over Manchester differs from the atmosphere over a Yorkshire moor; the atmosphere over the table-lands of Central Asia differs from the atmosphere over the Indian Ocean; the atmosphere over the plains of Lombardy differs from the atmosphere in the Upper Inn Valley; and although these differences are not so considerable as to actually interfere with the maintenance of human life, they may be sufficient, at certain times and under certain circumstances, to materially influence the health and well-being of men.

The chief conditions of the atmosphere that affect the climate of a place are these : (1) its temperature ; (2) its movements (winds); (3) its amount of aqueous vapour ; (4) its electrical conditions ; (5) its purity ; and (6) its density or pressure.

We shall find on consideration that the following are the chief influences which determine the climate of a place :

1st and chiefest is its distance from the equator ;

2, its adjacency to, or remoteness from, seas or other large tracts of water ;

3, its elevation above the sea ;

4, the prevailing winds ; and

5, strictly local influences, such as configuration and inclination of surface, nature of soil and vegetation, absence or presence of plantations, cultivation, aspect, protection, drainage, population, manufactures, etc.

The most important factor in the determination of climate is undoubtedly *temperature,* and the temperature of a place is, with certain qualifications which will immediately be stated, dependent on its distance from the equator.

Those countries are hottest upon which the sun's rays fall vertically, or almost vertically, and the warmest climates are therefore those of the inter-tropical regions. The further we go from the equator the more and more obliquely the sun's rays fall on the surface, and for that reason the same amount of heat-rays are spread over an increasing extent of surface, while they also have to traverse a greater mass of air. The nearer we approach the poles, the less the amount of heat received from the solar rays, and these climates therefore are the coldest.

The influence of distance from the equator is well shown in some European countries in the difference of time in the flowering or ripening of widely-distributed plants ; at Naples the elm comes into leaf at the beginning of February, at Paris not until late in March, and in the centre of England not until the middle of April. In the south of Italy ripe cherries may be gathered about the beginning of May, in northern France and central Germany at the end of June, but not in England usually till three or four weeks later.

Were it not then for counteracting influences we might say, that, universally over the surface of the globe, *temperature is regulated by latitude.* But this statement is by no means of universal application ; indeed, places which are on the same parallel of latitude rarely enjoy exactly the same temperature, and in some instances there are wide divergences.

If, for example, we compare London and Labrador, which coincide in latitude, we find that while in both the summers are mild, in England the winters are not very cold, while in Labrador they are excessively so.

By means of numerous thermometric observations taken at various parts of the world, at different times and seasons, it has been possible to construct maps, upon which lines are drawn through all places having the same temperature at the same season, or the same average annual temperature, and these *isothermal lines* as they are called show the general distribution of temperature over the globe.

Isothermal lines, or lines of equal temperature, are named after the degrees of temperature they express, e.g., "the isotherm of 60 degrees" means the line drawn through all the places on the map which have the average annual temperature of 60 degrees Fahrenheit.

Now it is found that these isothermal lines, drawn round the globe, instead of following the parallels of latitude, bend up and down, and these bendings are determined by the place of continents and oceans. They are noticed to be least irregular over the wide expanse of ocean in the southern hemisphere, while they show the greatest deflections across North America, the Atlantic, Europe, and Asia. It is thus seen that temperature is more uniform and more directly dependent on latitude in the oceanic parts of the globe than in the continental, or where the oceanic and continental come together as they do in the basin of the Atlantic.

If we take the isotherm of 50 degrees Fahrenheit (mean annual temperature) viz., that passing through London, it continues in nearly a straight line westward to the coast of Wales, it then bends southwestward, and crosses to the west coast of Ireland. If we trace it to the opposite side of the Atlantic, we shall find that in crossing the ocean it bends much further south, and reaches the coast of America, near New York, so that the mean annual temperature of London and New York is the same though New York is as far south of London as Madrid.

On reference to a map of isothermal lines, it will be seen that these divergences of belts of equal heat from the same parallels of latitude are determined by the manner in which the great areas of land and sea are grouped.

" Land gets sooner heated by the sun's rays than the sea, and also gives off its heat again sooner. The sea, though it does not get so hot as the land does, retains its heat longer, and is enabled by virtue of its liquidity and motion to diffuse it. Hence the influence of the sea tends to mitigate both the heat and the cold of the land. Its warm currents heat the air resting on them, and so give rise to warm winds which blow upon the land, while its colder waters in like manner temper the air, which reaches the land in cooling breezes, or it may be in cold, damp winds and fogs. Thus, in the basin of the North Atlantic, a warm ocean current, called the Gulf Stream, issues from the Gulf of Mexico, and, augmented by the surface-drift of warm water which is driven onward by the prevalent south-west winds, flows across the Atlantic to the shores of Britain, and even of Spitzbergen. It brings with it the supplies of heat which make the climate of the west of Europe so much less cold than it would naturally be. On the other hand, an icy stream of water coming out of Davis Strait, brings a chill to the coasts of Labrador and Newfoundland. The ocean, therefore, by its cold currents is depressing the temperature in America along the same latitudes, where in Europe, by its warm currents, it is raising it." *

Dr. Carpenter, however, thinks that the Gulf Stream exerts less influence in this way than what he calls "the general oceanic circulation," consisting in the main of an under-flow of cold Polar water from the Polar regions towards the equator, and an upper-flow of warm, equatorial water from the equator towards the poles. Warm water is in this way constantly carried along the surface from hot to temperate and cold regions. This surface stream on reaching the Polar regions is gradually cooled, and then sinks to the bottom and returns as the under-flow of Polar waters to the equator. This deep, cold current, however, comes to the surface in different places, and gives rise along certain shores to cold currents and bands of cold water. The Mediterranean, being an inland sea, is free from the influence of the Polar currents, and has, even at its greatest depths, a temperature of 54 to 56 degrees Fahrenheit; while in the Atlantic, outside the Straits of Gibraltar, the water, at the same depths, is only 36·5 to 37 degrees Fahrenheit.

Again, the influence of a large tract of land situated in high latitudes lowers the temperature below what it would be if the same regions were occupied by

---

* Geikie's "Physical Geography," p. 61.

sea, because it allows of the accumulation of vast
masses of snow and ice; and a similar tract of land in
low latitudes, by exposing a broad, motionless, and
quickly-heated surface to the tropical rays of the sun,
gives rise to a far higher temperature than would be
observed over a tract of ocean in the same region.   It
thus happens that the isothermal lines for June,
representing temperatures below the freezing point,
extend a considerable way southwards, over northern
Asia, and also over Greenland and North America;
but in crossing the ocean between these regions
they bend considerably northwards.   On the other
hand, if we look at the isothermal lines for July, we
shall see that over the equatorial parts of America
and the greater parts of Africa and southern Asia the
lines of 80 degrees swell out considerably, and include
a much larger area than where they traverse the ocean.

But we must not conclude that two places which
have the same average *yearly* temperature have the
same climate.   It has been found that Reykjavik, in
the south of Iceland (64 degrees 40 minutes north
latitude), has a mean yearly temperature of about 38
degrees Fahrenheit, and that Quebec has one of about
40 degrees; but in July the mean temperature of Quebec
is 70 degrees, and of Reykjavik, 51 degrees, and in
January, the mean temperature of Quebec is 12
degrees, while that of Reykjavik is 30 degrees; so
that while the south of Iceland in winter is often
without frost, in Quebec it is intensely cold, with 20
degrees of frost, whereas in summer in Quebec it is
19 degrees warmer; this is due to the circumstance
that Canada is chilled by the cold land and sea lying
to its north-east, while Iceland is warmed by the ocean
current which flows along its shores all the year round.

Again, Dublin and Munich have nearly the same
annual means of temperature, viz., 48 to 49 degrees
Fahrenheit; but the winters at Munich are 9 degrees
colder than in Dublin, and the summer about 6
degrees warmer.

In order then to compare the climate of different

places on the surface of the globe, it is necessary not only to know the mean annual temperature, but also how the temperature is distributed through the different seasons, and maps are prepared for this purpose, with isothermal lines showing the distribution of temperature for each month or season.

"Range maps" have also been constructed, enabling us to contrast different localities with regard to the extent of variation, or range, of temperature which they experience in the year, which is a very important factor in the determination of climate.

These "range maps" show that the regions of very great annual range are all situated within the land areas of the northern hemisphere.

Near Jakutsk in Siberia there is a district which has the enormous range of 100 degrees Fahrenheit. Jakutsk itself, in the hottest season, has a temperature of 65·8 degrees Fahrenheit, and in the coldest −44·9 degrees. Northumberland Sound in America has a range of 75 degrees, from 36·7 to −38·6 degrees.

By far the greater portion of the surface of the globe "enjoys a very uniform temperature throughout the year, the range nowhere exceeding 20 degrees Fahrenheit. This region extends nearly up to the 60th parallel on the west coast of North America, and beyond it in the neighbourhood of the Faroe Islands in the Atlantic. Almost the whole of the sea surface of the globe belongs to it, as well as the major portion of South America and South Africa."*

These "range maps" show that:

(a) The range increases from the equator towards the poles, and from the coast towards the interior of a continent: (b) the regions of extreme range in the northern hemisphere coincide approximately with the districts of lowest temperature in winter: (c) the range is greater in the northern than in the southern hemispheres: (d) in the middle and higher latitudes of both hemispheres, with the exception of Greenland

---

* Scott's " Meteorology," p. 341.

and Patagonia, the western coasts have a less range
than the eastern : and (*c*) that in the interior of ,
the continents the range in mountainous districts
diminishes with the height above the sea.

On inspection of such a map, the influence of the
sea in moderating extremes of climate is unmistakable,
" but even more decidedly do the agencies of prevailing winds and prevailing currents show their
effects.   Where the prevailing winds are westerly, the
so-called anti-trades, and where the ocean currents are
flowing from the equator into higher latitudes, the
cold of winter is mitigated, and the curves of equal
range bent polewards.   On the other hand, on the
eastern coast of America, and even more so on that
of Asia, the prevailing winds are northerly and cold,
so that the temperature in winter falls very low, while
the summer is comparatively warm, and the contrast
between the opposite seasons is very marked."*

We thus see how temperature is regulated, and
the effect of distance from the equator modified by
the *distribution of sea and land.*

The temperature of the air is also regulated, and
the effect of distance from the equator modified, by
the *elevation of a region*, by its height above the sea.
It is a matter of common observation, that in ascending mountains the air gets cooler and cooler the
higher we go; "in ascending a mountain from the
sea-level to the limit of perpetual snow, we pass through
the same series of climates, so far as temperature is
concerned, which we should do by travelling from the
same stations to the polar regions of the globe ; and
in a country where very great differences of level
exist, we find every variety of climate arranged in
zones according to the altitude, and characterised by
the vegetable productions appropriate to their habitual
temperatures." †

It may be taken as a somewhat rough average,

* Scott's " Meteorology," p. 342.
† Herschel's " Physical Geography," p. 249.

that the decrease of temperature with elevation is about 1 degree Fahrenheit for every 300 feet ascent; so that *height above the sea* is an important element in climate. There is an apparent exception to this observed in very cold, calm weather, when, in a valley, the cold, heavier air sinks to the lowest level, and the warmer and lighter air rises on the hill sides, so that a residence on a slight eminence, or on the side of a hill, secures immunity from the greatest severity of the frosts. Evergreens suffer less in such situations than in low-lying bottoms.

Before we consider the important influence of the movements of the atmosphere, i.e., of the prevailing currents or winds on climate, it will be desirable to consider shortly the influence of atmospheric pressure or density, and of atmospheric humidity, as it is on them that the movements of the air are dependent.

The use of the barometer to foretell changes of weather is founded, as everybody knows, on the fact that variations in atmospheric pressure give rise to winds and storms, and all those movements of the air upon which weather depends.

But if we ask, in the next place, what causes these variations of atmospheric pressure, we shall be told that they are greatly affected, first by temperature, and secondly by aqueous vapour.

We can readily understand how temperature acts. Air when heated expands, when cooled it contracts; therefore cold air is heavier than warm air, and warm air ascends while cold air descends.

" The ascent of warm air must necessarily diminish atmospheric pressure. When a broad tract of the earth's surface, such for instance as the centre of Asia, is greatly heated by the sun's rays, the hot air in contact with the ground rises and flows over into the surrounding regions. Hence the atmospheric pressure is lowered there during the hot months of the year."

The influence of the presence of water-vapour on the pressure of air is due to the circumstances that

water-vapour is very much lighter and has less pressure
than air ; so that air saturated with vapour of water is
lighter than dry air, and the difference in weight increases with the temperature, because the warmer the
air the more water-vapour can be dissolved in it.
For instance, a cubic foot of perfectly dry air at 32
degrees Fahrenheit weighs a grain and a quarter more
than a cubic foot of air saturated with vapour at the
same temperature, whereas a cubic foot of perfectly
dry air at 80 degrees Fahrenheit weighs six and a
half grains more than air saturated with moisture at
the same temperature.

"The vapour which rises so abundantly from sea
and land into the atmosphere diffuses itself through
the air, pushing the air atoms aside, and having far
less weight and more elasticity than the air, it necessarily lessens the density of the atmosphere, or, in
other words, lowers the atmospheric pressure."*

The amount of aqueous vapour in the air is constantly varying, the addition of a large volume of
vapour to the atmosphere lowers its pressure and
causes the mercury to fall; the removal of this vapour
by its condensation into rain for instance, restores the
pressure and the mercury rises.

"How it is that these changes in the volume of
vapour in any part of the atmosphere are brought
about is still unknown. It is well ascertained, however, that they determine movements of the air. When
sudden and extensive they are accompanied by
storms of rain and wind ; when less violent they still
show their influence upon the winds and weather."†

Every surface of water exposed to the air gives off
vapour into the atmosphere so long as it remains un-
saturated ; when the air reaches its point of saturation,
evaporation ceases, and the higher the temperature of
the air, the greater its capacity, as we have seen, of
absorbing moisture. Wind greatly favours evaporation, for it blows away the vapour from the surface

* Geikie's "Physical Geography," p. 55.   † Ibid.

of the water as it is formed, and brings other and drier air to absorb and carry off the fresh supplies.

It is clear, then, that more vapour of water must be added to the air during the day than during the night, and during the summer than in winter, during a brisk wind than in still weather, and in far greater amount in warm tropical regions than in temperate or cold ones.

The water which passes into the air in the form of this invisible vapour is condensed, and reappears in such visible forms as dew, rain, snow, clouds, and mists.

The presence of this water-vapour in the atmosphere has a very important influence on the temperature of the surface of the earth. It surrounds the earth with a kind of cloak, sometimes invisible, sometimes in the visible form of cloud, and it serves to protect it from the too great intensity of the solar rays during the day, and from the too rapid loss of heat by radiation into cold space during the night, when the influence of the sun is removed. "If it could be removed for a little from around us, we should be burnt up by day and frozen hard at night."

It is well known that when water passes into the state of vapour it absorbs a considerable amount of heat which is rendered *latent* or imperceptible, and that when the vapour is again condensed into water, the latent heat is again given out and becomes sensible. "Every pound of water which is condensed from vapour liberates heat enough to melt five pounds of cast iron." When then, in nature, the vapour of the air is converted into water on a large scale, the temperature of the air is thereby considerably raised. Rainfall is dependent on the amount of aqueous vapour which passes into the air, and its subsequent condensation, it is therefore usually greatest in tropical regions, where the amount of evaporation is greatest, and it diminishes in amount as the temperature falls toward the poles. This rule is, however, subject to important exceptions, dependent on the distribution of sea and land, and on the great aërial currents.

As condensation is more active over land than over sea, the rainfall also is greater over land than sea, and over the northern hemisphere, much of which is land, than over the southern hemisphere, most of which is water. Most of the vapour of the atmosphere being furnished by the ocean, it follows that the condensation of vapour into rain upon the land is greatest at the coast-line, so that the sea-board of a country may be rainy while its interior is comparatively dry. Rainfall is much influenced by the form of the surface ; as mountains act as condensers, they are therefore much wetter than plains. " Places which lie in the path of any of the regular air-currents are wet when they cool the current, and dry when they warm it. Hence winds blowing towards the equator, since they come into warmer latitudes, are not usually wet winds; but when they blow towards the poles they reach colder latitudes, and are chilled and therefore rainy.

" Some of these laws are well illustrated in the British Islands, where the rains are chiefly brought by the south-westerly winds, which have come across the Atlantic. The coast-line facing that ocean is more rainy than the east side looking to the narrow North Sea. In the former part of the country, away from the hills, the amount of rain which falls in a year would, if collected together, have a depth of from 30 to 45 inches. On the east side, however, the average annual rainfall does not exceed 20 to 28 inches. Where the western coast happens to be mountainous, an excess of rain falls ; hence the wetness of the climate along the north-west coast of Scotland and in the lake district of England, where the annual rainfall ranges from 80 to 150, and sometimes even more than 200 inches."*

Owing to the enormous evaporation which goes on between the tropics, a constant stream of vapour is rising into the upper regions of the atmosphere, and

* Geikie's " Physical Geography," p. 77.

becoming chilled there, falls in the form of heavy and frequent rains.

If a high mass of land, in these rainy regions, lies in the path of the warm, moist air-currents, the rainfall is still greater. In India, where the south-west monsoons, laden with vapour from the Bay of Bengal, encounter the Khasi range of hills, which stretches across the course taken by these winds, the annual rainfall amounts to from 500 to 600 inches ; for these winds as they spread up the hills into higher and cooler air, have their moisture at once precipitated as rain.

On the other hand, a tract of country lying behind a high mass of land upon which moist winds blow may be almost rainless. This is the case with the country on the east or lee side of the range of the Western Ghats in India, for these mountains, lying in the pathway of the warm, moist monsoons from the Indian Ocean, interrupt a heavy rainfall reaching on the hill tops to 260 inches annually, while places at the foot of these hills have only 26½ inches.

So again in Peru, owing to the high chain of the Andes precipitating all the moisture which blows from the east across the continent of South America, the winds which descend upon it are so dry that rain is almost unknown. "Another and much greater rainless tract is seen in the desert regions from North Africa across Arabia and far into the heart of Asia. In these parts of the earth's surface, the dry sandy soil is raised to an intense heat during the day. There is little or no water to evaporate, the hot dry air ascends, but the winds which blow in upon the deserts cannot deposit any moisture, for, instead of being cooled, they are heated and dried up in the ascending currents."

When it happens, as it does in some countries, that in one part of the year the wind blows in one direction, and in another part of the year in the opposite, we shall find that these periodical winds are usually accompanied with rain when they blow from warm to colder regions, and with dry weather when they blow from a cold to a warmer region, therefore in these countries rainy seasons alternate with dry ones.

Great irregularity of rainfall is characteristic of north-western Europe, and of temperate arctic regions generally ; but as a rule there is more rain at the end of autumn and in winter than in summer.

So far as the health of human beings is concerned, the cleansing effect of rain on the air is important, for a downfall of rain washes the air and carries away suspended and other impurities which, in dry weather, are blown about and diffused by the aërial currents.

The influence of prevailing winds on climate has already been mentioned, and it has been shown that the movements of the air are produced by differences of pressure. The following is the law which governs the direction of these movements : *Air always flows in spirally from areas of high pressure into areas of low pressure.* It is clear that this must be the case, as low pressure indicates a deficiency, and high pressure a surplus of atmosphere. The column of air is heavier in the latter case than in the former, consequently obeying the universal law of gravitation, the heavier column must necessarily flow out at the base to supply the deficiency in the lighter one.

A familiar and instructive illustration of the in-fluence of temperature in producing currents of air can be observed on the sea-coast of any country, where the days are warm and the nights cool. The surface of the land during the day, under the influence of the sun's rays, becomes much warmer than that of the sea, and thus heat is communicated to the air resting on it, which becomes hotter than that resting on the sea. The hotter air on the land expands, its pressure is diminished, and being lighter it ascends, while the cooler air from the sea streams in towards the land to take its place ; thus a light breeze, blowing from the sea on to the land, is developed, which increases in force as the day advances. This is the familiar *sea breeze!* As the sun disappears it dies away, for as soon as the sun is set, the land parts with the heat it has absorbed during the day by radiation into space, and it does so much more readily than the sea does; thus the air over the land becomes cooler, and therefore denser than that over the sea, and this denser air moves towards the sea to take the place of the ascending lighter current now flowing upwards from its surface; and thus a *land breeze* arises. This increases in force, as the difference in temperature between the air over the land and that over the sea increases by continued radiation from the former; this again dies away towards morning, when the sun's heat again becomes felt on the land, and the temperature of the air over land and sea becomes equalised.

A somewhat similar illustration may be observed in mountainous districts; during the day the air on the mountain sides, heated by the sun's rays, ascends, and a breeze blows up the mountains from the valleys; while at night, the cold, heavy air on the mountains flows down as a cool breeze into the valleys.

But it is in the tropical regions that the great influence of temperature and aqueous vapour in originating aërial movements can be studied on a large scale. "The air round the tropical belt of the globe is constantly receiving an enormous volume of water-vapour from the ocean. From this great influx of vapour, as well as from the high temperature along that belt, there is a continued ascent of heated and humid air into the higher regions of the atmosphere, and a consequent in-draught from north and south. It is called the belt of equatorial calms, a somewhat misleading name, because, although there are no constant winds, the air is in ceaseless disturbance as it gets heated and streams upwards." This hot, moist air expands and cools as it ascends, and loses much of its moisture, which becomes condensed and falls as rain.

When it reaches the higher regions of the air, it divides, one part flowing northward, the other southward, these upper currents moving in an opposite direction to that flowing in at the surface. They travel towards the poles until, reaching the belt of high pressure, they descend and reach the surface in temperate latitudes; there it streams partly back along the surface to the equator, and partly towards the pole. This is the fundamental aërial circulation of the globe.

If the earth were at rest, instead of rotating on its axis, these currents would flow uninterruptedly directly north and south; but, "by the rotations of the earth from west to east, the direction of these winds is modified, so that in the northern hemisphere the main current, blowing from the north, is transformed into a north-eastern current, and in the southern hemisphere the main wind from the south becomes a south-eastern wind (trade winds); while the main current in the upper regions, flowing from south to north in the northern hemisphere, appears as a south-west wind, and that from north to south in the southern hemisphere, a north-west wind. From the meeting of the main currents flowing in opposite directions, many intermediate winds result."* Thus it happens that in Britain and the west of Europe the prevailing winds are the familiar south-west winds.

The large masses of land in the northern hemisphere interfere considerably with the regular distribution of the aërial currents as they are observed over the broad unbroken expanse of ocean presented by the southern hemisphere. "In January the high and cold table-lands of Central Asia become the centre of a vast area over which the pressure of the air is high. Consequently, from that elevated region the winds issue on all sides. In China and Japan it appears as a north-west wind. In Hindostan it comes from the north-east. In the Mediterranean it blows from the east and south-east. But in July matters are reversed, for then the centre of Asia, heated by the hot summer sun, becomes part of a vast region of low pressure, which includes the north-

---

* Hermann Weber's "Climate and Health Resorts."

eastern half of Africa and the east of Europe. Into that enormous basin the air pours from every side. Along the coasts of Siberia and Scandinavia it comes from the north. From China, round the south of the continent to the Red Sea, it comes from the Indian Ocean, that is, from south-east, south, and south-west. Across Europe it flows from the westward. Hence, according to the position of any place with reference to the larger masses of sea and land, the direction of its winds may be estimated."*

Monsoons is the Arabic name (meaning any part or season of the year) given to the summer and winter winds on the shores of the Indian Ocean. We have seen that the air is drawn in towards the centre of Asia in summer, and flows out from the centre in winter; so in India the winter wind is the north-east Monsoon, which corresponds to the north-east Trades of the North Atlantic and North Pacific Oceans; whereas the summer wind is the south-west Monsoon, which is a complete reversal of the natural course of the Trade Wind, owing to the enormous indraught caused by the low summer pressure over Asia. On the Chinese coast the winter wind is a north-west Monsoon, and the summer wind a south-east Monsoon. Something similar occurs in North America, for in the Southern States the winter wind comes from the north-east, the summer wind from the south-west.

There are certain well-known *local winds* which blow in certain countries, and which have received local names, of which the following are the chief. If these winds come from a tract over which the pressure is high and the temperature low, to a region where the pressure is lower and the temperature higher, they come as cold blasts condensing the humidity in the air of the warmer region into torrents of rain. But if the area of low pressure surrounds hot desert regions, like Africa and Arabia, it draws out

* Geikie's "Physical Geography," chap. xi., "The movements of the air."

towards it the hot air lying over these burning sands ; such a wind is that known in Italy as the *Sirocco*—a hot, moist wind, causing extreme languor to men and animals. In Spain it is known as the *Soleno*, and it sometimes brings with it to this country, across the Mediterranean, fine hot dust from the African deserts. This wind from the desert is known in Africa and Arabia as the *Simoom*, and it sometimes blows with such violence as to whirl up clouds of sand in which whole caravans have been buried. A similar hot wind is encountered on some parts of the West Coast of Africa (Guinea), in December, January, and February, blowing from the interior out to sea ; this is the *Harmattan*. In Egypt there is a similar wind called *Khamsin* (fifty), a hot and very dry wind laden with fine sand ; it prevails for about fifty days in spring. The *Fohn* is a warm and dry wind frequently met with in the north-eastern cantons of Switzerland, and is generally very much dreaded on account of the physical and mental depression it produces. (Its prevalence during the past winter, 1884–85, at Davos, has been very prejudicial to many of the winter visitors there.) Some difference of opinion exists as to its source and origin, but it certainly blows on the Italian side of the Alps as a warm and southerly wind, laden with moisture, owing to its contact with the surface of the Mediterranean. Its moisture becomes precipitated in its passage over the Alps, and it is felt in Switzerland as a warm and dry wind.

The numerous names and other significations given to local winds on the Mediterranean coasts will be referred to in the chapter on the Riviera.

Winds often serve a useful purpose in distributing temperature and moisture. A wind blowing from a warm or mild region raises the temperature of the district to which it comes. The prevailing west and south-west winds of Great Britain, for example, are warmed by their passage over the Gulf Stream, and keep our climate much milder than, from its latitude, it ought to be. But when a wind blows from a colder to a milder region it produces a depression of temperature; thus it is that the winds that blow from the vast cold expanse of elevated land in central Asia westward into Europe, make the weather in winter and spring colder and drier there than when west winds blow.

Winds also distribute the moisture of the atmosphere, and if it were not for them the condensed vapour of the air would be discharged upon the same area from which it had been evaporated; but the winds convey vapours from the sea, and these become condensed on the land, so that, speaking generally, the wetness or dryness of any place will depend on the direction from which its prevalent winds come. If they come over a wide and warm expanse of sea, they will bring moisture, if they come from the interior of the continent they will be dry, and hot or cold according to the temperature of the regions from which they come.

The preceding considerations show how climate, *generally*, is influenced: (1) By distance from the equator; (2) by distance from the sea; (3) by elevation above the sea; (4) by the prevailing winds; and (5) by less general local conditions; and that in estimating the climate of any particular place, we have to take into consideration *its temperature*, not only its annual mean and annual range, but its seasonal and even its daily variations; its *rainfall*, and not only its annual amount, but the manner of its precipitation, whether in short and sudden torrents, or prolonged over considerable periods, and its distribution in the different seasons and months; its average atmospheric humidity, amount of cloud and mist; its *elevation* above the sea; the direction of the *prevailing winds*, during the different seasons of the year; and, finally, *local conditions* such as the presence or absence of protecting hills, forests and mountain chains, the shape and position of the ground, its relation to adjacent masses of water, sea, lake or river, the nature of its soil, whether porous and absorbent or the reverse, cultivation, vegetation, population, etc., etc.

It will also be seen from the foregoing that the elementary division of climates into *insular or oceanic* and *continental* is well founded. Owing to the fact that water absorbs heat more slowly, and parts with it more slowly than land does, the ocean and other large masses of water act as store-houses of heat which they absorb slowly during the hot seasons, and give out again slowly during cold seasons to the air lying on their surface; it follows, therefore, that the climates of the sea and of the coasts of land are much more moister and more equable than those of the interior of continents, and in proportions as places are distant from the sea, their climates become more extreme. "An insular and oceanic climate is one where the difference between summer and winter temperature is reduced to a minimum, and where there is a copious supply of moisture from the large water-surface. A continental climate is one where the summer is hot, the winter cold, and where the rainfall is comparatively slight." This difference is "well brought out by the fact that such evergreens as the Portugal laurel, aucuba, and laurustinus grow luxuriantly even in the north of Scotland, while they cannot withstand the severe cold of the winter at Lyons."

Little more need be said on these points. With regard to that important element of climate, the *prevailing winds*, we have seen that winds which come from a cold region are cold, and those from a warm region warm. Sea winds are usually moist, those from the land generally dry. Sea breezes are not subject to the same extremes of temperature as those from the land. The vapour they contain serves to cool the heat of summer and diminish the cold of winter ; whereas winds from the interior of a continent are usually hot and enervating in summer, bitterly cold and dry in winter. We have also seen that winds which come from lower into higher latitudes, or from warmer to colder regions, deposit their moisture in the form of rain and are therefore wet winds, while those which come from higher to lower latitudes, or from cold to warm regions, are dry winds.

Of *local* influences, one of the most important is the nature of the soil. Where the ground is wet and marshy the mean temperature is lowered, for the water absorbs and conveys downwards the heat which would otherwise be retained on the surface and warm the soil; when such ground is properly drained the mean annual temperature is found to rise. This rise of temperature from efficient drainage has been known, in our own country, to raise the annual average as much as 1·5 to 3 degrees Fahrenheit, which is as great a change as if the ground had actually been transported 100 to 150 miles further south.

A sandy desert presents the greatest extremes of climate, " for while the dry surface readily absorbs the sun's heat so as to rise even to 200 degrees Fahrenheit during the day, it cools rapidly by radiation, and during a clear night may grow ice-cold."

The presence or absence of vegetation is also an important local element of climate. When the surface is covered with vegetation the heating effect of the sun's rays on the soil is necessarily diminished, at the same time that the radiation of heat, during the night, from the surface into space is hindered, so that the soil is neither heated nor cooled as much as it would be if it were bare ; and as leaves never become so hot as the soil, the presence of vegetation is an equaliser of temperature. The influence of a large forest on the local climate is, therefore, to moderate both the heat of day and the cold of night.

The adjacency of lakes and other large inland surfaces of water also exercises a similar equalising effect. As the temperature of the water on the surface becomes lowered by the cold of winter, this cold surface water descends to the bottom, and the deeper, warmer water rises to the surface; as this becomes cooled in its turn, it sinks and allows another portion of warmer water to take its place. The temperature of the air lying over the water is thus raised above that lying on the adjacent land, and the colder, heavier air over the surrounding land flows down on to the surface of the water, displacing the warmer air and becoming itself warmed in turn.   Thus it is that deep lakes, which do not freeze over in winter, "serve as reservoirs of warmth to keep the temperature of the surrounding ground higher than that of places only a short distance away."   The reverse happens during summer, when the water cools the air on its surface and so lessens the heat of the locality.

We have seen how the neighbourhood of hills and mountains may act on local climates, either by serving as a screen or protection from prevailing winds, or by cooling the upper air and precipitating rain, or in causing local currents of air moving alternately up and down valleys, or in generating cold gusts of wind which rush down from the hills on to the plains.

There are also local influences affecting the *composition* of the air, which have an undoubted influence on the well-being of men.   It is well known that the air on the sea-coast and in the open country is purer and richer in oxygen than it is in the densely crowded districts of populous cities; but especial attention has been given to the amount of *ozone* in the atmosphere of a place—which is a condensed and more active form of oxygen—as a test of its salubrity. It is found to be absent in sick-rooms, and in the neighbourhood of substances undergoing decomposition; it is less abundant in the interior of large towns than in the open country ; it occurs in greater amount over green fields and woods, than over sterile plains or dusty roads; on the sea-shore, than on inland plains ; on the tops of mountains, than in valleys or level tracts.   Ozone possesses great powers of oxidation and disinfection ; hence its presence, in relatively large proportions, is an indication of the freedom of the air from substances prone to decomposition.

It has also been said (Professor Binz), that it gives a soporific property to the air, and on that account many persons sleep better on the sea-shore and in the mountains where the air is rich in ozone. The contrary is however the case with some persons.

The presence or absence of floating organic particles in the air has, no doubt, a great influence on its salubrity. Dr. Angus Smith has calculated that in Manchester the air that a man breathes in ten hours contains 37,000,000 of spores! Whereas, the air over the sea and in high mountain regions is usually remarkably free from these organic impurities.

Various attempts have been made to classify climates with more or less want of success; for, in most instances, what has been gained in precision has been lost in accuracy.

We may, however, accept as well-grounded, the fundamental division into:

A.—Sea or insular climates.

B.—Inland or continental climates.

The suggested subdivision of climates into (a) humid climates, and (b) dry climates, rather indicates approximately the characters of the climate of different places, than affords a basis for accurate classification, and the following division is partly comprised in the foregoing: (*a*) climate of plains; (*b*) climate of altitudes.

No less imperfect is the classification founded on the annual distribution of temperature into (1) hot, (2) cold, and (3) temperate climates, although it has a certain practical value, when qualified by a knowledge of the seasonal and diurnal variations of temperature.

A.—*Sea or insular climates* are represented by the climates of the open sea, of small and moderate sized islands, and of sea-coasts. These have much in common, although they, of course, vary according to latitude and to those other determining conditions of climate which we have considered.

They agree in possessing the following characters:

1. Their atmosphere is usually freer from organic and inorganic impurities than that of inland plains.

2. Owing to the constant evaporation from the surface of the sea, their atmosphere is comparatively moister than that of inland regions and the amount of atmospheric humidity is less variable.

3. Their temperature is more equable than that of inland climates, there is less difference between summer and winter and day and night temperatures.*

B.—*Inland or continental climates* differ from sea and coast climates in being less equable and more exposed to extremes, to great heat in summer and severe cold in winter. The east coasts of continents usually share in these extremes.

The difficulty attending the further subdivision of climates is at once seen when we consider that the suggested classifications of climates into (a) humid and (b) dry climates very nearly coincides with the preceding, *humid* climates being usually sea or insular or coast climates, and dry climates being ordinarily continental or inland climates. There are of course exceptions determined by local conditions. Then again in the division into (*a*) climate of plains and (*b*) climate of altitudes, it is obvious that the first of these will coincide with sea or coast climates if the plains are adjacent to the sea, and with continental climates if situated inland.

Mountain climates, or the climate of altitudes, is fully considered in the next chapter ; and other practical questions connected with climate will be discussed in subsequent chapters.

* In the next chapter the character of sea and coast climates is entered into more fully.

CHAPTER II.

SEA OR MOUNTAIN?

A STUDY OF THE ACTION OF SEA AND MOUNTAIN AIR.

THE results of recent investigations as to the relative influence and value of sea and mountain climates as remedial and invigorating agencies are of much interest. The restorative properties of sea air have long been fully appreciated, although regular and periodical migration to the sea-shore is a custom of modern origin. The popularity of mountain health resorts is, however, of quite recent date, and much has still to be learnt from careful observation and experiment as to the exact nature of the influences at work in them, and the precise limits of their application.

This is not a question of narrow professional import, but it is one of those practical physiological studies upon which educated persons may desire, and may be expected, to form just and correct ideas.*   It is, I believe, a somewhat prevalent notion that sea and mountain air are widely different in their mode of action; that they are, as it were, the extremes of climatic influences. This, however, is not the case. There is much that is common to both of them in their action on the human organism.

---

* This chapter was originally published in the form of an article in "The Fortnightly Review."

The results, indeed, of precise experimental observation on this subject are perhaps a little at variance with what we might, at first sight, have been led to anticipate. An attempt to determine experimentally the difference in the action of sea and mountain air was made by Professor Beneke, of Marburg, in 1872.* He had already established, by observation and experiment, that exposure to the air of the North Sea (his observations were made in the isle of Norderney) produced an appreciable acceleration of the nutritive changes in the nitrogen-containing tissues of the human body. In more simple language, it helped us to "throw off the old man," to get rid of our old material and to put new stuff in its place. By what precise means it led to so desirable a result he had not been able to satisfy himself. Was it the abundance of ozone in the air? Was it due to the influence of the strong reflection of light from the sea? Or was it simply a stimulating psychical effect? The phenomena observed were not sufficiently accounted for by either or all of these suggested influences. It occurred to him that he might establish some basis for a satisfactory explanation of these results if he could ascertain the relative proportion in which bodily heat was lost, in a given time, in sea air and in inland air. Experiments on the human organism itself were of little avail for exact observations, since they must inevitably be complicated by the heat-regulating processes within it. He therefore constructed the following simple apparatus, by which the loss of heat from a heated body, under various external conditions, could be observed :

A thermometer was suspended in a glass flask, into which water at a temperature of 50 degrees Centigrade was introduced, and then it was ascertained how long, under various external conditions, it took for the water to cool from 45 to 35 degrees. The influence of

* "Deutsches Archiv für Klinische Medicin." March, 1874.

clothing in interfering with the loss of heat was also
tested by enveloping the flask, first with shirting, then
with linen and flannel, and finally with shirting and
a double layer of flannel. The observations with this
apparatus were made, first in a closed room in the
island of Norderney, then outside the house in the
midst of the village, and then on the shore of the
island ; and these were compared with like observations in a closed room in Marburg, and on a terrace
in the professor's garden there. All these observations
gave the same result, viz. that in equal or even higher
temperatures of the air, the flowing-off of heat occurred
much more rapidly on the sea-shore than inland ; a
circumstance which Professor Beneke refers, first, to
the high degree of saturation of sea air by moisture,
and secondly, to the intensity of the currents of air on
the sea-shore. And he infers that the beneficial influence of the North Sea air on the human organism
is due, in great part, to the increased loss of heat it
occasions from the surface of the body. In answer to
the objection that the same effect would be produced
by a cold bath or by exposure to air of a low temperature anywhere, he rightly replies that the peculiar
effect of the sea air is, that it withdraws heat in a
more gradual and continuous manner, that its currents gently stimulate the surface, and thus a steady
restoration of the heat lost is produced without causing
any great tax on the reactionary forces of the body,
so that weakly persons may be exposed with perfect
safety, for hours together, to this cooling, and, at the
same time, reconstituting process.

The next point the professor desired to ascertain
was, how the loss of heat from the apparatus described
above would be affected by exposure to mountain air
at different altitudes, and accordingly he made a
series of observations at the following places : On the
Schienige Platte, near Interlaken, 5800 feet above the
sea, the temperature of the air ranging from 9·5 to
13 degrees Réaumur, it took 91·5 minutes to produce
the same loss of temperature which was brought about

in 53 minutes, temperature of air 13 degrees Réaumur (in 35 minutes during a storm), on the sea-shore of Norderney; on the Wengern Scheideck, 6370 feet above the sea, the temperature of the air ranging from 5 to 7 degrees Réaumur, the same amount of cooling took 68·5 minutes; on the great Scheideck, 6036 feet, the temperature of the air ranging from 5 to 8 degrees Réaumur, 90 minutes. The next three points of observation were lower. They are well-known health resorts. On the terrace of the hotel at Bürgenstock, on the Lake of Lucerne, 2900 feet above the sea, temperature of the air 7·5 to 8·5 degrees Réaumur, the same loss of heat was produced in 73 minutes. At Engelberg, 3109 feet, temperature of air 10 to 10·5 Réaumur, it took 69·25 minutes. At Seelisberg, 2336 feet, temperature of the air 11·5 to 12·5 Réaumur, 94·5 minutes. The last observation was made on the Rigi Staffel, 5048 feet, temperature of the air 7 degrees Réaumur—*a violent storm*, he says, was raging, such as one only expects to find on the sea-coast—and the same amount of cooling took 64 minutes.

Professor Beneke thus establishes the fact that heat is lost from the selfsame apparatus more slowly on the tops of mountains than on the shore of the North Sea ; and this notwithstanding that on the tops of the mountains the temperature of the air was almost constantly lower, a circumstance which would have led us to expect a more rapid loss. He tells us also that his observations were made at times when there was a considerable amount of moisture in the air, so that the slower loss of heat could not be referred to the dryness of the air, nor to the lesser intensity of the currents, for a violent storm was blowing during the observations on the Rigi Staffel. It remains to be determined whether it is due to the rarefaction of the air—whether rarefied air is a much worse conductor of heat than air on the sea-shore.

These observations appear to justify the following inference : Since the activity of tissue-changes will

correspond with-the lóss of heat, the greater the loss
of heat the greater will be the activity of change of
tissue, i.e. the greater the stimulus to nutritive changes.
Hence in mountain air these nutritive changes are
comparatively much less active than on the shore of
the North Sea. And Professor Beneke's practical conclusions are that individuals in whom the processes of
tissue-change do not require hastening are, *cæteris
paribus*, better off on mountain heights than on the
sea-coast. Highly irritable, nervous organisations,
people who, as we say, take too much out of themselves, profit more by mountain than by sea air. For
those, on the contrary, who have no tendency to
nervous irritability, and who are in a condition to
bear the increased stimulus to tissue-change, sea air
is a more powerful restorative agent. Hence the
greater proportion of scrofulous persons and those
exhausted by overwork, who retain some activity of
the digestive organs, should prefer the sea-side.

But although these general conclusions of Professor
Beneke's are probably in the main correct, there are
many other considerations to be attended to in determining the relative value, in individual instances, of
sea and mountain air. I have, however, thought it advisable to call attention, at some length, to these really
valuable observations and suggestions of Beneke, as
they are almost the only experimental researches that
have been hitherto published on this interesting and
important practical question.

I shall now proceed, in the first place, to consider
in detail what are those properties of sea air to which
it owes its special influence on the human organism.
The presence of ozone in sea air in greater proportion
than in the air of inland plains is well established.
This is a property which it shares with mountain air.
Its greater abundance on the sea-coast depends, in all
probability, on the influence of sunlight, which is one
of the most important sources of ozone. Vegetation
is also a source of ozone, and it is therefore found in
excess in forest air; where, therefore, we find pine-

forests on the sea-coast, as at Arcachon and Bournemouth, we may look for an unusual excess of this
hygienic agent.   Experience has thoroughly established the fact that where the amount of ozone in the
air is constantly high, there we almost invariably find
a high degree of salubrity.   It purifies the air by
destroying injurious gases, and especially by determining the oxidation of decomposing organic substances.   It promotes nutrition and blood-formation
by supplying to the respiratory organs a most active
form of oxygen.   The excess of ozone in sea air is,
therefore, one of its most important properties, as it is
also one of the most important properties of mountain
air.

Another hygienic property which sea air shares
with mountain air is the absence in it of organic dust.
This applies with especial force to the air of the open
sea, or on small islands, or to points of land standing
well out into the sea.   If people build a large town
on the sea-coast, which becomes densely populated,
organic impurities will tend to accumulate over the
thickly-inhabited area ; and when the wind blows off
the land such impurities may be wafted to a little
distance off the coast.   But as the sea presents an
ever-moving fluid surface, no impurities in the shape
of organic dust can rest upon it, so as to be again
blown about, in mischievous activity, with every fresh
breeze.

Equableness of temperature is another characteristic of sea air, and one to which it owes much of
its beneficial influence in many cases.   In this respect
it is contrasted with the air of elevated regions, in
which the diurnal variations of temperature are often
very considerable.   The temperature of the sea-coast
is warmer in winter and cooler in summer than that
of inland districts.   This admits of easy explanation.
In the first place the rapid cooling of the surface
of the land by radiation into space, after the sun has
gone down, is checked by the amount of moisture
in the air.   The aqueous vapour which is abundant in

sea air absorbs the heat given off from the soil during nocturnal radiation, and acts as a kind of screen to retard the loss of heat in this way. Hence great variations between the day and night temperatures are very rarely observed at the sea-side.

"Wherever the air is dry," says Professor Tyndall, "we are liable to daily extremes of temperature. By day, in such places, the sun's heat reaches the earth unimpeded, and renders the maximum high; by night, on the other hand, the earth's heat escapes unhindered into space, and renders the minimum low. Hence the difference between the maximum and minimum is greatest where the air is driest. In the plains of India, on the heights of the Himalaya, in Central Asia, in Australia, wherever drought reigns, we have the heat of day forcibly contrasted with the chill of night. In the Sahara itself, when the sun's rays cease to impinge on the burning soil, the temperature runs rapidly down to freezing, because there is no vapour overhead to check the calorific drain." It is a matter of common observation that, in the interior of continents, where the rainfall is small, the heat of summer and the cold of winter are greater than at or near the coast.

During the heat of the day the air over the sea is always cooler than that over the land; for the surface of the land gets rapidly heated and communicates its heat to the superjacent strata of air; but "when the sun's rays fall on water they are not, as in the case of land, arrested at the surface, but penetrate to a considerable depth," so that water is heated much more slowly by the sun's rays, as well as cooled more slowly by nocturnal radiation than the land. Moreover, the evaporation which is always going on at the surface of the sea, and going on rapidly where the sun's rays are powerful, carries away some of the heat of the surface-water, and helps to keep the air in contact with it cool.

Much of that feeling of agreeable *freshness* in the air at the sea-side during hot weather is due to currents of air produced by this inequality in the heating and cooling of the atmosphere on the land and over the sea.   As the day advances and the land becomes heated by the sun's rays, it heats the air on its surface, which thus becomes lighter and ascends, while the cooler and heavier air lying on the sea flows in to take its place, and so a refreshing sea breeze is generated. During the night the land is rapidly cooled, especially if the night be clear, by radiation into stellar space, and the air lying on it is cooled also, and thus becomes heavier than the warmer air over the sea, and so it happens that in the morning and early part of the day a gentle breeze is found blowing off the land towards the sea.

But the influence of the sea in equalising the temperature of the air is exercised in another very interesting manner.   "Over the surface of the ground slanting to the sea-shore the cold currents generated by radiation flow down to the sea, and the surface-water being thereby cooled sinks to lower depths. In the same way, no inconsiderable portion of the cold produced by radiation in all latitudes over the surface of the ocean and land adjoining, is conveyed from the surface to greater depths."

On account of this equableness of temperature, oceanic climates—the most equable of all climates— are said to afford almost absolute immunity from colds.*   It is only on board ship that such a climate in its perfection can be found.   A very near approach to it, however, may be obtained on such very small islands as, for instance, the isle of Monach, about seven miles to the west of the Hebrides, and fully exposed to the prevailing westerly winds of the Atlantic.   The mean January temperature of this island, which is nearly in the latitude of Inverness, is 43·4 degrees Fahrenheit, or 1·8 degree higher than

---

* It has also been observed that, after a sea voyage, on landing, there is much greater proneness to take cold, and extra precautions are necessary to guard against it.

the mean of January at Ventnor. On the other hand, the mean temperature of July is at Monach 55 degrees, and at Ventnor 62·6 degrees, so that in January Monach is 1·8 degree warmer than Ventnor, in summer it is 7·6 degrees cooler.*

But these two characteristics of sea air—an equable temperature and a high degree of saturation with moisture—are soothing rather than bracing properties, and if it were not for the currents of air induced on the sea-coasts they might be found actually relaxing, and this is no doubt the case in warm and cloudy weather on our own south-western shores. In these respects, therefore, sea air offers a great contrast to mountain air.

The same is the case with the next property of sea air I propose to consider, viz., its density. The absolute density of sea air is of course greater than that of the air at any higher level, and it must therefore contain bulk for bulk more oxygen, and it follows that in breathing sea air we take more oxygen into the lungs in a given time than in the air we breathe at places above the level of the sea; that is, supposing in both cases we breathe with equal frequency and equal amplitude. But it does not necessarily follow because an absolutely larger quantity of oxygen exists in a given volume of sea air than in the same volume of mountain air, that more oxygen, on that account, is taken into the blood at the sea-side than on higher ground. In the first place, the oxygen may be, for aught we know, in a more active form in mountain than in sea air; its chemical energy may be greater, and therefore the nutritive changes dependent on respiration may be accelerated, though the air be thinner and poorer in its absolute quantity of oxygen; or, in the second place, the respiratory act may be so much increased in frequency on the mountains, that although less oxygen is taken into

---

* See article "Climate," in the "Encyclopædia Britannica." Monach is a very small island with very few inhabitants!

the lungs at each breath, yet much more may be received into the organism in a given time. Moreover, if we compare the density of sea air with the density of the air inland, at places situated only a few feet above the sea level, as, for instance, the greater part of London, the difference would be so insignificant as really to merit very little consideration.

But disregarding, for the present, the absolute density of sea air, a more important point to be attended to is the great and frequent variations of barometric pressure met with on the sea and on sea-coasts. Now it has been shown by careful experiment that all rapid variations in atmospheric pressure increase the activity of the circulating and respiratory organs, and that the perfection of organic life depends on these alternations of excitement and repose. We are justified, then, in assuming that rapid changes in the barometric pressure are more favourable to vital functional activity than its relative stability.

It has also been shown that the barometric variations at the sea-side, besides being greater in amount than inland, occur with far more regularity, a circumstance which is regarded as tending to promote the accommodation of the organism to its new conditions.

These, then, are the most important properties of sea air: (1) Excess of ozone; (2) excess of aqueous vapour and equability of temperature; (3) great purity and absence of organic particles; (4) maximum density and great but regular variations of barometric pressure. Of minor importance are the presence of saline particles suspended in the air, which, of course, vary greatly in amount, according as the sea is calm or agitated, and probably exercise a mildly stimulating effect on the respiratory mucous membrane. The small amount of iodine and bromine diffused in sea air may not be without a real influence on some organisms.

Leaving, for the present, any further investigations into the effects of sea air and its usual concomitant sea-bathing, I propose in the next place to examine, also in detail, the characteristic properties of mountain air. And here, at the very outset of our inquiry, we come upon a very remarkable contrast. There was no need to define what we meant by sea air, although its effects, as I shall have to point out hereafter, may be greatly modified by circumstances of locality. But are we always sure what we mean when we use the term mountain air? In Scotland and Wales we speak of mountain air at a few hundred feet above the sea, considerably below the level of the towns of Lucerne or Geneva. In Germany we hear of mountain air at 1200 and 1500 feet above the sea, and in the Engadine at 6000 feet, in Mexico at 12,000! Now if we think only of one quality of mountain air, viz., its rarefaction, it is quite clear that we must be using the same term to express very different things. But if we are thinking only of the *general* bracing effects of mountain air we may find these, no doubt, at very various elevations, and we may even find them in great perfection at comparatively low levels. An open plateau in a temperate climate at an elevation of 2000 or 3000 feet above the sea will certainly possess a more bracing air than a close valley in a hot climate at twice that height.

If we confine our attention to the continent of Europe we may take the Upper Engadine (about 6000 feet) as the extreme limit of a permanently inhabited, and perhaps habitable, mountain district. (The village of Cresta in the Aversthal, 6295 feet above the sea, is reckoned the highest in Europe.) For all practical purposes of comparison we may take an elevation of 6000 feet as the limit in one direction of a habitable European mountain climate, and in the other direction such elevations as Heiden, above the Lake of Constance, 2660 feet; Glion, above Montreux, 2900 feet; and Seelisberg, 2400 feet, on the Lake of Lucerne. Places at a lower elevation than these, although they may have many advantages as health resorts, can scarcely be admitted into the category of mountain climates. Of localities such as these, then, ranging between 2000 and 6000 feet above the sea level, we have, within tolerably easy access, a great number to choose from ; while there are a few, for exceptional needs and for short periods of residence, between 6000 and 8000 feet.

There seems good reason to believe, as I shall hope to show presently, that at higher elevations than these the air reaches a degree of rarefaction which is inconsistent with the maintenance of vigorous health.

Diminution of atmospheric pressure is, then, one of the chief properties of mountain air, and the relative proportions of this diminution must necessarily, *cæteris paribus,* have much to do in determining the hygienic character of any particular mountain station, and its suitability to different individuals. It has been calculated that at an elevation of 2500 feet we lose about one-eighth of the atmospheric pressure, at 5000 a sixth, at 7500 feet a fourth, and at 16,000 a half.

Another important property of mountain air is its lower temperature. It is a very well known fact that the temperature of the air diminishes in proportion to the altitude. From observations made in the Alps of Switzerland the medium loss of temperature was 1 degree Centigrade (= 1·8 degree Fahrenheit) for every 520 feet of elevation during summer, and for every 910 feet during winter. Whence it follows that the tops of mountains are relatively much warmer in winter than in summer. It has, however, been pointed out that there are "extraordinary modifications, amounting frequently to subversions of the law, of the decrease of temperature with the height," owing to the circumstance that " the effects of radiation will be felt in different degrees and intensities in different places. As the air in contact with declivities of hills and rising grounds becomes cooled by contact with the cooled surface, it acquires greater density, and consequently flows down the slopes and accumulates on the low-lying ground at their base. It follows, there-

fore, that places on rising ground are never exposed
to the full intensity of frosts at night ; and the higher
they are situated relatively to the immediately surrounding district the less they are exposed, since their
relative elevation provides a ready escape downwards
for the cold air almost as speedily as it is produced."
Hence a southern slope at a considerably greater
elevation may have a higher night temperature than
a neighbouring plateau.   "On the other hand, valleys
surrounded by hills and high grounds not only retain
their own cold of radiation, but also serve as reservoirs
for the cold heavy air which pours down upon them
from the neighbouring heights."   And at the numerous
meteorological stations in Switzerland it is observed
that "in calm weather in winter, when the ground
becomes colder than the air above it, systems of
descending currents of air set in over the whole face
of the country.   The direction and force of these
descending currents follow the irregularities of the
surface, and, like currents of water, they tend to
converge and unite in the valleys and gorges, down
which they flow like rivers in their beds.   Since the
place of these air-currents must be taken by others,
it follows that on such occasions the temperature of
the tops of mountains and high grounds is relatively
high, because the counter currents come from a greater
height and are therefore warmer."   So the "gradual
narrowing of a valley tends to a more rapid lowering
of the temperature, for the obvious reason that the
valley thereby resembles a basin almost closed, being
thus a receptacle for the cold air-currents which
descend from all sides.   The bitterly cold furious
gusts of wind which are often encountered in
mountainous regions during night are simply this
out-rush of cold air from such basins."*

Considerations such as these are of the greatest
importance in determining the hygienic character of
any particular mountain health resort.

* Article " Climate," "Encyclopædia Britannica." New edition.

The question of the humidity or dryness of
mountain air is one not easy to resolve. The air on
the summits of high mountains is no doubt drier than
the air at lower levels. But at intermediate levels,
considerations other than those of altitude alone
determine the relative humidity or dryness of the
atmosphere ; so that each mountain station must, to
a great extent, be judged of by itself with regard to
this very important point. Perhaps, as a general rule,
one may say that the higher the locality the less rain
falls ; but, on the other hand, we have to face the
startling fact that twice as much rain and snow fall
at the St. Bernard and St. Gothard stations as at
Geneva ! Much will, however, necessarily depend on
the configuration of the ground, as well as its aspect.
A mountain ridge facing the direction from which
moist winds habitually blow will condense their
moisture and precipitate it in the form of rain or
snow on its sides, or on the valleys or plains at its
base ; while more remote summits of the same
mountain chain and the higher mountain valleys at
their bases may be thus protected and screened from
heavy and prolonged rainfalls.

Thus the moist Atlantic winds blowing against the
western ranges of Scotland and Cumberland determine
the great rainfall in these regions ; and the town of
Santa Fé de Bogota in the Andes, at an elevation of
8600 feet, is visited with almost incessant rain, owing
to its situation at the foot of a mountain on the sides
of which the warm trade winds of the South Pacific
Ocean become cooled, and condense their moisture.

"Ces phénomènes de pluie et d'humidité excessive," says M. Jourdanet, "observés en différents
points élevés, ne détruisent nullement la réalité
habituelle de sécheresse des altitudes. Ils sont la
conséquence exceptionnelle de conditions topogra-
phiques desquelles résultent, sur une localité, l'arrêt
tourbillonnant et l'ascension sur les flancs des montagnes de vents chauds et humides qui condensent
leurs vapeurs en pluie par le refroidissement."

The presence or absence of vegetation will also exercise a determining influence as to the relative humidity of the atmosphere. We must, therefore, bear in mind that certain topographical conditions will frequently induce, in stations of considerable altitude, a moister atmosphere than is found in the neighbouring plains. But if we consider the effect of altitude alone, it is easy to understand how the air of elevated regions must be, *cæteris paribus*, drier than that of lower situations.

In the first place, the lower the atmospheric pressure the more rapid is the process of evaporation, and hence the boiling-point of water is 28·3 degrees Fahrenheit less on the top of Mont Blanc than at the sea level.

Secondly, the energy of the sun's rays, and therefore their drying effect on the atmosphere, is greater the less the thickness and density of the layer of air they have to traverse. The slope of the soil, the absence of vegetation at great heights, and the greater intensity of the aërial currents all tend to promote dryness of the atmosphere.

Lombard * appears to think that we may distinguish two zones in mountain climates—an upper or dry zone, and a medium or humid zone; their limits varying greatly according to latitude, aspect, and configuration of the soil. For European climates he considers the dry zone to extend from about 3500 to 4500 feet upwards ; and the humid zone, where the air is moister than it is in higher or lower regions, to extend from an inferior limit of from 1600 to 2000 feet up to 3500 or even 4500 feet. It is difficult to see the value of a distinction which has such ill-defined limits.

Mountain air differs then from sea air in three main particulars : (1) In its diminished density; (2) in its lower temperature ; (3) in containing less humidity. The temperature is not only lower than that of sea

---

* "Les Climats de Montagnes."

air, it is also less equable.   Owing to the clearness of
the air, the absence of moisture, and the energy of the
sun's rays, very great differences between the day
and night temperature are constantly found at great
elevations.   There is but little aqueous vapour in the
air to prevent nocturnal radiation into stellar space
from the surface of the soil, greatly heated during the
day by the solar rays ; thus there is usually a rapid
fall of temperature when the sun goes down.   In
summer a difference of 40 to 50 degrees Fahrenheit
between the day and night temperatures will sometimes be registered.   There is often also a very great
difference between the sun and shade temperatures
during the day.

Mountain air resembles sea air in containing an
excess of ozone, in its freedom from organic and
other impurities, in being cooler than the air of inland
districts, and in the fact that its monthly and annual
variations of temperature are less than on inland
plains.

The study of mountain climates has hitherto
taken the form, chiefly, of an investigation into the
physiological effects of diminished atmospheric
pressure on the human organism.   Since different
individuals are very variously endowed with the
power of accommodating themselves to altered
external conditions, it is not to be wondered at
that some discrepancies are to be found in the
statements of different observers as to the effects
upon themselves and others of alterations of atmospheric pressure.   Even different animals seem to
possess very different degrees of sensitiveness in
this respect.   The cat appears to be the most
sensitive of animals in this particular ; it cannot
exist at an elevation above 12,000 or 13,000 feet.
Attempts to acclimatise it at Potosi, a town in Bolivia,
about 13,000 feet above the sea, have failed.   At this
elevation it is said to be attacked by very remarkable
tetanic fits, commencing, at first, as slight irregularities
of muscular movement, as in St. Vitus's dance, and

gradually becoming stronger and stronger, inducing the poor animals to make violent leaps, as if they wished to climb up the rocks or the walls of the houses; after violent efforts of this kind they fall exhausted with fatigue, and expire in a convulsive seizure. In the town of Mexico, about 7300 feet above the sea, efforts to introduce the cat, M. Jourdanet tells us, have been more successful. He mentions the attempt of a French lady, who imported a couple of white Angoras. He says: " They rapidly lost their habitual gaiety. They bred, however, but their young family was reared with difficulty, many of them dying in their earliest infancy (drowned, so to say, in rarefied air l). Those who survived had a dejected appearance, not the gay and lively aspect natural to kittens. Most astonishing thing of all, they were all of them deaf." The long-suffering dog, however, abounds in Mexico, and before the conquest the natives used to eat them.*

Jourdanet maintains that persons who are not accustomed to a rarefied atmosphere begin to suffer inconvenience when they attain an elevation of between 6000 and 7000 feet. Most of those who have reported their experiences of mountain ascents in Europe (I am not, of course, alluding to mountaineers *in training*) have not experienced any noticeable inconvenience until they reached nearly 10,000 feet. Soldiers going to Himalayan stations at 7500 feet complain at first of shortness of breath, and have a quicker and more feeble pulse; but these effects are temporary. Of the serious effects of exposure to the highly rarefied air of very considerable elevations we have most valuable evidence in the records of the balloon ascents of Mr. Glaisher. Acceleration of the pulse was one of the first effects noted. At 16,000

---

* There appears to be no getting out of reach of that enterprising little animal, the flea. We are assured that on the passes of the Himalayas, at an elevation of 18,000 to 19,000 feet, he is to be found in great activity.

feet it had risen from 76 to 100. Between 18,000 and 19,000 feet both Mr. Glaisher and his companion suffered from violent palpitations with difficulty of breathing ; then their lips and hands became of a deep blue colour. As they continued to ascend their respiration became more laborious. On another occasion, at 27,000 feet, Mr. Glaisher became unconscious. The attack came on with indistinctness of vision, inability to move arms or legs, though he could move his neck ; then he lost his sight completely, though he could still hear his companion speak, but he could not answer him. Then he became wholly unconscious. He also describes a feeling of nausea, like sea-sickness, coming on at great elevations.

The following are the various symptoms that have been recorded by many different observers as occurring during the ascent of lofty peaks or on elevated plains. Great loss of muscular power, palpitations, quick and laborious respiration, bleeding from the nose or gums, drowsiness, severe headache, nausea and vomiting, great thirst, mental depression, enfeebled senses, and impaired memory. The superficial veins become distended, the face pale and bluish. These symptoms were aggravated by exertion and mitigated by rest. Another significant symptom, reported on good authority, both in mountain and balloon ascents, is *increasing coldness of the body* beyond what would be accounted for by the lower temperature of these elevations.

It seems certain, then, both from the evidence of such actual observations as I have referred to, and from the experimental observations of M. Bert in the laboratory of the Collège de France, that when the rarefaction of the air reaches a certain degree, the due oxygenation of the blood is interfered with, and we get symptoms developed which point to oxygen-starvation, and to obstruction in the circulation through the lungs. In M. Bert's experiments it appeared that slight degrees of diminution of atmo-

spheric pressure did not lessen the affinity of the
aërial oxygen for the blood corpuscles; but when
that diminution approached or reached one quarter
of the whole atmospheric pressure, perceptible disturbances ensued.

M. Jourdanet,* who gives a full account of M.
Bert's experiments, concludes that the oxygenation
of the blood is not injuriously affected by residence
at an elevation below 6500 feet. Above this elevation, he believes the respiratory functions become
disturbed, and the due oxygenation of the blood is
interfered with. He proposes to restrict the term
"mountain climates" to places not exceeding 6500
feet in altitude, and to higher regions he gives the
title of "climats d'altitude." Moreover, he maintains
that those who live all their lifetime at great elevations, as, for example, the natives of the various towns
on the high plateaus of Mexico, Bolivia, and Peru,
are by no means striking examples of health and
vigour. They are, according to his experience,
especially prone to suffer from anæmia and the disturbances of health associated therewith — pallor,
breathlessness, palpitation, vertigos, dyspepsias, and
neuralgias! Lombard also tells us that the monks
of St. Bernard, after several years' residence there,
present various signs of anæmia, and these are
occasionally so grave as to necessitate a removal to
the plains.

Not less important than its rarefaction is the
dryness of mountain air. Dryness of the air has an

---

* I refer to his elaborate treatise published in 1875, with the
title "Influence de la Pression de l'Air sur la vie de l'Homme,"
a work in two large and profusely-illustrated volumes, which
would have been much more valuable than it is had it been less
diffuse. Much that Dr. Jourdanet writes is from personal
observation, as he resided for many years in the mountainous
regions of Mexico. But what can he know personally of the
"Séparation des hommes au pied de la Tour de Babel," of which
he presents us with an engraving, "d'après les indications de
l'auteur"?

important influence on the activity of the bodily
functions.   These "are facilitated," says Mr. Herbert
Spencer, in some interesting remarks on this head,
"by atmospheric conditions which make evaporation
from the skin and lungs tolerably rapid. . . . If
the air is hot and moist the escape of water through
the skin and lungs is greatly hindered ; while it is
greatly facilitated if the air is hot and dry.   Needful
as are cutaneous and pulmonary evaporations for
maintaining the movement of fluids through the
tissues, and thus furthering molecular changes, it is to
be inferred that, other circumstances being alike,
there will be more bodily activity in the people of
hot and dry localities than in the people of hot and
humid localities. . . .   The evidence justifies this
inference.   The earliest recorded civilisation grew
up in a hot and dry region—Egypt ; and in hot and
dry regions also arose the Babylonian, Assyrian, and
Phœnician civilisation."   He further points out that
from the "rainless district extending across North
Africa, Arabia, Persia, and on through Thibet into
Mongolia, have come all the conquering races of the
old world. . . . .   These races, widely unlike in
type, and speaking languages deemed as funda-
mentally distinct, from different parts of the rainless
district have spread as invaders over regions relatively
humid.   Original superiority of type was not the
common trait of these races; the Tartar type is
inferior as well as the Egyptian.   But the common
trait, as proved by the subjugation of other races,
was energy.   And when we see that this common
trait in races otherwise unlike had for its concomitant their long-continued subjection to these special
climatic conditions—when we find, further, that from
the region characterised by these conditions the
earliest waves of conquering emigrants, losing in
moister countries their ancestral energy, were overrun
by later waves of the same races, or of other races
coming from this region—we get strong reasons for
inferring a relation between constitutional vigour and

the presence of an air which by its warmth and dryness facilitates the vital actions."

But mountain air is not only drier than sea air and the air of inland plains, it is also colder. Now this lowering of temperature tends, to a certain degree, to compensate for the deficiency of oxygen dependent on its elevation. For instance, in a given volume of air at 1400 feet above the sea, at a temperature of 32 degrees Fahrenheit, there is as much oxygen as in the same volume of air at the sea level at 60 degrees Fahrenheit. So that such virtues as are lessened in mountain air by its elevation are, in part, restored by its coldness. And this leads me to speak of what I have always believed to be an important modification of mountain air. I mean the air in mountain districts that is found on the surface of vast glaciers. The contact of an enormous refrigerating mass, such as an extensive glacier is, with the lower strata of the air over it, has, I take it, two necessary effects upon that air. First, it makes it drier than the air over the adjacent country, because it must tend to condense whatever aqueous vapour there is in the air on to its surface, where it remains frozen. Secondly, it must exercise a certain amount of condensing effect on the air itself—on these strata in immediate contact, or very close to it—so that we breathe thicker, denser, richer air on a glacier than we do on the land near it, at the same elevation. Thus the air over a glacier may be compared to a can of milk turned upside down—in which the cream accumulates at the bottom instead of at the top. Whoever has walked much on glaciers, in elevated districts, must have noticed that they breathe with increased freedom, and with less effort as soon as they get well on to the glacier. Some have thought this simply a psychical effect; but I think I have observed it again and again when it was impossible to associate it with anything other than a purely physical influence. I have, therefore, great confidence in the restorative and tonic effect of glacier air for persons who retain a fair amount of muscular strength and activity ; and I consider the adjacency of a great glacier, of tolerably easy access, a great recommendation to a mountain health resort.   This is one of the advantages which belongs to Pontresina, in the Upper Engadine.   The great Morteratsch glacier is within about an hour's walk of the village ; and after the little climb that is necessary in order to get on it, there is a vast field of glacier easily traversed in all directions, extending for miles, and rising very gently along the whole distance until the broken part of this immense ice-stream is reached.

Having thus considered in detail the properties of sea and mountain air, having noted in what particulars they agree and in what important points they differ, we are now prepared to approach the consideration of the following highly practical questions : Who should go to the mountains? who should go to the sea ? and who should go to neither ? I should like to answer the last question first.   I believe there is no greater mistake made than that very general one of sending *all convalescents* to the sea-side, except the still greater one of actually embarking them on a sea voyage !   It arises from the very natural desire to hasten convalescence after acute disease.   I am now speaking exclusively of convalescents from acute diseases.   But these unwise attempts to hasten convalescence are the very frequent cause of serious relapses.   In the general debility which follows a fever or an acute inflammation all the organs share—the organs of nutrition, the secretory, the circulatory, the eliminatory organs, are all feeble and unable to do much work without exhaustion.   If an attempt is made to over-stimulate them, if an appetite is induced before digestive power has been regained, a feverish state is frequently re-excited, and the very effort that has been made to hasten recovery retards it.

Sea and mountain air are alike too stimulating and exciting for such cases.   They arouse to premature activity when the organism can strengthen itself only by absolute repose. "How *poor* are they that have not patience" was never so applicable as to cases such as these. Pure, unexciting country air, in a locality where the patient can be thoroughly protected from cold winds, and where he can "bathe in the sunshine or slumber in the shade"—that is the safest and best place for the invalid to slowly, but steadily, regain health after severe acute disease. Sea or mountain air may, however, be needed later on to promote recovery from the chronic affections which occasionally follow acute ones, and then sea air is probably the more appropriate of the two.

Speaking generally, those who seek health in high mountain districts should be capable of a certain amount of muscular activity. Those who suffer from great muscular debility as well as general exhaustion, and who need absolute or almost absolute repose, are unsuited for mountain climates. Such climates are too rigorous, too changeful, too exciting; and the persons to whom I now allude, when they find themselves in the cold, rarefied, exciting mountain air, feel out of place and become chilled, depressed, and dyspeptic. One also finds such persons amongst those whose desire for mental activity is somewhat in excess of their mental power, especially when this is combined with a feeble physique; or amongst those who incessantly and heedlessly work a strong though not exceptionally vigorous brain. Such persons need for a time much repose, and they will find renovation with repose *by* the sea, or, still better, in a yachting trip *on* the sea.

There are others, however, who, with vigorous frames and much actual or latent power of muscular activity, become mentally exhausted by the strain of incessant mental labour, anxious cares, or absorbing occupations. Mental irritability usually accompanies this exhaustion, great depression of spirits, with unrest of mind and body. These are the typical cases for the mountains. The stimulus and object which they afford to muscular exertion; the bracing atmosphere rousing the physical energies and re-awakening the sense of powers unimpaired and unexhausted ; the soothing effect of the quiet and stillness of high mountain regions, and the absence of the human crowd—all these influences bring rest and renovation to the over-worn mind.

It is important to remember that the same individual may, at different times and under different conditions, be differently affected by sea and mountain air. If he happens to be the victim of an irritable and exhausted nervous system, the result of overstrain, he will, probably, be benefited by removal to the mountains ; if, on the other hand, he should be slowly recovering from chronic disease, and especially from certain surgical maladies, or after surgical operation, where the processes of tissue-change require hastening without necessitating any activity in the patient himself, then he should go to the sea.

Sea air is better suited than mountain air to persons who cannot bear great and sudden changes of temperature, as is the case with most of those who suffer from grave chronic maladies, as well as with many others. If, however, it should turn out, as suggested by Professor Beneke, that rarefied air is a bad conductor of heat, we can readily understand why a high degree of cold at a great elevation should exercise a much less injurious and depressing effect on the animal organism than the same degree of cold at the level of the sea.

A certain morbid sensitiveness to cold, or rather to " taking cold," is often greatly lessened by a residence in the bracing rarefied air of -elevated localities, and the same good effects are also to be obtained by such persons from exposure to a bracing sea air, especially if accompanied by sea-bathing.

Speaking within very wide limits, mountain air is less suitable to persons advanced in years than sea air. The very stimulus to muscular exertion which mountain air produces is to persons much past middle life often a pitfall and a snare. *Qui va doucement, va loin,* is especially applicable to this period of life, and the state of feverish activity which is sometimes induced in aged persons in the mountains is not by any means for their good.*

We must not forget to consider that the effects of sea air vary very much with locality. The very bracing effects which Professor Beneke observed in the isle of Norderney would not be found, for instance, in the often warm moist air of some parts of our own south-western coast. The former locality would, no doubt, be frequently visited by the cold, dry, Continental east winds. The watering-places on our east coast enjoy a much more bracing and less humid atmosphere than those on our west and south-west coasts, and those on the north coast of France and Belgium have a drier air than either. The warm, moist, and soothing, but relaxing climate of Penzance and Torquay suits admirably many persons with irritable, hyper-sensitive mucous membranes, to whom the air of Cromer or Lowestoft would be unbearably exciting and irritating. To many the air of Bournemouth proves particularly soothing; the air here is, I believe, drier than in the more southern coast-towns, chiefly on account of the absorbent nature of the soil; moreover, it does not lack a certain bracing quality at certain seasons.

The Undercliff (in the Isle of Wight), Hastings, and St. Leonards are cheerful and sunny spots, more bracing than the resorts farther west, and not so bracing as those farther east. Brighton possesses a very bracing stimulating sea air; a much too-decided sea climate for many delicate persons, whom it often renders bilious and dyspeptic. Those who know Brighton well assure me that the higher part, viz.

---

* I met a well-known statesman in Switzerland a few years ago, who with characteristic wisdom and discretion informed me that he was going to visit a locality where he could *look at* a mountain he had once *climbed*. There are many who with much less vigour than this gentleman possessed, when they get into the neighbourhood of mountains, are not content, as they should be, with simply contemplating them.

what is called Kemp Town, is much more bracing and healthy than the central and western parts. One of the most eminent medical men at Brighton assures me that he has often had to remove patients from the central to the eastern part of Brighton to promote convalescence or recovery from illness, and with the happiest results. Eastbourne and Folkestone are excellent quarters, both for sea air and sea-bathing. Eastbourne, in many respects one of the most agreeable and bracing of our south-coast resorts, with its magnificent promontory of Beachy Head, has the unfortunate drawback of being greatly exposed to the east winds, which are especially obnoxious during the spring months. Some of the local medical authorities maintain, however, that if Eastbourne is somewhat more exposed to the east than other rival resorts on the south coast, this is more than compensated for by its smaller rainfall and greater amount of sunshine. Folkestone is especially bracing on its east and west cliffs, where, at a considerable elevation above the sea, the air is less charged with moisture, and when the wind blows off the land it comes fresh across the fine open downs behind the coast. Dover is a good and convenient bathing station. Still farther east, but maintaining something of a southern aspect, we have Ramsgate, and then, with a more decidedly easterly aspect, Margate.* These two last are most valuable bracing health resorts, the air there possessing important tonic properties. On the drier and more bracing east coast we have Lowestoft, Yarmouth, Felixstowe, Cromer, Filey, Scarborough, Whitby, Redcar, Saltburn, and others.

The health resorts on the French and Belgian coasts on the other side of the Channel possess a drier and brighter air than our own coast towns, and, as

* It deserves to be borne in mind that during the prevalence of north-easterly winds, which often blow for weeks together in spring, at Margate they blow directly from the sea, whereas in some other resorts on our own coast they are land winds, and persons exposed to them get no " sea breezes " !

they are very accessible, they offer decided attractions to those persons, and they are very numerous, who find advantage in breathing a drier air than can be obtained in Great Britain. Boulogne, Dieppe, Trouville, Fécamp, and especially Etretat, are favourite French resorts; the latter is exceedingly picturesque, and has the advantage of being a place of much more simple and quiet manners than its neighbours. On the Belgian coast, Ostend is the best known and most popular watering-place, but Blankenberghe, a fishing village about three hours by rail from Bruges, is, for many reasons, much to be preferred. It has a finer promenade on the sea-shore, and the life there is or used to be much more retired and simple.

Having thus briefly attempted to characterise a few of the principal resorts where sea air and sea-bathing can be enjoyed, I must next pass rapidly in review a few of the chief mountain stations on the Continent; and first of all, in importance and usefulness, I would mention the Engadine. The Upper Engadine, as a mountain health resort, must be regarded as typical and unique, so far as Europe is concerned. It is a wide valley, running for many miles at a nearly uniform level of from 5000 to 6000 feet above the sea, and bounded by mountains of comparatively moderate height. By its considerable elevation and its peculiar geographical position, it is removed in a great measure from the regions of cloud and mist. By its peculiar geographical position I mean its remoteness from the lower-lying districts which are nearest it. From the north it is approached by a road which, starting in the Rhine valley at an elevation of over 2000 feet above the sea, pursues for nearly fifty miles a steady ascent, the only notable descent in the whole way being at Tiefenkasten (2900 feet), thence for nearly thirty miles the ascent is unbroken till the northern barrier of the valley is surmounted. On the south it is separated from the Val Tellina, about thirty miles distant, by a vast mountain range covered with ice

E 2

and snow, and reaching an elevation of nearly 14,000 feet, and which protects it, to a great extent, from the mists rising from the southern plains of Italy. Towards the east it stretches for thirty miles without descending 1000 fect, and there it is again separated from any lower level by a high mountain range, which forms a striking and grand eastern boundary to the Upper Engadine. Its only vulnerable point, speaking climatologically, is towards the west, or rather south-west, where it descends somewhat suddenly into the Val Bregaglia; this descent continues steadily to-wards the south-west, till it reaches the Lake of Como. It is from this quarter that nearly all the clouds and rain come that visit the Engadine. It is to its remoteness from the lower levels that the Engadine owes its peculiar and characteristic mountain climate; and it is on this account that persons fail to find the same bracing effects at the same, or nearly the same, elevations elsewhere.

I shall have to return to a consideration of the Engadine and its several health resorts—St. Moritz, Pontresina, etc.—in subsequent chapters.

The *Davos* valley is another mountain station in the Grisons, about 5000 feet above the sea. It is situated to the north of the Engadine, with which it runs parallel, at a distance, as the crow flies, of probably not more than twenty miles. To the lover of the picturesque it can offer few attractions com-pared with those of the Engadine, only a few hours distant; while the fact that it is the special resort of several hundred consumptive patients would, in itself, deter many from making it a resting-place. It has a kurhaus, which has been carefully fitted up and adapted with baths and douches for the systematic treatment of chest affections. I shall have to speak more at length of Davos as a health resort when I come to treat, in a subsequent chapter, of high moun-tain valleys, as sanitaria for consumptive patients.

Bormio, 4300 feet above the sea, at the foot of the Stelvio, on the Italian side, and at the head of the

Val Tellina, is also in the immediate neighbourhood of the Engadine, and a pedestrian starting from Pontresina can cross over the mountainous path which extends from the Bernina road to Bormio in one long day. Carriages have to make a long detour by Tirano in the Val Tellina.

Bormio has long been known for its thermal springs; its climate is milder and more equable than that of the Engadine, owing partly to its southern aspect and partly to its lower elevation. It serves, therefore, admirably as a refuge for those who find the cold of the Engadine too severe, and its climate generally too exciting. Much, however, cannot be said as to the beauty of its situation. The country around has a barren and unattractive aspect, and the Baths of Bormio have a background of reddish, hot-looking, bare mountains, of uniform sugar-loaf form. But it is close to very beautiful scenery; for it is only seven miles through the picturesque Val Furva to the Baths of *Santa Catarina*, 5700 feet above the sea, a mountain station very nearly as high as St. Moritz, and, like St. Moritz, possessing a strong chalybeate spring. This is an exceedingly beautiful spot, surrounded by a semicircle of magnificent snow-covered summits. From its position on the southern side of the Alps, and from its being enclosed by an amphitheatre of lofty mountains, its climate is no doubt much less bracing than that of the Upper Engadine. Were it not for this circumstance, and its greater distance from home, Santa Catarina would certainly become a formidable rival to St. Moritz.

The Baths of *Tarasp* and the picturesquely situated village of *Schuls*, about a mile from the baths, as well as the little hamlet of Vulpera, still nearer the baths, ranging from 4000 to 4800 feet above the sea, are also in the vicinity of St. Moritz, being situated in the Lower Engadine, about six hours' drive from Samaden. Here the climate is much milder than in the upper valley of the Inn. The waters of Tarasp have virtues and a deservedly high reputation of their own, quite independent of the mountain climate; but Schuls and Vulpera offer excellent resorts for those who need less decidedly bracing treatment than is to be found at St. Moritz, while they also afford convenient intermediate resting-places for those delicate and sensitive persons who may desire to avoid either a too sudden approach to, or a too sudden descent from, the rarefied air of the Upper Engadine. (Of the special merits of Tarasp as a health resort I shall have to treat in detail by-and-by.)

The villages of Bergün and Molins, the one on the Albula Pass and the other on the Julier, are the places, perhaps, most commonly selected for the purpose of breaking this ascent or descent. But apart from the fact that the latter certainly is hardly sufficient of a break, they neither of them afford quite that amount of comfortable accommodation which invalids require for a three or four days' stay. Bergün is most picturesquely situated, and if the village were improved in cleanliness and a really good hotel esta-blished there, it ought to prove an attractive resting-place *en route* to or from the Engadine.

A new resource has been provided within the last few years for breaking conveniently the descent from St. Moritz, by the opening of an hotel in one of the most beautiful parts of the Val Bregaglia—the Hotel Bregaglia—situated at Promontagno between Vico Soprano and Castasegna, and between 1000 and 2000 feet higher than Chiavenna, the station to which persons hitherto had to descend in going from the Engadine towards Como and North Italy. This is a real gain to invalids going south.

Monte Generoso, situated between the lakes of Lugano and Como, and usually ascended from the town of Mendrisio, has lately become justly popular as a health resort, since Dr. Pasta has established a comfortable hotel a few hundred feet below the peak, which is between 5000 and 6000 feet above the sea. This, for its elevation, is a comparatively mild

mountain situation, and better suited on that account
to highly sensitive organisations, while the beauty of
the scenery it commands can scarcely be surpassed.
It is a most convenient locality for persons coming
north, after wintering in the south of France or Italy.

Of other frequented European mountain stations
a very brief account must here suffice. Of very bracing
health resorts over 6000 feet, the following are the
best known : The hotels on the Æggisch-horn and
the Bell-Alp, the former ascended from Viesch, the
latter from Brieg in the Rhone valley, in the midst
of the grandest mountain scenery, and close to the
great Aletsch Glacier ; there is also an hotel on the
Rieder-Alp, between these two ; the Riffel Hotel,
facing Monte Rosa and the Matterhorn ; and the
hospices of the Great St. Bernard, of the St. Gothard,
of the Bernina, and of the Grimsel passes.

Of bracing, but less extreme mountain climates,
ranging between 5000 and 6000 feet above the sea,
the following may be mentioned : The Baths of San
Bernardino, on the southern side of the pass of that
name, where there is a chalybeate spring. Its southern
aspect moderates the rigour of its high mountain
climate. Mürren, beautifully situated above Lauterbrunnen ; the Rigi Scheideck ; the village of Zermatt ;
Panticosa, on the southern slope of the Pyrenees, in
the province of Aragon, a few hours from Les Eaux
Chaudes ; here there are alkaline and other springs ;
this fact and its moderate temperature have made
it a resort for consumptive ¡patients, chiefly from
Spain.

Of milder and less exciting mountain climates we
have a great variety to choose from : ranging between
4000 and 4500 feet we have Comballaz in the Val
des Ormonds, about three miles above Seppey ; the
Baths of Leuk, at the foot of the Gemmi ; Weissenstein, a ridge of the Jura, near Solothurn, a station for
the goats' milk and whey cure, commanding a very
fine view ; the village of Andermatt, on the St.
Gothard road ; the well-known Kaltbad on the Rigi ;

Barèges, in the Pyrenees ; and the town of Briançon in Dauphiné, and many others.

Of those between 3000 and 4000 feet I may name Beatenberg, over 3500 feet, in an admirable situation above the right bank of the Lake of Thun ; Gurnigel, also over 3500, a frequented sulphur bath not far from Berne ; Courmayeur, nearly 4000 feet ; Grindelwald ; Engelberg, a favourite mountain resort at the foot of the Titlis, and near glaciers, with an equable, fresh, and tonic climate, and whey and goats' milk cure ; Château d'Oex in the Simmenthal ; Chaumont, overlooking Neuchâtel, and easily approached from that town—like most of the Jura stations, it is more exciting and bracing than other localities of the same elevation.   Sainte Croix, also near Neuchâtel, on a declivity of the Jura chain, is about 3600 feet.   St. Cergues, also in the Jura, is a village built in a gorge looking east, at the foot of the Dôle.   It is much frequented for its bracing climate, but it is considered too irritating for delicate impressionable persons. The climate of the Jura chain is said to be generally colder and more humid than that of the Central Alps.

As examples of very mild and slightly tonic mountain climates between 2000 and 3000 feet above the sea there are Glion and Les Avants, near Montreux ; St. Gervais, in a very favourable position, near the valley of Chamounix ; Seelisberg, in a protected and mild situation above the Lake of Lucerne ; Gais, Weissbad and Heiden, in the canton Appenzell, above the Lake of Constance ; these three are exceedingly pleasant, quiet resorts, out of the way of the beaten track, and excellently well suited for those who need repose and quiet in a pure and moderately bracing air.

Finally, whether we seek health in the mountains or by the sea, in either case we shall find change— that change which is the type of life and the condition of health ; that change which is rest.   And who shall estimate the moral, as well as physical, refreshment we gain by changing the sordid routine of city life, the "greetings where no friendship is," for the contemplation of the solemn moods of Nature, whether in sea or mountain?   Looking on these eternal realities, in the grandeur of their calm repose or in the majesty of their roused anger, we recover that sense of proportion which we are so prone to lose—our sense of the relative proportion of the individual to the whole. Or, if we need no such stern remindings, we may seek changeful Nature in her gentler moods in the soft woodland shade, and there, amidst the perfume of flowers, the songs of birds, and the murmur of the trees, we may, as well as by the sea or on the mountain, recover health of mind and body as we—

"Draw in easier breath from larger air."

CHAPTER III.

TOWARDS the end of the last chapter I mentioned several places which are well known as sea-side resorts; in this chapter I propose to give some additional information about some of these, and to give an account of others which are not mentioned there.

To begin with our own coasts. If we start from the mouth of the Thames and travel *westward*, we encounter successively the following more or less popular marine health resorts: Herne Bay, Westgate-on-Sea, Birchington, Margate, Broadstairs, Ramsgate, Dover, Folkestone, Sandgate and Hythe, Hastings and St. Leonards, Eastbourne, Seaford, Brighton, Shoreham, Worthing, Littlehampton, Bognor, Hayling Island, Southsea, the Isle of Wight, Bournemouth, Swanage, Weymouth, Lyme Regis, Sidmouth, Exmouth, Dawlish, Teignmouth, Torquay, Dartmouth, Plymouth, Falmouth, and Penzance. If now we turn round the Land's End, and continue along the west coast of England, we notice, that with the exception of St. Ives, New Quay, and Bude, which are but little known, an almost entire absence of such resorts along the north-west coast of Cornwall. We then come to the north coast of Devon, and notice, first of all, the picturesque fishing village, Clovelly; next comes Westward Ho, Ilfracombe, Lynmouth, Lynton, then Weston-super-Mare. Now if we cross the Bristol

Channel to the Welsh coast, we shall again find but few places of popular resort. Tenby is perhaps the best known ; then much further north Aberystwith, Aberdovey, Barmouth on the west coast, and Llandudno and Rhyl on the north-west. Then there is New Brighton close to Liverpool, and still further north Southport, Blackpool, Morecambe and Grange.

Leaving Scotland for the present, we now cross over to the east coast of England. We find on the Yorkshire coast Redcar, Saltburn, Whitby, Scarborough, Filey ; then coming southward, passing by the Lincolnshire coast, we reach, on the Norfolk coast, Hunstanton, Cromer, Yarmouth, Lowestoft, and on the Suffolk and Essex coasts, Felixstowe, Harwich, Walton-on-the-Naze, Clacton-on-Sea, back again to Southend.

Some account will be given of each of these in the order named, and an attempt will then be made to arrange them in groups according to local climate.

Herne Bay is the nearest sea-side resort to London on the Kentish coast. It is sixty-three miles from London and eight from Canterbury, and is reached in rather less than two hours by the Chatham and Dover Railway. It is situated on a fine bay on the estuary of the Thames, and like the neighbouring resorts, Westgate and Margate, it has a north-eastern aspect, and its climate, like that of those places, is tonic and stimulating. There is a good promenade, a mile long, where the baths are, and it has an iron pier. The town is but partially built, as the place has never acquired that amount of popularity which it was originally expected to obtain, nor has it been able to compete successfully with its more fortunate neighbours.

Birchington-on-Sea, which has recently come into notice as a sea-side resort, is sixty-nine miles from London and only three-and-a-half from Margate. It is situated on elevated ground and is bounded on the sea-coast by bluff cliffs, indented by Westgate Bay and Epple Bay. The village, about three-quarters of a mile in length, lies on the old London and Canter-

bury Road. It is a quiet resort, better suited than
Margate to those who are seeking rest and retirement.

Westgate-on-Sea has grown rapidly into popularity
as a health resort. It is only two miles from Margate,
of which parish it once formed a part. Good roads
have been made and sea walls built round the curves of
St. Mildred's and Westgate Bays, forming two promenades over a mile in length, from which by steps the
sands or cliffs are easily accessible. Gardens also
have been made on the cliffs, where there is a marine
drive two miles in length. The air is bracing and
pure, and the water supply is pure, abundant, and
continuous. The sea-bathing is good.

Westgate is a salubrious and convenient sea-side
resort for families with children who require good
bracing sea air and pleasant, quiet sea-bathing.

Margate is one of the most bracing of our sea-side
resorts, and, being within two hours' railway journey
of London, and within four or five hours by river
steamer, it is naturally and justly a very popular place
with the natives of the metropolis. Although often
associated in men's minds with the neighbouring
resorts on the south coast, it must not be forgotten,
in estimating the special qualities of the climate of
Margate, that its aspect is not to the south, nor to
the east, but to the north or north-east. Whoever
will take the trouble to look at a map of this part of
the coast of the Isle of Thanet will see that Margate
has a decidedly northerly or north-easterly aspect.
In this respect it differs entirely from the neighbouring
coast town of Ramsgate, which is so placed in an
indentation of the coast-line as to look south-east.
From this difference of aspect it not infrequently
happens that the local weather differs considerably in
these two places, only a few miles apart, for a storm
may be raging at Margate which is scarcely felt at
Ramsgate. It happens also from this northerly or
north-easterly aspect of Margate that during the
prevalence of the north-easterly winds of spring it is
one of the very few conveniently-accessible sea-side

resorts where during that season pure *sea* air can be obtained. For the prevailing north-east winds blow directly over the North Sea and the northern portion of the British Channel on to Margate and the line of coast of which it forms a part, whereas during the same season the prevailing winds at the resorts on the southern coast are land-winds, and blow off the land out to sea, driving off, as it were, the sea air— hence, I take it, the great value which Margate air is known to possess in scrofulous affections.

It has a chalk subsoil, and the ground quickly becomes dry after rain. The water supply is abundant and pure.

The best residential part of the town is on the Cliff, where the Cliftonville Hotel stands ; but the cheaper hotels and lodging-houses are on the lower ground, near the harbour and pier.

The drainage of the higher ground is better than that of the lower level, which is not all that could be wished.

Margate is too windy for cases of chest disease, except in the summer months, but it is especially beneficial to cases of scrofulous disease in children and to those of convalescence after surgical operations. This fact has led to the establishment there of that excellent institution, "The Royal Sea-Bathing Infirmary, or Royal National Hospital for Scrofula."

The air of Margate is excellent for promoting the progress of slowly healing wounds and ulcers, and it is valuable in cases of debility from inherited feebleness of constitution. It is also said, from its dryness, to be good for rheumatism.

Its autumn climate is often very fine. "In November, when London is foggy and the country damp, Margate rejoices in a clear, bright, and dry atmosphere."

Margate is celebrated for its sands, which are very extensive owing to the shallowness of the water. They are therefore the delight of children and those who do not care for a plunge into a deeper sea.

Broadstairs is a quiet, unpretending little sea-side town, which has maintained its character for simplicity and retirement, notwithstanding the temptation to become otherwise which is afforded by the neighbouring towns of Margate and Ramsgate, between which it lies. It is especially the resort of young children, who can bathe and amuse themselves all day long on its sheltered sands.

"For people who want absolute quiet, combined with bracing and healthful air, Broadstairs is the watering-place most within reach. The climatic characteristics of the place are the same as at the other resorts in the Isle of Thanet. The subsoil is of course chalk. The greater part of the district is from 120 feet to 150 feet above the level of the sea. The Local Board are now carrying out a system of drainage, the outfall for which is taken about one-and-a-half miles from the town to a point nearly opposite the North Foreland, where two different currents meet and take the sewage out to sea. Water of a high degree of purity is supplied by a private company. During last year the death-rate for Broadstairs was equal to 16.4 per 1000, that for the chief zymotic diseases being only 1.1. The inhabitants of this quiet little semi-rural watering-place, in which town and country characteristics are blended, are very zealous in their crusade against infectious diseases, and insist upon prompt isolation of such cases whenever they happen to be introduced into the district."*

Ramsgate, unlike Margate, has a south-eastern aspect. It is somewhat warmer than Margate, and is more protected from northerly and north-easterly winds. Its air is, however, bracing and tonic, and like Margate, it has fine sands for bathing.

The harbour and town lie between two cliffs, the east and west, where most of the visitors reside.

---

* I am indebted for the above, and for valuable information concerning some of the following sea-side resorts, to the excellent and instructive reports published in the Holiday Number of the *Medical Record*.

The large hotel on the East Cliff, with its comfortable and complete arrangements for sea baths of all kinds, has made that part of Ramsgate (St. Lawrence-on-Sea) especially popular. Close here is the station of the Chatham and Dover Railway, by which Ramsgate may be reached in two hours from London. There are steamers also from London Bridge in six hours. The harbour, enclosed by two piers, is full of activity; and in the season small steamers ply between here and Sandwich, Deal, and Dover. Pegwell Bay is within an easy walk, and Canterbury, with its Cathedral, is only a few miles by rail. In the summer months a four-horse coach runs between Canterbury and Ramsgate and Margate; while there is another four-horse coach from Canterbury to Folkestone, thus placing a fine drive through the attractive country between Ramsgate and Folkestone at the command of the visitors at either resort.

Dover, one of the first places to attend to its sanitary arrangements, has many merits as a health resort. Its climate is, for the most part, dry, tonic, and invigorating. It is not, however, to be recommended at all seasons of the year. It is very cold in January, very windy in March, and very hot in July. It is usually pleasant in May and June, and again from August to the end of October. The winter is often mild up to January. The cases that are said to do well at Dover are those of phthisis in its earlier stage; of bronchial catarrh in young people, with sensitive nervous systems and languid circulations; of dyspepsia, with loss of appetite, depression of spirits, and irritability of temper; of chronic diarrhœa from residence in tropical climates; of insomnia; of scrofulous disease in children. Most of the houses at Dover are built with a southern or south-eastern aspect, and are exposed to the direct influence of the sea breezes blowing from the Straits. It is protected by the chalk hills behind it, to some extent, from winds coming from the north, the north-east, and the north-west. Its subsoil is chalk, but most of the houses are built upon the beach, i.e., on shingle, flint, and sand overlying the chalk. The soil is, therefore, porous, and rain rapidly drains off the surface. The water supply is good and pure, and is derived from two communicating wells sunk in the chalk to a depth of 226 feet. The sewage is discharged into the sea at low water.

Folkestone, with its excellent service of tidal trains, and situated so conveniently as it is on the great highway of Continental traffic, possessing, moreover, attractions in itself of no mean order, has naturally become one of the most popular of health resorts, especially amongst the upper and travelling classes. Folkestone is an instance, as Dr. H. Weber has pointed out, how different parts of the same locality may vary in climatic character. " The houses situated close to the beach are almost entirely sheltered from the north, so that the air feels, on hot and sultry days, very close, and is, on the whole, less bracing, while those on the west cliff receive the air from all sides, so that one always has a sensation of freshness. Furthermore, the houses near the beach are much more under the influence of the sea, and in consequence, some persons become more easily bilious when staying there than on the cliff, which is fully 50 feet above the sea." *

The town is built on a lofty, porous cliff of greensand, and seen from the sea, or from the cliffs to the east of the harbour, it has a most picturesque appearance. The "Lees," on the west cliff, form a fine extensive promenade, high above the sea, commanding a vast sea-view up and down the Channel ; Shakespeare's Cliff, and on clear days, the French coast between Calais and Boulogne, are striking objects ; while a constant succession of steamers and sailing vessels passing through the Straits of Dover give animation to the scene. The open chalk downs behind the town, and the many attractive country

* Hermann Weber's " Climate and Health Resorts."

roads, afford abundant resources for horse and carriage exercise. The water supply is good, pure, and abundant, and the drainage has been most carefully provided for. The death-rate for the last ten years is given as 16·4 per 1000. Its annual rainfall is said to scarcely exceed twenty-five inches, which is very small for a town on this coast ; " this is owing probably to the water-laden clouds not breaking and discharging their contents until they meet the range of chalk hills some distance inland."

The climate is considered suitable to many forms of chest disease, to cases of early phthisis, to cases of chronic catarrhal tendencies in the over sensitive and scrofulous, to some forms of asthma. It is also highly useful in cases of depressed nervous tone, with irritability, sleeplessness, loss of appetite, and hypochondriacal tendencies, also in cases of protracted convalescence after attacks of acute disease.

There are many distractions for the visitors to Folkestone in the summer ; the presence of military bands, the activity of the port, with its daily arrival and departure of Continental traffic, the hospitality of the residents, etc., make it eminently a bright and cheerful resort.

One drawback, from the point of view of sea-bathing, is the rather rough shingly beach ; but baths of all kinds, including a large tepid sea-water swimming-bath, can be obtained at the Bath Establishment.

Sandgate is about a mile and a half west of Folkestone ; it stretches for some distance along the coast, with a fine sandy beach in front, affording great facilities for bathing, and a range of hills behind, affording protection from the north.

Shorncliffe Camp lies on a plateau above, and on the north side of the town. There are pleasant walks and drives towards Folkestone, or Hythe, or Shorncliffe, and the promenade in front of the sea forms an agreeable and sunny lounge. It has the same character of climate as the lower part of Folkestone, but it is quieter and less expensive, and the sea-bathing is decidedly better.   There is a sea-wall and parade between Sandgate and Hythe, which is two or three miles further west, and has the same climate and character as a health resort as Sandgate.

Hastings and *St. Leonards* have a complete southern aspect, and they are protected to the north and north-east by high cliffs, and the hills at the back of those towns ; but there is little or no protection from winds blowing due east, these enfilade the promenade along the shore with a force and violence at times almost unendurable.

These towns are said to have a remarkably equable temperature, both in winter and summer, and on that account they are considered suitable to cases of pulmonary disease.   The air is warmer along the shore than on the hills behind the town, where it is said to be cool during the warmest summer months. Pulmonary invalids are recommended to choose the former, and convalescent patients, suffering from debility and want of tone, the latter situation.

The soil is porous and sandy, and the rain that falls is rapidly absorbed, so that the air is free from the humidity which might arise from evaporation of water from the surface.   Hastings enjoys great immunity from fogs, and its relative proportion of sunshine is large.   The system of drainage, and for disposal of sewage, is excellent.   "All the sewage of the town is conducted into a well-constructed and efficiently-ventilated sewer, which runs along the sea-front and has a fall from opposite the St. Leonards archway (its highest point eastwards and westwards), and discharges at each end into capacious intercepting tanks.   The sewage is retained in these tanks to be let out at the east end of the town at low water, and is thus carried away from the town up channel for five hours upon the flood tide and at the west end two hours after the ebb tide, and is thus carried down channel away from the town.   By this means the sea-front is kept free from all impurities, and rendered cleanly and wholesome for bathing.

"The water supply is derived principally from springs and wells in the surrounding hills and valleys: it is of good character and quality, the slight excess of chlorides being due to the condition of the soil which is peculiar to sea-coast towns. In October 1883 a further supply of bright, clear, well-water was opened for the use of the inhabitants; this supply alone providing 500,000 gallons of good water daily.

"The value of Hastings as a health resort is seen in the lowness of its annual death-rate, which is equal to an average of 16·66 per thousand persons living. These are most satisfactory figures, and would be more so were it not for the large number of persons in advanced stages of disease who come to Hastings too late to receive benefit from its salubrious climate, and who die here. These average 17·43 per cent. of the total number of deaths recorded each year, and materially increase the natural death-rate of the town. The zymotic death-rate, too, is low, averaging 1·31 per thousand."

Hastings has a very fine pier which is peculiar in having a large covered pavilion in which excellent concerts are given daily. On the esplanade there are also very convenient glass-covered partitioned shelters of great service to invalids. There are numerous places of beauty and interest to be visited in the neighbourhood; the excursions to Fairlight and Ecclesbourne Glen are most popular, while more distant objects of interest are afforded by Battle Abbey, Pevensey and Hurstmonceux Castles, Bodiham, Rye, etc.

Owing to their well-known advantages as winter resorts, Hastings and St. Leonards are almost deserted in the summer and autumn, and most undeservedly so, for they are no hotter than adjacent places.

Eastbourne as a health resort has grown into

popularity with unexampled rapidity. This is no doubt greatly due to the wise liberality and foresight of the chief landowner of the town, the Duke of Devonshire. The roads and streets have been skilfully planned and laid out on a uniform system, so as to secure abundance of space, free ventilation, and the picturesque planting of trees throughout the streets of the town. Moreover, the streets and other residential parts of the town are not all huddled together close to the shore as in some resorts, but spread out over a considerable tract of land extending towards the magnificent downs behind the town for three-quarters of a mile or more from the sea-shore. This is a great advantage, for it enables the physician to make use of Eastbourne for many patients who are not comfortable when close to the sea, and for others to whom the exciting effect of strong sea-air might prove injurious, while the pure, bracing air of the downs, which are so readily accessible from the inland part of the town, might be advantageous.

Eastbourne is situated on the coast of Sussex, between St. Leonards and Brighton, at a distance of sixty-five miles from London; its train service is good, and the fast trains accomplish the journey in little more than an hour and a half.

The old town of Eastbourne (meaning that it is situated on the East—*burne*, a stream) is a mile inland ; the new town, which has naturally been built towards the sea, has a south-eastern aspect, like the coast. There is a level frontage to the sea, admirably laid out, with three parallel promenades at different elevations, extending from the pier westward for nearly three miles till it reaches the steeply rising ground reaching to the magnificent promontory at Beachy Head. There is an extensive sandy beach, affording admirable sea-bathing, and gentlemen are permitted, in the early morning, to bathe from the pier. The promenades are provided freely with seats and chairs and protected lounges for the delicate, and at night it is lit by the electric light.

The existence of the Devonshire Park and Gardens proves an attraction to many—to some on account of the excellent concerts provided there twice a day, and to others on account of the numerous lawn-tennis courts, which attract some of the best players in England. In connection with the same establishment there are hot and cold swimming and other sea-baths.

There are many riding and driving excursions to be made from Eastbourne; but its chief attraction will always be Beachy Head, and the fine stretch of downs extending inland from this grand cliff. There is probably no finer air or a more perfect sea-view to be found anywhere than on Beachy Head. Standing up in sheer perpendicular white cliffs to a vast height above the sea, this precipitous coast-line extends, with only one dip, at Birling Gap, for several miles westward till it reaches Seaford. As a background there is a vast expanse of breezy downs, extending for miles inland in the direction of Lewes. There is a good restaurant, established by the proprietors of the Queen's Hotel, on Beachy Head, which is a great convenience to pedestrians, as it is an hour's up-hill walk from the centre of the town to this its, to my thinking, chief attraction. To the east of the town the country is flat and unattractive, consisting of long reaches of sand, which enclose Pevensey Bay. Pevensey Castle, about five miles distant, is the chief object of attraction on this side.

The sanitary arrangements are as complete as they can well be made, and an abundant supply of pure, soft water is obtained from wells sunk through chalk into the green sandstone. Every house has a constant supply. The sewers are freely ventilated, and the sewage flows into the sea at a point three miles away to the east, where the tides carry it away without the chance of its returning to the bay. The death-rate is low, the annual average for the last three years being 14·18 per 1000.

During the summer I have known phthisical

patients do extremely well at Eastbourne, especially if they begin their stay by living away from the sea.

Cases of torpid scrofula, of slow convalescence from surgical operations or injuries, cases of anæmia and general want of tone, cases of depressed function, nervous or digestive, are all suited to this place. I am informed that in houses built to the west of the town, and with rising ground between them and the east, the winter temperature is not unpleasantly low, and that they get more than the average amount of sunshine. But those parts of the town which are unprotected from the east suffer much from the prevailing winds in spring.

Seaford is nine miles west of Eastbourne, the range of chalk cliffs and downs, of which Beachy Head forms the culminating point, lying between them. It is three miles from the port of Newhaven, which connects it by rail with London.

The air is pure and bracing, and the fine downs stretching up to Beachy Head, which is six miles distant, are readily accessible. The climate has much in common with that of Eastbourne, except that the great mass of cliff, which intervenes between the two places, affords a great protection from the east; it is, however, fully exposed to the south-west.

Seaford is now provided with a fine sea-wall and promenade, which preserve it from inundations of the sea, from which it formerly suffered severely.

Brighton is too familiar a resort to need detailed description. No sea-side place is so accessible to those who live in London. Its strong sea-air, coming from a wide open sea-board, is most invigorating to many, but for others it is too strong and irritating; and it is important to recognise a form of dyspepsia with torpidity of liver which becomes developed in certain constitutions at Brighton—as well as in some other sea-side places—and does not disappear so long as the patient remains there. Brighton is also not at its best when the north-east spring winds prevail, for then the wind is off the land, and the smoke, etc., from the town is blown down over the esplanade, and, instead of pure sea-air, we breathe not very pure land-air. South-south-west and westerly winds bring the best air to the shores of Brighton.

But the climate of Brighton, as was mentioned in the last chapter, differs considerably in different parts of the town. The houses in the King's Road and Brunswick Terrace, for example, are much more under the influence of the sea than those high up. near Montpellier Square; and the high east cliff (Kemp Town) is much more bracing, and has altogether finer air than the low western side, which is, however, more sheltered, and far more frequented and popular.

Under the east cliff, and extending along its whole length, a fine summer promenade has been made, which is completely sheltered from the north and north-west, but the east and north-east winds are felt there keenly; this is the so-called "Madeira Walk." It would be far more useful than it is if three or four glass-covered seats, or some similar protection for pedestrians, were placed along the walk, for, at present, when east or north-east winds are blowing, it is dangerous for delicate persons to rest on the open seats there.

"The soil of Brighton is exclusively chalk. No river runs near it within half-a-dozen miles; there are neither water-courses, stagnant pools, nor marsh-land in its vicinity, and it enjoys, therefore, almost absolute freedom from noxious vapours or dampness. In the spring the wind· is variable, in summer it comes most frequently from the west, in autumn and winter from the south-west—from the sea." It has a good and abundant water supply.

The sea-bathing is good, the water is clear, and the bottom is sandy. There are numerous private bathing establishments, where hot and cold sea-water baths can be obtained; as well as Turkish baths and *massage*. The South Downs, which stretch along the back of the town, are a great and valuable resource for healthful exercise.

Brighton is especially serviceable in cases of retarded convalescence, especially after surgical operations, and also after some acute febrile maladies. It is useful in cases of anæmia, and general loss of tone, induced by over-work, by chronic illnesses, or by other depressing agencies. It is of value in giving vigour to delicate young people, especially when of scrofulous constitution, during the most trying periods of rapid growth and development. Its bracing sea-air and sea-baths are also beneficial in diminishing that sensitiveness of skin and mucous membranes upon which the very prevalent tendencies to catarrhal and rheumatic affections depend.

Shoreham, or *New Shoreham*, "is a pleasant little town at the foot of the Southdown, on the river Adur, about six miles from Brighton and fifty-six miles from London. The town is excellently drained, and has a constant supply of water, which has been certified by a public analyst to be very pure. The death-rate for 1882 was 12·37 per thousand, and for 1883 14·53 per thousand. For the last four years the mean death-rate has been 13·83 per thousand, and the deaths from zymotic diseases have been only 0·88 per thousand. The air is very invigorating, and at the same time soft : it is suitable for persons suffering from chest affections. Excellent sea-bathing on a fine sand and beach is easily accessible by ferry-boat or bridge."

Worthing is a few miles to the west of Brighton, and fifty-two miles from London. It is a much more quiet resort than its popular neighbour, and enjoys much the same climatic character. The town is clean and well laid out, and the drainage is said to be good. It is a very suitable station for sea-bathing, as the sands are firm and good, and, with the exception of Brighton, it is the most accessible resort from London. There is an esplanade facing the sea about a mile long, composed chiefly of lodging-houses ; there is also a good pier, where, as usual, a band plays twice a day. It is surrounded by agreeable country, with some charming rural scenery.

Littlehampton, on the Sussex coast, is sixty-two miles by rail from London, sixteen from Worthing, four from Arundel, and eleven from Chichester. It is a much frequented resort, on account of its mild and salubrious air, its comparative retirement, its excellent sea-bathing, firm and clean sands, and its pleasing rural scenery. When the tide is low the sands are left firm and dry, so as to form a perfect paradise for young children. The town is situated at the mouth of the river Arun, the traffic from which is considerable. There are also steamers from this port to Honfleur three times a week, as well as to Jersey in the summer. The water supply and drainage are good.

Littlehampton is an excellent resort for families and children requiring sea change in the summer months.

Bognor, on the Sussex coast, is about sixty-three miles from London, thirty miles west of Brighton, and twenty-five east of Portsmouth. It is a quiet resort, and having a firm, clean, level, sandy beach, is well adapted for sea-bathing. The air is pure and mild, and the vegetation luxurious: "native and exotic plants" flourish, we are told, "at the very verge of the sea." The elevated downs lying behind it afford a protection from the north winds. The system of drainage has been improved, and the death-rate is said to be low. Most of the houses depend upon wells for their water supply. There is a fine pier 1000 feet long by 18 feet wide.

Bognor is surrounded by some very fine country. Chichester is only seven miles distant, Goodwood about ten miles, and Arundel Castle and Park also about ten miles.

Hayling Island has only recently come into repute as a sea-side resort, and for those who like to get away from the human crowd and from the beaten track, it has much to recommend it. It is reached by a short branch of the London, Brighton, and

South Coast Railway from Havant, the journey from London taking a little less than three hours, or it can be approached from Southsea by crossing the ferry at Fort Cumberland, from which it is distant about half-a-mile. Numerous good villas and a commodious hotel have recently been built there. Pure air, fresh breezes from the open sea, excellent sands and good bathing, exceedingly well suited to children, and the absence of a fashionable crowd, these are the attractions of Hayling Island.

"Hayling Island is ten miles square in extent, with a south-west aspect of about five miles; commanding scenery of great beauty and ever varying interest. The sea front of Hayling is about five miles in length, the beach sloping very gently, and presenting a continuous stretch for the most part of firm sand." To the west there is the Isle of Wight, with Bembridge Bay and the town of Ryde; in front the open sea; in the interior, "well-sheltered and old-fashioned farms, with green-hedged lanes winding among them. Nowhere can such country luxuries as abundant cream, new laid eggs, pure and well churned butter, fresh vegetables, fine fruit in its season, and plump poultry, be more safely reckoned upon. Along the whole southern coast there can scarcely be named a finer tract of sands than that afforded by the lovely bay—it is the place of all places for children."

Southsea has many attractions. It is within two-and-a-half hours, by a good train service, from London, and within half-an-hour of the Isle of Wight. It has a fine common facing the sea, a good pier, excellent hotels and lodging-houses, and, from its adjacency to a great military and naval arsenal, constant social activity and interests, and frequent military and naval displays. A number of charming excursions by land or sea can be made in fine weather. The beach is shingle, and is not so pleasant as sand for bathing upon.

As a winter residence it has been highly commended by some. A six years' resident writes: "We have much less rain than inland, and the ground, except in one low-lying part, rapidly dries when a fall does occur, the soil being gravel and shingle. In the winter house-rent is very moderate." The death-rate is low. One of the medical authorities of the place informs me that "more people come now in winter who have delicate chests, and they do well here if they live on the Common. Speaking roughly, our average temperature is, in winter, 5 or 6 degrees Fahrenheit higher, and in summer 5 or 6 degrees Fahrenheit lower than at Kew. In 1884 our lowest temperature was 27 degrees Fahrenheit, and our highest 84 degrees Fahrenheit. There is a good system of drainage, with a pumping station which discharges the sewage nearly three miles from the town. This will be soon even more improved by the building of large reservoir tanks, which will secure the discharge of the sewage at all times of tide and in all weathers. The sewers are well ventilated, and we have an efficient medical officer of health and a Contagious Diseases Hospital. There are now excellent baths built by the town, with a large swimming bath supplied with sea water and warmed in winter."

Southsea brings us to the neighbourhood of the *Isle of Wight.* This island is about twenty-three miles in length from east to west, and about thirteen miles wide at its widest part from north to south. A range of high chalk downs stretches right across the island near the middle from Bembridge Down, near the extreme eastern part of the island, to the extreme western part. There are other chalk downs at the extreme south of the island.

In 1881 the population amounted to 64,489. Its chief places are Newport, Ryde, West and East Cowes, St. Helens, Sandown, Shanklin, Ventnor, Yarmouth, Alum Bay, and Freshwater. Newport, the capital, is situated in the centre of the island, and does not, therefore, concern us in this place.

Cowes, chiefly associated with yachting meetings,

is divided by the river Medina into East and West
Cowes. Of these West Cowes is the most resorted
to by visitors, and is the most fashionable: its houses
are less crowded together. From the west bank of
the Medina there rises a steep, well-wooded hill, and
beautiful villas are built over it. In the old part of
the town the streets are narrow and dirty. The soil
is clay, with some gravel at the upper parts. During
the yachting season, and especially in the regatta
week, Cowes is crowded with visitors, but it really
possesses little interest as a health resort. Its aspect
is due north.

Ryde is a pleasure resort rather than a health
resort, and is the most fashionable place in the island.
Its long pier is a well-known and popular lounge, and
military bands play there during the season. It is a
great centre for yachting and boating, and for excursions into the interior and around the island.

The town occupies the face of a hill which slopes
mainly in a north-easterly direction, but partly also
in a north-westerly direction. It lies on the steep
north-easterly slope, and is much exposed to the
north-east winds. The most elevated part of the
town is 155 feet above the sea. The highest part of
the whole district, at its boundary, is 193 feet. The
subsoil at the lower and northerly parts is clay, with
here and there gravel and sand, and a little stone.
At the higher parts the subsoil is brick-earth with
gravel and some sand. In some places at the top of
the town a running sand, full of water, is found at a
depth of six or eight feet from the surface. The
town is regularly built, and, with the exception of the
High Street, the streets are on the whole of a good
width and airy. The tide runs out a long way,
leaving exposed a great breadth of wet sands, which
are muddy, and sometimes offensive to the smell near
the shore. The whole town is now completely and
well and efficiently sewered. The water supply of
the district is derived from waterworks belonging to
the Corporation, about four or five miles off, where

the water is derived from a deep well in the chalk hill, and from some springs on the opposite slope. Owing to its north-easterly exposure, and its open and comparatively unprotected situation, Ryde is only adapted for a summer resort, and for this purpose it is cool, cheerful, and attractive.

St. Helens is the name of a district stretching for about three miles to the east of Ryde. The slope and summit of the hill at its western extremity are largely built over, and two villages along the shore— *Spring Vale*, about a mile, and *Sea View*, about a mile and a half from Ryde—are developing into little towns. The village of St. Helens, the houses of which surround a village green, is situated on a hill at the extreme end of the district. Between this and Ryde some of the ground is marshy.

Sandown is almost the first station on the Isle of Wight Railway after leaving Ryde, from which it is distant about fifteen minutes.

It is built on the shore of a fine open bay, which affords great facilities for bathing and boating. The town is delightfully placed on the slopes of a low hill, and these stretch with a gradual fall towards the sea at the south-east. To the north-east a sea-wall protects the low marsh land from incursions of the sea. The streets, which are many feet above the sea level, are wide and well laid out, and the houses have gardens attached to them. The subsoil of a great part of the town is clay, but the westerly and upper third is built on sand. The town is well drained, and the sewage discharged into the sea at a long distance from it. Public walks and pleasure grounds have been laid out at great expense.

Some very perfect Roman villa remains are to be seen in the immediate neighbourhood. The surrounding country is attractive, and excursions to various parts of the island can readily be made from it.

Sandown is a pleasant, quiet, picturesque, sea-bathing resort, much frequented in the summer by families with young children.

Shanklin is about five minutes by rail from San-down, and is situated in a valley stretching inland from the shore of the Bay of Sandown. Like San-down, it has a firm sandy beach. The air is fresh, and pure, and bracing on the higher ground. Shanklin Chine is a beautiful rocky glen covered with luxuriant verdure. The walk from Shanklin to Bonchurch is considered one of the most beautiful in the island, passing, as it does, through the romantic Chine of Luccombe.

The houses are built on the sloping sides of a hill which, at its highest part, is 200 feet above the sea. It has for the most part a sandy subsoil. The water supply is said to be inadequate, and efforts are being made to remedy this.

Shanklin is a resort of great beauty and attractive-ness ; its sea-bathing is excellent, and the excursions which can be made from it, both by sea and land, are most varied and delightful.

Ventnor is situated at the most charming part of the southern shore of the island, and is built in a series of terraces on the wooded rocky slopes of the beautiful and celebrated Undercliff. It is sheltered from the north and east winds by a steep range of limestone rocks, and, owing to its protected situation, it has become a popular resort for pulmonary invalids ; on this account it has been selected as the site of the National Hospital for Consumption. This hospital is built in blocks, each for twelve patients; the patients' rooms all face south ; each patient has a separate sleeping room of from 1500 to 1750 cubic feet, and in the newest block, 2000 cubic feet! Each six patients have a separate sitting-room 3000 cubic feet square, and each block has an additional room of 3000 cubic feet, in which the patients take their breakfast, tea, and supper. The male and female patients dine apart in large rooms at the end of their blocks. In the new block there is an arrangement for the supply of 5000 cubic feet of fresh air per patient per hour at a temperature of 62 degrees Fahrenheit.

There is a large open piece of ground between the Hospital and the sea where patients can sit or take exercise in the open air. Very good results are, as might be expected, obtained there.

Besides the winter season, which begins in November, Ventnor also has a summer season from June to September. It has good sea-bathing in deep, clear water.

"Ventnor is situated upon a part of a huge land-slip from the chalk hills at the southern part of the island. The subsoil, therefore, upon which the town stands is of a heterogeneous nature, consisting of chalk, upper greensand (firestone), and gault, so indefinitely arranged in relation to each other that it is often impossible to predicate upon what sort of rock it is on which any particular part of the town rests, or what would be found at any particular depth on digging down into the ground. Behind the esplanade is a natural hollow or excavation termed ' The Chicken Pit,' where the subsoil is gault or ' blue slipper,' as it is locally termed. Houses erected here are much frequented by visitors on account of their protected situation."

It has been pointed out (H. Weber) that Ventnor has "varied climatic characters" in its different parts, some being close to the sea, some thirty to sixty feet above it, while some are more exposed to the sun's rays, and for a longer time than others ; and some, again, are nearer the shelter of the rocks, while others are in more open and airy situations.

The climate of Ventnor partakes of that of the whole Undercliff, the name given to the curious land-slip which forms a kind of terrace six miles in length on the south-eastern coast of the island, stretching from near Bonchurch to Black Gang Chine. This district enjoys protection from the north, north-east, north-west, and west. The rain that falls is rapidly absorbed by the chalk and sandstone rocks, so that the ground, which is almost wholly rocky, is generally dry. There is plenty of space for out-of-door exercise, and the warmth of the sun is reinforced by reflection from the cliffs, and in some parts by reflection from the sea. Its elevation of fifty to sixty feet above the sea prevents those irritating effects which some suffer from when on a level with it.

The climate then is mild and equable, but at the same time fairly dry and tonic, and is adapted to cases of early phthisis and to those who have reason to fear they may become phthisical, to chronic catarrhal throat affections, to scrofulous and anæmic and debilitated persons, and to convalescents from acute diseases.

Yarmouth, on the north-west coast of the island, and *Freshwater* and *Alum Bay* on the western, come under the influence of the Atlantic, and in the two latter especially fine breezy sea air can be enjoyed in comparative quiet and seclusion during the hot months of summer.

Bournemouth is well known as a winter resort for persons with delicate chests and others, and it is coming into vogue also as a place for summer visitors. Indeed, there is no reason why a place of *winter* resort at the sea-side should necessarily be unfit for *summer* resort, as is popularly supposed, for the possession of an equable climate, which renders such a place warmer than inland districts in winter, often also renders it cooler than those places in summer; and it has been said of Bournemouth that "the highest temperature registered on the hottest days of the year are often eight, nine, and ten degrees below those of several summer resorts celebrated for their coolness in hot seasons."

Yet there may be a source of fallacy lying concealed in a statement of this kind, for the cool and refreshing climate of any summer resort depends to some extent on the variations between day and night and sun and shade temperature; so that, although one place may during the day register a maximum temperature some degrees lower than that of another place, yet if its nights are comparatively warm and

there is less difference between day and night temperature, this equability will be a positive drawback,
and such a climate will in hot seasons seem relaxing
compared with another where there is a considerable
fall of temperature during the night, for this helps to
mitigate the effects of the oppressive heat of the
day.

As a winter resort Bournemouth has many advantages. The houses chiefly occupied by visitors
are situated on the higher ground to the east and
west of the sheltered depression, occupied by the
older part of the town. Many of the villas are built
in the midst of the pine trees, which here cover a
considerable tract of the sandy soil. These pine
forests not only afford a certain protection from
winds, but they also exert a salutary influence on the
atmosphere, and embalm the air with their aromatic
exhalations. The soil, composed of sand and sandstone, is dry and absorbent, so that the rain falling on
the surface rapidly drains away, and the atmosphere
is left drier than it would be were it retained longer
on the surface of the ground by a less readily perme-
able soil.

There is fair shelter from the north and north-
east, and, to a less extent, from the east winds ; but
the surrounding hills are low, and the protection they
afford must be at times very imperfect. Bournemouth
is, however, much exposed to the west, south-west,
and south winds. There is more shelter from cold
winds on the eastern than on the western cliff.
Compared with Torquay it is somewhat less sheltered,
but is drier and more bracing. It has a moderate
rainfall, and a medium degree of atmospheric humidity.

It would be an error to regard Bournemouth as a
perfect winter climate, which, indeed, it is far from
being.

As there is no esplanade or promenade by the sea,
with covered seats, as at St. Leonards, and as the
place consists chiefly of private houses, scattered
through the pine forest, it is found very dull by

invalids who go there without their families, and
without friends there. Owing, no doubt, to the porous
nature of the soil, the presence of pine trees with their
exhalations, the shallowness of the bay, and the
elevation of the residential part above it, the smell of
the sea in the air is absent to a remarkable degree.

Chest cases are said to do better at Bournemouth
during the spring than at other places, but such a
general statement is most difficult to prove or
disprove.

The adjacency of many interesting and picturesque
places of resort ought to add to its popularity as a
summer resort. Steamers run daily, in fine weather,
to the Isle of Wight, Weymouth, and Swanage, and
the beautiful scenery of the New Forest is readily
accessible.

"The smooth hard sands, and the almost uniform
height of the tide, permit the bathing to be peculiarly
safe and agreeable."

Swanage, a branch line to which has recently been
completed, is the chief place in the " Isle of Purbeck,"
and is reached in about three-and-a-half hours by fast
train from London. It can also be approached by
steamers from Poole, Bournemouth, or Weymouth.
Its situation is charming. It lies quite open to the
south-east, and is pleasantly cool in summer. It
commands extensive and varied views of the coast of
Hampshire and the Isle of Wight, which is only
fifteen miles off. It has smooth and fine sands with
excellent bathing. There is a good pier, and the
steamboats land and take up passengers from it. Not
the least of its attractions are the quiet and comparative
seclusion it offers.

Charles Kingsley speaks of it thus: "At the east
end of the Isle of Purbeck is a little semi-circular bay,
its northern horn formed of high cliffs of white chalk,
ending in white isolated stacks and peaks, round
whose feet the blue sea ripples for ever. . . . The
southern horn is formed by the dark limestone
beds of the Purbeck marble. A quaint old-world

village slopes down to the water over green downs, quarried, like some gigantic rabbit-burrow, with the stone workings of seven hundred years. Landlocked from every breeze, huge elms flourish on the dry sea-beach, and the gayest and tenderest garden flowers bask under the hot stone walls. A pleasanter spot for summer sea-bathing is not to be found eastward of the Devon coast than Swanage." The old town consists almost solely of one long narrow street of gray stone-roofed houses, stretching up the hill which forms the southern horn of the bay. There are good lodging-houses on its outskirts, and the Royal Victoria Hotel has a good reputation. To the north the cliffs are very precipitous, and rise to the height of nearly 600 feet above the sea. The surrounding country is very interesting, and a variety of agreeable excursions can be made into it; especially to Corfe Castle, St. Aldhelm's Head, Lulworth Castle, etc.

Swanage is particularly appropriate as a sea-side resort to persons requiring quiet and soothing sur-roundings in a cool and somewhat sedative atmosphere, and opportunities for exercise in the open air amidst pleasing rural scenery.

Weymouth, distant four hours by rail from London, is a popular resort on the Dorsetshire coast, which here runs out towards the south, and bending a little eastward forms a wide, open bay which looks to the east. A projection of the shore divides this bay into two parts, Weymouth Bay and Portland Road. The new part of the town is built along the curving shore of the bay, and commands a fine view of the coast to the east as far as St. Aldhelm's Head. There is a fine esplanade and a pier of stone and wood, and these afford good promenades by the sea.

· There is some peculiarity about the tides of Wey-mouth. "There is generally a secondary tide—a slight flow and reflux—which takes place after the lowest ebb," and is locally known as the *Gulder*. Steamers leave Weymouth daily for the Channel Islands and for Cherbourg, and in the summer small steamboats run to Lulworth, Swanage, Bournemouth, and the Isle of Wight. The reputation of Weymouth as a sea-side resort dates from the middle of the last century. Its chief attractions are its wide, open bay, and its smooth level sandy shore affording great facilities for sea-bathing. The Isle of Portland is about four miles distant, and can be visited by steam-boat or by rail.

Lyme Regis, on the Dorsetshire coast, is five miles and a half from Axminster station, which is five hours and a half from Waterloo by fast train. It is most picturesquely situated, being built on the slopes of a hollow on a wild, rocky coast. Miss Austen said of it: "He must be a very strange stranger who does not see charms enough in the immediate vicinity of Lyme to make him wish to know it better." It is sheltered from the north and east winds, so that its winter climate is mild, while the sea breezes make it cool and fresh in summer. There is good bathing on pleasant sands. The place is quiet, and there is not much in the way of social distractions, but the surrounding scenery is charming.

Sidmouth is the first sea-side resort of any consequence we come to on the south coast of Devon. It is between six and seven hours from London by rail (S.W.R.). It is situated at the mouth of a valley running at nearly right angles to the coast, and enclosed by lofty hills terminating seawards in the sheer precipices of Salcombe and High Peak, about 500 feet above the sea. The coast view from the beach is admirable, owing to its situation in the centre of the Great Bay, which is bounded to the east by the Isle of Portland, and to the west by Start Point. "The characteristic features of the sea view are the blood-red cliffs, which rise to a height of about 500 feet above the beach." The air is mild and fine, and at times moist and relaxing. Numerous interesting excursions can be made amongst the neighbouring hills and valleys.

Exmouth, between ten and eleven miles (railway) from Exeter, is a popular watering-place on the Devon coast. *Budleigh Salterton*, five miles from Exmouth, is much more sheltered, and has a warmer climate, as it is placed in a narrow valley, and is well protected from cold winds.

The situation of Exmouth is a fine one. It stands on the slope of a somewhat steep hill at the mouth of the Exe, and commands not only a fine coast view, but an extensive range inland over the cultivated country in front of it, and the barren moors in the distant background. The Haldon ridge, at an elevation of 800 feet, is about eight miles off, and forms a great feature in the landscape. There are beautiful promenades, especially on the Beacon Walks and along the Strand, which is bounded by a sea wall. The sands are extensive, and afford excellent facilities for bathing.

Dawlish is about midway between Exmouth and Teignmouth. It is a pleasant and popular resort, with many attractions to recommend it. It is in a sheltered valley on a sunny coast, with neat, well-built houses, some facing the sea, others built round public gardens, and it has very picturesque rural scenery surrounding it. It has a bright and cheerful aspect, which is not a little enlivened by the railway running across the mouth of the valley close to the sea. The line here skirts "the very edge of the sea, and, piercing several headlands, has a fine effect, especially when a train is approaching."

There is a fine esplanade by the side of the line of railway. There is safe and pleasant bathing on beautiful and extensive sands. The climate is, perhaps, as mild as Torquay and as equable, and living is quieter and less expensive. The cliffs of the bay are of bright red sandstone, and some of them assume fantastic forms.

The hill of Little Haldon is only about two miles distant. From its summit, 818 feet above the sea, a fine view is obtained, commanding the mouths of both the Exe and the Teign.

Dawlish can certainly be commended as a charmingly quiet, and yet bright and cheerful resort, with a mild climate, cooler in the hottest part of summer than most of the sea-side resorts near London, and with a sedative and at times somewhat humid, relaxing atmosphere. It is well suited to delicate invalids and children who do not tolerate a too bracing sea-side climate.

Teignmouth is a pleasant town on the south coast of Devon, only three miles from Dawlish, and fifteen from Exeter. It is situated on the estuary of the river Teign, Shaldon being on the opposite side. Like Dawlish and Exmouth, it faces a minor depression in the west of that large bay, which is bounded on the east by the Bill of Portland, and on the west by the Start Point ; adjacent to this depression in the coast and west of it is the much deeper indentation known as Torbay.

Teignmouth looks south and south-east, and is protected to the north by somewhat distant hills, which at Waldon Heath, two miles from the coast, rise to an elevation of 800 feet above the sea.

The surrounding scenery, like that adjacent to most of the coast, is exceedingly picturesque and varied, the rich red soil is most productive, and the luxuriant vegetation bears witness to the comparative mildness of the climate and the fertility of the soil.

The climate of this district is characterised by milder winters and cooler summers, by greater equability of temperature, than is found inland or further east.

January appears to be the coldest month in the year, and *July* the hottest.

The following are some meteorological facts, founded on an average of ten years' observations, applying to these two months, and which afford a very fair criterion of the climatic elements which are found on this part of the south coast.

January—Mean temperature	.	. .	41·8 Fahr. (shade).
Highest maximum temperature			58·0
Lowest minimum	,,		15·7
Mean of all maxima	,,		46·8
,, minima	,,		36·8
Humidity, 9 a.m.	.	. . .	88·0
Mean total rain in inches.		. .	3·40
,, number of wet days		· .	16·0
July—Mean temperature	.	. .	61·8
Highest maximum temperature			84·9
Lowest minimum	,,		45·2
Mean of all maxima	,,		68·5
,, minima	,,		54·1
Humidity .	.	. . .	76·0
Mean total rain in inches		. .	2·33
,, number of wet days		. .	13·0

March is the driest month, with a mean of twelve wet days and a rainfall of 1·98 inches; the next driest is May, with also twelve wet days and a rainfall of 2·3 inches. The wettest month is October, with eighteen wet days, and a rainfall of 4·77 inches.

There are 180 wet days, distributed pretty equally over the whole year, and thirty-seven inches of rainfall annually.

The defects of this climate are very evident from these data, viz., a high degree of atmospheric humidity—a large proportion of wet days—often a thick and cloudy sky, which, together with its equability of temperature, form a decidedly relaxing climate, modified somewhat by the tonic effect of the sea air.

Torquay is one of the most popular and fashionable winter resorts in England. It enjoys a magnificent situation, facing one of the finest bays in any part of our coast. It is encircled by hills, which shelter it from the north and north-west, and, to some extent, from the north-east.   North-east winds are, however, sometimes felt there severely, as was the case this last spring.

The hill to the north is called the Braddens, that to the west, covered with firwood, Waldon Hill, and that to the east, Park Hill.   The villas stretch up these hills at various elevations above the sea, up to 450 feet.

The air of Torquay is said to be drier than at any other place on this coast ; but different parts of the town differ greatly, the nearer the sea the more sedative and relaxing the air, the higher up the hills the more bracing it becomes.

The luxuriant growth of sub-tropical plants testifies to the general mildness of the climate.   Its climate is no doubt very equable, more so even than that of Teignmouth, but, like other neighbouring resorts on this coast, there is frequently a great deal of humidity in the air.   The rainfall is considerable, and the number of rainy days very great.   If, for instance, we compare the rainfall and the number of rainy days at Torquay and at Hyères, we find at Torquay the rainfall is 39·60 inches, at Hyères it is 27 ; while the rainy days at Torquay are 200, and at Hyères, 63.

It is claimed for Torquay that it is much cooler in summer than the sea-side resorts further east.   " In addition to the cooling effects of the sea, there are always, even in the hottest of summers, cool breezes on the sides and tops of the hills."   Bathing, boating, yachting, and sea-fishing, can be pursued with great advantage at Torquay, and beautiful excursions into the surrounding picturesque country, extending even to the wild hills of Dartmoor, can easily be made.

" The water supply is the purest supply of any town in the British Isles, on a par even with that supplied to Glasgow from Loch Katrine.   It is collected from one of the moorland streams sixteen miles from Torquay, the supply is inexhaustible, the reservoirs hold three hundred millions of gallons, and the constant service system is in force.

"A few years ago an entirely new and complete system of sewerage was constructed at a cost of over eighty thousand pounds, and which allows the sewage to flow away uninterrupted by tidal action, and at a safe distance from the town. The sewers and drains are well ventilated, and receive an amount of flushing from the numerous streams of water which have been turned into them, such as probably no other town can boast of."

The town appears to be exceedingly healthy, and the death-rate is low. The cases which appear to do well at Torquay are those of chronic bronchitis in old people, some forms of chronic phthisis, with tendencies to catarrhal attacks, some irritable throat affections ; also young and delicate children during the trying periods of rapid growth and development, and some elderly people, who find themselves more comfortable in a moderately mild and soft climate rather than in a dry and bracing one.

Torquay can now be reached in a little over five hours from London, by the Great Western express trains.

Salcombe, about eighteen miles west of Dartmouth, and nine miles west of the Start Point, lies on a protected inlet in the southernmost part of the coast of Devon. It is scarcely at all known, although by its position it is so sheltered as to be one of the warmest coast towns in Britain.

"The myrtle and other tender plants clothe the shores ; the lemon, orange-tree, and aloe flower in the gardens, but beyond the protecting influences of this ridge on the coast, the country consists of bare, bleak hills."

Notwithstanding its exceptional protection and warmth, it is too "cabined, cribbed, confined," to become a popular resort.

Falmouth, on the south coast of Cornwall, is beautifully situated on the shore of one of the finest harbours in England. The town itself is not very attractive, but from the ramparts there are grand views of the adjacent coast, and a great variety of interesting excursions about this picturesque coast can be easily made. Although so far from London—nine hours by express train—its reputation as a sea-side resort is rapidly growing, especially as a residence for invalids during winter and spring, on account of its mild climate.

Many houses and handsome detached villas have been built on the heights above the old main street of the town.

Penzance is the westernmost of sea-side resorts on our coasts. Its great drawback is its distance from London, but there is one fast train in the day which is timed to accomplish the journey between Paddington and Penzance in nine-and-a-quarter hours.

The town is built on the west side of Mounts Bay, so that it has a south-east aspect. It commands a fine view across the bay, and the long, low line of coast to the Lizard. It is sheltered by a lofty plateau on the west and north, but it is fully exposed to winds from other quarters. Notwithstanding this exposure to the winds, its temperature is remarkably equable, like that of the Scilly Islands, forty-two miles further west. These possess "the most equable temperature in the British Islands, if not in all Europe." At St. Mary's, Scilly, the mean winter temperature for four years was 47·9 degrees; the mean maximum, 50·5 degrees; the mean minimum, 44·5 degrees; the mean daily range, 6 degrees; the mean monthly range, 18·7 degrees; the mean relative humidity, 89 degrees; number of rainy days, 107; rain in inches, 17·13; winds from north and east, 56 days; from south and west, 90 days.

Penzance is 5 degrees warmer than London in winter; 1 degree warmer in spring; 2 degrees warmer in autumn; and 2 degrees cooler in summer. It has rather a large rainfall, 43 inches, and a large number of wet days.

There are many interesting excursions to be made from it both by land and sea.

For persons whom a mild, equable, somewhat

humid climate agrees with, Penzance is exactly suited. It is sedative, and the prevalence of winds from the sea must also confer upon it a certain tonic element.

Turning round the Land's End to the north coast of Cornwall, we pass the at present but little known sea-side towns of St. Ives, New Quay, and Bude, and, reaching the north coast of Devon, the first of the picturesque resorts on this coast we come to is Clovelly. This charmingly romantic fishing village can scarcely be called a health resort, and must not detain us now.

We soon, however, reach *Westward Ho.* This is a comparatively new resort twelve miles east of Clovelly. The coast immediately around it is flat and not very interesting, but very beautiful scenery is not far distant. It is a quiet retreat, and some might even think it dull compared with its fashionable neighbour Ilfracombe. It has, however, fine bracing air, and a long reach of tolerably firm sands. Its golf club will be an attraction to some, the ground upon which this game is played, the "Northern Burrows," being said to be "only surpassed by 'the Links' of St. Andrew's, and Musselborough."

Ilfracombe, on the north coast of Devon, offers great attractions to the lovers of the picturesque, situated as it is in a part of England unrivalled, perhaps, for the beauty of its scenery.

It affords remarkable facilities for sea-bathing. At one spot several acres of beach are enclosed by a sea wall, and separate portions appropriated to ladies and gentlemen. It has also a very fine swimming-bath, filled at every tide.

At present it is almost exclusively a place of summer resort, the season beginning with July and ending in September; but its winter climate has also been recommended for invalids who can bear a certain amount of bracing wind, as its winter temperature is only half a degree lower than that of Torquay, while it has less rain.

The promenades and excursions are numerous. "The most favoured resort is the Capstone Parade, which surrounds the base of the hill of the same name, and is provided with numerous seats, some of which are sheltered, even in a gale of wind.   The hill itself is an abrupt knoll, clothed with turf, to the summit of which are several winding paths.   The best possible view of Ilfracombe is to be obtained from this point. Another much-frequented walk is the Tors Walk, westwards along the cliffs, which affords an extensive view of the Bristol Channel, and of the varied stretch of coast from long, low Killaye on the east, to Bull Point, with its lighthouse, on the west, and the little bay of Lee between them.   Of the many delightful excursions from Ilfracombe perhaps the most enjoyable is that by coach to Lynton and Lynmouth (seventeen miles) along the splendid coast road to Combe Martin.   The scenery passed through on this journey is perhaps the most beautiful of its kind in England.   Other trips which should not be missed are those to Chambercome and back by Stele (two-and-a-half miles).   There are many other enjoyable trips to be taken, notably one by steamer to that charming spot Covelly."

Ilfracombe is six hours and a half by rail from London.

Lynton and *Lynmouth,* only seventeen miles by coach-road from Ilfracombe, are amidst the most exquisite scenery of the North Devon coast.   Lynton is built on hilly ground, at an elevation of 430 feet above the sea, while Lynmouth lies on the sea-shore, at the mouth of the Lynn.

" The air of Lynton is bracing but warm—lofty hills sheltering it on the east, west, and north.   The temperature of Lynmouth is warmer, and at all times very equable.   Lynton has the disadvantage of being separated from the sea by one of the steepest of hills, which is traversed by zigzag paths ; but the grand prospects that it offers, of which perhaps the loveliest is that of the gorge through which the East Lyn flows, more than compensate for this drawback. There is no flat sandy beach, and the place for bathing is rocky, and limited in extent, while the rocks themselves, though grand in the extreme, are dangerous. The spring tides, too, run very swiftly here, and those who are unable to swim, or who are not adepts at the art, should be cautious how they venture into the water. Thus the district is hardly the one to which to take young children, nor, owing to its hilly nature, is it altogether suitable for invalids or weakly persons. The one way to see the locality is on foot, for not only is carriage travelling expensive, but those who cannot walk will miss some of the most charming features. To those, then, who are in the enjoyment of good health, and who go out of town in search of quietude and repose, not of excitement— for the amusements of the place only comprise a couple of reading-rooms and fishing—to such it would be difficult to discover one single drawback in Lynton and Lynmouth. The drainage arrangements and the water supply are alike excellent. The district is most frequented in July and August; by the close of September it is almost deserted. The points of interest are as beautiful as they are numerous. The Valley of the Rocks and Watersmeet (the junction of the East and West Lyn) may be said to excel in exquisite loveliness. One of the nearest rambles is to Glenlyn, along the beautiful course of the West Lyn. As to whether Lynton or Lynmouth is the more convenient point of departure, it may be said, generally speaking, that Lynton is better situated for expeditions westwards, while Lynmouth offers greater facilities for the exploration of the glen scenery to the east. Lynmouth, however, has this advantage, that lodgings are cheaper there, and, with the sea-side and river in closer proximity than at Lynton, is usually preferred as a place of residence."

There are several routes from London. That by London and South-Western Railway to Barnstaple, and thence by coach to Lynton, is probably the best. Another route is by the Great Western Railway to Minehead, from which there is a coach to Lynton.

Weston-super-Mare and the adjacent resorts, Mine-head and Clevedon, on the shore of the Bristol Channel, have the disadvantage of large muddy sand-fields at low water, and their climate is rather relaxing than bracing. Weston is, however, greatly frequented for sea air and sea-bathing by the inhabitants of some of the large towns in the west of England, Bristol, Bath, etc., and by people from the Midlands. The season lasts from July to October, when there are from three to four hundred lodging-houses occupied by visitors. The sanitary condition of some of these would appear to be anything but satisfactory.

" The town is sewered, and is provided with water derived from springs which appear near the surface in a vein of dolomite conglomerate at the foot of the Worle Hill, about a mile to the eastward of the town. A scheme has been started for building over the locality of the springs, an arrangement which must be viewed with considerable apprehension, as affording imminent risk of pollution ; and it behoves the authorities to take efficient steps to prevent any danger to the water supply of the town. The supply is said to be plentiful and wholesome. From the springs the water is pumped up to a reservoir, whence it passes by gravitation to the town on the continuous system.

" Some of the houses are, however, dependent upon wells for their drinking water, though the number of cheap houses is rapidly decreasing. Considering that the town is traversed by drains and sewers of a more or less imperfect character, and that the soil is polluted by soakage from surface nuisances, such wells must be regarded as affording a dangerous water supply. There are no means of systematic flushing of the sewers; a matter of especial importance, seeing that many of them are laid very flat."

Of the lodging-houses the medical officer of health reported in 1883, "the sanitary arrangements in numbers of these houses are such as could never ·be sanctioned by the local board in the case of houses about to be built."

Passing now to the opposite coast,

Tenby is the most important sea-side resort on the south coast of Wales. It has a decided sea climate, mild and fairly dry. It has the disadvantage of being a long way (274 miles by rail) from London, a journey which is accomplished in about nine hours by fast train. It is thus described by Dr. D. A. Reid : " The town stands upon the western side of Carmarthen Bay, on the south-west coast of Pembrokeshire, a little to the north-east, and protected by Caldy Island. It is built upon the point and north-eastern margin of a rocky peninsula, composed of mountain limestone, rising to nearly 100 feet above the sea. On the North Cliff a new suburb is springing up, from whence an uninterrupted view of land and sea for many miles can be obtained. The margin of the peninsula on which the old town is built is concave, and includes within its wooded cliffs the bay, harbour, and pier, which are sheltered from the open sea by a rocky projection connected with the main peninsula by a low narrow neck, and crowned with the ruins of the ancient castle, under which is the rocky islet of St. Catherine, notable for its caverns, containing number-less zoophytes, sea anemones, star fishes, etc., and also for a large and formidable fort erected about fifteen years ago for coast defence. The new part of the town is on the southern side, and includes a hand-some row of houses, forming the esplanade (over-looking the south sands, which extend for nearly two miles westward), Victoria Street, South Cliff Street, etc. The sands, which cannot be surpassed by any watering-place in the kingdom, are most extensive on this side. The two tracts are separated by the Castle Hill, the property of the Corporation, which is laid out in convenient paths for promenade and well provided with seats. The houses are, as a rule, well-built and substantial, though some few are more adapted for a summer than a winter residence. In

the larger lodging-houses the drains are ventilated to
the roof, and every precaution taken to insure perfect
sanitation.   The water supply is obtained from three
or four sources, collected in reservoirs, and conveyed
by pipes direct to every house and cottage, remaining
at full pressure by day and night.   The reservoirs are
situated on high ground, respectively 130 and 70 feet
above the highest level of the town, and are fed by
springs of very pure water.   The drains are ventilated
and supplied with flushing chambers at proper intervals.   The greater portion of the sewage is conveyed
into the sea at a considerable distance from the
bathing places, and as the currents at the point of
discharge are very strong, everything is carried at
once into the Atlantic.   A small proportion is conveyed to a piece of land a mile from the town, and
there used to fertilise some fields.   Situated as Tenby
is, on ground from 20 to 100 feet above the sea level,
there is a good fall to all the drains, and, with proper
flushing arrangements, no sewage is likely to remain
in them."

The population of Tenby in 1881 was 4733.   In
the season this number is increased by several
thousands of visitors.   Tenby has its summer and
winter season.   The summer visitors are those who
seek the recreations afforded by the sea-side, and who,
on pleasure bent, delight in rambling on the shore,
bathing, and making excursions to the old castles and
other objects of interest in the locality.   As a winter
resort Tenby is becoming more and more popular as
a residence for invalids suffering from pulmonary and
cardiac affections.   The soft mild air, impregnated as
it is with iodine, the absence of frost, and the sheltered
situation, render it peculiarly suitable for such cases.
As an instance of the mildness of the climate, it may
be mentioned that during the whole of the year 1882
the thermometer never registered a lower temperature
than 32 degrees; this was on December 10th.   The
highest temperature registered for the year was 90
degrees on August 9th.   There is, the writer believes,

no recorded instance of epidemic disease existing in Tenby.

"Since Tenby has had a separate return the mortality has ranged from 14·2 to 8·1 per 1000, thus comparing most favourably with other health resorts. In the year 1882 only fifty deaths occurred, and three of these were visitors. This gives a death-rate of exactly 10 per 1000 of all ages. If, however, we exclude twelve persons over seventy years of age and eight under one year, the mortality will stand at 6 per 1000. One death was registered of a lady aged one hundred years and four months.

"Chief among the attractions of the town must be placed the beauty of the town itself and its surrounding scenery.

"But not only does Tenby amply repay the visitor by its own beauties, but it is a convenient centre from which to make excursions, for those so inclined, to such objects of interest as Caldy and St. Margaret's Islands, Lydstep, with its wonderful caverns, St. Govan's Head and Chapel, the Huntsman's Leap, Bosheston Mere, the Sunken Forests, Stackpole Court, the seat of the Thane and Earl of Cawdor, the Stack Rocks, with their myriads of sea-birds (notably the *guillemot*), the old castles of Pembroke, Manorbier, Carew, Llawhaden, Amroth, Cilgerran, and Picton, and also the Royal Dockyard at Pembroke Dock, Milford Haven, and the ancient and beautiful cathedral of St. David's.

"The bathing at Tenby is singularly good and free from danger. The sands are hard, gradually shelving, and beautifully clean. On the north shore the beach is more pebbly, and perhaps safer, because more sheltered. To the more timid bather it is to be preferred. There are excellent machines on both shores. Boating is a popular amusement during the season, and there are comfortable and safe pleasure-boats always to be hired, as well as rowing-boats for those who prefer them. Sea fishing may be had all the year round. There is also trout fishing within a few miles."

H

At Gumfreston, a mile and a half from Tenby, there is a chalybeate spring, which is said to " resemble that of Malvern in its purity, and that of Tunbridge Wells in the quantity of iron it contains, exceeding all other chalybeate waters in Great Britain in the large quantity of carbonic acid it holds in solution."

Aberystwith, on the shores of Cardigan Bay, is one of the chief resorts on the west coast of Wales. It can be reached by fast train from Euston in eight hours, *via* Shrewsbury and Welshpool. It is sheltered by hills from the north and east, but it is exposed to the west, south-west, and north-west. Its climate is somewhat mild and humid, but it is said to prove both soothing and invigorating to those whose nervous systems have become irritable and enfeebled by over-work. It has a mild and equable temperature in winter, 8 degrees warmer than some of the nearest inland towns. The beach is pebbly, and slopes off gradually to deep water, and is considered very safe for bathing, though not, of course, so pleasant as sand. The drainage is good, and the water supply, which is abundant and pure, is derived from a lake near the top of Plynlimmon, fifteen miles from the town.

A few miles to the north of Aberystwith, on an arm of the bay, is *Aberdovey*, a small sea-side resort of recent growth, with a southern aspect and mild climate.

Another neighbouring resort on this coast is *Towyn*, with very firm safe sands for bathing. It is rapidly becoming popular.

Barmouth is, like Aberystwith, on the shores of Cardigan Bay, but some little distance further north. It has much the same climate, the same protection from the north and east, and the same exposure to the west. Owing to the nearness to Barmouth of some of the finest of the mountain scenery of Wales, it has become of late years a favourite summer resort, but it has the disadvantage of large muddy sand-fields at low water. Its winter climate is very equable.

It is said to be dry and mild, and suitable as a winter residence for pulmonary invalids, but we are not aware of the existence of reliable meteorological statistics extending over some years, upon which only ought such an opinion to be based. The soil is sandy, and the rainfall is soon absorbed.

"The town is situated on a rocky eminence, facing southwards, and the houses are mostly built in terraces one over the other on the slope of this hill. A system of drainage with an outfall into the sea at a distance from the place has recently been completed, and is reported to act well. The water supply is derived from a reservoir about a mile from the town, and is soft and clear. The bathing is very good. The beach is sandy, with no shingle, and the descent into the water is gradual. A variety of fine views may be obtained from an arm of the sea called the Mawddach, which is navigable for upwards of seven miles above the estuary at Barmouth, and is much resorted to for boating purposes. The best time to visit Barmouth is in the early summer. In August the heat is too great, and the glare from the omnipresent sand too strong for perfect comfort."

Beaumaris, on the Menai Straits, fully exposed to the east, has fine bracing air and beautiful sea and mountain scenery; the prospect is further enlivened by the frequent passage of shipping. There are numerous walks and rides in the neighbourhood. There is a steamer runs between Beaumaris and Liverpool in summer, calling at Llandudno. Some of the finest scenery in North Wales is within an easy distance.

Llandudno, on the north coast of Wales, occupies an exceptional position, and enjoys an exceptional climate, owing to the protection it receives, chiefly from the Great Orme's Head, rising to an elevation of 678 feet above the sea, which shelters it from the west, north-west, and north. It is also protected by a lower range connected with the Little Orme's Head, to the south and east.

H 2

The town lies in a valley formed by these two elevations, and this valley opens at both ends to the sea—to Llandudno or Orme's Bay on the north, and Conway Bay on the south-west.   Owing to this position, all the winds to which it is exposed have come over the sea, and as a rule have had their temperature raised thereby ; they are, however, often violent and sometimes rather cold.

As a summer resort Llandudno is highly and justly popular on account of the freshness of its atmosphere, its two beautiful bays, and the admirable facilities for bathing enjoyed there in almost all weathers.

"The East or Llandudno Bay, which extends all the way between the Great and Little Ormes, and forms a circular arc of two miles in extent, is at present the only place used for bathing purposes. For such purposes it is well adapted, because of its freedom from dangerous currents, the firmness of its sands, the complete absence of sewage, and such gentle shelving of the foreshore as to permit sea-bathing at almost any state of tide."

An efficient system of sewerage has been carried out at great cost, and a good water supply is obtained from two lakes thirteen-and-a-half miles from the town on the western side of the Conway river.   The supply is unlimited, and the service continuous, the water itself is exceedingly pure and soft.

"The Orme's Head Marine Drive is one of the great attractions of the place.   From end to end this is about five miles in length.   There is probably no drive equal to it in the kingdom for its uniqueness and beauty.   Its air—a fine combination of sea and mountain air—is exceptionally pure and exhilarating, and the picturesque views and varied scenery it commands can scarcely be rivalled."

Llandudno has pretensions also as a winter resort, especially for invalids who can support a certain amount of windy weather.   Its annual rainfall— 32·36 inches—is somewhat below the average rainfall of England and Wales, which is 35 inches. The mean humidity of the air in winter is 82 degrees, so that it has a drier winter atmosphere than some of our south-west coast resorts; this is partly due to the porousness of its subsoil, which is mostly gravel and sand.

During the five winter months (November to March), from observations extending over five years, it would appear that the mean winter temperature is 43·5 degrees; the mean maximum, 48·3 degrees; the mean minimum, 38·8 degrees; the mean daily range, 9·5 degrees; the mean monthly range, 28·3 degrees. North and east winds blow on 44 days, south and west winds, 107 days. There were 83 rainy days, and a rainfall of 15·96 inches.

Nothing, perhaps, could show better the character of our winter climate than the fact that a place claims to have attractions as a winter resort, where, on an average of five years, there have been 83 wet days out of 151—more than one-half!

Rhyl is only twenty-four miles west of Liverpool, and six-and-a-half miles east of Llandudno. It has little that is attractive, beyond its pure sea air and firm and extensive sands, but it is a popular and convenient sea-bathing resort for the inhabitants of Liverpool and adjacent towns.

When the weather is clear there are fine views from the beach of the Great and Little Orme's Head, and some of the Welsh mountains.

New Brighton, at the mouth of the Mersey, and close to Birkenhead, with a north-west aspect, looking partly out to sea, and partly over the opposite Lancashire coast, is one of the most accessible sea-side resorts for the inhabitants of Liverpool, by whom it is much frequented. The town is built on ground which rises considerably above the sea level, and commands extensive views, both over land and sea. The beach is sandy, and affords good bathing. Public baths, including a very fine swimming-bath, have been erected, and a winter garden is attached.

A sandy beach and a wide stretch of sand-hills, about seven miles in length, connects New Brighton with Haylake, another sea-side resort, at the mouth of the Dee. The drainage of New Brighton is good, and a complete system of sewerage is in operation. The death-rate is low, 12·21 per 1000 in 1881.

The climate is decidedly bracing, as the situation is much exposed to winds. It is found very suitable for some convalescents, especially after surgical operations; and there is a large female convalescent home there; but the want of protection from prevailing winds must be much felt in winter.

Southport, on the Lancashire coast, is a popular sea-side resort for some of the large towns of the north of England, as it is only eighteen miles by rail from Liverpool, and thirty-seven from Manchester.

"It lies between the estuary of the Mersey and the mouths of the Ribble, at a distance of about eighteen miles from the entrance to the former. Throughout almost its whole extent the sea-border here presents a continuous range of sand-hills, upon the outer or western side of which there is a broad belt of level sand covered with water at high tide, but left bare during a considerable portion of every twenty-four hours. Inland from the sand-hills, the country is flat for a distance of several miles, but then rises with very agreeable undulations.

"The Marine Promenade, including the recent extensions north and south, now measures no less than 2600 yards in length. Raised well above the level of the sands, excellently paved and asphalted, and bordered by handsome terraces with hotels and other large buildings, few sea-side roadways of the kind supply a more inviting place for exercise. The coast of Lancashire has always been noted for the beauty of its sunset views, and these are certainly nowhere witnessed to greater advantage than at Southport.

"The pier is one of the longest in England (1465 yards), and at the extremity is expanded into a platform of 180 feet in length and proportionately wide. The water goes out a great distance, but leaves behind it broad and firm sands.

" Amongst the other attractions of the place are the Atkinson Art Gallery and Free Library and the Winter Gardens, which have an area of about nine acres, and include an aquarium, pavilion, and one of the finest conservatories in the United Kingdom. The town possesses a public park of thirty acres, near which are the Churchtown Botanical Gardens of twenty acres. The Convalescent Hospital and Sea-bathing Infirmary at Southport is now the second in the kingdom in extent, being surpassed only by the institution at Margate."

The climate of Southport seems well suited to the various classes of invalids who require the tonic and alterative influence of a bracing sea air. Owing to the sandy nature of the soil, the atmosphere is fairly dry, but the "changes of temperature during the same day are frequently considerable," so that the statement made by the same authority that " Southport enjoys a remarkably *equable* climate" must need modification. It is scarcely necessary to say that great diurnal variations of temperature are inconsistent with the possession of "a remarkably equable climate!" The same writer states that cases of asthma do extremely well there.

Blackpool, on the Lancashire coast, facing the Irish Sea, has become a very popular resort in the north of England. It lies about midway between the mouth of the Ribble and Morecambe Bay. It is only sixteen miles from Preston, and is also readily and quickly reached from Manchester and Liverpool. From London it is about six hours by London and North-Western Railway.

Blackpool has no protection to speak of, and lies fully open to the westerly winds which prevail on this coast. It has a fine sea-port, bracing air, and miles of good sands for bathing. It is a good centre for excursions, and a variety of entertainments are provided in the place itself for the amusement of visitors.

It is said that the easterly winds of spring are felt less trying here than further south.

"The promenade and carriage drive runs for a length of nearly four miles, and from it a series of remarkably fine views can be obtained. Walking or driving towards the north, the Isle of Man, which is sixty miles distant, can be frequently seen. The coast on the further side of Morecambe Bay is also a special feature in the landscape. The busy town of Barrow-in-Furness comes out in its foreground, and the whole range of those Lancashire and Cumberland mountains, the lofty peaks of which look down on the enchanted spots which give the Lake District its name and fame, make up the background. Turning to the south, the ships that 'come and go' to the famous port of Liverpool are outlined on the horizon. Beyond their track looms up the Great Orme's Head, and beyond that the wild Welsh hills. In fact, the whole coast line from Anglesea to the mouth of the Mersey is distinctly visible. There are few walks in England from which so rare and so varied a view can be obtained. There are at Blackpool, in addition to this splendid road, the two piers, both of which are extensively used for promenade purposes. Fine, well-fitted, and fast steamers, in charge of skilful commanders, ply every fine day in the summer to Southport, Liverpool, Morecambe, Barrow (for Furness Abbey), Douglas (Isle of Man), and Llandudno.

"As to sanitary matters, all that can be done to make them perfect is being done. The borough is well drained and sewered. The sewers are carried far out to sea, and discharged at a spot where it is practically impossible for them to produce any deleterious effect.

"The water supply, which is gathered upon the hills beyond Garstang, some eighteen miles from Blackpool, is excellent in quality and abundant in quantity. The rainfall is very moderate, and may be said to average thirty-two inches per annum."

Blackpool is pleasantest and quietest in May and June, and the first half of July, so also in the latter half of September and October. From the middle of July to the middle of September it is crowded with visitors.

Morecambe and *Grange*, both on the coast of Morecambe Bay, have considerable merit as sea-side resorts for those who inhabit the great manufacturing towns in this part of the north of England. Morecambe commands fine views of the bay and the Lake hills. It has a good pier, from which extensive views of the Lake mountains can be obtained in clear weather, and beautiful walks along the shore.

Grange is rising into popularity as a sea-side resort, on account of the beauty of the adjacent scenery and the protection it obtains from the lofty crags around it, which renders its climate a mild one and suitable for winter residence. Coaches run daily during the season to Lakeside, at the foot of Windermere, and return the same day.

Crossing to the eastern coast of England, the most northern of the watering-places on the Yorkshire coast is *Redcar*, which faces a fine open sea, and has extensive sands; but it is too near the manufacturing town of Middlesborough to make it an attractive resort to visitors from the south.

Saltburn-on-the-Sea, only a few miles from Redcar, enjoys a fine position on the coast between it and Whitby. Two picturesque wooded glens run down to the sea; a portion of these is enclosed, forming pleasure grounds, for entrance to which a small sum is charged. There are beautiful sands for bathing, and the fine, bold, lofty cliffs afford admirable promenades and fine sea-views. There is a lift for ascending from the sands to the cliff. Saltburn is very picturesquely situated, the town is new and well-built, and offers very considerable attractions as a bracing, healthy, sea-side resort.

Whitby is situated about midway between Salt-burn and Scarborough, and is reached in between six and seven hours from London by Great Northern Railway. It depends for its prosperity greatly on its fisheries, especially on its take of herring. It is also famous for the manufacture of jet ornaments, the material of which is found in the neighbouring cliffs. Iron shipbuilding also goes on there.

Whitby has of late years grown greatly in favour as a pleasant, bracing, summer resort, and on the West Cliff, where the visitors mostly reside, a spacious building has been erected with accommodation for concerts, balls, and theatrical representations, with refreshment rooms, billiard and reading rooms, and a fine promenade where a band performs twice daily. Lawn tennis courts are also provided.

The sea view is enlivened by the frequent passage of large ships close to the piers, of which there are two—east and west.

Boating is a favourite amusement, and regattas are held at Whitby in the season. There is good sea-fishing, and there is also good salmon and trout fishing in the river Esk. Whitby can boast of a public library such as is rarely found in a provincial town. It has been in existence for more than a hundred years, and contains a valuable collection of books.

The sea-bathing is very good; the sands generally are firm and smooth; and although the town sewage passes from the harbour into the sea, the constant "set" of the tide towards the south carries all the sewage-charged water in the direction of the East Cliff; consequently the bathing-ground, which is immediately in front of the West Cliff, is at all times perfectly pure and untainted. There is a boat, in charge of an experienced seaman, in attendance daily, to prevent accidents to the bathers.

The water supply is excellent—pure and abundant. It is conveyed from springs at Goathland, about ten miles distant.

There are many pleasant villages and woods within a short distance of Whitby. Three miles north is Mulgrave Castle, the seat of the Marquis of Normanby. About ten miles from Whitby is the fishing village of Staithes, where the celebrated navigator, Captain Cook, was apprenticed to a grocer.

Robin Hood's Bay, a village of considerable antiquity, lies to the south, and a few miles up the Esk is Egton, with the fine woods of Arncliffe, and the picturesque Beggar's Bridge.

There are many excellent lodging-houses on the West Cliff. The two best hotels are the " Royal " and the "Crown." August to November is considered the best season for Whitby. July is sometimes rainy, and the spring months are to be avoided. Its fine weather comes late in the season. The following description of Whitby is from " Harper's Monthly : "

" Whitby is much better suited to those who want relief from the pressure of an overworked and aggressive civilisation than Scarborough, and is, to our own mind, the most picturesque little town in Great Britain. . . . Let us take our first look at Whitby from the summit of the East Cliff—one of the two promontories between which the river Esk enters the German Ocean. On both sides of us is a precipitous line of coast, with bristling cliffs, washed by a boiling surf in some places, and in others fringed with a narrow beach, on which gigantic moss-covered boulders are piled. The sea itself melts in the extreme horizon. The ground at the summit is uneven, and ends in a precipice. Looking to the east and north, the embattled cliffs and the restless sea fill the views. Looking to the west, we see the river cleaving the valley, with the town built on both sides of it. The two characterising colours of the picture are red and blue. One house rises above another, apparently supported by the cornice of that below it ; the floor of one seems to be the roof of the other. The roofs are peaked and gabled, and dormer-windowed, with tall chimney-pots shooting up from them; nearly all of them are sheathed with crimson tiles, which, with the lazy blue smoke drifting over them, are the things that give colour to every picture . of Whitby. The colour and architecture are both foreign. The cold gray look of the usual English village on the coast is substituted by a delightful warmth and richness. Leading down from the summit of the East Cliff to the town is a curving flight of 196 well-worn steps, up which the worshippers come on Sunday to the old parish church which stands at the head of the cliff, surrounded by a full crop of gravestones, with the sea behind it. It is a very old building of the early Norman period; and the interior, with its undecorated oak and skylights in the low roof, is more like the cabin of a ship than a church. On the cliff also are the beautiful gray ruins of St. Hilda's Abbey, which are the crowning glory of Whitby."

Scarborough, the most popular of northern sea-side resorts, is situated in the North Riding of Yorkshire, at a distance of five hours and a half from London by Great Northern Railway.

The greater part of the town is built upon a site elevated more than 100 feet above the level of the sea, surmounting a range of precipitous cliffs. The Castle Hill on the north-east of the town rises to an altitude of 285 feet, it juts out into the sea, and serves to divide the town into two parts, the North Bay and the South Bay. It is on the borders of the latter that most of the attractions of Scarborough are situated.

Oliver's Mount, more to the south, is, at its summit, 500 feet above the sea.

Scarborough, it will be seen, is healthfully situated, and as it is well looked after, from a sanitary point of view, by the local authorities, it has high claims to favourable consideration as a tonic and exhilarating sea-side resort. For the health-seeker, pure and simple, it is perhaps a little too fashionable, too gay, and at times a little noisy; but these defects, from one point of view, become attractions from another, and render this town particularly delightful to the large class of persons who cannot enjoy life without a certain daily provision of lively distractions.

The death-rate from consumption there is said to be unusually low—1·7 per 1000 against 2·4 for the whole of England. Its mean annual temperature is 46·7 degrees Fahrenheit, or 42·6 degrees for the first 6 months, and 50·6 degrees for the last 6 months. The mean annual rainfall for the last 5 years was 31·67 inches. In 1883 the mean temperature was 48·5 degrees, and the rainfall 27·99.

Mr. Haviland says: "Its peninsular position ensures it an agreeable temperature, as proved by the small annual and daily range of the thermometer; whilst the bracing character of its air gives life and tone to those who breathe it."

The water supply is good, and comes from springs nearly four miles from the town, and the sewage arrangements are very satisfactory. There are fine sands for bathing, and there are chalybeate and saline springs at the Spa, of which Dr. Macpherson says: "They are not very powerful agents, but the visitors to the place are so numerous that some are tempted to make use of them. The wells are surrounded with handsome masonry, and are on the Esplanade. There is no question that they prove useful in dyspepsia, and sometimes in chlorosis, their action being supported by the fine sea air of the place."

Scarborough possesses, perhaps, the finest marine aquarium in the world. There are a constant succession of varied and high class amusements. The walks and drives in the neighbourhood are numerous and attractive.

Filey, seven miles from Scarborough, has the same fine bracing climate as the larger town without its noise and excitement. It stands on an elevated position above the sea, has excellent bathing and many interesting walks in its neighbourhood. The death-rate is reported to be very low, and for those who desire a healthy and quiet, but bracing sea-side resort, without the gaiety of a fashionable populous town, Filey will prove more attractive than Scarborough.

Hunstanton, at the mouth of "The Wash," is a popular sea-side resort for the adjacent towns of the Eastern Counties. It stands near a fine cliff, a mile long, and sixty feet high at its highest point. It commands an extensive sea-view, and there is a firm sandy beach for sea-bathing. It shares in the dry bracing climate of this coast.

Cromer, on the Norfolk coast, about twenty-two miles from Norwich, with which it is connected by railway, has long been a favourite resort with those who seek for a quiet holiday in good sea air, apart from the fashion and movement of a popular wateringplace. It stands high, sheltered by wooded hills, surrounded by pleasant inland scenery, and commanding a view of the bay called the "Devil's Throat," from the heavy and dangerous sea continually rolling in upon it. Many landslips have occurred on this part of the coast, where the sea has gradually encroached greatly on the land. The esplanade affords a pleasant walk by the sea, the inroads of which are checked by its stone wall. There is good bathing on a sandy shore.

The fossil remains found in the Cromer cliffs have made the neighbourhood most attractive to geologists.

There are many interesting walks of various lengths inland and along the coast.

Great Yarmouth, on the coast of Norfolk, facing nearly due east, is 108 miles from London, rather more than four hours by fast train. There is also regular steamboat communication with London during the summer season. It is one of the great seats of the herring fishery. It has fine and extensive sands for bathing, and the sea air is fresh and bracing. It is, however, at certain seasons of the year much exposed to violent winds, which blow the sand from

the surface into the air, and make out-of-door exercise
unpleasant.

There are a great number of cheap lodging-houses,
which are filled during the summer months by visitors
chiefly from the poorer districts of London.

Lowestoft, about ten miles from Great Yarmouth,
on the Norfolk coast, is the most popular sea-side
resort in this county. Like Yarmouth, it looks nearly
due east, and has good bracing sea air; it is, however,
quieter than Yarmouth, and is resorted to by a better
class of visitors, indeed it may be regarded as the
special sea-side resort of the neighbouring inland
districts.

It has a fine harbour, a good pier, and a good
beach, partly sand, partly shingle, with excellent
bathing. It has a public park with fine views of the
sea. Its adjacency to the Norfolk Roads makes
the neighbourhood attractive to yachting men and
anglers.

It has good hotels and lodging-houses.

"*Felixstowe*, upon the Suffolk coast, about twelve
miles from Ipswich, has a southerly aspect, and its
merits as a health resort are deserving attention.
The air is bracing and dry, and its cliffs being sixty
feet above the sea level, there are some well-sheltered
spots, which are suitable for the treatment of chest cases
even in winter. The weather in most seasons is then
generally delightful until the end of February. The
colder months of the year are March, April, and
perhaps May, the prevalent east winds are then,
as elsewhere, trying, although the high crag cliffs
much shelter the houses along the under cliff. The
water is very pure, though hard, and contains some
iron derived from the iron pyrites in the red crag.
The bathing is safe and excellent in every respect.
There is one first-rate hotel, 'The Bath,' pleasantly
situated, and its beautiful grounds are much sheltered.
Felixstowe has long enjoyed a great reputation with the
people of the district, many of the leading inhabitants
having residences there. A Convalescent Home, on

a large scale, has also been established. The medical men of the surrounding neighbourhood think highly of Felixstowe as a health resort, and many eminent members of the profession at a distance have been impressed with its healthiness. The late Doctors Lankester and Tilbury Fox went year after year to Felixstowe; the former remarked upon the abundance of ozone prevalent in the air, which exceeded that of any other sea-side place of his acquaintance.

"Felixstowe, until recently, was inaccessible by rail, it is now within two-and-a-half hours from London, and is rapidly increasing. The warrens at Felixstowe are used as Golf Links. One of the largest Golf Clubs in England has established itself here." *

Harwich, well known as an important port for the steamers of the Great Eastern Company, and as an ancient fishing town, is also a great resort during the sea-bathing season for excursionists from London, who come in great numbers by steamboat and by rail. *Dovercourt*, which is a kind of suburb of Harwich, has better sea-bathing. It has fine firm sands, and is altogether a pleasant place to reside in.

Walton-on-the-Naze and *Clacton-on-Sea*, both near one another, on the Essex coast, are rising into popularity as good, bracing sea-side resorts, providing facilities for bathing on a sandy shore, and con-veniently accessible by steamboat or by rail from the east end of London. They are cheaper than the more fashionable resorts on the south coast.

Southend, on the north bank of the estuary of the Thames, finishes our survey of the watering-places on our English coast. It has the merit of being readily accessible, and is a convenient resort for the inhabi-tants of the crowded districts at the east of London. Though at the mouth of the river, its adjacency to the eastern coast obtains for its climate a certain bracing character, and it has extensive sands for bathing and for recreation.

* Dr. Elliston of Ipswich has favoured me with the foregoing description of Felixstowe.

The foregoing sea-side resorts on the English coast admit of a somewhat rough classification, according to their situation. Those on the west coast are warmer and moister than those on the east coast, owing to their contact with a warmer sea, and owing to the prevalence of comparatively warmer and moister winds. In winter, when this difference of temperature is chiefly manifested, there may be an increase of from 3 to 6 degrees Fahrenheit of warmth on the west coast over that on the east. But from January to May this gradually decreases, and as summer advances, owing to the sun heat and the prevalence of warm winds from the continent on the east coast, the conditions of temperature may be reversed.

It has also been noticed, especially in the resorts on the Norfolk and Suffolk coasts, that in autumn, often up to November or December, owing to the comparative absence of easterly winds, and the less influence of westerly and south-westerly gales over the eastern counties, that the weather is warmer and drier and sunnier than at any other part of the year, and than the weather on the west and south-west coasts.

The east coast towns are much drier than those on the west or south-west, the easterly winds to which they are exposed being continental and dry winds, and blowing over a comparatively narrow expanse of sea.

It follows that the climate of the sea-side resorts on the east coast are, generally speaking, drier and more bracing than those on the west and south coasts, and less suited for residence in winter, and especially in spring, during the prevalence of easterly and north-easterly winds ; whereas their superior dryness and sometimes even warmth make them valuable in summer and autumn.

Of course the further north we go on the east coast, the colder and more bracing the climate of the sea-coast becomes, except in localities which enjoy some exceptional amount of shelter, or from indentations of the coast-line some unusual aspect.

A great drawback to the resorts on the west and south-west coasts is their large rainfall, which is spread over a great number of rainy days, and the associated humidity and dulness of their atmosphere. These conditions are the opposite of bracing, and these places may be said to possess a sedative and somewhat relaxing climate, modified, no doubt, in certain places by strictly local conditions of shelter from prevailing winds, nature of soil, etc.

On the west coast, as on the east coast, latitude makes itself felt in lowering temperature as we proceed north, but less so owing to the warmth of the Atlantic currents.

The resorts on the south coast have been grouped into three main sub-divisions, according as they approach its eastern or western extremity.

Those situated furthest east—from Margate to St. Leonards—are considered to partake somewhat of the character of the east coast resorts; to be drier and more bracing than those further west, while they are warmer in winter than those east of them.

Those furthest west—from Weymouth to the Land's End—are especially moist and warm (in winter), more equable both in winter and summer, and for these reasons especially sedative and relaxing.

A third or intermediate group, extending from Eastbourne to Weymouth, and including the Isle of Wight, are fairly warm and dry, less relaxing and moist than those further west, less bracing than those further east.

This classification is by no means free from defects, as strictly local conditions constantly come into play, greatly determining the bracing or relaxing influence of particular places.

Of popular sea-side resorts in Ireland and Scotland, the following are the chief:

Queenstown, in the Cove of Cork, has a southern

aspect. It is built in terraces on the side of a hill rising from the sea, completely open to the south, and well protected from the north. Its climate is remarkably mild and equable, if somewhat humid and relaxing.

Its mean annual temperature is 51·9 degrees.
    „   winter   „   44·2 „
    „   spring   „   50·17 „
Its annual rainfall is 34 inches.

It would seem to be somewhat warmer than Hastings, Bournemouth, and Ventnor, and to have about the same temperature as Torquay. It has a good sandy beach for sea-bathing.

Glengarriff, on Bantry Bay, has also a southern aspect, and enjoys considerable protection from cold winds. It has a mild and equable climate, like that of Queenstown, and is surrounded by beautiful scenery. The coast around is very interesting and romantic. It merits to be better known and more frequented as a winter resort for persons with delicate chests.

The following are frequented bathing resorts in summer :

On the east coast, *Kingstown*, near Dublin ; *Howth*, on the north side of the Bay of Dublin ; *Bray*, on the coast of Wicklow ; *Dundrum* (on the bay of that name), *Rosstrevor, Newcastle, Donaghadee*, and *Holywood*, all in County Down, afford facilities for sea-bathing, with a mild and humid climate.

Portrush, County Antrim—the " Brighton of Ireland," near the Giants' Causeway—and *Port Stewart*, County Londonderry, are on the north coast, and are considered to have fine air and a more bracing climate.

On the west Atlantic coast, there are *Bandarem*, on the Bay of Donegal, *Kilkee*, County Clare, and *Kilrush*, at the mouth of the Shannon ; these have good sands for sea-bathing, and a fine healthy climate, though, of course, much exposed to the Atlantic westerly gales, and at times much covered with fog and mist.

In Scotland, *Rothesay*, on the west coast, the

capital of the Island of Bute, at the mouth of the
Firth of Clyde, is popular both as a winter and
summer resort. The climate is mild; the scenery is
very fine, and the bathing is good. The island, which
is only eighteen miles in length, and four or five in
breadth, is almost surrounded by the lofty hills of the
opposite coast. Its climate is characterised by great
equability; the temperature rarely falls below freezing
point, and rarely rises above 70 degrees Fahrenheit.
The difference in temperature between Glasgow and
Rothesay is sometimes as much as 10 or 15 degrees.

There is very little accurate information to be
obtained as to other resorts on this coast, but there
are probably many with almost similar advantages
of mildness and equability. One of these—

Ardrossan, on the coast of Ayrshire, has been lately
rising into popularity. It is frequented chiefly for its
sea-bathing. It has cool, humid, and equable summers,
and mild winters, with rather much rain, however.

Other small resorts on the same coast are Wemyss
Bay, Millport, Largs, Tunellan, Dunoon, etc.

The following are the chief sea-side resorts on the
east coast :

North Berwick, twenty-two miles from Edinburgh,
at the mouth of the Firth of Forth, is a pleasant
sea-side resort in summer, and is much frequented by
visitors from Edinburgh and adjacent towns for sea-
bathing on its excellent sands. It also has extensive
Links, for golf-players, between the sea and the town.
It has a fine sea view, enlivened by the constant
passage of ships. Its climate is tonic and invigorating.

St. Andrews, on the coast of Fifeshire, forty-four
miles by rail from Edinburgh, is situated on the small
bay of St. Andrews. From its position on the bay
it is exposed to the north-east winds, which are the
prevailing winds in spring. For fairly strong persons,
however, the climate is bracing and healthy. The
city stands on a high cliff or rock, forming a peninsula
jutting into the North Sea, between the bay on one
side and the Burn of Kinness, or Nether Burn, a

small stream skirting the town, on the southern and eastern sides. To the north-west of the town stretch the celebrated "Links"—uneven downs formed by the sea. These are nearly two miles in extent, and there are similar downs south-east of the town. St. Andrews is the head-quarters of the old Scottish game of "golf," and for the sake of this game, as well as for fine bracing sea air, it is much resorted to in the fine months of summer. There are many antiquities and objects of interest in this ancient town, and there are the ruins of an old castle on a cliff on the bay, the foundation of which dates back to the thirteenth century.

Stonehaven, on the coast of Kincardineshire, sixteen miles south of Aberdeen, is very picturesquely situated near the mouth and on both sides of the river Carron, and is popular as a bathing-place. It is a pleasant, cheerful, summer resort, with fine bracing sea air. It has a railway station.

Aberdeen.—There are magnificent sands at Aberdeen for sea-bathing, on the coast just to the north of the port, but they are at an inconvenient distance from the city, and there are no houses or bathing establishment close at hand.

Nairn is the county town of Nairnshire, and is situated on the left bank of the river Nairn, near its confluence with the Moray Firth, about eighteen miles north-east of Inverness. It is much frequented for sea-bathing on account of its good sands and its accessibility. It has also been recommended as a winter resort, as its climate has been stated to be much warmer owing to local conditions than might have been expected from its latitude. Precise information and meteorological details are, however, wanting.

Having now passed in review the chief sea-side resorts on our own coasts, I propose in the next place to describe briefly the chief resorts of this kind which are frequented on the Continent.

And in the first place I shall speak of those which

have most interest for us, viz., the sea-side resorts on the other side of our Channel, beginning with the French resorts.

Passing from east to west the first we encounter is *Dunkirk*, an old Flemish town of 33,000 inhabitants, situated in the midst of barren downs. It is an important fortress and garrison town, as well as a commercial port.

As a sea-bathing place, Dunkirk is chiefly frequented by the native population and a number of visitors from the neighbouring country. The *Établissement des Bains* is about two miles from the town, with which it is connected by omnibuses running frequently to and fro at low fares. The Casino des Dunes is a fine and commodious building, and the sands are good and extensive.

Dunkirk has a large number of English residents, life there being cheap and pleasant.

There is a railway from Calais to Dunkirk, and also a diligence, the distance by road being thirty miles. There is also communication with London by steamboat, three or four times a week.

Calais is not attractive as a sea-bathing station, the beach is shingle, and the resources for visitors very limited.

Boulogne-sur-Mer is one of the most popular French watering-places. The old town stands on a hill on the right bank of the river Liane, and is connected with the new town, lower down and nearer the port, by the Grande Rue, a handsome thoroughfare, with good shops. The railway station and steamboat pier are on the left bank of the river, which is crossed by three bridges. The streets, which are regular and wide, are filled by a busy crowd; and the traffic, especially near the port, is at certain hours of the day very considerable. The great attractions of Boulogne are the fine sandy beach and splendid Casino and *Établissement des Bains.* The gardens and terraces overlooking the sea are a favourite promenade in summer evenings, as well as the pier and jetty.

During the season there are races, regattas, *fêtes de nuit* with illuminations and fireworks, and other entertainments provided by the Municipality. The theatre, where good operatic performances are given, is one of the best in the provinces.

Although life at Boulogne is rather more expensive than at other places on the coast, it is easy to find good accommodation at very moderate prices.

The climate is fairly dry and bracing ; drier and sunnier than on the opposite English coast.

Le Crotoy, a few miles west of Boulogne, is a small fishing village at the mouth and on the right bank of the Somme, opposite Saint-Valéry. It has of late years grown into popularity, and is now resorted to greatly by a number of Parisians on account of the excellent sea-bathing on its fine sandy beach. The port is shallow, but is well sheltered against violent sea winds.

A fine *Établissement des Bains* has recently been erected, with an extensive terrace facing the sea, and on the other side a large garden. It has all the usual conveniences of such an establishment.

There is an extensive sandy beach, occasionally exposed to rather stormy currents. The surrounding country is flat and uninteresting.

It has much the same climate as Boulogne, but it is much quieter.

Saint Valéry-sur-Somme, on the opposite bank to Le Crotoy and at the mouth of the Somme, is connected by rail with Abbeville, from which place it is distant about forty minutes. There is a fine railway viaduct across the bay, 4500 feet in length.

Saint Valéry is really composed of two towns, La Forte, near the port, and St. Valéry on the top of a hill, surrounded in part by the old ramparts. It is a clean, neat, agreeable town, with many ancient monuments of interest. Its well-sheltered port admits vessels of 500 tons. William the Conqueror is said to have sailed from this port when he invaded England.

The bathing here is not so good as at Le Crotoy, as the water is generally sandy, except at high tide, and the *Établissement* is not as comfortable as it should be.

Saint Valéry, compared with neighbouring resorts, has not many attractions.

Between *Saint Valéry* and the next well-known resort on this coast, *Le Tréport*, there are two picturesque little fishing villages, *Ault*, or *Le Bourg d'Ault*, and *Cayeux-sur-Mer*. These are now provided with *Établissement des Bains*, casinos, and a few hotels and *pensions*.

Le Tréport is one of the most frequented resorts on the coast of Normandy. It is about two hours from Longpré, a station between Amiens and Abbeville. It is at the mouth of the river Bresle, and consists of an old town half-way up the cliff, and a new town extending to the beach, where the *Établissement des Bains* and many elegant villas and châlets have been erected.

A picturesque view of the town and neighbouring villages, with the forest of Eu in the background, is obtained from the cliff on the west side of the *Établissement*. A long range of cliffs stretches northwards towards Ault and Cayeux.

Bathing (at high tide on shingle, at low tide on sand) is carried on in three different reserved spaces ; that on the right being for ladies only, the next for families, and that on the left for gentlemen only.

Close by is a hydropathic and warm sea-bath establishment.

The beach is lined with elegantly-built houses, and is the favourite promenade.

" The harbour, although small, is frequented by a certain number of Swedish, Norwegian, and English vessels, and contains about sixty or seventy fishing smacks, the entire population being exclusively composed of fishermen.

" A few places of interest are within easy distance of Le Tréport. Mers, on the other side of the valley of La Bresle, has lately been resorted to as a bathing place. At the entrance of the village stands a small but well-appointed *Établissement des Bains*, and the beach is fine and convenient.

"About two-and-a-half miles distant from Le Tréport is the little town of Eu, which can be reached by rail or road. The historical château is the residence of the Comte de Paris, who, during the last few years, has repaired and much improved it. On the spot where it now stands Charlemagne had built a fortress in order to check the progress of the Norman adventurers ; this was destroyed in the fifteenth century, and the present castle was commenced in 1578.

"Queen Victoria was twice received at Eu by King Louis Philippe, in 1843 and 1845."

An excursion to the extensive forest of Eu should be made by tourists ; the finest landscape is that seen from the Mont d'Orléans, about six miles from Eu.

From Le Tréport to Dieppe the best means of communication is the fine road passing through Creil, Saint-Martin, and Graincourt. Coaches run twice daily, performing the journey—eighteen miles—in a reasonable time.

Dieppe is the most popular of sea-side resorts on this coast, and justly so. It is picturesquely situated between two ranges of hills, forming on both sides high white cliffs, visible from a considerable distance. The harbour divides the town into two parts, Dieppe proper on the west, and a suburb (Le Pollet) on the east, where the fishermen and sailors live.

Dieppe, during the bathing season, is full of life and gaiety.

"The beach is more than two-thirds of a mile in length, and extends from the ship-building yards to the foot of the cliff where the Château stands. As in all watering-places in France, the baths are under the superintendence of a physician-inspector. The Casino and *Établissement des Bains* are united in a handsome building of glass and iron erected in 1857, somewhat after the style of the Crystal Palace. The central pavilion contains a very tastefully-decorated ball-room and two halls reserved for concerts and other entertainments. Two wide, covered galleries on each side lead to the terminal pavilions, containing the various reading and conversation saloons, club rooms, billiard rooms, etc. The top of the galleries has been judiciously arranged as a terrace, from which there is a splendid view of the sea and the surround· ing country. A theatre has lately been added to the Casino ; it is an elegant house, with room for 800 people, and forms, during the season, one of the chief attractions of the town, the artists generally coming from the most renowned Parisian playhouses.

"Behind the Casino is a very fine, carefully laid-out, and well-kept garden, a special portion of which has been set apart for children. Numerous minor buildings have been erected in the grounds, such as a *café*, a shooting gallery, a gymnasium, and a very complete hydropathic establishment.

"Between the Casino and the sea stands an elegant covered platform, where open-air concerts are given, and two large marquees afford excellent protection against the rays of the sun.

"Lower down, close to the sea, are the bathing cabins, those for the ladies being on the west side.

"There is also an establishment, in the Rue de l'Hôtel de Ville, where warm sea-baths can be taken. It is open all the year round, and comprises reading and billiard rooms, together with a fine concert room, where entertainments are occasionally given."

The Jardin Anglais, between the town and the beach, the piers, and the Cours Bourbon, are the chief promenades in the town ; the latter is a fine promenade, with a central road, and two wide side avenues of chestnut, acacia, birch, and poplar trees.

The chief industries at Dieppe are lace-making and ivory-carving, for both of which it is celebrated.

The excursions into the beautiful country around Dieppe are numerous and interesting, and lovely walks abound in its neighbourhood. "Not far from

Pollet is a small hamlet called *Pays*, inhabited in summer by a few English and French gentlemen, who prefer it to 'busy and lively' Dieppe. Here reside during part of the season, Alexandre Dumas, the celebrated dramatist, and the Marquis of Salisbury, whose country house, Châlet Cecil, is one of the finest in this region."

Veules is a very picturesque little sea-side resort on the line from Paris to Fécamp. It is rapidly becoming popular, on account of its good beach for bathing and its charming situation. It is on the river Veule, in the valley of which there are lovely walks.

Another sea-side village, within two miles of Veules, named *Veulettes*, has recently been turned into a sea-bathing place, with a comfortable *Établissement des Bains* and good hotels.

Saint Valéry en Caux, a little further west on this coast, and about two-and-a-half hours by rail from Rouen, is situated at the end of a small bay between two high cliffs. It is a place resorted to greatly during the season by families desirous of a quiet and inexpensive life and good sea-bathing, which its firm, sandy beach affords.

Fécamp is an active sea-port, with one of the safest harbours on this coast. It is a town of 10,000 inhabitants, two-and-a-half hours from Rouen by express train. It is built in a picturesque, winding, narrow valley, and has a simple, quiet, and peaceful aspect. The abbey church has some interesting objects worth visiting.

The beach is of shingle, and a little distance from the town. There is a very grand *Établissement des Bains de Mer,* "one of the finest buildings of the kind ever constructed." It contains a very good hotel, theatre, concert-hall, and reading and recreation rooms.

"Warm sea baths are a speciality at Fécamp, where the custom is to line the bath with a thick layer of sea-weed. This has the effect of imparting to the water the peculiar properties of the weeds, so rich in bromine and iodine."

Good hotels and pretty villa residences are found on the cliffs behind the Casino.

A cliff, 300 feet high, rises to the north of Fécamp, with a chapel seven centuries old on its summit. It is visited by a number of pilgrims on the 25th of March. There are some curious grottoes in the cliff.

Yport, four miles west of Fécamp, is another of those small picturesque fishing villages on this coast which have of late years become converted into fashionable sea-side resorts. The beach (all sand), sheltered by two high cliffs, is somewhat small, and bathing can only take place at low tides. There are very few stations on the coast more pleasantly situated than this one, as on no other point are trees to be found so near the sea. The forest of Hogues slopes down almost to the very edge of the cliffs.

Not far from Yport is the pretty sand beach of *Les Petites Dalles,* generally resorted to by families.

Between Fécamp and Hâvre is *Etretat,* situated between two gigantic cliffs, 270 feet high, and renowned for the beauty and picturesqueness of its landscapes. It is one of the most frequented sea-bathing places on the Channel coast.

" It is unlike any of the other watering-places ; its aspect is striking and original in the highest degree. The ground, lower than the water at high tide, is protected by a barrier of fragments of rocks which has often been destroyed by the waves. An old custom, derived from this peculiar position, still prevails. On Ascension Day the local clergy, in a curious ceremony, including the benediction of the waves, bid the sea not to go beyond its limits, and to respect the little village.

" The history of Etretat is that of most sea-bathing stations. Formerly an insignificant village, it was one day 'discovered' by artists and literary men, who, for a long time, enjoyed undisturbed possession of the place, until the paintings of Lepoitevin and the novels of Alphonse Karr called attention to this hitherto *terra incognita.* The miserable huts and cottages soon had to make way for elegant villas, châlets, and hotels.

"In spite of the influx of visitors, artists and literary men are still in the majority at Etretat.

"The Casino, erected in 1852, on the highest and most central point of the beach, is small but comfortable. From the terrace, the view of the country around is really magnificent.

"Etretat is famous for its cliffs, which extend on both sides of the town and abound in geological phenomena of the most curious aspect."

The environs of Etretat are most charming, and abound in beautiful walks.

Le Hâvre, the chief port on the north of France, has also bathing stations in that part of the bay which extends from the North Jetty to *Ste. Addresse;* the most important are the well-known *Établissement Frascati* and that of *Ste. Addresse* at the foot of the hills. The beach is pebbly, and bathers require shoes.

Honfleur, on the left bank of the Seine, opposite Le Hâvre, can scarcely be called a health resort; its beach is muddy, and the sea-bathing is not, therefore, attractive to visitors, it has, however, many points of interest for the tourist. Steamers go from New-haven to Honfleur twice a week.

Villerville, six miles from Honfleur by a picturesque and shady road, skirting the sea, is a small bathing-place, rapidly rising in popularity amongst the Parisians, especially artists. It is unfortunately losing its simplicity, and has now an *Établissement des Bains,* but has not got a Casino! It has a fine sandy beach and a bracing climate.

Trouville and *Deauville* form practically but one town; they are five-and-a-half hours by express train from Paris. Small steamboats also ply between Trouville, Le Hâvre, and Honfleur.

Trouville is one of the most fashionable and pleasant watering-places on the coast. Originally a little fishing village, it has rapidly become a popular bathing station, where the members of the Parisian aristocracy repair during the summer months.

The situation of this town is particularly fine.   It stands at the mouth and on the right bank of the river Touques, at the foot of a pleasant hill studded with pretty villas and gardens.

A notable building is the splendid Hôtel des Roches-Noires, situated at the foot of the cliffs, in an admirable situation.   The Casino of Trouville, built at an expense of over £12,000, is composed of several detached pavilions, in front of which is a wide terrace, communicating with the beach by a flight of steps. It comprises four large saloons, reading, writing, conversation and card rooms, billiard rooms, and a set of apartments for the exclusive use of the members of the local Union Club.   The interior decoration of the Casino, and particularly of the "grand salon," is of great beauty and richness.

The bathing season extends from June 1st to the middle of October.   The beach, a very fine sandy one, is divided into three separate parts—the first on the left being reserved for ladies, the second for ladies and gentlemen, and the third for gentlemen only.   Trouville is also provided with a very good hydropathic establishment.

A fine stone bridge connects Deauville with Trouville.

The Casino of Deauville, situated between the sands and a fine garden, is composed of a central building with two wings, round which runs a glazed gallery. It contains numerous saloons and a theatre, and the usual reading, writing, conversation, and card rooms.

One of the chief attractions of this place is its racecourse.   The races take place every year during the first two weeks in August, and last three consecutive days.

The old village of Deauville is situated at about ten minutes' walk from the new town, on the top of a hill, from which there is a magnificent view of the Touques and the neighbouring country.

The country around is very picturesque and well wooded, and there is a number of interesting places to be visited, thus forming objects of excursion of great variety and picturesqueness. Two-and-a-half miles from Trouville, on the road to Villers, stands the picturesque village of Bénerville, at the foot of Mont Canisy (336 feet high), from the summit of which there is a splendid view of the sea, the Cap de la Hève, the lighthouses of Hâvre, Deauville, and Trouville, and the valley of Villers, whilst in the background, in a western direction, the cliffs of Auberville, and the low-lying coast of the littoral of Caen, are just perceptible in the distance. A great drawback to these places is the fashionable and expensive life that prevails there.

Five miles west of Trouville and Deauville is *Villers-sur-Mer*, which is now connected with the main line of railway from Paris to Caen. It is much quieter, less pretentious, and less given over to gaiety and folly than its fashionable neighbours, and it is less expensive. It is therefore a popular resort for families who come in great numbers in the summer months to enjoy its cool sea breezes and to bathe on its excellent sands. It is most charmingly situated at the opening of a wide valley, and has most agreeable and picturesque surroundings. A comfortable and elegant Casino has been built at the foot of the cliff.

The *Vaches Noires*, a curious undulating range of cliffs, extend from Villers to Houlgate. They can be visited either by a road near the church or by following the beach at low tide.

A few miles further west we come to *Houlgate* and *Beuzeval*, connected together by a street bordered with pretty houses. Houlgate, with its beautiful villas, splendid hotels, and admirably appointed Casino, is an aristocratic and fashionable bathing-place. The sands, sloping gently towards the sea, are lined with rows of bathing-machines or cabins ; and large *umbrellas*, firmly set in the ground, afford excellent shelter against the overpowering rays of the sun. Bathing at Houlgate is very convenient, and there is, besides, a very complete hydropathic establishment, where baths and douches of all descriptions can be taken at all times. The Casino of Houlgate is a luxurious and commodious establishment. During the season entertainments of a high order are given there.

Beuzeval, on the other side of the small stream which separates it from Houlgate, is essentially a quiet place, a family bathing station, where all the visitors know each other. It has no Casino, no entertainments, and a canvas tent does duty for a reading room ; but it is, nevertheless, a charming and most comfortable little town, where one can enjoy sea-bathing on a fine sandy beach, free from noise and excitement, whilst its proximity to Houlgate enables the residents who feel so inclined to share in all the amusements going on there, without experiencing the disadvantages inherent to fashionable watering-places.

Near Houlgate-Beuzeval is a fine promenade called the " Desert." It is a large amphitheatre produced by the fallen cliffs, and forming a succession of deep bays and valleys, covered in places with luxuriant vegetation, owing to the numerous streams which run through them, and sometimes presenting a strange appearance, a chaotic wilderness of rocks, curiously shaped, resembling turrets, gables, pinnacles and buttresses, thrown together in a most picturesque and imposing confusion.

Still further west, close to the little town of Dives, and on the left bank at the mouth of the river of that name, is

Cabourg, which, like so many of the places we have noticed on this coast, has been transformed in a few years from a little fishing village into a fashionable and frequented watering-place. Its great attraction is a fine sandy beach, four or five miles in length, where bathing can be indulged in at any state of the tide. It bids fair to rival Trouville in popularity. A pleasant, smaller, and quieter bathing-place, *Home-Varaville*, is only two-and-a-half miles from Cabourg, with which it is connected by a service of omnibuses; and many prefer it on account of its greater quiet and retirement.

A number of small watering-places succeed one another on the coast of Calvados, west of the mouth of the river Orne. The first of these is *Lion-sur-Mer*, a little distance from the sea, with a small bathing establishment. Next comes *Luc-sur-Mer*, about two miles from the former. It is frequented in summer by bathers chiefly from Caen. Here the celebrated belt of rocks—the rocks of Calvados—commences. Next we have *Langrune-sur-Mer*, with a small bathing establishment ; then, about a mile further west, *Saint Aubin-sur-Mer*, where, as most of the houses are on the beach, visitors can take their sea-bath from their own door—a plan both convenient and economical. Still three miles further is *Courseulles,* a small port at the mouth of the Seulles, with productive oyster-beds and a small bath establishment.

These and others adjacent are all pleasant, quiet sea-side resorts where good sea-bathing and pure sea air can be obtained, and where living is simple and inexpensive.

S. Vaast la Hogue is a small sea-port with a fine bay looking east, on the Normandy coast. Previous to the rise of Cherbourg it was the chief port of the Cotentin. It has good sea-bathing, and fine coast views can be obtained from the country around.

Barfleur, on the same coast and with the same aspect as the preceding, is only seven miles from it. It is about a mile from the extreme point of the Cap de Gatteville, where there is a fine lighthouse 271 feet above the sea.

Fifteen miles west of *Barfleur* is *Cherbourg,* the principal naval port of France ; it is situated in the centre of a bay at the north extremity of the peninsula of the Cotentin. It is nearly opposite Portsmouth, from which it is distant seventy miles. There is good and safe bathing on the sands to the end of the *avant-*

K

port and *jetée*, where there is a bathing establishment, with a Casino for balls, concerts, etc.   In front the Casino has a garden and fine terrace, with a beautiful view over the harbour and pier.

At *Cap de la Hague* the coast line turns again sharp to the south, and at nearly the end of this indentation of the coast line we find the sea-port of *Granville*, situated at the foot of a rocky promontory, projecting into the British Channel.   It is about twenty-five miles from Cherbourg by land.   It is a busy trading town, and many ships are built there.

The baths lie close under the cliff, and can only be approached through a breach in the rocks.   There is a fine expanse of smooth, broad sands, quite shut out from the town.   Instead of machines the bathers are enclosed in canvas cases, carried like sedan-chairs.

About two miles from Granville is *St. Pair*, a pretty little bathing-place, on a creek or bay, with excellent sands, and much frequented in the summer.

Cancale is three hours by sea from Granville, across the bay of St. Michel.   It is celebrated for its rocks and its oysters, and is beautifully situated on the east of the fine bay of Cancale, at an elevation from which a grand panorama is commanded.   To the south of the town is the port named *La Houle;* this is inhabited almost exclusively by the oyster fishers. The view from this port comprises not only the whole of the bay, but also, in the distance, Mount St. Michel and Mount Dol.

St. Malo is about six miles from Cancale, and there are omnibuses run daily between them.   There are steamers also twice a week to and from Jersey, in about three hours.   The town is built on a peninsula, connected by a causeway with the mainland.   The situation is very picturesque, and the town and its neighbourhood present many objects of interest to the visitors.   It is much frequented for its sea baths, as well as for the beauty and interest of its surroundings. The beach, which is covered with fine sands, descends gently to the sea. From the terraces of the Casino fine views are obtained, and, during the bathing season, concerts, balls, regattas, and horse-races, and fêtes of various kinds, are provided for the amusement of the numerous visitors.

It is a most attractive, healthy, and pleasant sea-side resort in summer.

Dinard is on the opposite side of the estuary of the Rance to St. Malo, and a steam-ferry plies hourly between them. The sea-bathing here is good, and some prefer it as a residence to St. Malo.

It is only ten or eleven miles—steamers make the journey by river—to the most romantically situated town of Dinan, amidst the most beautiful scenery in Brittany. Numbers of English families are settled there.

Paimpol, Tréguier, Roscoff, are three small towns on the coast of Brittany, situated in the midst of romantic land and coast scenery and associations, near the mouth of the Channel, too much out of the way of the "beaten track" to ever become popular resorts. The climate of Roscoff is exceptionally equable, as it is much under the influence of winds from the sea, and in consequence of which the soil is of extraordinary fertility, and early fruits and vegetables are sent from Roscoff in abundance to Paris and to the English and Dutch ports.

It must be remembered that the whole department of Finistère is much exposed to storms from the Atlantic, which bring a large rainfall, and at times a good deal of mist and fog.

Before passing round to the west or Atlantic coast of France, we must complete our review of sea-side resorts on the coasts *opposite* to our own, viz.: those to the north of the Straits of Dover, on the coasts of Belgium and Holland. These somewhat resemble in their climate the resorts on our own east coast. They are generally drier and more bracing than those further west, and have a less equable climate ; they are quite out of the reach and influence

K 2

of the warmer Atlantic winds and currents, and the continental winds which blow from the east and north-east are colder in winter and spring, and warmer in summer and autumn, than they are further west.

Between Dunkirk and Ostend, and but a short distance from the latter, is *Middlekerke*, with a fine sandy beach and excellent sea-bathing. It is rising in importance as a sea-side resort, and deserves a visit from those who think Ostend too lively and too expensive.

Ostend is one of the most popular sea-side resorts in Europe, commending itself to persons of nearly all nationalities who require, or who think they require, good sea-bathing in fine bracing air. The sands of Ostend are very extensive, and every facility for sea-bathing is afforded there. Being so popular and fashionable, it is consequently a very expensive resort, and only suited to persons who can afford to be somewhat indifferent to the cost of accommodation.

There are abundant amusements provided, and it is not without a suspicion of dissipation, while its undoubted character as a resort of pleasure distinctly detracts from its value as a health resort.

Blankenberghe, near Bruges, a little distance to the east of Ostend, on the Belgian coast, has a finer seaside promenade than Ostend, and even more extensive sands ; and it had, a short time ago, the advantage of affording a quieter life and more simplicity than its fashionable neighbour, but it has now become very nearly as gay and costly as Ostend.

Heyst, only four miles further east than Blankenburghe on this coast, affords quite as good sea-bathing as either of the preceding resorts, while the life there is much more quiet and simple. It presents fairly good accommodation, and is rapidly rising in repute.

Scheveningen, three miles from the Hague on the coast of Holland, is but little known or frequented by English families, yet some who have spent a season there are enthusiastic as to its charms. The hotel accommodation is good, but is lacking in refinement in some important details. The sea view is very fine, and the extensive sands are beautifully white and clean and afford the most admirable sea-bathing. "The gently-sloping beach makes it a paradise for children." The air has been described as "fine and elastic, with a softness in it which makes it delicious; it is said not to be bracing, but it is very healthy, and must be delightful to persons who do not like sharp winds The wind is sometimes very high, but never sharp."

It is certainly an advantage that at Scheveningen, if one gets weary of the monotony of the sea, the most interesting places of this interesting country, with their picture galleries, their antiquities, and their many objects of social interest, are within easy reach. The Hague, with its deer-park and promenades, its picturesque buildings, its galleries, its canals, is within two or three miles. It is only twenty minutes from the Hague to Delft; it is a short hour from Scheveningen to Haarlem, and twenty minutes from Haarlem to Amsterdam. It is necessary to remember that Holland is a rainy country, and that the rain comes on suddenly, making the possession of a waterproof a great comfort.

There are many more or less popular sea-side resorts on the German coasts of the North Sea, with good sea-bathing and fine, bracing, tonic sea air; but they are not likely to prove attractive to English visitors on account of their distance from home.

Some of these, like our own Heligoland, are islands in the North Sea near the mainland, as Borkum, Norderney, and Wangeroog; and some are on the adjacent coast, as Cuxhafen, at the mouth of the Elbe, Westerland-Sylt, on the west coast of Schleswig, Dangast, and others.

Heligoland, which belongs to England, is one of the most frequented of these. It is six miles from the mouths of the Weser and the Elbe, and is connected by regular steam communication with Ham-

burg and Bremen. It consists of a sandstone rock,
a portion of which, the foreland, is level, and another
portion elevated, the latter being reached by an ascent
of nearly 200 steps.

Heligoland, being a small island, has the advantage
of presenting an entirely insular or sea climate, almost
as completely so as in a ship anchored out at sea. From
whatever quarter the wind may blow it must come
over sea. Much the same applies to the island of
Norderney, where some of Professor Beneke's experiments, alluded to in the last chapter, were performed.

There are several sea-side resorts on the coasts of
the Baltic, which naturally possess but little interest,
except for those who live within easy reach of them.
As sea-bathing places they have this peculiarity, that
the water contains very little salt. The chief are
Colberg, Cranz, Doberan, Dusternbrook, Heringsdorf,
Swinemunde, Travemunde, and Warnemunde. These
are excellent cool summer resorts, most of them in the
neighbourhood of beautiful forest, but with little that
can be called truly "marine" in their climate.

Returning now to the west or Atlantic coast of
Brittany and France, we find there several sea-bathing
stations. The first we come to after passing Brest is
Douarnenez, usually reached from Quimper, from
which it is distant fourteen or fifteen miles. It is
pleasantly situated on a fine bay of the same name.
Its chief industry is catching and preserving sardines.
The beauty of its bay attracts a certain number of
bathers in the season, but the baths are situated in
the hamlet of Riz, more than a mile from the town.

Some twelve miles from Douarnenez is *Audierne,*
on an extensive bay, which is much exposed to the
fury of the Atlantic gales. Its shores are of fine sand,
with a few small scattered rocks, and afford good
bathing in calm weather. "The spot has the most
sublime grandeur, not surpassed by any scene of the
kind in France, and bearing comparison with the sea-
cliffs on the west coast of Ireland and the precipices

of a Norwegian. fjord. The sea around is almost always tempest-tossed, and the shore of the Baie des Trépassés, so called from the number of dead bodies washed upon it, is perpetually covered with wrecks."

Some few miles further south, in a rather deep indentation of the coast, is *Concarneau*, a town the inhabitants of which are devoted to the sardine fishery; but its chief point of interest is the possession of a great aquarium and marine laboratory for the study of marine natural history, which is open to all comers who are interested in this branch of natural sciences. The situation is a fine one.

Le Croisic, situated on a point of this coast just before we reach the mouth of the Loire, is a popular resort in summer for sea-bathing and for its fine sea air. It has the usual *Établissement des Bains*, and several lodging-houses.

There are two beaches from which baths are taken, one close to the *Établissement*, and another about half a mile distant. Omnibuses convey bathers from the town to both of these. There are two artificial embankments, which serve as promenades, by the sea. The rocks of the adjacent shore are worn into most curious and fantastic shapes, and are well worth visiting.

Many interesting excursions can be made from Le Croisic.

Pornic, near the mouth of the Loire, on the opposite side to Croisic, and on the shore of the Bay of Biscay, is prettily situated, and has some interesting objects in the neighbourhood. Steps cut in the rock connect the upper with the lower part of the town. It is much resorted to as a sea-bathing place by the inhabitants of the adjacent city of Nantes.

Royan, the next bathing station of note on this coast, is situated at the mouth of the Gironde, where it pours its waters into the Atlantic. It is a small sea-port town, twenty-seven miles from Rochefort. It has a bathing establishment, and is much resorted to in the summer for sea-baths. The surrounding country is flat and uninteresting.

Arcachon, which will be fully described in a future chapter,* is a popular summer resort for sea-bathing, especially by the inhabitants of Bordeaux and the neighbourhood, from which it is distant about thirty-five miles. It is situated on the south shore of a large land-locked basin or inlet from the Bay of Biscay. This is the Bassin d'Arcachon, which communicates with the open sea by a channel about two miles wide, which runs for some miles in a southerly direction before it opens westward into the bay. This protected basin with its shelving beach affords a convenient place for sea-bathing, especially for families, as some of the houses facing it have gardens running down to the beach, and its entire protection from the Atlantic rollers makes bathing much safer than at Biarritz.

Biarritz also will be described in another chapter. It is only necessary to say now that it is a very popular bathing station, and is thronged with Spanish visitors during the hot months of summer. It is too hot to be attractive to those who live further north.

It is not necessary to do more than mention the sea-bathing resorts on the coast of the Spanish peninsula, for they are only of local interest, and although they may appear as comparatively cool retreats to those who live in the adjacent inland towns, they are far too hot in summer to be healthful or attractive to English visitors. The coolest of these resorts are those on the Spanish coast of the Bay of Biscay, as they are the furthest north and have a northern aspect. The chief of these are San Sebastian, Santander, and Gijon. On the coast of Portugal there are Oporto and Lisbon ; on the Mediterranean coast of Spain, Cartagena, Alicante, Valencia, and Barcelona.

Sea-bathing can be had at nearly all the coast towns of the French and Italian Riviera. Via Reggia,

---

* See chapter on Winter Quarters.

between Spezia and Leghorn, is said to be better adapted than many of them for this purpose owing to the adjacency of large pine-woods, affording shade and shelter from the sun's heat.

Bathing on the Lido is very popular during the summer months with vast numbers of the inhabitants of the inland towns in central, eastern, and south-eastern Europe, who come to Venice and other ports on the Adriatic for this purpose.

Trieste also has a fine sandy beach, and is greatly frequented for sea baths during June, July, and August.

This completes our survey of sea-side resorts at home and abroad, and it certainly cannot be said that there are not enough to choose from.

CHAPTER IV.

THE ENGADINE AND ITS HEALTH RESORTS.

VACATION STUDIES IN THE UPPER AND LOWER VALLEYS OF THE INN.

THE JOURNEY THERE.

It is now more than fourteen years ago that I first spent an autumn vacation in the Upper Engadine; since then I have passed some part of eight other vacations in that invigorating climate. The means of getting there remain much the same now as they were fourteen years ago, save that there is now the alternative route by the St. Gothard, the Italian lakes and the Val Bregaglia.* The railway journey from Bâle to Zurich has been shortened by a new line which avoids the detour by Olten, and a new line has also been made along the west shore of the Lake of Zurich. At Zurich, if you are so disposed, you can quit the train and cross the lake to Raperschwyl by steamer, an agreeable change in hot weather. At Raperschwyl you join the train again, and in a short

---

* A correspondent sends me the following with regard to the St. Gothard route, which he thinks preferable, and he says takes less time. Leave London 10.15 a.m., Bâle, 7.30 a.m., Lugano, by steamer for Porlezza, 5.5 p.m. Train from Porlezza to Menaggio, 6.37 p.m. Next day steamer, 9.23 a.m., for Colico. Thence to the Maloja Hotel in the Engadine by carriage takes about nine hours.

time reach the beautiful Lake of Wallenstadt. The
railway runs along the shore of this lake in nearly its
whole length, and so enables the traveller to see its
wild and grand scenery to advantage. Soon after
leaving the shores of the Lake of Wallenstadt the
train reaches Ragatz, where most persons who have
not already seen the Baths of Pfeffers remain a few
hours in order to visit the remarkable gorge where
these hot springs have their source. Those who are
independent of public conveyances may find it convenient to sleep at Ragatz, where there are excellent
hotels. From Ragatz to Chur is only half-an-hour by
rail, and the carriage-drive from Chur to Tiefenkasten
makes altogether a very easy day, and conveniently
divides the long distance of nearly fifty miles by road
from Chur to the Engadine. Some prefer to break
the journey at Thusis, as they object to the noise made
by the river at Tiefenkasten as it rushes along its
stony bed, under the window of the hotel.

There are very fair hotels in the quaint picturesque
little town of Chur, and the Steinbock may certainly
be said to be "clean and comfortable." It is rather
noisy and bustling in the early morning hours, when
the various diligences are preparing to start for their
several destinations; and, as I have already said,
invalids will find Ragatz a quieter resting-place.

I never see Chur without thinking of Thackeray's
charming and characteristic description of it; nor
without wondering that, in this age of rapid change,
it should remain almost as accurate now as when he
wrote it.

"The pretty little city," he observes, in one of
the "Roundabout Papers," "stands, so to speak, at
the end of the world—of the world of to-day, the
world of rapid motion, and rushing railways, and the
commerce and intercourse of men. From the northern
gate the iron road stretches away to Zurich, to Basel,
to Paris, to home. From the old southern barriers,
before which a little river rushes, and around which
stretch the crumbling battlements of the ancient

town, the road bears the slow diligence or lagging vetturino by the shallow Rhine, through the awful gorges of the Via Mala, and presently over the Splugen to the shores of Como.

"I have seldom seen a place more quaint, pretty, calm, and pastoral than this remote little Chur. What need have the inhabitants for walls and ramparts, except to build summer-houses, to trail vines, and hang clothes to dry? No enemies approach the great mouldering gates; only at morn and even the cows come lowing past them, the village maidens chatter merrily round the fountains, and babble like the ever-voluble stream that flows under the old walls. The schoolboys, with book and satchel, in smart uniforms, march up to the Gymnasium, and return thence at their stated time. There is one coffee-house in the town, and I see one old gentleman goes to it. There are shops with no customers seemingly, and the lazy tradesmen look out of their little windows at the single stranger sauntering by. There is a stall, with baskets of queer little black grapes and apples, and a pretty brisk trade, with half-a-dozen urchins standing round. But, beyond this, there is scarce any talk or movement in the street. There's nobody at the book-shop. 'If you will have the goodness to come again in an hour,' says the banker, with his mouthful of dinner, at one o'clock, 'you can have the money.' There is nobody at the hotel, save the good landlady, the kind waiters, the brisk young cook, who ministers to you. Nobody is in the Protestant church (oh! strange sight, the two confessions are here at peace!)—nobody in the Catholic church, until the sacristan, from his snug abode in the cathedral close, espies the traveller eyeing the monsters and pillars before the old shark-toothed arch of his cathedral, and comes out (with a view to remuneration possibly) and opens the gate, and shows you the venerable church, and the queer old relics in the sacristy, and the ancient vestments (a black velvet cope, amongst other robes, as fresh as yesterday,

and presented by that notorious 'pervert,' Henry of
Navarre and France), and the statue of St. Lucius,
who built St. Peter's Church, opposite No. 65, Cornhill.
His statue appears surrounded by other sainted persons
of his family. With tight red breeches, a Roman
habit, a curly brown beard, and a neat little gilt
crown and sceptre, he stands, a very comely and
cheerful image.

"What a quiet, kind, quaint, pleasant, pretty old
town! Has it been asleep these hundreds and
hundreds of years, and is the brisk young Prince of
the Sidereal Realms in his screaming car drawn by
his snorting steel elephant coming to waken it? Time
was when there must have been life and bustle and
commerce here. These vast venerable walls were not
made to keep out cows, but men-at-arms, led by fierce
captains, who prowled about the gates and robbed the
traders as they passed in and out with their bales,
their goods, their pack-horses, and their wains. Is
the place so dead that even the clergy of the different
denominations can't quarrel? Why, seven or eight,
or a dozen, or fifteen hundred years ago—a dozen
hundred years ago, when there was some life in the
town, St. Lucius* was stoned here on account of
theological differences, after founding a church in
Cornhill."

The first evening I saw Chur the rain was pouring

* Dean Stanley observes that King Lucius is the reputed
"founder of the originals of many English Churches—St. Peter's
Cornhill, Gloucester, Canterbury, Dover, Bangor, Glastonbury,
Cambridge, Winchester. He it was who was said to have converted the two London temples into churches" (the Temple of
Apollo at Westminster, and the Temple of Diana at St. Paul's).
"Of him, too, the story is told how the British king deserted his
throne to become a Swiss bishop at Coire in the Grisons, where
in the cathedral are shown his relics, with those of his sister
Emerita, and high in the woods above the town emerges a rocky
pulpit, still bearing the marks of his fingers, from which he
preached to the inhabitants of the valleys, in a voice so clear
and loud that it could be heard on the Lucienstieg (the Pass of
Lucius) twelve miles off."—" Stanley's Memorials of Westminster
Abbey."

down in torrents, and the surrounding mountains were concealed by a dense covering of mist, while the swollen Rhine, hurrying along its shallow bed and overflowing its very low banks, looked on the whole a dirty, straggling, and untidy river. It was not pleasant to think of a drive of fifty miles over the mountains, into the Engadine, in weather like this. Happily, the next morning dawned on a very different day.

From Chur to St. Moritz is a day's journey by diligence. At four o'clock in the morning after my arrival I was roused from my slumbers in a somewhat sulky mood, for one does not get reconciled immediately to the early hours of the country, and I prepared to take my place in the diligence which leaves Chur for St. Moritz somewhere about five in the morning. I joined the crowd of passengers surrounding the vehicles, some going by the Splugen and the Bernardino into Italy, others bound as I was for the Engadine. A motley group we formed at this early morning hour—Swiss, Germans, French, Italians, Americans, English, mixed up together, and on the whole looking rather gloomy and taciturn, a sense of injury at having been disturbed so early was plainly depicted on the countenances of most of those who were of Anglo-Saxon race. Some were unmistakably invalids. Their languid, drooping appearance and pale faces marked them out as fellow-travellers seeking a renewal of health and vigour from the iron-waters and mountain air of St. Moritz. As it happened to be just the very height of the season, the number of passengers desirous of proceeding into the Engadine was unusually large, and, in addition to the ordinary diligence, vehicles of various kinds were pressed into the service. If, at this time of the year, you would see in comfort the fine scenery through which you will have to pass, it is a wise precaution to secure, by telegraphing some days beforehand, the *coupé* or still better the *banquette* places in the diligence. Those who fail to do so are compelled to accept such accom-

modation as the officials at Chur choose to provide.
It may be a place in the interior of the diligence, with
small chance of seeing to advantage the beauties of
the country ; or you may be packed into the inside of
a narrow omnibus, with scarcely room enough to move
a limb !*

There are two routes from Chur into the Engadine traversed daily by diligences—one by the
Albula to Samaden, the chief place in the Engadine,
about three miles from St. Moritz ; the other by
the Julier Pass to Silva Plana and St. Moritz. The
latter is the shorter route of the two for those in-
tending to stay at St. Moritz. The former is more
convenient for those who intend making Samaden or
Pontresina their head-quarters. The road as far as
Tiefenkasten used to be the same for both routes ;
but a most agreeable route from Chur to St. Moritz,
especially to those who have never visited the Via
Mala, is now taken by the diligences which follow the
Julier Pass. This route is somewhat longer and less
direct than the other by the Albula, as it goes round
by Thusis, which is situated at the entrance to the
Via Mala. But it takes you through the beautiful
scenery of the Schyn Pass, between Thusis and Tiefenkasten, which is equal to almost anything in Switzerland for grandeur and variety.

By the other route, in rather more than an hour
from Chur, you reach the pleasant village of Churwalden, 3975 feet above the sea. Churwalden is a
" cure-place" (kurort), for you are now in a country
of " cures." Switzerland has been called the " play-
ground" of Europe ; with almost equal justice it
might be· termed the " Kurhaus" of Europe. The
mineral springs of this country are indeed most
abundant—saline, chalybeate, sulphurous, and simply
thermal ; but it is not to its mineral sources merely
that Switzerland trusts in order to restore lost health,

---

* Those who travel with a private carriage should sleep at
Thusis, instead of Chur, and so shorten by three hours the next
day's tiresome drive.

and to bring back declining powers to the invalids of Europe ; it has also its mountain-air cure, its whey cure, its grape cure, and a facetious German fellow-traveller assured us that it was likely soon to add another to this list, in the form of a sandwich cure. So that it is no uncommon thing in this country to meet with an individual who, as Émile Souvestre has observed in one of his charming Swiss sketches, "après une *cure de bains* à Bex, faisait une *cure d'air* au Selisberg, pour continuer par une *cure de chaud lait* dans la Gruyère, et finir par une *cure de raisin* à Clarens."

But if Switzerland is *par excellence* the land of *cures*, this Canton Graubunden is *par excellence* the canton of cures ; and nearly every third village is a "kurort," and has its "kurhaus," its "kurliste," its "kurarzt," its "kurmusik," and finally its "kuristen," as those are termed who come to be cured.

Churwalden, then, is a kurort, a "Luft-kurort," or air cure, and has several hotels and *pensions*, and a kurhaus, and baths, and a kurarzt, and all that is necessary for the equipment of a "cure-place." This spot has really many attractions ; it is within an easy distance of Chur, and it is most pleasantly situated in a mountain valley, running from north to south, watered by the Rabiosa, and rich in woods and meadows. Its climate is represented as being extremely mild, its summer temperature moderate and very little variable ; and, unlike many other high mountain valleys, it is singularly free from extremes of heat and cold ; so that during the summer months it has no frosts and no oppressive heat, no fogs, very little damp, no snow, a remarkably small rainfall, and but few storms. Interesting walks and excursions abound in all directions. It seems to answer excellently well as a temporary resting-place for persons who are recovering from the effects of exhausting diseases, and who are too weak to undertake a longer journey to a more distant or higher mountain-station, for which also it serves well as a preparation.

It is certainly to be recommended that weak and delicate persons should not proceed at once from the lower health resorts to a high mountain valley like the Upper Engadine, but should remain for a short time in a moderately elevated situation such as Churwalden, as an intermediate step between the relaxing temperature of the low valleys and the somewhat rigorous air of elevated localities, such as St. Moritz or Davos.* The special curative means which are provided at Churwalden are cows' and goats' milk, whey, and wild strawberries, and conveniences for a regular course of baths. It is especially celebrated for its excellent white Alpine honey, which gained the prize medal at the Exhibitions of Paris and Berne. Woods of pine and birch cover the hills around. The season begins on June 1, and ends on September 30.

Between Churwalden and Lenz the road is not very interesting; but at Lenz there is a fine view extending in nearly all directions, looking down on Tiefenkasten, and including the commencement of the two routes into the Engadine and the road through the Schyn Pass from Thusis. All these routes used to converge at Tiefenkasten; but the new Albula road, instead of descending to this village, now keeps on a higher level, and passes through Alveneu, where there are sulphur springs and a bath establishment; and near this village a post-road branches off from the Albula route and leads to Davos.

Alveneu is a good stopping place for those who are proceeding by the Albula route to Samaden or Pontresina. It divides conveniently the long carriage drive, and there is fair hotel accommodation to be obtained there.

Bergün, eight miles farther on, is also a convenient stopping place, and is most picturesquely situated, but the hotel accommodation used not to be very good.

* Two or three days, however, are quite long enough.

But as we intend to follow the Julier route we must descend to Tiefenkasten.   Here, formerly, the two roads into the Engadine diverged, that by the Albula to the left, and that by the Julier to the right. At Tiefenkasten, and at the adjacent hamlet of Solis, there are mineral springs of some local repute, but Solis is more remarkable for its bridge, which is situated very boldly at a reputed height of 1458 feet above the stream of the Albula, which rushes along a narrow and rocky bed at this awful depth from the road.

I have already observed that to the traveller who is bound for St. Moritz the pass of the Julier affords a somewhat shorter journey than that by the Albula. The latter route can be taken in returning from the Engadine, so as to avoid going twice over the same ground.   For my own part I would say, as M. Michelet did in choosing between the two routes : " Je préfère les grandes voies historiques où l'humanité a passé."   Such an "historical road" is the pass of the Julier.   That it was frequented by the Romans is very certain, since Roman remains have been discovered in several places in the Oberhalbstein Valley, at the head of which there is also another important Roman road, the Septimer Pass, leading into the head of the Val Bregaglia.

But it was believed that long before the Romans occupied the province of Rhætia this road was frequented by the earlier Celtic population of the country, and the name of the pass, instead of being derived from Julius Cæsar, would seem to owe its origin to the name of one of the Celtic gods—"Jul."

In the Middle Ages, merchants, pilgrims, and crusaders toiled over this road on their way to Venice, which was then the gateway of the East ; and ruins of mediæval castles may still be seen in many parts of the valley.

The Oberhalbstein Valley is entered immediately on leaving Tiefenkasten ; it is nearly twenty-four miles long, and its scenery, as well as that of the side valleys, which open into it, is very fine.

Villages and hamlets are seen dotted about the mountain-sides; old towers and ruins appear on the heights; and for every village there is a little church with its pointed spire. Most of these churches are very picturesque, and, considering the size of the villages or hamlets to which they belong, even ambitious in their character. They are built after the Italian style, and are frequently adorned externally with curious frescoes, the work of itinerant painters. The inhabitants of this valley are Roman Catholics. Sometimes one sees a church perched up apparently in a most inconvenient and inaccessible position, where churchgoing in bad weather must be an arduous if not a perilous undertaking.

Below the villages there are cultivated fields, and above them, amidst the mountains, extend woods and green meadows; and above these again, appear the jagged mountain summits, closing in the landscape on every side.

At Mühlen, or Molins, a village about thirty miles from Chur, and eighteen from St. Moritz, there is usually a little halt. This is the place often selected as a resting-stage by those who are not equal to making the entire journey from Chur to St. Moritz, or from St. Moritz to Chur, in one day, and who do not wish to break the journey at Tiefenkasten. It serves admirably for this purpose, not only on account of the excellent accommodations to be found at the inn there, but also because it is placed in the centre of the finest scenery of the Oberhalbstein Valley, and is surrounded by some of the highest peaks in this district.

The road, gradually ascending, reaches, in about six miles from Molins, the last and highest village in the Oberhalbstein—Stalla or Bivio. Here the two roads divide—the high road over the Julier Pass lying due east; that into the Val Bregaglia, over the Septimer, mounting south-south-west. For about two hours from Stalla the carriage winds slowly up a very well-made road, which leads in numerous zigzags to the summit of the Julier Pass.

The air here is keen and bracing, and you are reminded of the observation made by the author of the "Regular Swiss Round," that "many thousand Roman noses have been pinched blue with the sharp air of the Julier."

The pass of the Julier is a scene of mixed grandeur and desolation. The most perfect silence and stillness prevail on every side, broken only by the cracking of the whips of the postboys. No tree, no blade of grass, nothing but huge stones hurled down from the decaying mountains around, with here and there a patch of snow.

The summit of the pass is formed by a flattened ridge extending between two high granite mountains —the Piz Munteratsch and the Pulaschin. A little lake lies on the highest point of the pass, and near it there are two curious roughly hewn granite pillars, about three or four feet high, one on each side of the road, concerning the history of which there has been much discussion.

Their erection has been erroneously ascribed to Julius Cæsar, but many consider them to belong to a much earlier date, and to derive their name Julius from that of the Celtic sun-god Jul, and that they are Celtic sacrificial stones ; but such sacrificial sites else-where (and there are many in Western Switzerland and in the interior of France) have a very different appearance. Older writers describe only one pillar ; then it appears there were three. Murray states that recently, when some alterations were making in the road, one of these columns had to be removed, when some Roman coins were found at its base, bearing the name of Tacitus and the date of A.D. 275. Of all the passes of the canton this is the earliest free from snow, and it is also said to enjoy a remarkable immunity from avalanches.

The climate of the Oberhalbstein Valley, which ends in the pass of the Julier, is reported to be, on the whole, more inclement than that of other valleys lying equally high, because of its exposure to the

north wind. Above Roffna there is but little cultivation. A great part of the soil is rocky and dry,
and unfavourable for Alpine farming. The population
is scanty ; the inhabitants are for the most part tall
and strong, and renowned of old for a brave and
resolute demeanour. They speak the Romansch
dialect.

The road begins to descend immediately after
passing the Pillars of Julius, and soon the green
meadows and blue lakes of the Upper Engadine, with
the pretty village of Silva Plana, come into view—a
somewhat sudden but pleasing transition from the
wild and desolate scene which is left behind. Very
beautiful indeed did this part of the valley appear to
me when first I saw it from the Julier Road. The
clear, deep blue, unclouded sky; the rich green
meadows, lit up with the soft and subdued light of
the late afternoon sun ; the wooded mountain-sides,
topped by the eternal snow-clad summits of the
Bernina chain ; the clear, still lakes, whose unruffled
waters reflected, as in a mirror, blue sky and wooded
mountain, clustering village and ice-clad summit,
composed a scene which made one feel glad that, for
a few weeks at least, one had exchanged the hot and
dusty streets of noisy London for the calm beauty
and fresh breezes of this quiet Alpine valley, much
quieter fourteen years ago than it is now !

The diligence sweeps rapidly down the steep zigzags that lead from the summit of the pass to the
village of Silva Plana. In a quarter of an hour
more it reaches the picturesquely situated village of
Campfer, with its noted hotel, the Julierhof, renowned
for its comfort and excellent food ; here many of the
visitors to the baths of St. Moritz wisely reside, and
by so doing avoid the lowlying hotels around the
Kurhaus, or the toil uphill to those in the not very
pleasant village of St. Moritz. Campfer is just as
near the Kurhaus and the baths as St. Moritz, it is
connected with them by an excellent level road, as
well as by a most charming foot-path through woods

and meadows. On leaving Campfer the road continues on the left bank of the river, and we soon observe the Kurhaus of St. Moritz, placed in the meadows at the western extremity of the St. Moritz lake, looking "like a union house in a fir-plantation." In a few minutes the diligence turns off from the high road in order to cross to the opposite bank of the river, and soon it draws up on the gravel drive in front of the Kurhaus.

ST. MORITZ.

Fourteen years ago St. Moritz was far other than it is now, although even then it had begun to assume the aspect of a fashionable watering-place. But hotels and *pensions* were few and accommodation limited, and the struggle for rooms which followed a long day's journeying over the mountains was most trying both to mind and body. I well remember how anxious we all felt, when we were drawing near the end of our journey, as to what chance we had of being comfortably housed for the night. We had been warned in London, in Paris, at Bâle, at Zurich, at Chur, and indeed at every stopping-place on the road, that St. Moritz was full, full to overflowing. And so it proved to be. The Kurhaus was quite full, and during the delay of the mail-coach there, some, desirous of learning their fate, hired an ein-spanner (a light carriage with one horse), and dashed on towards the village, determined to secure apartments if possible at the Kulm Hotel or perish in the attempt. Others had engaged rooms weeks ago, and had even received replies assuring them that apartments would be retained for them ; but even this was not sufficient to allay the anxiety which the reports we had heard on the way had aroused. A painful expression of anxious uncertainty as to what might be our fate, for that night at least, was visible on all our countenances. In this unhappy condition we drove along the road bordering the pretty little St. Moritz lake, and in a very short time reached the village of St. Moritz, situated about a mile from the Kurhaus. Descending at the Bureau de Poste, confusion and disappointment soon became

apparent amongst the numerous arrivals. Unprotected females in ones and twos, as well as in larger
and more formidable groups, were making violent
and as it seemed unsuccessful attempts to force an
entrance into the chief hotel of the place ; others
were wandering about in a state of disconsolate
uncertainty, having received some kind of promise of
accommodation, but almost fearing to discover what
that accommodation might prove to be. Some of
them were manifestly invalids, and, unfortunately,
upon those not unfrequently the least attention was
bestowed, so that very uncomfortable apartments fell
to their lot.

It is hardly necessary to say that invalids should
always be careful to secure a promise of good accommodation some time beforehand wherever they go.
But St. Moritz at that moment was the new health
resort, and invalids were rushing to it in great numbers
without much inquiry and without good advice. Now
things have changed completely. St. Moritz, instead
of having but one good hotel and one good *pension*,
as was the case then, is covered with huge hotels and
lodging-houses, to the entire destruction of its seclusion
and the great impairment of its picturesqueness.
Moreover, the peculiar character of its climate and the
proper uses of its mineral springs are much better
understood now than they were then, and the many
victims of grave incurable maladies that found their
way to St. Moritz fourteen years ago are no longer
permitted to commit so serious a blunder.

Many invalids at that time were obliged to put up
with whatever accommodation they could obtain, no
matter how rough. They were turned away from the
Kurhaus, they were turned away from the only good
hotel in the place, and after wandering about the
village in search of a resting-place, they perhaps finally
obtained shelter in a room that was little better than a
hayloft.

The hotel-keepers then were no respecters of persons. I remember one very cold evening in August I

was requested to see a foreign lady of rank, who had brought an introduction to me from her physician in Paris. She was in delicate health, and travelling with another lady as a companion, and accompanied by a maid and man-servant. I found them placed in a wretched little room, which I had to approach by a ladder, and which was placed over a hayloft. The man and the maid-servant were on the landing. Whatever food they needed had to be brought from the nearest hotel. This lady had engaged rooms some time before her arrival ; but as she did not appear on the precise day she had named, they were not kept for her. Things have altered greatly since then. Besides the Kurhaus, the situation of which has always been objected to by many—it is built in flat marshy meadows, at the end of the lake of St. Moritz, about a mile from the village, and about three hundred feet lower than its highest part—the principal and indeed the only decent hotel at St. Moritz at that time was the Engadiner Kulm, kept by Herr Badrutt, and built in the most agreeable and commanding part of the village, overlooking the lake. It still remains the most popular hotel with the English, although there has never been a time that complaints have not been made about the food there. But Herr Badrutt, though a pillar of the Republic, and, I have no doubt, a democrat outside his own house, is a severe autocrat within it, and knows how to suppress disaffection with a strong hand and a stern countenance.

At present there are about ten hotels and twice as many good *pensions* and villas at St. Moritz.

One ought, of course, to feel grateful for all this increased accommodation, but at the same time it is impossible not to feel some regret at the departure of that simplicity of life which had lingered so long in this neglected corner of eastern Switzerland.

I like to remember, though with a certain sadness, that when I first visited St. Moritz, being unable to find a room in the hotel, I was provided with one in the village, which was very comfortable, save for the odours which penetrated it, and which appeared, from their nature, to proceed jointly from a cow-house and a cheese-store. It was then the custom in the Engadine, even in houses which had a very showy appearance externally, to devote a part of the ground-floor to the accommodation of the live-stock—cows, goats, fowls, etc.—and to stores of various kinds; so that emanations from the occupants of the ground-floor had to be tolerated by those who occupied the upper storeys. And I lived in the odour of cows for two days, till I received intimation that the room destined for my reception in the hotel was vacant, and I at once took and kept possession for the remainder of that season at St. Moritz of a very pleasant little room in the new wing of the hotel, with two windows, one looking over the bright green lake and on to the wooded mountain-side opposite; the other overlooking the Kurhaus, and up the valley to the great Cima di Margna.

The Upper Engadine scenery is by no means seen at its best from St. Moritz, and the farther we go eastward, along the valley, the less interesting it becomes. A writer in "Fraser," who evidently knows Switzerland well, speaks in somewhat disparaging terms of the Upper Engadine.

He says: "The upper valley of the Inn is one of the very few Alpine districts which may almost be called ugly. The high bleak level tract, with monotonous ranges of pine-forests, at a uniform slope, has as little of picturesque as can well be conceived in the mountains. Even in the great peaks, there is a singular want of those daring and graceful forms, those spires, and domes, and pinnacles which give variety and beauty to the other great mountain masses." This is true enough of that part of the valley which extends from Samaden to the commencement of the Lower Engadine, but it scarcely does justice to the neighbourhood of St. Moritz; while around Campfer, Silva Plana, and Sils, the scenery of this valley, with its chain of Alpine lakes, is as beautiful as it is unique.

For those who are able to walk well, there are beautiful and interesting mountain excursions of all degrees of arduousness. "The frequented and inhabited valleys of this region," writes an "Edinburgh" Reviewer,* "are so high above the sea-level, that the visitor is already on a platform of from 5000 to 6000 feet, even in his inn. Hence he starts from a 'coign of vantage' for ascents, and is invigorated by sleeping in the purest and keenest air. He must not be surprised to see snow falling in the middle of July, or to experience a cutting blast after sunset which would do no discredit to December in Britain. We have been well-nigh chilled to death on an August evening while gazing at the Norwegian-looking lake of St. Moritz, from the village of the same name, which itself stands at an elevation of 6000 feet above the sea.

"From this explanation it will be at once understood that mountaineering in this region becomes comparatively facile in relation to some particular summits. The finest and most accessible of all the famous views is that from the Piz Languard, and it is hard to refrain from enthusiasm in adverting to it. From this summit of 10,715 feet above the sea-level, if one has but a cloudless sky—an exceptional advantage, which we enjoyed in two ascents—spreads out to the eye, certainly one of the most extended, if not actually the most comprehensive, circle of snow-mountains to be seen from any one Alpine observatory of equal altitude.

"The distinguishing feature of this view is the multitude of snow-crowned summits on all sides, rather than the massiveness of many on one side, although it includes the imposing Bernina range. Every kind and shape of mountain seems to rise up and roll away into dim distance and indistinct azure.

* July, 1869.

A whole day would be too brief to count and identify
the several peaks, and the two or three hours of clear
morning sky, ordinarily permitted, seemed to fly away
like minutes.   A vast and varied relief mountain map
is perhaps the best verbal description of the view.   If
all the world's icy kingdoms and the glories thereof
were to be imagined as visible from one pinnacle,
assuredly the Languard would be its nearest representative.   From the tops of Mont Blanc and Monte
Rosa we embrace a far larger circle, but from neither
one a panorama so distinct and so appreciable in
details as that from the Languard.   The altitude is
sufficient to command a vast view, but not so extreme
as to dwarf the visible mountains.   Every candid
mountaineer will confess that a height of about 10,000
feet is best adapted to a panoramic view.   We may
see more from a greater altitude, but we distinguish
less definitely.

" Pontresina or Samaden, the latter having a capital
inn, is the spot we should specially commend to
tourists of moderate physical ability, who desire new
and grand Alpine views from points of ordinarily
possible attainment.   Some visitors, indeed, regard it
with particular, and perhaps overweening, partiality ;
yet its future wide popularity may be safely prophe-
sied.   Not its least attractions are its remarkably
invigorating air, and the old Valtelline wine there to
be drunk.   Some will delight themselves in its rich
flora ; and the little green-shuttered windows of
several of the natives of these villages and towns are
filled with bright flowers.   Those who desire ice-work
without danger may easily walk over a great portion
of the Morteratsch glacier, towards the bend of which
a very grand near view of the Bernina mountains is   .
obtained.   Near the foot of the glacier is one of the
most beautiful of waterfalls ; while from the roadside,
amidst rocks and firs, a view of the glacier is gained,
which never fails to elicit enthusiastic admiration.
The easy walk up the fine Roseg valley presents
striking wood and rock views, with a beautiful termi-

nation of glaciers; while athletic mountaineers have before them excursions, including every degree of difficulty, up to the ascent of the highest peak of the Bernina, a height of 13,294 feet, which is of difficult and laborious attainment. The Piz Morteratsch, which is 12,316 feet, is a safe and, comparatively, not difficult expedition, while the view is probably nearly as impressive as that from the highest peak."

The village of St. Moritz is the highest in the Upper Engadine, and the projecting ridge of rock upon which it is built is on this account termed the "Engadiner Kulm." The highest part, where the Kulm Hotel stands, is placed at an altitude of 6100 feet above the level of the sea, the general bed of the valley being about 300 feet lower. Green grassy meadows descend in gentle slopes from the village to the north-west shore of the beautiful little lake, the St. Moritzer See, which stretches across the valley to the wooded foot of the Piz Rosatsch, a huge mountain mass, which rises steeply on the opposite shore of the lake, its base covered with larch and pine trees, and its summit overhung by a great glacier mass. Herr Badrutt's hotel is built so as to command one of the finest views in this part of the valley. Standing on the terrace of the hotel, directly opposite, rises the aforesaid Piz Rosatsch; to the right the view stretches far up the valley towards the Maloja Pass, overlooking the Kurhaus and its grounds, catching a glimpse of the series of lakes which extends up to the head of the valley, and forms one of its greatest beauties; and finally the eye rests on the gigantic outline of the Piz Margna, always a conspicuous object in this direction. Turning to the left, the view is extremely beautiful: the rich green meadows at the eastern end of the lake, with the dark pine woods, which thickly cover the gentle elevation, concealing the village of Pontresina on the other side of it, form a pleasant foreground; while directly behind Pontresina rises the great mountain-range which lies to the south-east of the Bernina Road, and amongst which the pyramidal summit of the Piz Languard is a striking object. Still farther to the left the rest of the upper valley of the Inn opens out—the white villages, the winding river, the somewhat bare-looking mountain-sides—at the end a grand mountain-range, whose wonderfully bold and rugged summits seem to close the valley in this direction, and they do indeed form the natural boundary of the Upper Engadine. Few mountain masses that I have ever seen look more beautiful than this, when it is lit up by the rays of the setting sun. There is a variety and a warmth of colouring, and a grandeur of outline about its rocky masses, which have at certain times seemed to me more striking than some of the finest snow-covered summits.

Behind St. Moritz, to the north-west, rises the Piz Nair, a mountain very easily ascended; and continuous with this, towards Samaden, rise the neighbouring summits of Piz Padella and Piz Ot. In the opposite direction, towards Campfer, is seen the triangular pyramidal summit of the Piz Munteratsch, a mountain which rises on the eastern side of the Julier Pass.

The village of St. Moritz itself is not attractive. It is built in an irregular, untidy, scrambling way, along the hillside, with narrow dirty streets, and a terribly rough and jolty pavement. It is the least tidy and neat of all the Engadine villages I have seen. Some decent-looking houses have been built on the outskirts of the village, for the accommodation of visitors, and a pretty little Catholic church, and a house adjoining for the priests, have been erected just beyond the Kulm Hotel, on the road towards Samaden. An English church, the site for which was given by Herr Badrutt, and the foundation-stone laid in 1868 by the Archbishop of York, has been built in a convenient situation, about midway between the village and the Baths. Our countrypeople are indebted for this edifice mainly to the exertions of the Rev. A. B. Shettell, the English chaplain at St. Moritz, who, besides founding the chaplaincy there, was chiefly instrumental in procuring the erection of this church, and has been unremitting in his efforts to improve and beautify it since its completion.

But to turn from matters general to matters medical.

Most of the visitors to St. Moritz come for the purpose of going through the "cure" there. This consists in drinking the waters and bathing in them.

These waters are mildly chalybeate, the amount of iron contained in them being comparatively small, as will easily be seen, by comparing them with other well-known chalybeate springs :

	Carbonate of Iron.
St. Moritz	0·25
Spa	0·37
Tunbridge . . . . :	0·39
Pyrmont	0·42
Schwalbach	0·64
Orezza	0·80

It is here seen that even the Spa water contains a larger proportion of iron than the strongest of the St. Moritz springs, while the Orezza water contains nearly four times as much. The water of the Orezza spring is also abundantly impregnated with carbonic-acid, and I believe, as it becomes better known, it will be largely used. The spring itself is unfortunately situated in the island of Corsica, and on that account it is not likely to be much drunk at its source ; it is, however, exported in considerable quantities, and has long been in general use in Paris.

The following is the detailed official analysis of the water from the two sources drunk at St. Moritz. There is also a third source, but that has not yet come into use for drinking. One of these is termed the "Paracelsus," and the other the "Alte Quelle."

IN 1000 GRAMMES.

Gaseous Constituents.

	Alte Quelle. Grammes.	Paracelsus. Grammes.
Carbonic acid . . .	2·5485	2·5220
Nitrogen	0·0047	—
Oxygen	0·0015	—

Solid Constituents.

		Alte Quelle. Grammes.	Paracelsus. Grammes.
Carbonate of lime	. . .	1·0460	1·2832
,, magnesia .	. .	0·1911	0·2412
,, iron .	. .	0·0327	0·0454
,, manganese	. .	0·0059	0·0059
,, soda .	. .	0·2694	0·2935
Chloride of sodium	. .	0·0389	0·0404
Sulphate of soda .	. . .	0·2723	0·3481
,, potash	. .	0·0164	0·0205
Silica . . .	. .	0·0381	0·0495
Phosphoric-acid .	. .	0·0004	0·0006
Alumina	. . .	0·0003	0·0004
Bromine, Iodine, Fluorine .	.	traces	traces
Total of solid constituents	.	1·9113	2·3287

The water is strongly charged with carbonic-acid, which makes it sparkling and pleasant to drink. It also has a distinctly chalybeate taste, and is of rather low temperature — 3·5 degrees Réaumur, equal to about 40 degrees Fahrenheit.

A non-medical writer remarks of the St. Moritz waters : * "They are delicious — far too nice for medicine, though they are said to perform great cures. They combine the finest flavour of the best soda and seltzer water iced. There is a keen refreshing edge to them which spreads all over your being, and sharpens you up at once."

Practically these waters may be regarded as containing a small quantity of iron, about three grains of the carbonate in a gallon of the stronger source, and a considerable amount of carbonate of lime, about eighty grains in a gallon, held in solution by an abundance of carbonic-acid.

The presence of this large amount of carbonate of lime, in the absence of any appreciable amount of aperient saline constituents, interferes somewhat with the usefulness of this water in many cases where the use of a chalybeate is indicated.

* The author of " The Regular Swiss Round."

The history of these mineral springs is of most respectable antiquity, although their celebrity in our own country is of recent date.

About eight-and-twenty years ago, an English clergyman staying at Schwalbach heard of the existence of a very similar spring at St. Moritz, and wishing to exchange the somewhat sultry summer atmosphere of the former place for the more bracing influence of mountain air, came for the first time into this now popular watering-place, and found only two English people there; and indeed, so far as our own country-men were concerned, it was at that time practically unknown as a health resort.

Germans, Swiss, Italians, had long known of the virtues of its waters or of its mountain air, or of both together, and annually came in considerable numbers to go through the regulation cure; but it is only within the last sixteen or seventeen years that the place has been at all well known to English physicians. A book, by Mrs. Freshfield, entitled "A Summer Tour in the Grisons, and in the Italian Valleys of the Bernina," which was published in 1862, was one of the first publications which drew the attention of English tourists to the, at that time, unfrequented upper valley of the Inn.

It would seem, from recent discoveries, that the Romans were acquainted with these springs, and made use of them during their occupation of the country; but the earliest written account of them belongs to the sixteenth century. In the year 1539, Theophrastus Paracelsus, a physician of Hohenheim, wrote thus: "Ein *acetosum* fontale, das ich für alle, so inn Europa erfahren hab, preiss, ist im Engadin zu Sanct Mauritz, derselbig lauft im Augusto am sauristen; der dessel-bigen Trankes trinket, wie einer Arztnei gebürt der kann von Gesundheit sagen." Paracelsus is, in con-sequence, held in great honour at St. Moritz, and the principal well is named after him. Since 1680 the waters have been exported, and so long ago as 1703 the springs were frequented by Germans, Italians, and Swiss. At one time the waters were collected in the hollowed-out trunk of a huge larch-tree, which was discovered during the excavations made in 1852, for the purpose of cleaning out and utilising the new source, and improving the supply from the old one. The workmen were not a little astonished to come upon a great wooden tube, which further exploration proved to be formed of the trunk of a larch-tree artificially hollowed out. This was evidently the ancient conduit of the mineral source.

It was cleared of the earth and rubbish with which it had become filled, and the explorers were delighted to perceive that the clear and sparkling mineral water gushed forth in nearly ten times its former quantity, and of purer quality; since the ancient conduit had served, and had been doubtless designed especially, for the purpose of protecting it from contamination and dilution by the common surface water.

About fifty years ago (1817) there was only a little pavilion or shed built immediately over the spring, and there was no kind of shelter provided for the drinkers. Three little "Trinkhalles" were subsequently built, and these lasted up to the year 1832. In that year a sort of *Établissement des Bains* was built, consisting of baths and a drinking-house, with a few bedrooms—simply, however, for resting in, after taking the baths. At this time all the patients lived at St. Moritz, and had to take a twenty-minutes' walk before they reached the baths.

In 1854 a company hired the spring of the commune for fifty years, on the condition that they should build a large bath establishment, with a dwelling and boarding-house in the immediate neighbourhood of the spring. This building, commenced in 1854, was finished in 1859; and in 1866 the large and convenient Trinkhalle, which extends from the Alte Quelle to the Paracelsus Quelle, was added. There is accommodation in the Kurhaus for about three hundred guests; the bedrooms are many of them very small, but there are spacious and comfortable apartments for those who can afford to pay a high price for them. The dining-room is very large and handsome, and capable of seating three hundred persons. There are also *salons,* and reading, coffee, billiard, and smoking rooms. Besides the Kurhaus there are many good hotels now built, conveniently near the baths and springs.

I must now enter into some medical details as to the proper use and applicability of the St. Moritz cure.

Fourteen years ago St. Moritz was a watering-place of such recent growth, so far as our own country-people were concerned, that at that time much doubt and misconception existed amongst English physicians as to the precise character of the illnesses in which the " cure " at St. Moritz might most appropriately be recommended. Their difficulties were further increased by the reception of most conflicting accounts from different patients who had been sent there. Some returned assuring them that, of all places in the world, St. Moritz was the most delightful, the most invigorating, and health-restoring they had ever known. Others gave a most gloomy account of it : they came back no better, or even worse, than when they departed ; they had shivered all through August, they were never warm except in their beds, they had been nearly starved for want of decent food, and they had been bored to death by the unutterable dulness of the place. Since then things have greatly changed. I may perhaps be allowed to have had a small share in providing data for the better discrimination of suitable cases to be sent there : but there are now few English physicians of eminence who have not visited St. Moritz themselves, and seen and judged for themselves of its merits and uses. The accommodation for invalids, which was at one time very indifferent, is now so good as to leave little to be desired.

Before I state my own views, I will quote briefly the opinions of the Swiss physicians as to the nature of the cases in which a residence at St. Moritz is likely to be beneficial or otherwise, as well as the method of drinking and bathing which they were in the habit of prescribing.

In the first place, conditions of general debility are mentioned, whether arising from constitutional tendency, or from overwork, or from attacks of exhausting diseases ; and especially those debilitated states which often occur in females from special causes.

In cases of slow recovery from the effects of typhus, typhoid, and intermittent or malarious fevers. In certain cases of delicacy of the skin, with a disposition to take cold, to rheumatic affections, and to exhausting perspirations.

In chlorosis of young people and in anæmia generally—that is to say, in all cases of blood degeneration —and especially in those cases of chlorosis which are accompanied with cardialgia (heartburn). When there is great irritability of the stomach, with a tendency to eructations, it is recommended that the water should be mixed with whey.

In chlorosis associated with nervous irritability— a most distressing and unfortunately common condition—in addition to drinking the waters, the cool carbonic-acid baths are said to be especially calmative. The anæmia which follows as a result of repeated hæmorrhage is treated successfully by the waters and baths combined.

Nervous affections, associated with weakness of the circulating organs, or with general debility, of which the following are especially named : Migraine, pain in the eyes, palpitations, uterine neuralgia, hypochondriacal conditions.

Certain special forms of weakness, when unconnected with disease or injury of the spinal cord. Such cases improve greatly at St. Moritz.

Many derangements of the digestive organs, cardialgia, slighter forms of chronic gastric catarrh, the atony left after vomiting of blood, when all inflammatory conditions have passed away.

In hæmorrhoids, when considerable loss of blood has led to great exhaustion, and when they are unconnected with any organic disease.

Certain maladies peculiar to the female sex.

In vesical catarrh, worms, atonic gout, scrofula.

In convalescence after all severe diseases.

Such are the numerous and various disordered conditions which the Swiss physicians state will be benefited by a sojourn at St. Moritz and by taking the waters there.

They also mention certain diseases in which this "cure" is counter-indicated. They are—Convulsive, nervous disorders in plethoric and vigorous constitutions, excitability of pulse and disposition to general plethora, frequent throbbings in the breast and head, threatened apoplexy, hæmoptysis, organic diseases of the heart, tendencies to active hæmorrhage, all forms of cancer, epilepsy, and violent hysterical conditions in plethoric persons.

The routine of drinking and bathing prescribed by the physicians of the place used to begin between six and seven in the morning. At this early hour groups of promenaders, wrapped up warmly and braving the chilly morning air, might be seen briskly pacing up and down the gravel paths in the ornamental ground in front of the Kurhaus. A select band of instrumentalists—the so-called "Kurmusik"—endeavoured with feeble strains to enliven the monotony of the occupation.

Ever and anon the promenaders disappeared into the Trinkhalle to take their dose of water. This is a convenient building, in which the waters are dispensed. A boy stands at a pump and pumps up the water as it is required. On the walls of the pump-room are racks of pigeon-holes, containing the glasses of the patients, and to each pigeon-hole the name of the drinker is affixed. Near the pump-room you ascend by a few steps to a covered promenade, surrounded with comfortable lounges, and provided with reading and work tables. There are also several stalls' for fancy articles on one side of this covered space. Those who are ordered to drink from the weaker "Alte Quelle" have to carry their glasses with them to the end of a long corridor, where the water of this well is pumped up in a similar manner.

A fee of twelve francs has to be paid on inscribing one's name on the Kurliste, two francs of which go towards the maintenance of the Kurmusik.

Patients are ordered to begin with one or two glasses, and increase to four or six glasses daily, and to walk for a quarter or half-an-hour after each glass.

Later in the day, generally after breakfast, the baths are taken. These are regarded as an essential part of the cure, and a course of twenty-five baths is said to be necessary in order to give them a fair chance. The baths are heated by jets of steam, which pour through two perforated pipes into the baths. By this means the water can be heated to a considerable degree without any loss of carbonic-acid or precipitation of oxide of iron. By this arrangement, too, a great number of bathers can be furnished with baths at the same time. If hot water simply were added, the carbonic-acid would escape rapidly, and the iron would be precipitated. A threefold system of pipes leads into each bath, one for the conveyance of steam, one for the mineral water, and a third for common water, for the purpose of cleansing the baths. They are generally ordered to be taken at a temperature of 26 degrees Réaumur at the commencement, and gradually reduced to 21 degrees, or even lower. Twenty minutes to half-an-hour is the time for remaining in the bath.

The water supplied in the baths is derived from the weaker spring, the "Alte Quelle," the "Paracelsus" or "Neue Quelle" being the one more generally used for drinking. But, owing to the small supply of water compared with the number of bathers, the water of the spring is considerably diluted before it reaches the baths. The baths themselves are also constructed with the view of economising the supply of water as

much as possible. They are simply long and narrow
wooden boxes, just large enough to receive the body
of the bather. An unusually stout and tall man
would find himself straitened for room in one. These
boxes are covered in by a movable lid, which fits
round the neck of the patient, so that one's head
appears outside the box, while all the rest of one's
body is shut in.

The bathers are accommodated in small wooden
compartments, separated from one another by wooden
partitions, and arranged on each side of a long corridor.
As the steam by which the baths are heated is allowed
to escape freely into these corridors, they constantly
become filled with a moist, hot, close, and unwholesome atmosphere, to remain in which for more than
half-an-hour must be very injurious to most delicate
persons, especially as they enter from and pass out
into a thin dry air, often *very many* degrees lower in
temperature than the air of the bath-rooms.

These arrangements for bathing do not compare
favourably with such admirable ones as are found at
baths like Schwalbach, Schlangenbad, Aix, etc.

Let me now proceed to make what I hope I may
call a rational inquiry into the physiological and
therapeutic action of these waters when drunk and
when used for bathing purposes.

The effect of immersion in water charged with
carbonic-acid, a few degrees below the temperature of
the body, is not disagreeable. The heat of the surface
of the body sets free the carbonic-acid of the layer of
the water in immediate contact with it. This accumulates on the surface in minute bubbles, so that the
whole of the skin from head to foot, as well as every
little hair, becomes covered with sparkling beads. of
gas. As the temperature of the layer of water in
contact with the body rises (and, in order to favour
this, one is particularly cautioned not to move in the
bath, but to remain perfectly still), the bubbles of gas
expand, and at last part from the skin, and escape at
the surface of the water. As each little bubble of gas

is set free from the skin it imparts to it a slight tingling
effect, comparable only to the effect of an intensely
feeble galvanic current; and doubtless its effect on
the surface is of this nature. The result is that the
skin becomes red and congested. If the bubbles of
gas be swept off the skin, they do not reappear.

The question is—Can these baths have an active
curative influence? Or are they, as some suggest,
ordered for the purpose of filling up the time of the
patients, and also as a source of profit, since a franc
and a half has to be paid for each bath? That they
may do *harm*, I fear there can be no doubt. There
are, however, cases in which these baths have a
peculiarly soothing effect, since it seems certain that
prolonged immersion in warm water, rendered somewhat astringent by earthy matters held in solution,
exerts an extremely calming influence in cases of
exalted sensibility and nervous irritation ; while, in
other cases, the stimulating effect of the carbonic-acid
on the skin is not without a beneficial influence.

Some persons state that the effect of these baths
upon them is to produce a state of intense exhilaration
and excitement. But then these are persons of highly
excitable nervous temperament. To some the effect
of the bath seems scarcely appreciable. The feeling
is agreeable, but the exhilarating effect certainly does
not surpass, probably scarcely equals, that of an
ordinary cold sponge-bath at home.

But if the good these baths do be in many cases
problematical, the harm they occasion in some instances
is by no means doubtful.

The mere fatigue which the taking these baths
occasions in some cases of great general debility and
exhaustion, and the lowering effect on a very weak
circulation of remaining for so long a time as twenty
or thirty minutes in a bath of any temperature, as
well as the breathing, at the same time, the hot
steaming atmosphere of the bath-rooms, produce, in
many instances, as I have had occasion often to
observe, a low feverish condition, which it has taken

some time to recover from. Especially is this the case in feeble persons advanced in years, who come out of the bath exhausted, chilled, and uncomfortable, and then walk slowly, or perhaps are driven, through an unusually cold air; for it would not matter so much if the baths were taken only on warm fine days—but such days, in a bad season, are "few and far between," and the baths are taken daily. I say nothing of the absorption of iron by the skin, which is not now seriously maintained by any persons of authority.

The best practical rule with regard to bathing at St. Moritz is this : if, after a bath (not necessarily immediately after, but shortly after), you feel more vigorous and refreshed, continue with them, they are doing you good; but if, on the contrary, after each bath you feel languid and fatigued, or feverish and uncomfortably excited, leave them off, they are not doing you good.

Next, as to the routine of water-drinking. It may doubtless be very wholesome for persons who suffer only from imaginary illnesses, or for those who simply have to shake off the effects of the unwholesome excitement of the habitual life of our large European capitals, or for those habitual dyspeptics to whom an entire revolution in their mode of existence serves as a useful alterative—it may be very well for all such persons to be made to get up at six o'clock in the morning, and drink three or four or more glasses of cold water before breakfast, with free exercise in the open air. But to certain cases of real illness which used to be sent to St. Moritz, much more commonly than they are now, such a *régime* was absolutely hurtful. There is, doubtless, a tendency in these watering places to treat all cases alike; whereas, my own experience pointed out, in the most forcible manner, the necessity of great discrimination in the course to be prescribed for different cases, and I am glad to have an opportunity of stating that the practice of the resident physicians, both at St. Moritz and elsewhere,

since they have had more experience of English
patients and more intercourse with English physicians,
has been much modified from what it was fourteen
years ago.

To most persons of languid circulation, with a
deficient blood supply, and with greatly reduced
muscular power, either on account of slow recovery
from severe illness, or from depressing chronic ailments, or exhausting discharges, the effect of a cold
douche to the stomach, before any food has been
taken, will very constantly interfere with the digestive
process for the rest of the day, and give rise to distressing feelings of weight and flatulence in the
stomach, as well as troublesome headaches.   These
are common consequences of the practice of drinking
the waters in the early morning.   I need scarcely
observe, that there are many other constitutions and
conditions in which this process proves refreshing and
invigorating.   I am only pointing out the absolute
necessity of classifying and discriminating different
cases of illness.

Patients who have suffered in the way I have just
pointed out used to be told by the local physicians to
take the water slightly warmed, in order that some of
the carbonic-acid might be dissipated, the excess of
which, they said, was the cause of the headache.   I
could not agree with this opinion, because precisely
similar feelings of discomfort followed the drinking of
water of the same low temperature and at the same
period of the day, even though it contained little or
no carbonic-acid.

The real explanation of these symptoms is doubtless to be found in the shock which the repeated
ingestion of cold fluid gives—in systems, be it remembered, not prone to vigorous reaction—to the branches
of the pneumogastric nerves distributed to the
stomach, as well as to the great sympathetic ganglia
and their ramifications, which are placed in the immediate vicinity of this organ.   A remarkable corrobo-
ration of this view was afforded me one day at St.

Moritz in the person of a gentleman who, having arrived only the night before, walked down the next morning to the spring and drank off a glass of this very cold water, when he instantly fell down faint on the floor in my presence. This was an extreme example of the effect of the shock of the sudden ingestion of a draught of unusually cold water to a person not in a vigorous state of health.

Fourteen years ago I pointed out as the result of observations at St. Moritz, and these observations have been corroborated by other English physicians since, that persons with feeble and impaired digestive powers, associated with a weak circulation, and especially when these conditions were due to chronic, functional, or organic diseases, could not drink a quantity of cold fluid before breakfast, or remain for a *prolonged* period in baths of any kind, without experiencing considerable augmentation of the distressing symptoms from which they suffered, and, indeed, it was very singular that they should ever have been recommended to attempt any such rigorous process of treatment.

I have formerly seen patients at St. Moritz so utterly fatigued and exhausted by the early rising, and the drinking and bathing which followed, as to be quite incapable of any further exertion for the rest of the day. Moreover, it is no new observation, that repeated immersion in baths for longer than a few minutes at a time has a very enervating influence in many conditions of general debility; and repeated confirmations of this experience also came under my observation at St. Moritz. On the other hand, in conditions of nervous excitement or irritability, the baths doubtless have a soothing effect; but even in such cases it is questionable whether they should be long persevered in. As to the suggestion that there is some particular reason why a certain fixed number of the baths should be taken, be it eighteen, or five-and-twenty, or any other number, greater or less, all that can be said is that it is often good to have some *definite* plan to follow, but there is no particular efficiency in any particular number.   Yet it is curious to consider that this tendency to believe in the efficacy of repeating a process a particular number of times is deeply rooted in the human mind.   "Go and wash in Jordan *seven* times" was the direction of the prophet to the leper!

It was remarkable to me to observe the heroic perseverance with which some persons pursued the *régime* they had been directed to follow, notwithstanding the irresistible conviction that they were losing and not gaining ground.   But then they had come all the way from England on purpose!   At length some became really ill; and then the very immediate improvement in health which was observed to follow in many instances the giving up the bathing and drinking, left no manner of doubt in my own mind that the *chief* health influence in the place was the pure, clear, bracing mountain air.   I must not be understood to say that I think the bathing, and especially the drinking the water, useless; on the contrary, I believe them to be of great value in many instances: all I wish to insist upon is, that many, very many, persons used to be sent to St. Moritz who would have done better without any of the waters and without any of the baths; and there are others to whom the waters may be useful, but who are not strong enough to rise at the early hour of six, or to drink them before breakfast.   For all such the hours of eleven and five are the best for drinking.

Since I first knew St. Moritz it is remarkable how many persons have taken to the hour of 5 p.m. for drinking.   It is an excellent hour; after the afternoon walk the glass of water is refreshing, and taken just about an hour or so before dinner it does not disturb digestion, and tends to promote appetite.

There are two methods of judging of the therapeutic action of the St. Moritz mineral springs.   One is by a process of rational deduction from the known effects of the ingredients which enter into their com-

position. The other is by the direct observation of
their effects on those who drink them. Both are
valuable; but the former is necessarily the more
exact of the two, since in the latter method the effects
observed may be due to other causes than the employment of the waters, and notably to the influence of
the mountain air, and the entire change in the habits
of life of the patient.

I shall limit myself, then, at present to the inquiry
as to the probable physiological action of these waters
on the human economy. It is not unimportant, in
the first place, to consider what the effect may be of
drinking daily a large quantity of water simply, apart
from the mineral substances which it holds in solution,
especially in the cases of persons unaccustomed to
the use of pure water as an ordinary beverage. This
is a part of the inquiry very commonly omitted, yet it
cannot be doubted for a moment that the admission
of from one to two pints of an influential physical
agent like water into the alimentary canal every day,
in opposition to ordinary habit, must have a very
decided influence on the health of the body. Let us
imagine, for instance, a typical alderman, or a chair-
man of a city company, or even a diner-out of more
moderate capacities; let us imagine, if we can,
the amount of food and drink, of turtle, salmon,
whitebait, venison, champagne, burgundy, sherry, port,
hock, etc. etc. etc., in excess of their requirements, in
actual hurtful excess, taken by such persons between
Christmas and midsummer of every year. Conceive
the residue of all these excesses lurking about the
human body, in the blood, in the secretory organs, in
the crypts and corners of the alimentary tube, and
consider also how little pure water has found its way
along the same channel during the same period. And
yet it is necessary to perfect health that the interior of
our bodies should be washed and made clean as well
as the exterior.

Put such a person on a daily dose of a pint and
a half of pure water, for three weeks or a month;

restrict him, at the same time, to a moderate and limited supply of food, and make him take plenty of active exercise in the open air; the effect will assuredly be to eliminate from the body, from the alimentary canal; from the blood, through the secretory organs, a certain amount of waste or hurtful material, the retention of which in the body might prove in time provocative of chronic functional derangement, or even of organic disease.

The necessity for an occasional course of this kind applies to most persons, except the most abstemious and self-denying. In the course of a year the blood may well get somewhat tainted with the products of impure food and faulty digestion, and the alimentary canal may also be the better for a little flushing, to clear away the accumulated *débris* which its ordinary action has not removed; and the blood will itself be purified by the continued absorption and elimination of a considerable quantity of an important solvent like water. Water is rapidly absorbed into the blood, and also rapidly discharged from it. In its rapid passage through the blood and the vessels of the glands, by which it is finally thrown off, it will carry away, dissolved in it, substances which were harmful to the economy, and which only needed an excess of this solvent in order to be eliminated by the channels of excretion.

But when the water that is drunk contains saline aperient salts, then its depurative action is more remarkable. In that case we have the advantage not only of the solvent action of the water, but also of the stimulating action of the compounds it contains, on the organs of elimination, and especially on the glands of the alimentary mucous membrane. It is for this reason that we send those, whom we know to have lived freely, to flush their alimentary canals, their main-drainage tubes, as it were, at such springs as those of Carlsbad, Homburg, or Tarasp. The most satisfactory results constantly follow this cleansing process. The advantage of such purgation in aiding the operation of the intellectual and imaginative faculties is not perhaps generally known or admitted; yet we have no less considerable authority than that of Dryden and of Byron on this point. Dryden says: "When I have a grand design, I ever take physic and let blood; for when you would have pure swiftness of thought and fiery flights of fancy, you must have a care of the pensive part; in fine, you must purge the belly." And Lord Byron observes: "The thing that gives me the highest spirits (it seems absurd, but true) is a dose of salts; but one can't take them like champagne," and we have far more ancient testimony to the same effect. We are told that "Carneades, one of the most famous disputants of antiquity, was accustomed to take a copious dose of white hellebore, a great aperient, as a preparation to refute the dogmas of the Stoics." *

There is, therefore, one consideration common to all the springs at which we send our patients to drink, and that is the quantity of water we thereby induce them to consume. In those cases where we desire also to exert some stimulating effect on the secretory organs generally, and especially on the liver, and on the intestinal mucous membrane, we certainly ought not to prescribe a course of the St. Moritz springs, and for the following reason: The St. Moritz water is constipating, and it has this effect in virtue of the very (comparatively) large amount of chalk which it contains—nearly ten grains to the pint. Many other mineral springs contain as much chalk as those of St. Moritz, but then very many of them contain also a considerable quantity of aperient salts, which counteract and overcome the constipating effect of the carbonate of lime, which then acts beneficially as a simple antacid.

The only secretion which the St. Moritz water increases is that of the kidneys, which of course *must*

---

* Disraeli's "Curiosities of Literature," under "Medicine and Morals."

be increased by the ingestion of a large quantity of fluid, none of which passes off by the bowels. It is on account of this tendency of the St. Moritz waters to check the secretions, especially those of the liver and the intestinal canal, that persons suffering from chronic hepatic congestion, from obstinate constipation, and from visceral obstructions generally, are often made excessively uncomfortable when they attempt to follow the course prescribed there. The dryness of the mouth and throat complained of by some while drinking these waters is referable to the same cause. On the other hand, this astringent and bracing constituent proves of considerable use in cases of a different type ; this is especially observable in those persons whom we designate as of leuco-phlegmatic temperament ; pale, lax-fibred, languid people, who commonly suffer from relaxed mucous membranes, chronic mucous discharges, and sometimes from passive hæmorrhages ; a tendency to constipation may also exist in such cases, but then it is due rather to atony of the muscular fibre of the alimentary canal than to a deficiency of secretion.

Cases of chronic diarrhœa, not of a dysenteric character, but rather dependent on an irritable condition of the intestinal mucous membrane, and often associated with excitability of the nervous system, are treated successfully by these waters. The abundance of carbonic-acid in the St. Moritz springs is doubtless advantageous. There need be no great mystery however as to its action. Water charged artificially or naturally with carbonic-acid forms the commonest of beverages throughout Europe. Such waters are grateful to the palate, and refreshing and invigorating to the system generally. In irritable conditions of the gastric mucous membrane they act as a sedative to that organ ; we can therefore readily believe that cases of gastric catarrh, with acid eructations, are relieved by drinking these waters, containing as they do the alkaline carbonate of lime and the sedative carbonic-acid.

Carbonic-acid acts also in an important manner as a solvent; it holds in solution the oxide of iron and the large quantity of lime in these waters; it promotes their absorption into the blood. This brings me to the consideration of the most important ingredient in the St. Moritz springs, viz. the iron. Here arise two practical questions: First, why is iron so essential to us? and, secondly, why do we go all the way to St. Moritz for it? To answer the first question is very easy, but it is not quite so easy to answer the second.

It has long been known that when persons become pale, from loss of blood or other causes, the natural colour may often be restored by the administration of medicines containing compounds of iron, or even the metal itself reduced to fine powder. This is the observed fact; now for the reason. The blood is the nutritive fluid of the body. In countless streams of liquid living food, it flows through nearly every tissue of the body. The precise manner in which the food passes from the blood into the tissues of the body and becomes incorporated with them is not clearly known to us. But the blood not only conveys nutrient matter to the tissues of the body, it serves also as the medium for the conveyance away, from the tissues, of the substances formed by their decay and disintegration—a process which is continually going on. Every one of the finest bloodvessels of the body not only brings a supply of food to the portion of tissue which surrounds it, but it acts also as a minute drainage pipe, to carry off the waste products of the life and activity of that particular portion of tissue. We can easily see from these considerations how essential it must be to the health of the body that the integrity of the blood should be maintained. It must neither be defective in quantity nor quality.

There are many circumstances, which I need not enumerate, but notably the direct loss of blood by hæmorrhage, that may have the effect of altering the

N

quality and diminishing the quantity of that fluid.
When this is the case, it becomes the duty of the
physician to attempt to increase its quantity, and
to restore its quality.   One of the most important
changes which the blood undergoes in the way of
deterioration is the loss of a portion of what we
call its *red corpuscles*—minute microscopic cell-like
bodies, of which there have been calculated to be
five millions in a cubic millimetre of human blood,
and that rather more than ten millions of them
would lie on a space one inch square.   Now it would
appear that these little red corpuscles are probably
the instruments or agents of nutrition ; that they
are the busy workers in the blood, building-up
and pulling-down with equal industry, while they
themselves die, and are replenished continually.
Chemical examination shows that the metal iron
is an essential constituent of red blood-corpuscles ;
and, therefore, a certain supply of iron is needed for
their constant formation.   For ordinary purposes we
obtain a sufficient supply in our usual food, the flesh
of animals, etc.; but when an extraordinary supply is
required for the rapid manufacture of much-needed
red corpuscles, then we have to give iron in the form
of medicine.   It is on this account that, in medical
language, we call iron a blood restorative.   In those
cases in which it is necessary, the demand for it is
imperative, and nothing else can supply its place.
That it acts by restoring the red blood-corpuscles we
have abundant proof : in the first place, we constantly
see the colour and freshness return to pale faces and
blanched lips during its administration ; and in the
second place, we have carefully examined the blood
with the microscope in such cases.   In less than a
month the quantity of red corpuscles has been found
to increase from 50 to 76 parts in 1000 parts of blood,
while the patient has been taking iron medicines.
Yet the amount of iron in the whole body is but
small.   According to some physiologists, there are
about 100 grains of metallic iron in a man weighing

eleven stone. A sentimental Frenchman, having lost
a friend to whom he was much attached, caused the
body to be burnt, and having extracted the iron from
its ashes, converted it into a mourning-ring, which he
wore in memory of the dead !

Although there is only this small quantity of iron
in the human body, we occasionally find great difficulty, notwithstanding all the varieties of iron medicine at our disposal, to induce the system to take up
and appropriate any portion of that which we are
constantly supplying it with. This brings me to the
second question, viz., Why do we send patients all the
way to St. Moritz, to get a few additional grains of
iron, when we can supply it so abundantly at home ?
The truth is, that in some instances the power of
assimilating the ordinary compounds of iron appears
to be absent, and we send patients to St. Moritz not
for the sake of the iron in the water merely, but that
they may recover the power of assimilating this substance. This object the mountain air and the active
out-of-door life in many cases effect.

There is also another reason. Of all the compounds of iron, the carbonate of the protoxide, when
held in solution by carbonic-acid, is the one most
easily assimilated, and it is in this form that iron
exists in most mineral waters. I have already pointed
out that the St. Moritz water is unsuited to some
persons who require iron, on account of its constipating rather than aperient character, and it is a matter
of the commonest experience that iron salts are useless, and produce discomfort in many cases, unless
combined with laxatives. This is especially the case
in *commencing* an iron course.

Now the mineral waters of Tarasp, and those of
Kissingen, supply this combination of chalybeate and
aperient ; and it is for this reason that it is often advantageous to begin the iron course at a place like
Kissingen or Tarasp, and finish it at St. Moritz. It is
also useful while drinking the waters at St. Moritz to
take an occasional aperient in the form of a bottle of

the Tarasp water or a seidlitz-powder, or some other simple medicine.

But the quantity of iron in the Paracelsus spring is so small—only about the third of a grain in a pint—that when any great deficiency of iron in the blood exists, a large quantity daily for a long time is needed to supply this; and I have already pointed out that in great debility of the digestive organs, so much fluid cannot be absorbed without considerable disturbance of the general health. In such cases I have seen the plan of adding some of the ordinary preparations of iron to the iron-water answer admirably.

I must again remark that I am particularly anxious not to be understood as laying down general laws—that is quite impossible in an investigation of this kind ; but I am endeavouring to make clear why it is that the St. Moritz course is so successful in some cases and not in others. To sum up : The cases that appear to derive most marked improvement from the waters and air of St. Moritz are that very numerous class of lax-fibred, leuco-phlegmatic, hysterical women, who commonly suffer from chronic mucous discharges, or passive hæmorrhages, or other functional diseases. The improvement which is observed in some of these cases is rapid and remarkable. Cases of nervous irritability and nervous depression in both sexes, arising from over-work or over-excitement, or from merely constitutional tendency, often derive very considerable benefit from following the course pre-scribed for them there.

Cases of chlorosis and anæmia in young women, who fail to improve under chalybeate medicines at home, as well as those distressing nervous conditions which so commonly accompany the climacteric period of middle age, are often very remarkably benefited by a short residence at St. Moritz. Certain persons of sanguine and bilious temperament do not get on so well there. The very dry and stimulating air and the astringency of the waters are not favourable to such persons. In cases of hepatic derangement, the waters of St. Moritz generally do harm, since they tend to arrest rather than promote secretion.

Cases in which there is pulmonary emphysema are not adapted to this great altitude. The air seems too thin, as it were, to satisfy their respiratory requirements, and they are consequently unable to take as much physical exercise as they could in a less rarefied atmosphere.

I have yet to speak of St. Moritz from a purely climatological point of view, and this is a most important part of the subject. The general question of the influence, beneficial or injurious, of the climates of different parts of the world on healthy or diseased human beings, is one of the greatest interest to the practical physician. It is one, also, in which the welfare of the general public is very intimately concerned, for many valuable lives have been prolonged by a wise and judicious removal from the action of climates which were clearly imperilling their existence.

The question, then, of a cure by climate is one which constantly arises in medical practice, and it is often the only one which leaves any hope of success. The importance therefore of sound and correct views as to the exact nature of the climate of our chief health resorts is obvious.

The investigation, however, of the beneficial influences of the different places that have been put forward as "climate cures," is by no means free from difficulties ; and not the least of these is found in the unfortunate but unavoidable circumstance, that the personal interests of a great number of individuals become more or less intimately associated with the success or non-success of the localities in question. So that we must always be prepared to meet with a certain amount of exaggeration, and a certain amount of concealment. We are further constantly embarrassed by the over-brilliant descriptions of the constitutionally enthusiastic, as well as by the dismal accounts of the constitutionally discontented.

All observers agree that one of the chief characteristics of the climate of the Upper Engadine is very sudden and great diurnal variations of temperature. The thermometric variations in the same day are often so very considerable that in summer a temperature below freezing-point will be registered, and on the same day a temperature of from 40 to 50 degrees above freezing-point ; while a westerly wind in the winter will cause the thermometer to mount from — 13 degrees Fahrenheit to + 42 degrees Fahrenheit ! —a range of 55 degrees. These sudden changes are admitted by the resident physicians to induce, even in the acclimatised, attacks of inflammation of the lungs, of pleurisy, of chronic rheumatism, and of catarrhal fever.

The early morning is generally cold and damp, as there is a strong dew-fall ; but the damp fogs which are common in the lower Swiss valleys are almost unknown at this great elevation. The midday is often very hot, as the sun's rays act very powerfully, owing to the perfect clearness of the sky, and the thinness and purity of the air. The evenings again are cold ; but on some few nights in the height of summer, when the south wind comes over the mountain-passes from the plains of Italy, the air becomes positively warm.

On such a night, at the end of August, we picnicked on the hillside, between St. Moritz and Campfer, till eight o'clock in the evening, and afterwards basked in the brilliant moonlight on the terrace of the hotel until after midnight. These still summer moonlit nights in the Upper Engadine are full of a calm still beauty almost unearthly.

Speaking generally, there is in the Engadine a short and temperate summer, and a long and very cold winter. Formerly very exaggerated ideas were current as to the cold of the Upper Engadine in summer. Persons who had only spent a day or two there, or who may have remained longer, but who had encountered an exceptionally cold season, brought back with them accounts of the rigour of its summer climate, which a longer experience could not fail to have modified. I have known invalids who had passed some portion of nearly every day in July and August reclining on a couch in the open air, on one or other of the terraces of the Kulm Hotel. Of course on many days in order to do this it was necessary to wrap up warmly.

From a series of meteorological observations, continued over many years, at Bevers, near Samaden, by M. Krättli, the following facts have been taken :

The mean annual temperature of the Upper Engadine is 36·5 degrees Fahrenheit. The mean for the three summer months of June, July, and August, 50·8 degrees Fahrenheit ; for the three winter months of December, January, and February, 17·5 degrees Fahrenheit; for the three months of spring, March, April, and May, 35·4 degrees Fahrenheit ; for the three months of autumn, September, October, and November, 37·8 degrees Fahrenheit. The two extremes of temperature observed by M. Krättli (in 1854) were − 25·8 degrees Fahrenheit, or 57·8 degrees of frost in February, and 79·7 degrees Fahrenheit in July. In November and December there are usually thick fogs, but the three first months of the year are generally calm and clear, and the sunsets and sunrises are said to be most magnificent.

For five months in the year the snow covers the ground to the depth of two or three feet, and the lakes are covered with ice several feet in thickness ; snow occasionally falls in summer. M. Binet Hentsch states that he saw the valley white with snow on the 2nd of July, 1857, and that on the 3rd of August, 1858, he had broken ice on the banks of the Inn; and I have myself seen the valley covered with snow on the 4th of August : but an occurrence of this kind is usually followed by a spell of very fine bright weather.

The extreme dryness of the air renders the cold in winter less insupportable than it otherwise would be. The normal barometric pressure is considerably diminished, owing to the greater rarity of the air at this elevation : it ranges between twenty-four and twenty-five inches.

The relative amount of ozone in the air is much increased.

St. Moritz is certainly more favourably situated than most of the other villages of the Upper Engadine ; it is sheltered on the north and north-west by the Julier chain of mountains, and on the east by a wooded elevation which projects as a spur from the mountains at the back, and so forms the eastern boundary of the St. Moritzer See. The declivity upon which it is built has also a southern aspect ; but the drawback to nearly the whole of the Engadine is the great height and nearness of the mountains bounding it to the south, so that they intercept a great deal of the sun's light and heat during some hours of the day.

Of the salubrity, then, of the climate of St. Moritz in summer and in fine weather, there can, I imagine, be no difference of opinion. The air is perfectly pure, clear, dry, and bracing. There is an absence of that oppressive heat, even in the hottest weather, which makes the lower valleys almost unendurable ; for wherever there is shade in the Engadine there is also coolness. The freshness of the air, moreover, induces an increased capacity for muscular exertion, and the author of " The Regular Swiss Round " mentions that he has known some people come there "who have been so indisposed as to feel scarcely able to make the journey from London to Paris, and after a time have been able to make a twelve hours' excursion on the glacier."

This statement goes more to the root of the matter than the writer of it probably thought at the time he penned it. I believe this kind of climate is especially useful to those who have been *strong ;* but by some accident or other, such as over-work, or ill-ness, or trouble, have become weak: to those who possess a latent power of reaction. I do not think it is so advantageous to the essentially weak person, to those to whom "twelve hours on a glacier" always has been, is, and ever will be an utter impossibility.

For the same reason the Engadine does not suit persons advanced in years, unless they retain considerable bodily vigour. Show me an aged person whom the Engadine suits well, and I know there are such, and I should be disposed to conclude at once that he possessed a naturally vigorous constitution.

But much of the benefit that is derived from a short residence in an elevated region like that of the upper valley of the Inn is due, I conceive, to the *alterative* influence which it exercises on the human organism. "Alterative" is a term we constantly have to use for want of a better one; we apply it to any agency which produces a change in, or "alters," by the way of improvement, the action of the bodily organs in a manner which we cannot otherwise explain. Now in passing from the sea-level to an elevation of over 6000 feet, we must *alter* in a very essential manner the conditions of our lives. Our circulatory organs, our respiratory organs, our secretory organs, are working under altered conditions. This alterative action is, of course, assisted by the coincident change from possibly unwholesome to wholesome food, from weary hurtful work to refreshing rest, and from insufficient exercise to bodily activity.

It will be observed that I have said a *short residence;* by that I mean from three to eight weeks, and I think it undesirable, in many instances, to prolong a stay in the Engadine beyond this period. Most of the visitors to this Alpine valley do not wish to become acclimatised there. They have to return to the sea-level, and live there for the rest of the year. We know from experience that it is not wise to continue an alterative course of treatment too long, or repeat it too frequently; for either it will lose its effect altogether, or, worse, it may do harm.

The Swiss physicians have, I presume, observed

this, as they recommend that, after three weeks, if it be deemed advisable to prolong the "cure," that the patient should go down into Italy or into the Tyrol for a short time, and return and complete the course subsequently.   I doubt not that this is sound advice, though it may be inconvenient to follow.

The experience of most of those who have passed several seasons consecutively at St. Moritz, has been that after their first season they have left immensely benefited; after their second season, considerably benefited, but less so than after their first; after their third season still benefited, but decreasingly so.

The conclusion to be drawn from all this is, that you may place yourselves under very altered conditions of life, with advantage, for a time, but it is better not to continue to do so too long, or repeat it too often.

It is wiser, therefore, to go to St. Moritz every other year than every year, or to go for two years and remain away for two years.   And yet ·it is quite possible that there may be persons to be heard of who obtain benefit from a visit to St. Moritz *every* year; for the human constitution most provokingly resists any attempt to make it universally conformable to general rules.

St. Moritz as a resort in winter will be referred to in the next chapter.

OTHER RESORTS IN THE UPPER ENGADINE.

What is most characteristic about the Upper Engadine is its great extent as well as its great elevation. Nowhere else in Europe is there a valley of the same elevation, and of the same magnitude, and with the same number of permanently inhabited villages.

Here there are nearly thirty miles of broad valley and good level carriage road, traversed daily in various directions by postal diligences, at an average elevation equal to that of the Rigi Kulm; while some of its villages are situated at an elevation of over 6000 feet above the sea. The direction of the valley is from south-west, where it commences at the low pass of the Maloja, to north-east, where it terminates in the bridge Punt Auta; its natural boundary is, however, some three or four miles lower down.

Just beyond St. Moritz a ridge crosses the valley, leaving only a narrow gap through which the foaming waters of the Inn force their way, and forming a sort of natural division of the Upper Engadine into two nearly equal halves, which differ considerably in aspect and agricultural character. The upper half, viz. that between the Maloja and St. Moritz, is narrower, its mountain boundaries on each side are grander and wilder, and much loftier, and their summits are covered with extensive glaciers and snow-fields; while the floor of the valley is occupied by a series of small but beautiful lakes abounding in trout, and linked together by the stream of the Inn, which flows through them. In some parts the mountains, covered with dark-green pine forests, rise gradually in gentle slopes from the shores of the lakes, in others they rear themselves as wild rocky barriers precipitously from the surface of the water. The lower half, that which extends from the ridge above mentioned to the termination of the Upper Engadine, has a very different appearance. Here there are no lakes, the floor of the valley is much wider, and is occupied by broad stretches of meadow-land, through which the Inn quietly and tamely flows along. The mountains on each side are of lower elevation, they all rise in gentle slopes from the floor of the valley, and present no bold or striking features of form or outline.

In the upper half of the valley, viz. between the Maloja Pass and the Baths of St. Moritz, there are three well-known villages which are resorted to by visitors to this district in the summer months; these are Sils, Silva Plana, and Campfer. In the lower half of the valley, Celerina, Samaden, and Pontresina are the only villages which can be said to have any vogue as resorts for strangers. The village of Zuz has, however, recently been promoted into a *Luftkurort*, and must therefore be reckoned amongst the health resorts of the Upper Engadine. (A large Kurhaus has recently been built on the Maloja plateau itself.)

There is no more characteristic and picturesque village in the Upper Engadine than Sils ; or rather there are two villages of that name, one in a raw windy situation on the north side of the valley is called Sils Baselgia, and the larger and better-built village is termed Sils Maria—this is on the south side of the valley, in a protected situation, at the commencement of the ravine which leads into the beautiful Fex valley.

Sils Maria certainly commands some of the finest views in the Upper Engadine. The Silser See, close to which it is built, is the first and finest of the lakes encountered in this upper part of the valley, and, indeed, is the largest lake in the Alps at this elevation, being three miles in length and a mile in breadth at its broadest part. The mountains surrounding it, especially in the direction of the Maloja, are remarkably picturesque and varied; and some wooded promontories projecting into it from the southern shore close to Sils Maria add considerably to its beauty. The village of Sils Maria is 5880 feet above the sea, it is clean and well built, and possesses two very good hotels. It is especially well adapted to those who desire to lead a very quiet life, as it is far away from the more frequented spots, such as St. Moritz, Samaden, and Pontresina; and as it is well protected from the prevailing winds, it serves as a good summer station for consumptive cases.

Numerous pleasing excursions into the adjacent Fex valley, and to many picturesque spots along the wooded hills which adorn the southern shore of the lake, are quite within the powers of invalids; while the more robust can find abundance of severe work in the surrounding valleys and mountains. Few persons who spend a season in the Engadine fail to pay a visit to Sils Maria, or to explore, at any rate in part, the Fex valley, at the entrance of which it lies, a valley especially rich in specimens of the brilliant Alpine flora which characterises this region.

Silva Plana is a village of considerable size, with a large posting establishment, about three miles from Sils and five from St. Moritz. It is the next village you come to in descending the valley from Sils. It is very pleasantly situated at the foot of the Julier Pass, in the centre of the lake scenery of the upper part of the valley, having the Silva Plana lake on one side of it, and the Campfer lake on the other. It often has rather a bustling appearance, from the frequent arrival and departure of post carriages, the changing of horses, etc. etc., and this gives it somewhat of an air of unrest, as if everybody there was on the point of going somewhere else!

It is, however, a convenient stopping-place for tourists on their arrival in the Engadine either by the Julier Pass or by the Maloja from Italy; and it serves as a convenient centre for visiting the attractive scenery around. It has a large and well-appointed hotel, and one or two humbler *pensions*. There is also an English chaplain there during the season, and many families pass the whole summer very pleasantly and comfortably there, where they avoid the crowd of visitors at St. Moritz and at Pontresina, and get perhaps better accommodation at a little less cost. Silva Plana is most conveniently situated for making many of the popular mountain excursions in this district, and an office for guides has been established there, where an experienced guide can always be obtained. Nearly opposite the village a deep depression may be noticed in the mountain-chain forming the southern boundary of this valley. This is the "Fuorcla da Surlej," or Surlei Forcla. This is a moderately high pass (9042 feet) into the Roseg valley and to Pontresina. It is an interesting and easy walk of about seven hours, and is an excursion frequently made between Pontresina and Silva Plana or *vice versâ*.

To the right of this pass, and also nearly opposite the village, is the great mountain peak of the Piz Corvatsch, the rounded summit of which is covered with vast fields of snow and glacier. This is an easy mountain to climb, and is often ascended from Silva Plana in about four hours and a half; good mountain-climbers take less. There is also a path up the Piz Julier from this village. Besides these and other mountain excursions, there are many easy and interesting walks in the neighbourhood of Silva Plana. One of the longest and most attractive of these is across the bridge to the little village of Surlej, and then along the right shore of the Silva Plana lake to Sils Maria, by a recently-made foot-path. By means of this path it is now possible to walk the whole of the way from Sils Maria to the

Baths of St. Moritz through shady pine-woods, instead of along the dusty high-road on the northern side of the valley. The village of Silva Plana is 5900 feet above the sea.

The next village we come to is *Campfer.* This admirably situated village is two miles from Silva Plana. Its immediate surroundings are exceedingly picturesque, and it possesses most excellent hotel accommodation—the " Julier Hof," kept by M. Muller, having long and deservedly enjoyed the reputation of being one of the most comfortable and best-conducted hotels in the valley. It is a convenient abode for those visitors to the Baths of St. Moritz who would be at St. Moritz but not of St. Moritz ; it is as near the Kurhaus and the baths as the village of St. Moritz itself. The walk through the woods on the right bank of the river, between Campfer and the Kurhaus, is infinitely preferable for pedestrians to the hot, steep, and dusty road which leads from the latter to the village of St. Moritz; while both the upper road, which connects Campfer and St. Moritz, and the lower road, along the left bank of the river, command charming views of lake, river, and mountain scenery. The wooded and grassy mountain slopes around the village afford facilities for quiet rambles which are not to be found in the more frequented parts of the valley, while for exploring the attractive lake region of the upper part of the valley, it is far more conveniently situated than St. Moritz, Samaden, or Pontresina.

If it were not for the gregarious instincts of average humanity, Campfer would be as popular as St. Moritz ; it is perhaps fortunate for those who appreciate its attractions that it is not so. Campfer has an elevation of 5950 feet above the sea, and by the position of the surrounding mountain-ridges it enjoys considerable protection from the prevailing winds, its warm and sunny situation, and the many easily accessible walks around, rendering it a very suitable summer station for invalids.

It is an easy and beautiful walk up to the restaurant and summer-house called *Crest-alta*, which is finely situated on the top of a wooded promontory, which projects in the most picturesque manner from the south side of the valley into the Campfer lake. Another easy walk in the opposite direction is to the summer restaurant known as the "Alpina," situated among the green Alps, which cover the mountain slopes lying above the upper road which connects Campfer with the village of St. Moritz. The view of the chain of lakes of the Upper Engadine, with their wild mountain barriers, as viewed from the Alpina, is one of the finest things to be seen in the valley.

Passing now to the lower half of the Upper Engadine beyond St. Moritz, the first village we arrive at is Cresta, and a few minutes farther on we reach the village of Celerina, to which parish the adjacent village of Cresta belongs. Celerina is situated about midway between St. Moritz and Samaden ; but between St. Moritz and Celerina we have to descend in steep zigzags the high ridge which here stretches across the valley and forms a natural protection to St. Moritz from the north-east. Beyond Celerina the valley continues at an almost unbroken level to its termination. These two villages, Cresta and Celerina, are particularly neat, with their limewashed walls, green shutters, handsome old doorways, and windows filled with flowers, amongst which, some especially fine carnations are the envy and admiration of most visitors. There are some comfortable *pensions* in both villages, where quiet and modest accommodation can be obtained at a cheaper rate than either at St. Moritz or Samaden. Celerina is 5600 feet above the sea, rather more than 400 feet lower than St. Moritz Kulm, which is 6032.

Samaden, rather more than three miles from St. Moritz, has obtained a vogue, and is popular in spite of its extremely uninteresting situation. It is situated just at the spot where the Upper Engadine begins to be almost ugly. But a good hotel, and an obliging and clever landlord, have no doubt contributed much towards creating and maintaining its popularity. Then it has a certain prestige as the capital of the valley, and as the largest and most central village in it. It has long served also as a kind of reservoir for the reception of the stream of visitors waiting for accommodation at Pontresina and St. Moritz, being about equi-distant from both; and it has no doubt happened that many of these, finding themselves in good quarters at Samaden, have preferred to stay there, and so Samaden has become a considerable resort of the summer visitors to the Engadine. The road from the Val Tellina, which crosses the Bernina Pass, here joins the main-road of the Engadine. Indeed numerous post-roads converge at Samaden—two from the north, over the Albula and the Julier Passes respectively, one from the south-west coming up from the plains of Lombardy and the Italian lakes over the Maloja. Another also from Italy, as we have already said, from the south-east over the Bernina, and one from the east, connecting the Lower Engadine and the Tyrol with these other roads. Recently a Kurhaus has been established at Samaden, which is especially intended for the reception of winter visitors, and it is fitted up with bath and douche rooms, with properly heated apartments, corridors, and verandahs to adapt it to the purposes of a winter sanatorium for consumptive patients. For many years past one of the hotels at Samaden has been kept open for the same purpose through the winter, and a few invalids have generally remained there each year through the winter season.

There is one fine view from Samaden, and that is the view of the snow-clad summits of the Bernina group, which is well seen from the terrace of the salle-à-manger of the Bernina Hotel.

There are many pleasant and invigorating walks along the meadows which cover the lower slopes of the mountains, on the north side of the valley, between Samaden and Celerina; and on a warm summer day, when the high-road is exceedingly hot, dusty,

o

and fatiguing, the air on these grassy slopes, where there are good paths, 300 or 400 feet above the level of the valley, has seemed to me often very refreshing and grateful. The mountain-paths along these alps, between Celerina and Samaden, I have thought one of the pleasantest walks in the Upper Engadine; the bracing character of the air there being particularly noticeable. There are also pleasant walks through the pine-woods, on the same side of the valley but in the other direction, viz. towards *Bevers*. A pretty little English church has been built on a terrace behind the post-station, in a conspicuous and agreeable position.

But the most popular and interesting excursion from Samaden is that to the Bevers valley, at the entrance to which the village of *Bevers* is situated. This valley is one of the most picturesque and beautiful of the whole of this region, the deep green of its verdure, richly variegated with the choicest Alpine flowers, which are here found in great abundance, the stillness and repose of its shady retreats, its bubbling streams, as well as the stern grandeur of the mountain precipices that close it in at its deepest part, impart to it a charm of variety perhaps nowhere else to be found in this region. The Hon. Lionel Tollemache remarks of this valley, in an interesting essay on the Upper Engadine, which he published some years ago in the "Fortnightly Review": "No part of the Engadine impresses me nearly so much as the beautiful valley called Beversthal. In it the number of creeping firs is said to be almost unexampled, that of *Pinus cembra* is certainly very great. These with their dark foliage heighten the effect produced by this narrow valley, which is enclosed between high walls of steep and rugged mountains. It runs in a crescent round the back of Piz Ot, and altogether its aspect has a peculiar charm, a charm which a German writer declares to be unparalleled. Nor is it less to the ear than to the eye that this dim religious valley is impressive . . . . it is a sort of mountain *cul-de-sac*, wholly without traffic, and which the absolute stillness helps to make solemn and even death-like." Samaden is nearly 5700 feet above the sea. About three miles from Samaden, along the road which turns to the south-west and goes to the Bernina, lies the most popular of Engadine villages, *Pontresina*. Its situation is exceedingly picturesque, and it possesses several excellent hotels. It is, moreover, the most convenient station for exploring the high mountains, the valleys, and the glaciers of this portion of the Upper Engadine. It is close to the foot of Piz Languard, the Rigi of the Engadine, and it is about an hour nearer the glaciers of the Morteratsch and the Roseg than either St. Moritz or Samaden. It is out of the way of the patients and the doctors of St. Moritz, but it is in the way of the raw, blistered, be-spectacled faces, and be-roped bodies of Alpine climbers of all nations. It has been said to have a milder climate than St. Moritz. From its situation, in a wide open space, at the junction of two lateral valleys with the main one, it is much exposed to the direct rays of the sun for many hours during the day ; and around and near the village I have at times found the sun-heat greater and more unbearable than in almost any other locality I am acquainted with. But for the same reasons it is to be expected that the nocturnal cold would be greater than at St. Moritz, and it is so situated as to receive directly the cold gusts of air blowing down the Roseg valley. I have certainly experienced colder winds at Pontresina than at St. Moritz.

To those who are vigorous enough to devote themselves daily to mountain and glacier excursions, Pontresina is undoubtedly the best resort in the Upper Engadine, but I do not think it is so well adapted for the quieter life of an invalid as either Campfer, Samaden, or St. Moritz. It is certainly a convenience to those who need the extremely bracing and tonic influence of glacier air, to be tolerably near, as one is

at Pontresina, so large and accessible a glacier as that
of the Morteratsch.   The elevation of this village is
nearly as great as that of the highest part of St.
Moritz, being about 5950 feet above the sea.   Excursions of every variety of distance and difficulty can be
taken from Pontresina, and level walks and rambles
of much interest abound.   The walk through the
woods from Pontresina to the Meierei, a sort of farm-
house restaurant, close to the Lake of St. Moritz, is
one of the most popular and interesting, passing on
the road the beautiful little Statzer See.   Another
mile by a path along the southern shore of the St.
Moritz lake brings one to the baths and Kurhaus.

Most visitors to Pontresina make at least one excursion to the Hospice on the summit of the Bernina
Pass, a drive of about two hours; and this Hospice
has been occasionally resorted to as an *air cure* by
those who have felt that they needed an extraordinary
and exceptional amount of bracing!   It is more than
1500 feet higher than St. Moritz Kulm, and the air is
there wonderfully keen and cold, especially after sunset.
The only night I ever spent there—a clear moonlight
night at the end of August—the cold was very great,
and when one got into one's bed it felt iced; it seemed
to me that without the aid of a hot-water bottle, which
I placed between my shoulders to thaw the pillow and
upper part of the bed, one would have been chilled
into restlessness for the whole night.   The bracing
climate of the Bernina Hospice has, however, proved
very grateful to many.   The Hon. Lionel Tollemache,
in the article I have already quoted, speaks of "a delicate lady" who found it worth her while to go almost
daily from Pontresina to the top of this pass, a distance
of twelve miles, so as to breathe the fine air for a few
hours.   I was certainly surprised to find a gentleman
suffering with heart-disease, a disease in which a residence in high regions is considered to be specially
counter-indicated, apparently gaining benefit there.
To quote Mr. Tollemache again, who is really an

authority on the subject of the Bernina Hospice, as he must have spent altogether many months there, he says: "A few have found their stay at Bernina the turning-point after a long illness; and how enthusiastically do they now dwell on its abnormal combination of charms! In fact, they go to the Bernina to have the summer of their discontent made glorious winter; transformed indeed into a sort of expurgated edition of the English winter—the English winter without its damp, and the east winds without their pungency: differing also from the English winter in the deep-blue sky, and in the dazzling and enchanting brilliance of the sunlight. One drawback, however, there is to a long residence on this pass—there are absolutely no trees; unless, haply, we count as a tree the dwarf willow (*Salix herbacœa*), which rises barely two inches from the ground! . . . Perhaps, after all, the absence of trees is not an unmixed evil. The superiority of Bernina to Pontresina in point of bracingness is out of all proportion to the difference between the places in respect of cold. That superiority is, in great part, due to the extreme dryness of the Bernina air; and the dryness must be increased by the scantiness of vegetation . . .. Many wild-flowers grow there, including some not found at the lower elevation of St. Moritz." For quite exceptional cases and exceptional constitutions, the extremely rigorous bracing climate of the Bernina may, for a time, be suitable; but I am satisfied it is of extremely limited applicability as a health resort, even to those exceptional cases.

The only other village in the Upper Engadine that can be spoken of as a health resort is the village of Zuz, about eight miles below Samaden. An attempt has recently been made to establish it as a winter as well as a summer station, and a large hotel has been adapted for this purpose, with what success I have not heard. It has about the same elevation as Samaden, 5680 feet, and is raised somewhat above the floor of the valley at this part. The scenery around is pleasing, and there arc several excursions to be made into the adjacent valleys and upon the neighbouring mountains ; but the scenery is altogether inferior to that of the upper part of the valley, and is not likely to prove attractive to English visitors.

THE LOWER ENGADINE—THE BATHS
OF TARASP.

The least interesting part of the Upper Engadine is certainly that which lies between Samaden and the commencement of the lower valley of the Inn. The road is almost a dead level, there are no lakes to add to its picturesqueness, as there are above St. Moritz, while the mountains which rise on each side have a bare appearance, and present no very striking characters either of form or magnitude. On the whole it must be admitted that this part of the Inn valley is somewhat monotonous, although we are assured that those who have time to visit and explore the side valleys will find the region full of interest.

The Lower Engadine is very beautiful, more pastoral, more picturesque, more varied, than the upper valley. It is about a five hours' drive through for the most part charming scenery, from Samaden to the Baths of Tarasp, the chief attraction in the Lower Engadine for the investigator of health resorts. The Kurhaus consists of a large mass of buildings, almost buried in a narrow gorge between the bases of the mountains on each side of the valley. It can receive over two hundred guests, and affords excellent accommodation. It is built close to the principal springs, and the baths are within the building itself.

A small *dépendence*, called the Villa Kurhaus Tarasp, was built in 1876 at the west end of the Kurhaus garden, and is intended for the reception of families and others who may desire a quieter place of abode than the Kurhaus itself. A fine Trinkhalle, with covered promenade and rows of shops, was built in 1875–76, on the right bank of the river, close to the principal spring. Here the Kurhaus orchestra plays in the early morning, while the visitors alternately drink and promenade.

The information I have to give with respect to this health resort will naturally arrange itself under the three following heads : 1. The nature, composition, and therapeutic effects of its mineral springs. 2. The nature of its climate. 3. Some topographical and general details.

There are many mineral springs which rise very near each other in the neighbourhood of Tarasp— some alkaline and aperient, some chalybeate, and others in the Val Plafna, behind Tarasp, sulphureous. The springs to which Tarasp chiefly owes its repute are the saline-alkaline sources which rise close to the banks of the Inn, and in the immediate vicinity of the Kurhaus. Those that are used for drinking are situated on the right bank of the river ; there are others on the left bank which are used for bathing purposes.

The drinking springs are named the St. Lucius and the St. Emerita; they rise quite close together, and are almost identical in composition. The St. Lucius is somewhat richer in carbonic-acid. The following is the latest analysis of these two springs :

ANALYSIS BY PROFESSOR HUSEMANN, 1872.

In 16 ounces = 7680 grains, the carbonates calculated as bicarbonates.	St. Lucius Spring.	St. Emerita Spring.
Sulphate of potash	2·916	3.090
„ soda	16·131	15·912
Borate of soda	1·322	1·354
Nitrate „	·006	·006
Chloride of sodium	28·216	28·308
„ lithium	·022	·020
Bromide of sodium	·163	·165
Iodide „	·007	·007
Bicarbonate of soda	37·426	37·545

In 16 ounces = 7680 grains, the carbonates calculated as bicarbonates.	St. Lucius Spring.	St. Emerita Spring.
Bicarbonate of ammonium . .	·507	·504
,, lime . . .	18·800	18·772
,, strontia . .	·005	·005
,, magnesia . .	7·524	7·563
protoxide of iron .	·165	·163
,, manganese . .	·002	·002
Silica	·069	·070
Phosphoric-acid	·003	·003
Alumina	·002	·002
Barium, Rubidium, Cæsium, Thallium, organic matter . . .	traces	traces
Total amount of fixed matters	113·492	113·289
Free and half-free carbonic-acid gas, cubic inches . . .	76·17	75·42
Quite free ditto . . .	33·92	33·08

The carbonates are all contained in the water, combined with some carbonic-acid, in the form of bicarbonates; and in each pint of water there are nearly thirty-four cubic inches of free carbonic-acid. It will be seen that these springs contain a large amount of bicarbonate of soda—thirty-seven grains in a pint; also a considerable quantity of chloride of sodium (common salt), and a moderate amount of the aperient sulphates of soda and potash. Next in importance figure the alkaline earthy carbonates of lime and magnesia, and there is a small quantity of iron.

It is interesting and instructive to compare the composition of these springs with that of other well-known European spas which possess some properties in common with them; such, for instance, as the Grande Grille at Vichy (a simple soda spring), the Rakoczy at Kissingen (a common salt spring), and the Sprudel at Carlsbad, and the Ferdinands Quelle at Marienbad (both characterised by the amount of aperient sulphates they contain).

Such a comparison is made in the following table:

In a pint, i.e. 16 oz. = 7680 gr. the carbonates calculated as mono-carbonates.	Tarasp Lucius Quelle.	Carlsbad Sprudel.	Marienbad Ferdinands Quelle.	Kissingen Rakoczy.	Vichy Grande Grille.
Total solid matters.	93·8	42·4	70·6	85·6	90·8
Carbonate of soda ...	26·4	10·0	11·2	—	27·0
Sulphates of soda, magnesia, and potash ...	19·0	19·9	36·6	4·5	2·3
Chloride of sodium ...	28·2	8·0	13·1	44·7	4·1
Carbonic-acid gas, free and half-free—cubic in.	76·2	12·6	68·8	41·8	—
Carbonic-acid gas, quite free, ditto	33·9	3·9	—	27·3	13·1
Temperature in Fahrenheit degrees	44·0	164·5	50·5	51·2	102·5
Elevation in feet, about.	4000	1200	1900	600	800

It will be seen that the Tarasp springs contain nearly as much carbonate of soda as the Grande Grille at Vichy; as much of the aperient sulphates as the Sprudel of Carlsbad, and about half as much as the Ferdinands Quelle at Marienbad; while it contains about two-thirds the quantity of common salt contained in the Rakoczy of Kissingen. It is richer in free and half-free carbonic-acid than any of them. The high temperature of the Carlsbad and Vichy springs is, no doubt, a great advantage in some cases. But for tonic and bracing purposes the considerable altitude of Tarasp is an important condition.

We find, then, that the Tarasp water combines the alkaline character of the Vichy and the aperient character of the Carlsbad springs, while the presence of so large an amount of carbonic-acid renders it more agreeable as a beverage, and promotes the absorption of its constituent salts. In common with the Kissingen and Marienbad springs, the presence of an appreciable quantity of iron renders it also a blood restorative, while the quantity of chloride of sodium in it makes it a valuable stimulant to the organs of digestion.

There is scarcely a spring in Europe that is known to possess so many important qualities.

But Tarasp has the additional advantage of possessing several acidulous chalybeate springs of a tonic character. The one most commonly drunk is the St. Bonifacius spring. It is situated on the right bank of the Inn, about two miles distant from the Kurhaus. Recent analysis, however, of this spring shows that it does not contain nearly so large a quantity of iron as was formerly supposed ; and, what is also of great importance, that both the absolute and relative proportions of its constituents vary with variations in the rainfall, so that the spring is clearly not protected from the influence of surface drainage. This spring has a pleasant and but slightly chalybeate taste, and contains a large amount of carbonic-acid.

It has the drawback of containing a large amount of the constipating carbonate of lime.

There are several other iron springs in the district of Tarasp, but the only other to which I shall now call attention is the Wyh spring, which rises out of the ground, and when first I visited this region was quite unprotected and unenclosed, in one of the meadows on the hillside above the village of Schuls, and about twenty minutes' walk from the Kurhaus. This spring contains a considerable quantity of iron, and a very large amount of free carbonic-acid, which makes it one of the pleasantest mineral waters to drink I have ever tasted. At that time it was not much frequented. We found a tumbler hidden under an adjacent bush, which had been left there by a benevolent fellow-countryman for the use of those who might subsequently find their way to this delightful spring. We made use of it and returned it to its hiding-place. This spring in 1877 was provided with a new basin, and an " elegant little

wooden pavilion" erected over it. It is most picturesquely situated, and commands fine views in all
directions. The waters of this spring, which flow in
great abundance, are now carried off to the new
bathing hall at Schuls, where they become of some
importance as baths on account of the quantity of
carbonic-acid in them. Another spring near Schuls
and more accessible than the Wyh Quelle, is the
Sotsass Quelle; this spring contains very little iron
and is chiefly valued as an agreeable drinking water,
on account of its richness in carbonic-acid gas. The
analysis of these three springs is given in the following table:

In 16 oz. = 7680 gr., carbonates as mono-carbonates.	Bonifacius.	Wyh.	Sotsass.
Sulphate of potash	·589	·084	·038
„ soda	1·840	·087	·153
Chloride of sodium	·300	·016	·077
Carbonate of soda	6·184	·028	—
„ lime	15·286	9·467	7·970
magnesia	2·589	·648	·600
„ protoxide of iron ...	·141	·204	·098
Total of solid matters... ...	27·866	10·696	9·173
Free and half-free carbonic-acid in cubic inches...	57·26	48·42	47·93
Quite free carbonic-acid in cubic inches	32·82	38·37	39·49

I have now a few words to say as to the therapeutic
action of these waters.

It will not be difficult, bearing in mind the composition of these mineral springs, to analyse briefly
their action on the human economy.

The primary effect of these alkaline-saline springs
must be an antacid action on the gastric mucous
membrane. After absorption of the saline constituents, we naturally get increased activity of the
secreting organs, by which these salts are eliminated

from the blood, as well as increased activity in the
gastro-intestinal glands by which the digestive fluids
are secreted, and in this way they act as stimulants to
digestion. If the waters be taken early in the morning,
as is generally ordered, with gentle walking exercise,
their laxative effects are commonly obtained before
breakfast or immediately afterwards. The diuretic
action of the water is manifested later in the day. It
is further reasonable to conclude that when a considerable quantity of an alkaline-saline fluid is taken
into an empty stomach, from which it must at once
pass into the portal system of veins, and thence through
the secreting structure of the liver, that the condition
of this organ and the character of its secretion may
be greatly modified thereby. We know that sodium is
the base with which the acids of the bile are found
combined, and there can, I think, be but little doubt
that the internal administration of dilute solutions of
the carbonate of sodium exert an important influence
in modifying the character of this secretion.

One of the most useful remedies in cases of
jaundice associated with congestion of the liver, as
well as in cases of obstruction from biliary calculi, or
from the presence of inspissated bile in the duct of
the gall-bladder, is large draughts of warm water
containing carbonate of sodium in solution. What
is the Carlsbad water, which is found to be so efficacious in the relief of biliary concretions and in the
reduction of fatty and congested livers, but a warm
dilute solution of carbonate of sodium with some
additional saline purgative constituents ? Doubtless
the warmth of the water, in this instance, favours its
action. The Tarasp water is much richer in the
alkaline carbonate of soda, and very nearly as rich
in aperient salts. We are therefore quite prepared
to hear that these waters are found especially useful
in visceral congestions and obstructions, as, for
example, chronic affections of the liver, hyperæmia
of that organ, and the fatty liver arising from errors
in diet, certain cases of jaundice, and gall-stones ;

congestion of the spleen after intermittent fevers;
dyspepsias, especially those associated with acid
fermentation and with vomiting, and originating in
want of exercise, improper or irregular feeding, or
obstinate constipation; general corpulence; chronic
gout and rheumatism—conditions specially associated
with increased development of offending substances
in the blood or with defective action of the eliminatory
organs.   Tapeworms are said to be occasionally ex-
pelled during the use of these springs.   Some forms
of catarrhal affection of the kidneys and bladder, and
disposition to the formation of renal calculi, are also
relieved here, just as, indeed, they are by the use of
the Vichy water. Disorders of menstruation, associated
with a chronically congested state of the uterus and
ovaries, are reputed to be relieved in a very remarkable manner by a course of the waters at Tarasp.
Some cases of chronic eczema have also been greatly
benefited here.   It is further stated that persons
suffering from chronic laryngeal or bronchial catarrh
with hoarseness are relieved by drinking these waters,
and some physicians believe them to be useful in
the incipient stage of pulmonary tuberculosis.   It is,
however, very probable that the pure and invigorating
mountain air has more influence in producing the
improvement observed in cases of pulmonary disease
than the use of the mineral springs.

As to the quantity of water that is usually taken,
patients begin with two or three glasses (each holding
six ounces), and increase gradually to six or eight.
These are drunk in the morning before eating, at
intervals of a quarter of an hour, gentle walking
exercise being taken all the time.   It not unfrequently happens that persons with weak digestions
suffer considerable inconvenience from headache and
flatulence if they drink any quantity of cold water
before breakfast. When such is the case it is better to
take an early light breakfast, and drink two or three
glasses of the water two or three hours afterwards.
If necessary, a glass or two more may be taken late in

the evening. A common practice also is to warm the water before drinking.

It is very obvious that a powerfully alterative aperient and alkaline water should not be taken continuously for a long time. There are very few cases in which its use should be prolonged without interruption for longer than a fortnight; and many cases would probably derive much greater benefit from a combination of the saline and the chalybeate treatment than from either singly—the saline water for three or four days, then the chalybeate for a similar period, and so on. I am thinking especially of those cases of anæmia in females, associated with obstinate constipation and disorder of the generative system, in which we know, as a matter of common observation, that preparations of iron are rarely useful unless combined with an aperient, and that frequent aperients are injurious unless combined with some tonic remedies. Tarasp, with its saline aperient water on the one hand, and its bracing chalybeate spring on the other, seems to be exactly suited to such forms of disease.

In the next place I would desire to say a few words on the climate of Tarasp. This district, situated at an elevation of between 4000 and 4500 feet above the level of the sea, possesses all the invigorating characteristics of an Alpine climate, while it has the advantage of being much less severe and rigorous than that of the Upper Engadine. There are here fewer sudden changes of temperature, and an unexpected fall of snow in the summer months, by no means an uncommon occurrence at St. Moritz, is at Tarasp quite an exceptional event. The milder character of the climate is indicated by the much greater luxuriance of vegetation; rye and flax are extensively cultivated in this district, and fruit-orchards flourish near Schuls, while the local flora is exceeding rich and diversified.

As well as being milder, the air is not so dry and rarefied as in the Upper Engadine—a condition which occasioned frequently most uncomfortable excitement and irritation in many cases of functional nervous disorders which used to be sent in considerable numbers to these high Alpine stations. Unpleasant and troublesome herpetic eruptions about the face are very commonly produced in persons with delicate skins by the dry stimulating air of St. Moritz. These are not so frequently observed in the milder climate of the Lower Engadine. This part of the valley is almost entirely protected from the north and north-east winds, the prevailing winds being the south-east and north-west.

The snow, which accumulates to some thickness on the ground during winter, begins to melt here in April. The spring is short, as in all mountain districts, and in June the summer is so far advanced that on the 15th the bath season commences. This continues to the end of September, a month generally remarkable for constantly fine genial weather, the only drawback being the shortness of the days. The mean atmospheric temperature in the months of July and August ranges from 56 degrees Fahrenheit to 60 degrees Fahrenheit. The maximum and minimum temperatures noted in the same months were 82 degrees Fahrenheit and 37·5 degrees Fahrenheit. It will thus be seen that the climate of Tarasp especially commends itself to those cases in which it is thought desirable to try the influence of mountain air, without incurring the risk which certainly attaches itself to an exposure to the sudden changes of temperature, the highly rarefied air, and often the continuous cold of the Upper Engadine.

Also, on leaving the Upper Engadine, Tarasp offers an admirable intermediate point where patients may break the suddenness of their descent into the plains of Italy and Switzerland, which to many people proves very trying, and the presence of chalybeate springs of a somewhat similar character to those of St. Moritz offers facilities for continuing or prolonging the course of steel waters if it be thought desirable to do so.

This account of Tarasp would be very incomplete if I were to omit to add to the preceding remarks on its climate and mineral springs a few topographical and general details.

The quickest way of getting to the Lower Engadine from England is through Bâle to the Landquart station on the railway between Zurich and Coire—it is the next station but one after Ragatz. A little distance from this place, the beautiful and picturesque valley of the Prættigau opens. The road ascends through this valley to Davos, in the Davosthal, which is reached in about six hours by diligence or by posting. From Davos a carriage road crosses the Fluela Pass to Süs in the Lower Engadine, and by this route Tarasp may be reached in about seven hours from Davos, so that the whole journey between Landquart and Tarasp may be accomplished in one rather long day, and the entire distance between London and that place in three days, sleeping one night on the road at Ragatz or Landquart. But this would be too fatiguing for most persons. A better plan is to rest five or six hours at Bâle, and go on in the afternoon to Ragatz or Coire, and the next day go from Coire or Landquart to Davos. There is a post route between each of those places and Davos. Sleep a night at Davos, and proceed the next day to Tarasp.

I have heard it remarked that "Tarasp is in a hole," and an impression of this kind is certainly produced upon one when, coming from the Upper Engadine, the diligence drives down to the left bank of the river and deposits one at the Kurhaus, which is built at the bottom of the somewhat narrow gorge through which the Inn here flows. But although it has pleased the proprietors to build the *Établissement des Bains* at the bottom of this gorge in order to be close to the principal mineral springs, it does not follow, nor is it true, that "Tarasp is in a hole." Tarasp is the name of the district, and a spot more beautifully situated or with greater natural advantages it would be difficult to find. It forms a small plateau, around which

spreads a hilly country covered with meadows, corn-
fields, and wooded slopes in charming variety, crowned
on each side by mountain summits of singularly bold
and striking outline. Beautiful lateral valleys, the
sides of which are covered with shady and fragrant
pine-woods, penetrate deeply into the recesses of the
Alpine chain, and afford a great variety of singularly
attractive walks. For mountain climbers there is
every kind of work, from rugged hitherto unscaled
peaks to those lower points of view which may be
described, in guide-book phraseology, as "easily accessible for ladies;" while the largest glacier-fields in
this part of the Alps—the Vadret-Lischanna—can be
readily reached in a few hours. One of the great
charms of the place is the number of pleasant shady
walks over the wooded park-like hilly slopes in the
immediate vicinity of the springs, which can easily be
prolonged into more ambitious efforts as the strength
of the invalid returns.

Those who object to the situation of the Kurhaus
can obtain excellent accommodation at the village of
Schuls, which is admirably placed, and distant about
a mile from the springs. Schuls possesses several
hotels: the best known are the Old and New Belvidere.
It has also recently become possessed of a bath-house,
where baths can be taken just the same as in the
Kurhaus at Tarasp. But commend me especially to
the hamlet of *Vulpera*, where most assuredly the
Kurhaus ought to have been erected; it is placed in
a sunny and pleasant open spot about 300 feet above
the bath establishment, on the opposite bank of the
Inn, and from it a good zigzagged path leads in about
seven minutes to the mineral springs. Vulpera, when
first I knew it, contained but two or three homely
pensions, now it has a Bellevue Hotel, kept by
Fanconi, and several good *pensions*.

The village of Schuls, a most interesting spot, lies
at the base of the slate mountains on the left side of
the valley. It consists of two pretty little villages
named Ober and Unter Schuls. They are surrounded

by well-cultivated orchards, for in this comparatively mild climate fruit trees flourish greatly. High above on the mountain-side are green Alps, broken by stretches of pine-woods, while down by the banks of the Inn are grassy meadows. Schuls is protected by its position from cold winds, and the snow soon melts on its sunny heaths. The houses and buildings in the village are good, the old church stands most picturesquely on a high rock, and in olden times it offered a place of refuge and was often bravely defended. A fierce encounter took place on this spot in 1621 between the inhabitants and the invading Austrians. Men and women equally fought with the energy of despair; their courage however was unavailing against the overwhelming numbers of their assailants, although they repeatedly drove back the besiegers. At length the brave little band gave way, but not until the ground was covered with their own bodies and those of their enemies. Close below the church a bridge leads over the Inn to Vulpera and Tarasp, and then there is a second one over the Scarlbach, which here rushes forth out of a deep ravine.

The mountains on the right bank of the river present a very different formation from those on the left. Those above Tarasp are joined together by high *grats*, and from them short yet high ridges stretch towards the Inn. Nearly every one of these is surmounted by two massive pinnacles terminated by steep jagged points, presenting altogether a great variety of graceful and striking forms.

In front of this principal chain, a second runs from Tarasp towards Finstermunz, not inferior to the first in height. It begins with the Piz Pisog (10,427 feet) and ends near Martinsbrück in the Piz Lat. The two principal points of this chain projecting towards the Inn, are the Piz St. Jean and the Piz Lischanna. Behind the latter, and enclosed by fearfully wild rocky precipices, lies the Vadret-Lischanna, the single great glacier of this side of the valley, being nearly

six miles long and more than a mile and a half wide.
There are many other small glaciers amongst these
mountains, but no great ones, on account of the
steepness of the mountains' sides and the narrowness
of their ridges. From their summits deep ravine-
like valleys stretch downwards towards the Inn. The
principal of these are the Val Lischanna, Val Triazza,
Val Uina, and Val Assa. The Scarl valley, with its
lateral branches, insinuates itself deeply between the
two adjacent ranges of chalk mountains.

There is scarcely any portion of the whole Alpine
chain which presents a wilder or more broken outline than the mountains on the southern side of this
valley ; while it is rare to see such gigantic mountain
forms so close to one another, and at the same time
separated by such deep valleys. On this account it
is that their outlines appear so sharply cut. The Piz
Pisog rises directly to the south of Tarasp, in bare
gray dolomite walls, whose perpendicular and inaccessible sides lead to a slender-pointed snow-covered
summit. This mountain is one of the most imposing
of Alpine forms ; seen from Tarasp it appears as a
pyramid, while behind it lengthens out into a long
ridge studded over with many small low points, till
it reaches the upper part of the rocky valley, Zuort,
which is almost entirely occupied by glacier. The
district of Tarasp spreads out at the foot of this
grand mountain, and is surrounded on all sides by
a hilly country covered with green meadows and
patches of woodland in charming variety. Scattered
here and there are hamlets and single houses, which
constitute the parish of Tarasp ; westward lies the
hamlet of Fontana, commonly described as the *village* of Tarasp, and above this, on a high rock, the
old Castle of Tarasp. The white walls of this fine
old ruin are seen from far and wide, and form an
ornament to the whole valley. It is a conspicuous
object even from the height of the Fluela Pass. The
building is still in good condition. The view from it
is an exceedingly grand and varied one, extending

over the whole of the Unter Engadine valley. It once served as the seat of the Austrian governor of this district, and the Austrian eagle may still be seen on its outer walls. The principal church of the valley, in connection with a Capuchin convent, is situated at Fontana, on the bank of a beautiful clear lake of some extent, which affords good fishing. The whole of this region is remarkably beautiful. A new hotel and *pension* has been erected here, close to the little lake.

Tarasp was until 1815 an Austrian possession, and with its rocky fortress, proved very troublesome to the rest of the canton, particularly in time of war. The inhabitants, about three hundred souls, are Roman Catholics, and speak, for the most part, German. Above the lake there is a second plateau, separated from the lower one by a steep terrace, and on this plateau there is a second small lake, the Schwarze See. On the other side a stream called the Chlemgia flows through a fearfully deep gorge, its white foaming water eddying swiftly over black serpentine rocks.

The hamlet of Vulpera, which I have already alluded to as a charming spot for the residence of the visitors to the baths, is about half-an-hour's walk from the castle. It is about midway between the Château of Tarasp and the village of Schuls. It is built on a little plain above the Inn on its right bank. The river here flows over a deep rocky bed, close by the side of which the Kurhaus has been erected.

One of the most remarkable natural phenomena in the neighbourhood of Tarasp are the so-called Moffete. These are orifices in the ground, in the meadows on the way between Schuls and Fettan. Carbonic-acid gas streams out of these holes in sufficient abundance to suffocate any small animals that come near them. All vegetation is also destroyed for some distance round. We observed numerous dead insects, small birds, mice, and even a snake, scattered around these curious apertures in the soil.

Tarasp is a most convenient centre for interesting

excursions ; situated as it is within a few hours of the grand defile of the Finstermunz and the Austrian frontier, within a day's walk of the finest pass in the Alps—the Stelvio, and bounded directly to the north by the peaks and glaciers of the Silvretta group.

NOTE.—Since the completion of the Arlberg tunnel a convenient way of reaching Tarasp will be by rail to Landeck, and thence by diligence to Tarasp-Schuls in about eight hours. Diligences run twice a day during the season. This route follows the course of the river Inn, and passes through the interesting Finstermunz road and does not cross any mountain pass. There is a new and rapid train (express to Vienna) which leaves Bâle at 7.20 a.m., and Zurich at 9.55 a.m., and arrives at Landeck at 5 p.m.

APPENDIX TO CHAPTER IV.

SOME NOTES ON THE FLORA OF THE ENGADINE.

IT is scarcely possible to pass a season in the Engadine without taking some interest in those beautiful Alpine flowers which grow in such varied profusion on the surrounding mountain sides and in the lateral valleys. "Cette brillante flore des Alpes," says M. Rambert, "m'a poursuivi de son image : c'était comme une idée fixe, comme un de ces refrains qui durant des jours et des semaines se chantent d'eux-mêmes dans la mémoire et qu'on fredonne sans y songer." There is no district in the whole of Switzerland where one can study with so much ease the special characteristics of the vegetation of the Alps as in the villages of the Upper Engadine. Owing to the great elevation of this part of the valley of the Inn, numerous species of Alpine plants are to be gathered within a few feet of the door of one's hotel, which can only be obtained in other parts of Switzerland after a long and arduous climb ; so that without any laborious or dangerous ascent some of the rarest and most delicate of Alpine flowers may be collected even by children. Two localities in the Upper Engadine are especially rich in different species of wild-flowers : one is the Fex Valley, the entrance to which is placed close to the pretty village of Sils Maria ; the other is the Val Bevers, of the beauty of which I have already spoken. Another valley more distant, but celebrated also for the number and variety of Alpine species which can be obtained there, is the Val del Fain, which opens to the south-east of the Bernina road, a short distance before one reaches the summit of this pass. But even in the woods and meadows immediately surrounding St. Moritz numbers of beautiful wild-flowers may be gathered, and an hour's ramble along the hill sides in any direction will be rewarded with an abundant harvest of interesting species.

What strikes one most forcibly as a characteristic of these Alpine flowers is the richness and intensity of their colours, and their very large size compared with the scanty foliage and dwarf-like aspect of the plants which produce them. The purpose of this great development of the flowers seems to be to gain the full advantage of the brief but powerful sunshine, in order that they may ripen their seed as rapidly as possible.

Common British species, many of which may be found growing in these regions, appear, by the great size and intense richness of colour of their blossoms, to present quite different appearances from those with which we are familiar. This is especially noticeable with regard to the species of House-leek (*Sempervivum*), and Stone-crop (*Sedum*), which are very abundant on some of the hills in this neighbourhood, and present a perfectly dazzling variety and brilliancy of hues. The common Monkshood (*Aconitum Napellus*) grows here in some of the meadows in great luxuriance, and its large and stately racemes of blue helmet-shaped flowers look very ornamental.

One of the commonest of plants around St. Moritz is the mountain Arnica (*Arnica montana*), a composite plant, whose large, deep orange-yellow flower-heads are met with in all directions. It is thought to possess virtues of a healing nature, and the tincture of arnica is a common application to wounds and bruises in our own country, but whereas we make a spirituous extract of the root, the natives use an infusion of the flowers. The forests in the Engadine reach a greater height than in any other part of Switzerland, often extending to 6600 feet and upwards. They are composed almost exclusively of three species of coniferous trees—the larch, the pine, and the arolla or pinus cembra. The latter is the characteristic tree of the forests of the Engadine, and grows highest on the mountain sides. Its wood is remarkable for its durability, its agreeable perfume, the fineness of its grain, and the ease with which it is worked. Its cones are about three inches long and two inches broad ; the seeds are large and contain a sweet oily kernel of a pleasant taste, which in some parts is used as food.

The arolla is very rare in the other cantons of Switzerland. The deep sombre hue of its foliage and its stern majestic aspect contrast strongly with the elegant outline and delicate green of the larch. It is curious to notice the length of time it takes one of these trees in this elevated region to reach a certain development. As many as 250 consecutive woody layers, indicating as many years of growth, so close together as to be distinguished with difficulty, have been counted on the trunk of an arolla only two feet in diameter. This slowness of growth is explained by the short duration of the summer, and hence the renewal of these forest trees when once they are destroyed is very difficult. The Bergamesque shepherds, though they doubtless add somewhat to the picturesqueness of the landscape, and pay also a considerable sum to the commune, yet do great and irreparable damage to the young forest trees through the injuries inflicted on them by their flocks.

The following list of species of plants found in the neighbourhood of St. Moritz and Tarasp, does not pretend to be complete. I believe, however, it will be found to contain nearly all the more interesting and rare flowering plants. I am indebted to Dr. Killias for the names of the species that have been found in the valleys and on the mountains around Tarasp. The method I have adopted of arranging them under their natural orders has seemed to me the most convenient. In nearly every instance the familiar English name of the plant is given, and in many places the German name also. Wherever it has been convenient to do so, I have briefly mentioned one or two of the more striking characters of the plants in order to facilitate their recognition by beginners.

The letters S.M. refer to the neighbourhood of St. Moritz, the letter T. to that of Tarasp.

I. Natural Order: RANUNCULACEÆ.

Stamens, indefinite, hypogynous ; ovary, apocarpous.

RANUNCULUS MONTANUS (Mountain Buttercup). Germ. *Berg Hasenfusz.*

A small plant, often flowering when not more than three inches high ; its leaves are in compact tufts, of a dark shining green colour, and deeply cut, the lobes rounded ; the flowers are of a brilliant yellow, a little larger than those of the common buttercup, and covering the plant in a dense mass. S.M. Brail. (Spring.)

R. GLACIALIS (Glacier Buttercup). *Germ. Gletscher Hasenfusz.*

One of the highest of mountain plants, growing on the margins of glaciers and eternal snows. The dark, brownish-green leaves are smooth and deeply cut ; the calyx is covered with soft, shaggy, brown hairs ; the flowers are rather large, and the petals white, tinted with a dull purplish rose on the outside. S.M. Val Bevers, Alp Err, Piz Languard, Piz Corvatsch, Cima di Margna, and in many other localities. (June.)

R. ALPESTRIS (Alpine Crowfoot). Germ. *Alpen Hasenfusz.*

A very small and pretty species, three to four inches high, leaves roundish in outline, deeply cut, dark green and shining, flowers *pure white*, one to three on each stem. Grows generally in moist, rocky places, as the higher mountain pastures. S.M. Albula Pass. (August.)

R. ACONITIFOLIUS (Fair Maids of France).

A large variety, often growing to two or three feet in height, and having rather small white flowers ; leaves deeply divided, nearly down to the base. Grows in moist places in Alpine woods and valleys. T. on the Alps by Fettan.

R. PARNASSIFOLIUS (Parnassia-like Buttercup).

A beautiful species, with stems from two to eight inches high leaves entire, cordate or reniform, dark brownish green ; flowers one to twelve on each stem, beautiful pure white, with yellow centres. S.M. Val Bevers, Piz Err, Piz Padella.

R. RUTÆFOLIUS (Rue-leaved Buttercup). Germ. *Schenckblume.*

Leaves deeply cut, the radical leaves twice cut ; stem three to six inches high, bearing usually but one handsome flower, which is about an inch across, white, with an orange centre, and occasionally rose-tinted at the margin. S.M. Val Lavirum. (July.)

R. PYRENÆUS.

A very elegant species of buttercup; the flower is composed of beautifully delicate white petals, and measures about an inch across. The leaves are nearly linear and undivided. The whole plant is from three to four inches high. S.M. Alp-Otta, near Samaden, Piz Languard.—Agagliog's Rocks, Roseg Glacier.—T. Piz Arina, Piz Campatsch.

R. LAMIGINOSUS.

T. Adjacent valleys.

THALICTRUM ALPINUM (Alpine Meadow-rue).

A plant generally found at considerable elevations in mountain regions, it is common in the Highlands of Scotland. This is a dwarf species, stem from four to six inches high ; leaves compound ; the leaf-stalk twice divided into three or five branches ; the leaflets small and rounded with indented or lobed margins. The few drooping clustered flowers have no petals, and their four small petal-like sepals are almost concealed by the prominent stamens, ten to twenty, whose long anthers project beyond the calyx. T. Scarl valley.—S.M. Val Lavirum. (Summer.)

T. AQUILEGIFOLIUM.

A tall, handsome lilac variety, found near the path leading to the Morteratsch Glacier. S.M.

T. MINUS (Maidenhair Meadow-rue).

A species greatly admired on account of the resemblance of its foliage to that of the Maidenhair fern. It is a native of Britain, but by no means common, being found occasionally in Scotland and North-Western England. Its flowers are inconspicuous, the sepals of a pale greenish-yellow, with a pink tinge. Stamens numerous, with long narrow anthers. T. Adjacent valleys.

T. FŒTIDUM.

T. Adjacent valleys.

ANEMONE SULPHUREA (Yellow Wind-flower). Germ. *Schwefelgelbis Windröschen.*

Differing from the Alpine anemone only in the yellow colour of its flowers. It varies in size according to the altitude at which it grows, from four inches in height near the mountain summit, to two feet in the valleys. The plants of this genus have no petals, but their sepals are coloured and petal-like, and usually six in number. Their leaves are radical, and the flower-stalk is naked, excepting three leaves which form an involucre usually at some distance from the flower. In this species the leaves are large and much divided, and the sepals are covered externally with a soft down. S.M. Val Bevers. (June.) Val del Fain.

A. VERNALIS (Shaggy Pasque-flower). Germ. *Frühlings Windröschen.*

A very dwarfed species; flowers large and shaggy; the sepals covered with brownish silky hairs. It flowers early in spring, and is usually confined to very elevated positions. S. M. Val Bevers. (May.)

ACONITUM NAPELLUS (Monkshood).

A fine stately plant, found growing wild in great abundance in shady places in the meadows around St. Moritz. Its stem grows erect to three or four feet in height, and ends in a handsome dense raceme of dark blue helmet-shaped flowers. The leaves are dark green, and divided to the base into five or seven deeply cut linear-pointed segments. It is a medical plant of great activity; all parts of it are poisonous. On more than one occasion the root has been eaten for horse-radish with fatal consequences, although the two roots do not the least resemble one another.

A. PANICULATUM.

T. Adjacent valleys.

AQUILEGIA ALPINA (Alpine Columbine).

This is one of the grandest of Alpine flowers, measuring nearly three inches across. The flowers are much finer than those of our common garden columbines, of a showy blue colour, and having stamens which project beyond the petals. The stem varies from one to two feet in height; the leaves are compound, each leaf being composed of three primary divisions; each of these divisions being again divided into three lobes, and each lobe being deeply cut. The leaves are almost as elegant as the flowers. There is an excellent *variety* which has the centre of the flower white. S.M. The cliffs of Piz Err.

A. ATRATA.

Another species, found in the neighbourhood of San Moritz, with smaller flowers of a very much darker colour, brownish purple, and leaves much less divided.

II. PAPAVERACEÆ.

Flowers regular. Sepals, two, caducous; Petals, four; Stamens, numerous. Ovary, syncarpous.

PAPAVER PYRENAICUM (Alpine Poppy). Germ. *Mohn.*

A dwarf poppy about three or four inches high, with flowers of a beautiful orange colour; dissected leaves, with small pointed lobes, slightly hairy. A native of the higher Alps.—S.M. Val del Fain. Cambrena Glacier; near summit of Canciano Pass.—T. Val Plafna (July.)

III. CRUCIFERÆ.

Sepals, four; Petals, four; Stamens, six, tetradynamous (four long, two short).

ARABIS ALPINA (Alpine Rock-cress). Germ. *Gänsekraut.*

This simple little plant is commonly cultivated in our gardens at home. It is very widely diffused over the Alps. The stem is erect, with a few scattered sessile lanceolate toothed leaves, which are rather fleshy. Flowers in clusters, small; petals spreading; white or purplish. S.M. Common.

A. CILIATA. Smaller than the preceding; stem not so erect; leaves oblong; margins not toothed, but fringed with delicate hairs. S.M. Abundant.

A. HALLERI. S.M. Saluver near Celerina, also lake of Poschiavo. (June.)

A. CÆRULEA. A variety with purple flowers. T. Piz Minschun.—S.M. Val del Fain.

A. PUMILA. T. Piz Lat.

CARDAMINE PRATENSIS (Bittercress, Ladies' Smock). Germ. *Schaum-kraut.* A common British plant. Stem erect, about a foot high, pinnate leaves. Flowers pinkish purple, petals obovate and spreading. S.M. Common.

C. RESEDIFOLIA. A dwarf species with white flowers. S.M. Val Bevers. (June.)

C. IMPATIENS. A larger species; stem a foot high or more; on each side of the base of each leaf-stalk there is a curved appendage, embracing the stem, like a stipule. Petals very small and sometimes absent. T. Valleys adjacent.

C. ALPINA. T. Piz Minschun.

CAPSELLA PAUCIFLORA (Shepherd's Purse). A small mean-looking plant, with few inconspicuous flowers. T. Near the village.

DRABA AIZOIDES (Sea-green Whitlow-grass).

Found indigenous in Britain, near Pennard Castle, Swansea. It is a dwarf plant, not exceeding three or four inches in height. The leaves are three or four lines long, sessile linear, of a bright green colour, edged with stiff white hairs, and forming dense tufts at the base of the leafless flower stem, which bears about a dozen or more bright golden yellow flowers in an erect raceme. S.M. Piz Languard.

D. FRIGIDA. Germ. (*Kaltes Hüngerblümchen*).

A very dwarf species, with a few *white* flowers. S.M. Piz Languard.—T. Val Muschens, near Scanfs.

D. NIVALIS. The most diminutive of the genus. The leaves are covered with minute stellate hairs, which give them a whitish-green appearance. The whole plant when in flower is not more than two inches high. T. Piz Lat. (Very rare.)

D. WAHLENBERGII. T. Piz Minschun.

BISCUTELLA ALPESTRIS (Buckler Mustard). S.M. Val Bevers.

ERYSIMUM LANCEOLATUM. Common in Britain, related to the mustard; conspicuous yellow flowers, not unlike those of the wall-flower. S.M. Abundant.

E. STRICTUM. T. Valleys adjacent.

HUTCHINSIA ALPINA. A neat little plant with shining leaves, deeply cut into narrow lobes, and pure white flowers in clusters on stems about an inch high. S.M. Bevers. (June.)

H. BREVICAULIS. S.M. Piz Padella. (August.)

IV. CISTACEÆ.

Wiry stems; opposite leaves; regular, showy rose-like flowers. Sepals, three; stamens indefinite; hypogynous.

HELIANTHEMUM GRANDIFLORUM (Large-flowered Rock-rose).

Leaves and shoots downy and hairy; large handsome yellow flowers. S.M. Adjacent hills.

V. VIOLACEÆ.

The very characteristic Violet Family.

VIOLA ALPESTRIS.

A beautiful yellow violet, growing abundantly in the meadows round St. Moritz. The plant grows to six inches or more in height; its flowers are of a rich handsome yellow, and the three lower petals are striped with thin black lines.

V. CALCARATA (Spurred Violet). Germ. *Gesporntes Veilchen.*

An Alpine species, often found growing very abundantly at great altitudes. Its large purple flowers contrast remarkably with the dwarf character of the plant, its leaves being scarcely visible in the turf. It is distinguished by having stipules deeply divided into three lobes at the top, and by its habit of increasing by runners under the soil. S.M. Val Bevers. (July.)

V. TRICOLOR (Heartsease). Our common field pansy, but in this Alpine region with larger flowers and richer colouring of purple, white, and yellow. S.M. Val Bevers.

V. CANINA (Dog Violet). The common purple dog violet. S.M. Val Bevers.

V. PENNATA. T. Val Uina.

VI. CARYOPHYLLACEÆ.

Opposite entire leaves; regular flowers; definite hypogynous stamens; free central placenta.

DIANTHUS GLACIALIS (Glacier Pink).

A most lovely little plant, growing on the highest Alpine summits. Its tufts, just like short wiry grass, grow close to the ground, with leaves not more than half an inch to an inch long, and flowering stems from one to three inches high. This dwarf stem bears a large brilliant flower of the deepest and purest rose-colour, about an inch across or even more. The outer margins of the petals are slightly notched. S.M. Val Bevers, Piz Err, on the road from Muhlen to the lakes of Scalotta; Piz Padella.

D. SUPERBUS (Fringed Pink). Germ. *Stolze nelke.*

A large and handsome species, very abundant in the meadows around the lake of St. Moritz, easily known by its sweet fragrance, and by its petals being cut into strips or fringes for more than half their length. The flowers are of a lilac colour, and often present a dark spot or eye in the centre.

D. SYLVESTRIS (Wood Pink). S.M. Common.

D. DELTOIDES (Maiden Pink). Found in many parts of Britain. Flowers small, little more than half an inch across, but bright and pink-spotted or white. T. Valleys adjacent.

ALSINE RECURVA. S.M. Bernina Pass, on banks of Lago Bianco; Val Bevers; near summit of Canciano Pass.

A. SUBNIVALIS. S.M. Val Bevers.

A. ROSTRATA. T. On rocks of Tarasp and Ardetz.

CERASTIUM LATIFOLIUM (Broad-leafed Cerast). Germ. *Breitblatterisches Hornkraut.* A shaggy, dwarf, tufted plant, spreading freely, but not more than two inches high, covered with a silky down ; large white flowers. S.M. Piz Padella, near Samaden. (August.)

C. GLACIALIS (Glacier Cerast). Not so shaggy as the preceding species ; leaves narrower ; flowers large and white. S.M. Val del Fain.—T. Piz Minschun.

GYPSOPHILA REPENS. Germ. *Gypskraut.* Not a showy plant, growing to six or eight inches high ; glaucous leaves spreading in dwarf tufts. Flowers in cymose clusters, white, veined with rose, on thread-like stems. S.M. Albula. (August.)

ARENARIA BIFLORA (Two-flowered Sandwort). S.M. Cambrena Glacier and Val del Fain.—T. Val Floss.

LYCHNIS ALPINA. Germ. *Alpen Lichtnelke.*

This pretty dwarf lychnis is rare in Britain ; it has been found on the summit of Little Kilrannock, a mountain in Forfarshire, and also in lonely and high mountain gorges in Cumberland. It rarely reaches six inches in height ; leaves long and narrow, flowers pink in compact heads, petals narrow and deeply bifid. S.M. Piz Err, Val Lavirum, and Lago della Crocetta, on Bernina Pass. (July.)

L. FLOS-JOVIS. A showy plant, with showy purplish flowers. T. Near Guarda.

SILENE QUADRIFIDA. T. Scarl valley.

CHERLERIA SEDOIDES (Cyphel). S.M. The grat between Mont Pers and Piz Cambrena.

VII. GERANIACEÆ.

GERANIUM SYLVATICUM. The common wood geranium. S.M. Abundant.

G. DIVARICATUM. T. Valleys adjacent.

VIII. POLYGALACEÆ.

POLYGALA ALPESTRIS (Alpine Milkwort). Germ. *Alpenkreuz Blume.* S.M. Alpotta, near Samaden. (June.)

P. VULGARIS. S.M. Val Bevers.

P. AMARA. S.M. Val Bevers.

IX. RHAMNACEÆ.

RHAMNUS PUMILA. T. Valleys adjacent.

X. LEGUMINOSÆ.

Flowers papilionaceous (resembling the flower of the pea). Stamens, ten, in one or two bundles.

HEDISARUM OBSCURUM. A vetch-like creeping plant, with long handsome spikes of purplish violet flowers ; leaves pinnate with seven to nine pairs of leaflets, the stipules united opposite the leaves. S.M. By Stätzer See.

ASTRAGALUS ALPINUS (Alpine Milk-vetch). Germ. *Alpen Tragant.* Found native on a few of the Scotch mountains. An attractive plant with short close racemes of drooping flowers, bluish purple, or white tipped with purple, on rather long peduncles. Flowers rather more than half an inch long ; leaves pinnate with eight to twelve pairs of leaflets and an odd one. S.M. Alp Spignas.

A. ONOBRYCHIS. Flowers purplish crimson, supported on peduncles an inch or more longer than the leaves, which are pinnate and about four inches long. A very handsome species. T. Adjacent valleys.

A. DEPRESSUS. T. Near ruins of Steinsberg.

A. CICER. T. Valleys adjacent.

PHACA ASTRAGALINA. S.M. Val del Fain.

OXYTROPIS URALENSIS. Not uncommon in Scotland. Leaves and peduncles springing from root-stock ; pinnate leaves, each with from ten to fifteen pairs of leaflets and an odd one ; leaflets oblong or lanceolate and hairy ; peduncle about four inches long with a short dense spike of bright purple flowers. The whole plant is densely covered with short silky hairs. S.M. Piz Languard.—T. Piz Minschun.

O. LAPPONICA. Germ. *Lapplandischer Spizkiel.* Plant more glabrous than preceding ; leaves, springing from stem as well as stock, longer, and leaflets more numerous ; flowers smaller and fewer ; petals blue, marked with white. S.M. Val Bevers ; Albula. (June.)

O. CAMPESTRIS. Found in one locality in Scotland, the Clova mountains. Flowers pale yellow tinged with purple ; not so hairy as *O. uralensis.* T. Adjoining meadows.

COTUS CORNICULATUS (Bird's Foot Trefoil). A common British plant, with tufts of bright yellow flowers, often *red* at the upper part. S.M. Common.

HIPPOCREPIS COMOSA (Horseshoe Vetch). Common also in the south of England. A small plant with deep-yellow flowers in an umbel, like *Coronilla;* small pinnate leaves, smooth leaflets. S.M. Abundant in meadows.

CORONILLA VAGINALIS. T. Scarl valley and ruins of Steinsberg.

VICIA CRACCA (Tufted Vetch). Common in Britain. Flowers in one-sided many-flowered racemes, on peduncles rather longer than the leaves, of a fine bluish purple, and about five lines long. Leaves pinnate, ending in tendrils ; leaflets nearly linear. The plant climbs by means of tendrils to a height of two feet or more. S.M. Common. (July and August.)

V. GERARDI. T. Meadows adjacent.

TRIFOLIUM ALPINUM (Alpine Clover). Germ. *Alpen Klu.* S.M. Piz Munteratsch. (June.)

T. BADIUM. S.M. Val Bevers.

XI. ROSACEÆ.

Regular flowers, usually with indefinite stamens inserted on the calyx. In some apetalous genera (Alchemilla, etc.) the stamens are definite, but they are then distinguishable by their divided leaves and conspicuous stipules ; ovary apocarpous or spuriously syncarpous from adhesion of the tube of the calyx.

ROSA ALPINA.

This beautiful shrub is a true member of the Rose family and has no relation with the Rhododendron or "Alpenrose" of the Germans. It is celebrated for having no thorns, but it is said to produce them when it descends into the plains ! Its flowers are of moderate size and of the purest carmine ; they are brilliant but fleeting, and the corolla often fades and perishes on the same day that it expands. S.M. Rocks in the neighbourhood.

R. RUBRIFOLIA. T.

R. POMIFERA. T.

POTENTILLA ALPESTRIS (Alpine Cinquefoil). *Germ. Goldfarbiger Fingerkraut.*

Closely related to the P. verna, found thinly scattered over hilly districts in England and Scotland. This plant, however, is larger, sometimes nearly a foot high. The radical leaves are composed of five wedge-shaped divisions notched at the top, and of a bright shining green colour. The flowers are about an inch across, and of a bright yellow. S.M. Piz Surlei.

P. FRIGIDA. S.M. Piz Languard, Lago della Crocetta, near summit of Bernina Pass. T. Piz Cotschen.

P. ANSERINA. The common silver weed. The flowers in this region are unusually large, and of a very brilliant yellow.

P. TORMENTILLA. Another very common British plant, having small yellow flowers with four petals. S.M. Woods around.

P. GRANDIFLORA. A large species with fine yellow flowers, and leaves composed of three sessile leaflets. S.M. Val Bevers.

P. CAULESCENS. T. Ardetz, Scarl valley, etc.

DRYAS OCTOPETALA (Mountain Avens). **Germ.** *Dryade.*

A creeping plant, with crowded leaves in dense spreading tufts and large cream-white flowers; common in most Alpine districts, and found abundantly in the north of Scotland. It is an evergreen, its leaves are little more than half an inch long, deeply and regularly crenate, shining above, and white and downy beneath. The flower-stalks are erect, two or three inches long; calyx of eight segments, rather shorter than the petals, which are also eight in number. Numerous yellow stamens and numerous carpels, which, when they have ripened into fruit, have feathery appendages an inch long. S.M. Val Chamuera.—T. Its environs.

GEUM MONTANUM. A dwarf genus, with large handsome yellow flowers. S.M. Val Bevers.

G. REPTENS. Germ. *Raukendes Sagenskraut.* S.M. near Piz Ot, Piz Languard, borders of the Lago Bianco, Bernina Pass.—T. Piz Minschun, Piz Arina, etc. (June.)

ALCHEMILLA PENTAPHYLLEA. S.M. Surlei Pass.—T. Piz Grian.

COMARUM PALUSTRE or POTENTILLA COMARUM. S.M. Near Statzer See.

XII. ONAGRACEÆ.

Herbs with two, four, or eight stamens; ovary inferior.

EPILOBIUM ALPINUM (Alpine Willow-herb). *Germ. Weidenröschen.*

A dwarf species, seldom more than four or five inches high, with stems decumbent, and much branched at the base; leaves ovate, the upper ones nearly lanceolate and slightly toothed. The flowers are large in proportion to the size of the plant, they spring from the axils of the upper leaves, and form short leafy racemes. The petals are notched. The style ends in a club-shaped stigma. S.M. Val Bevers. It abounds in the Scotch Highlands.

E. FLEISCHERI. S.M. Val Bevers.—T. Adjacent meadows.

E. ROSMARINI FOLIUM. S.M. Near Morteratsch glacier, on left bank of stream.

XIII. CRASSULACEÆ.

Herbs with fleshy leaves, very regular flowers, and apocarpous or nearly apocarpous pistils. Carpels three or more.

There are several species of Sedum (Stone-crop) and Sempervivum (House-leek) in the neighbourhood of St. Moritz ; but they are chiefly those which are common in other localities.

SEMPERVIVUM WULFENII (Yellow Flowers). S.M. Val Roseg, near entrance.

XIV. RIBESIACEÆ.

RIBES ALPINUM. T. Adjacent valleys.

XV. SAXIFRAGACEÆ.

Herbs with regular flowers, definite stamens (as many or twice as many as the petals), syncarpous pistils, and two or more styles.

SAXIFRAGA CÆSIA (Silver Moss). Germ. *Murgrüner Stein-brech.*

A very dwarf saxifrage, found in small tufts, covering rocks and stones like a silvery moss. It is known readily by its extreme dwarf-ishness, and its three-sided keeled leaves, with white crustaceous dots on their margins. The flowers are white, about a third of an inch across, on smooth thread-like stems, from one to three inches high. S.M. Albula Pass, Val del Fain. (August.)—T. Scarl valley.

S. OPPOSITIFOLIA (Purple Saxifrage).

Low straggling tufts, seldom more than an inch high ; leaves small, crowded, obovate, opposite, and ciliate. Handsome, rather large, purple flowers, growing singly on very short stout branches, but often so crowded as almost to conceal the foliage. S.M. Val Bevers. (June.)

S. STELLARIS (Star Saxifrage). Germ. *Steinbrech.*

Leaves spreading, in tufts, rather thin, obovate or oblong, tapering at the base. Stems erect, three to six inches high, leafless but a small leafy bract under each pedicel. White starlike flowers, rather small, on slender spreading pedicels, forming a loose terminal panicle. Petals narrow and spreading. Capsule with two diverging beaks. S.M. Piz Gravatscha, near Bevers.

S. AIZOON.

Flowers greenish white, not showy. The leaves, in silvery rosettes, are small, oblong, and obtuse, rather stiff and leathery, and bordered with fine teeth. The flower stems, six to twelve inches high, are smooth below, but covered with glandular hairs on the upper part. S.M. Val Bevers.

S. BRYOIDES. A small tufted plant, with pale yellow star-shaped flowers. The foliage is almost moss-like in character ; the leaves very small and lanceolate, in dense nearly globular tufts. S.M. Val Bevers.—T. Val Floss.

S. ANDROSACEA. S.M. Val del Fain.

S. EXARATA. S.M. Val Bevers.

S. PLANIFOLIA. S.M. Albula Pass ; Alp Falotta, near Mühlen ; Cima di Margna.

S. CONTROVERSA. S.M. Val Bevers, Val d'Agnelli, Julier Pass, Val del Fain.

Q

S. SEGNIERI.   T. Val Tuoi, etc.—S.M. Mont Pers.

S. STENOPETALA.   S.M. Val d'Agnelli, Julier Pass. (Also many other species of this truly Alpine genus.)

PARNASSIA PALUSTRIS (Grass of Parnassus). Germ. *Herzblatt.* A plant with handsome white flowers, an inch or even more across the petals, having strongly marked veins. It is curious in having five perfect stamens and five imperfect ones, bearing, instead of anthers, a tuft of globular-headed filaments. S.M. Extremely common ; frequent also in Britain.

DROSERA LONGIFOLIA. A curious little plant, two to five inches high, with radical leaves only ; these are erect and elongated, broad at the extremity, and tapering towards the base ; the blades of the leaves are densely covered with red hairs, each of which is tipped with a drop of viscid fluid. The small whitish flowers are borne on a wiry leafless stalk. S.M. Near the Statzer See.

XVI. UMBELLIFERÆ.

Herbs with umbellate flowers, five epigynous stamens, and two styles.

ATHAMANTA CRETENSIS.   T. Scarl valley.

CAUCALIS DAUCOIDES (Small Bur Parsley).   T. Neighbouring valleys.

ASTRANTIA MINOR.   Germ. *Sterndolde.*   S.M. Val Bevers. (July.)

GAYA SIMPLEX.   S.M. Piz Languard, summit.

XVII. CAPRIFOLIACEÆ.

Shrubs or herbs with monopetalous epigynous corolla, epipetalous stamens, and three to five-celled ovary.

LINNÆA BOREALIS (Twin Flower).

An interesting little plant, named after Linnæus, who was very fond of it. It is a slender evergreen, which creeps or trails along the ground, and has small, opposite, broadly ovate, or obovate leaves slightly toothed at the top. Its flowers are in pairs ; they are fragrant, and of a delicate pale pink colour ; small and bell-shaped, they droop gracefully from the slender branches of a long bifurcated peduncle. In Britain it is very rare, being found only in the " firwoods of some of the eastern counties of Scotland, and in a single locality in Northumberland." S.M. In many localities, in the woods at the base of Piz Rosatsch, amidst the stones on the right-hand side of the entrance of the Roseg valley, etc.—T. Amongst the stones scattered through the forests of Tarasp, near Fettan.

ADOXA MOSCHATELLINA (Tuberous Moschatel). Germ. *Bisamkraut.* S.M. Alp Spignas. (June.) Common in Britain.

XVIII. VALERIANACEÆ.

Herbs with monopetalous epigynous corolla, stamens epipetalous, fewer than the corolla lobes. Fruit one-celled.

VALERIANA SUPINA.   Germ. *Niedriger Baldrian.* S.M. Val Bevers.—T. On Alps of Seesvenna.

XIX. STELLATÆ.

GALIUM MONTANUM.   S.M. Adjacent meadows.

XX. COMPOSITÆ.

Herbs with flowers in compact heads, five syngenesious stamens, a one-celled inferior ovary, with one erect ovule.

ASTER ALPINUS. Germ. *Alpen Sternblume.*

A pretty composite flower of a violet-blue colour, with an orange-yellow eye, not unlike a large blue daisy. The flower is about an inch and a half across. The leaves are narrow and slightly downy. S.M. Val del Fain, near the Bernina Pass ; Pontresina, rocks on right bank of stream ; twenty minutes from hotel.—T. Meadows adjacent.

ERIGERON ALPIORUM.

Distinguished from preceding (*Aster alpinus*) by the scales of the involucre being narrow, sharply-pointed, and violet-brown ; in the *Aster* they are green, blunt, and incurved. Flower-heads generally solitary on each stem. The outer florets are pink or purplish, very narrowly ligulate, forming a short coloured ray. The central tabular florets are few and of a pale yellow. Leaves lanceolate, tapering at the base. S.M. Summit of Piz Languard.

E. VILLARSII. S.M. Rocks on right bank of stream ; twenty minutes from hotel.—T. Near ruins of Steinsberg.

CHRYSANTHEMUM ALPINUM. Germ. *Alpen Wucherblume.* Like a dwarf variety of the *Ox-eye Daisy.* S.M. Val Bevers.

C. HALLERI. S.M. Salüver, near Celerina. (August.)

ARNICA MONTANA. S.M. Abundant in all parts of the neighbourhood.

DORONICUM SCORPIOIDES (Leopard's Bane). S.M. Val del Fain.

ACHILLEA MOSCHATA (Musk Yarrow). Germ. *Wild Frauli.*

An aromatic herb, from which is distilled a liqueur termed Iva. The leaves are alternate and very much divided pinnately into narrow segments. The whole plant is but a few inches high, terminated by a corymb of small flower-heads, with white rays and yellow disks. S.M. Val Bevers, etc.—T. Scarl valley, Val Tuoi, etc. (July.)

A. NANA. S.M. Piz Languard, banks of the Lago Bianco, Bernina Pass. —T. Piz Minschun.

ARTEMISIA MUTELLINA. S.M. Val Bevers ; Pontresina, rocks on right bank of stream ; twenty minutes from hotel.

A. SPICATA. T. Piz Minschun.

GNAPHALIUM LEONTOPODIUM (Cudweed, or Alpine Lion's Foot). Germ. *Edelweiss.*

A peculiar plant, said to grow in the line of perpetual snow, or even under the snow. The whole plant is covered with white velvety down ; it has very little sap, and can therefore be preserved for a long while. The kind of large paw, which spreads itself out on the top of the stems, and looks something like a corolla, is formed of a number of white down-covered leaves, amongst which the small flowers, collected together in little heads, are concealed. S.M. Val Bevers, Val del Fain, etc.—T. Scarl valley, Piz Minschun, Val Tasna, etc. (Very common.)

G. HOPPEANUM. Germ. *Ruhrkraut.* S.M. Piz Padella, near Samaden.

G. DIOICUM. Germ. *Hausiges Ruhrkraut.* S.M. Val Bevers.

G. SYLVATICUM. Germ. *Wald-Ruhrkraut.* S.M. Val Bevers.

G. CARPATHICUM. S.M. Summit of Piz Languard.

ARONICUM GLACIALE. Germ. *Schwindelkraut.* S.M. Piz Padella.

HIERACIUM ALBIDUM. (Hawk-weed). Germ. *Weissliches Habichtskraut.*
S.M. Val Bevers.—T. Val Tuoi. (August.)

H. ALPINUM (Hawk-weed). Germ. *Alpen Habichtskraut.* S.M. Albula.

H. GLACIALE. S.M. Lago della Crocetta, near summit of Bernina Pass.

H. AUREANTICUM. S.M Val Bevers, and on left bank of stream about a
mile from Morteratsch Glacier.

H. VILLOSUM. T. Valleys adjacent.

H. PULMONARIOIDES. T. Valleys adjacent.

CENTAUREA NERVOSA (Knap-weed). Germ. *Flockenblume.* S.M. Near
Samaden. (August.)

C. RHÆTICA. S.M. Val Bevers.

C. MACULOSA. T. Valleys adjacent.

SOYERIA HYOSERIDIFOLIA. S.M. Albula. (August.)

CREPIS JUBATA (Hawk's-beard). S.M. Val Lavirum.—T. Fimber Pass.

C. JACQUINII. Germ. *Pizzan.* S.M. Val del Fain (July.)—T. Pi Lat, etc.

SENECIO CARNIOLICUS (Groundsel). Germ. *Krautliches Kreuzkraut.*
S.M. Summit of Piz Languard. (July.)—T. Mountains adjoining.

S. ABROTANIFOLIUS (bright orange flowers). S.M. Val Bevers, and Val
del Fain.—T. Very common.

S. NEBRODENSIS. T. Adjacent valleys.

S. NEMORENSIS. S.M. Pontresina, rocks on right bank of stream;
about a mile from hotel.

SERRATULA RHAPONTICUM (Saw-wort). T. Alps by Fettan.

SAUSSUREA ALPINA. Found also on Snowden, and in the Scotch Highlands. S.M. Albula. (August.) Val del Fain.—T. Val Tuoi, Val
Uina, etc.

S. DISCOLOR. S.M. Val Bevers.

CARDUUS PLATYLEPIS (Thistle). T. Adjacent valleys.

CIRSIUM ERIOPHORUM (Plume Thistle). ⎫
C. ERISITHALIS. ⎬ T. Adjacent valleys.
TRAGOPOGON MAJUS (Goat's-beard). ⎭

XXI. CAMPANULACEÆ.

Flowers generally bell-shaped, but sometimes irregular, with five epigynous
stamens (anthers syngenesious in Lobelia) and a two to five-celled
ovary and indefinite ovules.

CAMPANULA CENISIA (Mont Cenis Harebell).
A plant growing at very great elevations ; light green compact
rosettes of leaves ; flowers solitary, blue, somewhat funnel-shaped,
but open and cut nearly to the base into five lobes. S.M. Albula.—
T. Piz Minschun.

C. THYRSOIDES. S.M. Val del Fain. The more common species of
Campanula are abundant.

PHYTEUMA PAUCIFLORUM (Rampion). Germ. *Armblütige Rapunzel.*
S.M. Alp Prunella. (July.)—T. Piz Linnard and Piz Minschun.

P. ORBICULARE. S.M. Val Bevers.

P. HUMILE. S.M. Val Bevers and Val del Fain.

XXII. ERICACEÆ.

The Heath Family.

:Shrubby plants with wiry stems ; flowers monopetalous ; stamens eight
or ten, epigynous or hypogynous ; anthers opening in two small
terminal pores.

RHODODENDRON FERRUGINEUM. Germ. *Rothbrannen Alpenrose.*

R. HIRSUTUM. Germ. *Zottige Alpenrose.*

This is the "Alpenrose" of the Germans. Both species grow as
low thick bushes, with tortuous and interlaced branches, having a
hard black bark, and bearing leaves only at their extremities. The
leaves of *R. hirsutum* are margined with a row of hairs. Those of the
other species are longer, perfectly smooth, and of a dark shining
green on the upper, but covered with a rust-coloured layer on the
under surface ; hence its name. · The branches of flowers spring from
the centre of the tufts of leaves. The calyx is scarcely visible. The
corolla of the *R. hirsutum* is rather larger and more open than that of
the *R. ferrugineum*, and varies from a pale rose to a bright purple
colour. The narrower and more compressed corolla of the other species
has a less brilliant hue ; it is also purple, but of a more sombre tint.
S.M. Val Bevers. (July.)—T. Environs of Vulpera.

.AZALEA PROCUMBENS. Germ. *Felsenstrauch.*

A low trailing shrub ; leaves small, smooth, and rigid, with the
margins remarkably rolled back ; flowers flesh-coloured, in short
terminal clusters or tufts. S.M. Alpotta, near Samaden.

PYROLA ROTUNDIFOLIA (Winter Green).

A plant with roundish leathery leaves only slightly toothed, and
erect stems, bearing long and handsome slightly drooping racemes of
pure white fragrant flowers, half an inch across, ten to twenty being
borne on a stem six inches to ten inches high. It is marked by its
long style, bent down, and at its extremity curved upwards. S.M.
Common on the hills around and by the Statzer See.

P. UNIFLORA (One-flowered Winter Green).

A very pretty plant, with ovate leaves, and bearing one large
elegant, white, highly fragrant flower. T. Adjacent valleys.—S.M.
By Statzer See. (July.)

P. CHLORANTA. T. Adjacent valleys.

ERICA CARNEA (Spring Heath). Germ. *Fleischfarbene Heide.* S.M.
Bevers. (April.)

ARCTOSTAPHYLOS ALPINA (Black Bear Berry). Flowers white, with a
purplish tinge. S.M. Common.

-CALLUNA VULGARIS (Common Heath). Germ. *Besenheide.* S.M. Val
Bevers. (Sept.)

XXIII.   PRIMULACEÆ.

Herbs with a monopetalous hypogynous corolla, five epipetalous stamens opposite to the corolla lobes, and a free central placenta.   The following are only a few of the many beautiful species of the genus Primula which abound in this region.

PRIMULA FARINOSA (Bird's-eye Primrose).   Germ. *Mehlige Primel.*

Found also in Scotland and the north of England.   The stems, from three to twelve inches high, are slender and powdery, and spring from rosettes of musk-scented leaves, having their under surface covered with a silvery-looking meal ; the flowers are small, in graceful clusters, of a lilac-purple colour.   S.M.   Val Bevers.   (June.)

P. INTEGRIFOLIA (Entire-leaved Primrose).

A very dwarf species, having smooth, entire, ciliated leaves lying quite close to the ground, and handsome rose-coloured flowers, with deeply divided lobes, large enough to obscure the plant that bears them.   S.M.   Val Bevers.   (May.)

P. VISCOSA (Viscid Primrose).

Leaves dark green, obovate or nearly orbicular, with closely-set teeth, covered with glandular hairs and viscid on both sides ; the flowers are of a rosy purple, with white eyes.   S.M.   Val Bevers (April), and rocks leading to Morteratsch Glacier.—T.   Val d'Assa, Val Uina, etc.

P. GRAVEOLENS.   S.M.   Piz Gravatscha.

P. LATIFOLIA.   A very handsome species, and sweet-scented, bearing from two to twenty violet flowers in a head, on a stem about twice as long as the leaves.   The leaves are large, and sometimes nearly two inches broad, toothed and covered on both surfaces with glandular hairs.   S.M.   Rocks leading to Morteratsch Glacier.

P. MURETIANA.   S.M.   Val Bevers (July) and Agagliogs Rocks, Roseg Glacier.

CORTUSA MATTHIOLI (Bear's-ear Sanicle).   Germ. *Bär Sanikel.*   T.   Adjacent valleys.   (June.)

ANDROSACE CHAMŒJASME (Rock Jasmine).

Leaves fringed and forming large rosettes ; flowers on stout little stems from one to five inches high, white at first, with a yellow eye, which changes to a deep crimson, the outer part to a delicate rose.   S.M.   Val Bevers.   (July.)

A. OBTUSIFOLIA.   Allied to preceding, but rosettes of leaves larger ; the leaves themselves oblong-lanceolate ; flowers two to five together at the summit of the stems, white or rose-coloured, with yellow eyes. S.M.   Val Bevers.

A. GLACIALIS (Pink Flowers).   Germ. *Gletscher Manschild.*   S.M.   Val Bevers, Piz Languard, Cima di Margna.

A. SEPTENTRIONALIS.   S.M.   Near Samaden.

TRIENTALIS EUROPÆA (Star Flower). Germ. *Siebenstern.*

A delicate plant, extremely rare in the Alps, with erect slender stems, not more than six inches high, bearing a whorl of five or six leaves, of different sizes, from one inch to two inches in length. From the centre of the whorl spring one to four flower stems, each supporting a small white or pink-tipped star-shaped flower. S.M. Alp Nova, near Pontresina, and on left bank of stream, about a mile from Morteratsch Glacier.

SOLDANELLA ALPINA. Germ. *Alpenglockli.* A most interesting plant, growing near the snow-line, having beautiful pendent pale-blue open bell-shaped flowers, cut into numerous narrow linear strips, three or four being borne on a stem two to six inches high ; the leaves are radical, small, leathery, shining, roundish or reniform. S.M. Val Bevers.

S. PUSILLA. Leaves kidney-shaped ; flowers not so deeply cut as in preceding species. S.M. Cambrena Glacier and Agagliogs Rocks, Roseg Glacier.

ARETIA GLACIALIS. S.M. Val Bevers.

XXIV. GENTIANACEÆ.

Bitter herbs with a hypogynous regular corolla ; the lobes twisted in the bud ; stamens alternate with the corolla lobes ; ovary usually one-celled, with parietal placentas.

GENTIANA VERNA. Germ. *Fruhlings Engian.*

One of the most beautiful, if not *the* most beautiful of Alpine flowers. Its leaves are small and leathery, ovate or oblong, and in dense tufts. The flower stems are so short that the flowers appear sessile on the tufts of leaves. Each stem bears a single terminal flower of a splendid bright-blue colour. The tube of the corolla is cylindrical, and nearly an inch long ; the limb is broad and spreading, with five ovate lobes and smaller two-cleft ones between them. S.M. Bevers, etc.

G. BAVARICA. A native of the high Alps, and resembling in its bloom the vernal gentian, from which it is distinguished by its smaller box-like leaves, of a yellowish green tone, and longer flower stems. The flowers are of the deepest and most brilliant blue. S.M. Val Bevers.

G. NIVALIS. A high Alpine plant, with slender erect stems two to four inches high ; sometimes only an inch in height, and bearing a solitary small blue star-like flower, having the same characters as that of the vernal gentian, but very much smaller. S.M. Meadows near Samaden, path leading to Roseg Glacier. (July.)—T. Piz Minschun.

G. EXCISA. S.M. Val Bevers. (June.)

G. BRACHYPHYLLA. S.M. Adjacent meadows. (July.)

G. OBTUSIFOLIA. S.M. Val Bevers. (August.)

G. CILIATA. S.M. Celerina. (August.)

G. CHARPENTIERI. S.M. Val Bevers.

G. PUNCTATA. S.M. Near Statzer See.—T. Val Tuoi.

MENYANTHES TRIFOLIATA (Buckbean). Germ. *Dreiblatteriger Bitterblee.* Flowers white, bell-shaped, deeply five-lobed, fringed inside with white hairs, tinged externally with red, in an oblong raceme. Calyx short, with rather broad green lobes. Leaves in a dense tuft, each having a long sheathing stalk, and three obovate leaflets. S.M. Bevers. (June.) And other species.

XXV. POLEMONIACEÆ.

POLEMONIUM CÆRULEUM (Jacob's Ladder).

A familiar garden plant, with pinnate leaves, and showy blue or white flowers. Monopetalous corolla, with a short tube and a broad, open, five-cleft limb. Stamens five, bearded at the base. Ovary three-celled. S.M. Albula (July), and path leading to Roseg Glacier. T. Near Ardetz, etc.

XXVI. BORAGINEÆ.

Rough herbs, with alternate leaves and regular or nearly regular monopetalous flowers ; stamens same number as corolla lobes (five) ; ovary four-lobed.

MYOSOTIS ALPESTRIS (Alpine Forget-me-not). Germ. *Vergissmeinnicht.* Found rarely in Scotland and the north of England. It has beautiful bright clear blue flowers with small yellowish eyes. "Inferior to no gem of the high Alps in vivid colour and beauty." S.M. Alp Signas. (June.)

M. NANA, or ERITRICHIUM NANUM. A very dwarf species, scarcely two inches high, but with comparatively large brilliant blue flowers. S.M. Summit of Piz Languard.

PULMONARIA AZUREA (Lungwort). Coarsely hairy radical leaves, nearly oval and acute at the apex, and marked on the upper surface with white blotches. Corolla blue, with a straight tube open at the mouth, without scales, and a spreading five-lobed limb. Flowering stems from six inches to a foot high. T. Val Tasna.

ECHINOSPERMUM DEFLEXUM. Has all the appearance and small flowers of a Myosote, but with triangular very rough nuts. T. Adjacent meadows.

XXVII. OROBANCHACEÆ.

Broom-rape family, consisting of parasitic plants, distinguished by the absence of green leaves. The floral characters are nearly those of the *Scrophularineæ.*

OROBANCHE CRUENTA. T. Near Schuls. (Rare.)

O. ARENARIA. T. Common.

O. LUCORUM. T. Very common.

XXVIII. SCROPHULARINEÆ.

Herbs with irregular monopetalous, generally personate corolla ; stamens fewer than the corolla lobes (except *Verbascum*) ; ovary two-celled, many-seeded.

VERONICA CHAMŒDRYS (Germander Speedwell). A well-known native plant. S.M. Val Bevers, etc.

VERONICA SPICATA. Flowers in long dense spikes, blue or pale pink. S M. Val Bevers. (July.)

V. PROSTRATA. A dwarf spreading plant, less than six inches high ; stems downy, flowers deep blue, sometimes rose-coloured and white It flowers very freely, so that the leaves are often quite obscured by the flowers. T. Adjacent valleys.

V. SAXATILIS. On Alpine rocks, often at great elevations. A dwarf, bush-like plant, with pretty blue flowers, striped with violet, with a narrow crimson ring near the bottom of the corolla, its base being pure white. S.M. Val Bevers.

V. BELLIDIOIDES. Germ. *Ehrenpreis.* S.M. Val Bevers.

V. OFFICINALIS. S.M. Val Bevers.

LINARIA ALPINA (Alpine Toad-flax). Germ. *Alpen Leinkraut.* A truly Alpine plant, forming dense, dwarf, silvery tufts, covered with bluish violet flowers, with two spots of intense orange in the centre of the lower division of each. S.M. Bevers. (June.)

PEDICULARIS INCARNATA (Red Ruttle). S.M. Val del Fain, Val Bevers. —T. Alps by Fettan.

P. ROSTRATA. S.M. Val Bevers.

P. RECUTITA. T. Val Tuoi.

P. TUBEROSA. S.M. Val Bevers.

P. ASPLENIFOLIA. T. Alps of Samnaun.

P. TAQUINII. T. Val Triazza.

EUPHRASIA LUTEA (Yellow Eye-bright). T. Near Naiss.

XXIX. LABIATÆ.

Herbs, with an irregular, usually two-lipped corolla ; stamens fewer than the corolla lobes ; ovary four-lobed ; style from between the lobes.

DRACOCEPHALUM RUYSCHIANA.
A showy plant with blue flowers more than an inch long, in rather close spikes at the summit of the stem. The leaves smooth, narrow, entire, and opposite. S.M. Near Samaden.

D. AUSTRIACUM. Larger flowers than preceding, in whorled spikes. Floral leaves velvety, trifid ; the leaves three or five-cleft with narrow segments.

CALAMINTHA ALPINA (Alpine Basil). S.M. Val Bevers.

SALVIA VERTICILLATA. T. Meadows adjacent.

GALEOPSIS VERSICOLOR (Large-flowered Hemp-Nettle).
' The flowers are large, yellow, with a broad purple spot upon the lower lip. T. Meadows adjacent.

XXX. ORCHIDACEÆ.

CORALLORHIZA INNATA (Coral Root).
Marked by its curiously toothed roots, which in figure resemble branched coral. T. Val Uina, etc.

EPIPOGIUM GMELINI. T. Scarl valley.

GOODYERA REPENS. T. Fir-woods around.

GYMNANDRIA ODORATISSIMA. T. Adjacent forests.

ORCHIS GLOBOSA. S.M. Val del Fain.

XXXI. LILIACEÆ.

Monocotyledons with parallel-veined leaves, petaloid, coloured perianths of six divisions, six stamens, and superior syncarpous, three-celled ovary.

LLOYDIA SEROTINA (Mountain Lloydia). Germ. *Graslilie.*

A delicate little plant, with a small bulb and two or three almost thread-like leaves, and a slender stem, about three or four inches high, bearing a single white star-like flower, about three-quarters of an inch across. The segments of the perianth marked inside with three longitudinal reddish lines and a small yellow spot at the base. S. M. Gravatscha near Bevers. (June.)

PARADISIA LILIASTRUM (St. Bruno's Lily).

A beautiful white lily-like flower, found generally in the warmer Alpine valleys. The half-pendent blooms, about two inches long, are of a clear and delicate white, each division faintly tipped with pale green. There are two to five flowers on each stem. S. M. Val Bevers.

LILIUM BULBIFERUM. S. M. Val del Fain.

GAGEA LIOTTARDI. S. M. Val Bevers.

ALLIUM SCHŒNOPRASUM (Chive Allium). Leaves very narrow, cylindrical, and hollow. Umbel contracted into a dense globular head of rather large purplish flowers ; perianth, segments three to four lines long, very pointed. S. M. Val Bevers.

ALLIUM VICTORIALIS. T. Val Tasna.

MAIANTHEMUM BIFOLIUM. Germ. *Zweiblätterige Schattenblume.* A graceful little plant, allied to the lily of the valley, from which it is distinguished by having its perianth divided to the base into four segments. It has two leaves to the stem, and a terminal raceme of two small white flowers. S. M. Bevers. (July.)

TOFIELDIA BOREALIS. S. M. Cambrena. (July.) T. Col de Joata.

ANTHERICUM LILIAGO. T. Adjacent meadows.

XXXII. JUNCACEÆ.

The Rush family are separated from the Liliaceæ by their scarious perianth, and their very minute embryo ; and from the sedges, which they resemble in general aspect, by possessing a complete six-leaved perianth.

JUNCUS TRIGLUMIS (Three-flowered Rush). Germ. *Z'balgige Simse.* S. M. Albula. (August.)

J. ARCTICUS. S. M. Sandy plain between Silser and Silva Plana See.

J. TRIFIDUS. S. M. Albula.

LUZULA LUTEA.

L. SPICATA (Spiked Woodrush). Germ. *Genährte Hainsimse.* S. M. Piz Ot. (August.)

L. SPADICEA and L. ALBIDA. S. M. Piz Spignas. (August.)

XXXIII. CYPERACEÆ.

The Sedge family are distinguished from the Rushes by the absence of any regular perianth, and from the grasses by their leaves being closed round the stem, by the want of an inner scale or palea between the flower and the axis of the spikelet, and by the simple, not feathery, branches of the style.

CAREX FIRMA. Germ. *Steifblätterige Segge.* S.M. Albula. (August).

C. VAHLII (Alpine Carex). In Britain, only in the Clova mountains of Scotland. S.M. Albula, Val Bevers, Piz Languard.

C. BUXBAUMII. S.M. Near Samaden.

C. MICROGLOCHIN. S.M. Val Bevers.

CAREX IRRIGUA. S.M. Albula.

C. CÆSPITOSA. Common.

ERIOPHORUM SCHOUCHZÈRI (Cotton Sedge.) S.M. Near Piz Corvatsch.

SCIRPUS TABERNÆMONTANI. T. Near Crusch.

S. ALPINUS. Germ. *Alpen Binse.* S.M. Val Bevers. (May.)

KOBRESIA CARICINA. S.M. Albula. (August.)

ELYNA SPICATA. S.M. Albula. (July and August.)

XXXIV. GRAMINEÆ.

Sheaths of leaves split in front. Innermost glume (palea) two-nerved.

STIPA PENNATA (Feather-grass). A very beautiful delicately feathered grass, not found in Britain. S.M. Val Bevers.—T. Rocks near Ardetz, etc.

KÆLERIA HIRSUTA. S.M. Near Samaden and Lago della Crocetta, near Bernina Pass.

PHOCA AUSTRALIS. Germ. *Bergliese.* S.M. Albula. (June.)

P. FRIGIDA. S.M. Near Samaden.

MELICA CILIATA. T. Near Schuls.

LOLIUM LINICOLUM. T. Near Schuls.

AVENA SUBSPICATA. S.M. Lago della Crocetta.

POA LAXA. S.M. Summit of Piz Languard.—T. Piz Cotschen.

FESTUCA SCHINCHZERI. T. Piz Minschun.

F. HALLERI. S.M. Summit of Piz Languard.

SESLERIA DISTICHA. T. Piz Minschun.—S.M. Piz Ot. (August.)

TRIGLOCHIN PALUSTRE (Arrow-grass). Germ. *Dreizahn.* S.M. Bernina.

AGROSTIS RUPESTRIS. S.M. Summit of Piz Languard.

ADDITIONAL SPECIES.

FUMARIA VAILLANTII. S.M. Bevers. Nat. Ord. Fumariaceæ.

ARMERIA ALPINA (Alpine Thrift). S.M. Near Muhlen. Nat. Ord. Plumbagineæ.

HIPPOPHÆ RHAMNOIDES (Sea Buckthorn).   T.   Adjacent valleys.   Nat. Ord. Elæagnaceæ.

LINUM CATHARTICUM (Purging Flax).   Germ. *Purpier Lein.*   S.M. Alpotta, near Samaden.   Nat. Ord. Linaceæ.

RUMEX NIVALIS.   T.   Adjacent valleys.   Nat. Ord. Polygonaceæ.

SALIX PENTANDRA (Bay-willow).   Germ. *Funfmassige Weide.*   S.M. Val Bevers.   (August.)   Nat. Ord. Amentaceæ.

S. GLAUCA.   S.M.   Albula.—T.   Val Lavinuoz.   Nat. Ord. Amentaceæ.

S. HEGETSCHWEILEREI.   T.   Val Uina and Val Triazza.   Nat. Ord. Amentaceæ.

TAMARIX GERMANICA.   S.M.   Val Bevers.   (June.)   Nat. Ord. Tamariscineæ.

GLOBULARIA CORDIFOLIA.   Nat. Ord. Globulariæ.   Pretty blue flowers in a globose head.   Inferior five-cleft monopetalous corolla.   Stamens four ; ovary with one cell and one ovule ; a representative of a small natural order allied to the Verbenaceæ.   S.M. Guardoval.

WOODSIA HYPERBOREA.   S.M.   Near the fall of the Inn, etc.—T.   Near Lavin.   Nat. Ord. Filices.

ASPLENIUM SEPTENTRIONALE.   S.M.   Rocks leading to Morteratsch Glacier.   Nat. Ord. Filices.

CYSTOPTERIS ALPINA.   T.   Val Uina.   Nat. Ord. Filices.

GRIMMIA INCURVA.   S.M.   The Grat between Mont Pers and Piz Cambrena, Piz Languard.   Nat. Ord. Musci.

G. APICULATA.   S.M.   Cima di Margna.   Nat. Ord. Musci.

BRYUM CINCLIDIOIDES.   S.M.   Septimer Pass.   Nat. Ord. Musci.

B. CULLATUM, and B. LUDWIGII.   S.M.   Banks of Lago Bianco, Bernina Pass, and Piz Languard.   Nat. Ord. Musci.

DICRANUM MUHLENBECKII.   S.M.   Alp Err.   Nat. Ord. Musci.

ENCALYPTA APOPHYSATA.   S.M.   Val Lavirum.   Nat. Ord. Musci.

HYPNUM JULACEUM.   S.M.   Val Lavirum.   Nat. Ord. Musci.

H. AIRRHATUM.   S.M.   Val Faller.   Nat. Ord. Musci.

ORTHOTRICHUM KILLIASII.   S.M.   Piz Languard.   Nat. Ord. Musci.

POLYTRICHUM SEPTENTRIONALE.   S.M.   Piz Languard, Septimer Pass. Nat. Ord, Musci.

LECANORA VENTOSA.   S.M.   Grat between Mont Pers and Piz Cambrena. Nat. Ord. Lichenes.

LECIDEA ARMENIACA.   S.M.   Cima di Margna.   Nat. Ord. Lichenes.

GYROPHORA ANTHRACINA.   S.M.   Grat between Mont Pers and Piz Cambrena.   Nat. Ord. Lichenes.

G. POLYMORPHA.   S.M.   Cima di Margna.   Nat. Ord. Lichenes.

ENDOCARPON INTESTINIFORME.   Near Mühlen.   Nat. Ord. Lichenes.

·CETRARIA NIVALIS, and C. CUCULLATA   S M.   Grat between Mont Pers and Piz Cambrena.   Nat. Ord. Lichenes.

A LIST OF COLEOPTEROUS INSECTS (BEETLES) OBSERVED IN THE NEIGHBOURHOOD OF TARASP.

(a) SPECIES FOUND IN THE VALLEYS.

Cicindela riparia.
Lebia crux minor.
Harpalus honestus.
— latus.
Cymindis angularis.   Near Fettan.
Carabus gemmatus.
— Neesii.
Boletobius pygmæus. Near Schuls.
Bryoporus rufus.
Omalium florale.
Oxytelus inustus.   Near Schuls.
Silpha alpina.
Onthophagus austriacus.
Aphodius obscurus.
— alpinus. Near Guarda.
— depressus.
Cetonia floricolor.
Ancylocheira rustica.
Dolopius marginatus.
Diacanthus melancholicus.   Near Fettan.
Athous vittatus.
Agriotes gallicus.
Cardiophorus musculus.
Pheletes Bructeri.
Cantharis abdominalis.
— Erichsonii.
Podabrus alpinus.
Ragonycha rufescens.
— testacea.
Telephorus albomarginatus.

Malthodes flavoguttatus.
Isomira hypocrita.
Oedemera virescens.
Toucartia squamulata.   Near Schuls.
Phyllobius psittacinus.
— atrovirens.
Otiorhynchus rhæticus.
— hirticornis.
— helveticus.
Cleonus cinereus.
Magdalinus violaceus.
Pachyta interrogationis.
— clathrata.
Cryptocephalus lobatus.
— interruptus.
Tetropium luridum.
Callidium violaceum.
Clytus Verbasci.
Oberea pupillata.
Toxotus cursor.
Pachybrachys hieroglyphicus.
Oreina intricata.
— venusta.
— speciosa.
— pretiosa.
Phratora major.
Longitarsus Verbasci.
Anoncodes rufiventris.
Malachius geniculatus.
Scymnus ater.   Near Schuls.

(b) SPECIES FOUND ON THE MOUNTAINS.

Cicindela alpestris.
Cychrus attenuatus.   Alps of Fless.
Nebria Germari.   Alps of Fettan.
Amara bifrons.   Alp of Urschai.
Pterostichus alpestris.
Bembidium glaciale.
Helophorus glacialis.
Argutor alpestris.   Val Urschai.
Philonthus micans.   ,,
— picipennis
Stenus filum.   ,,
Homalotha alpicola.   ,,
Byrrhus pulchellus.   ,,

Basytes obscurus.
— alpigroidus.
Geotrypes alpinus.
Anthaxias helvetica.
Corymbites rugosus.
Otiorhynchus maurus.   Val Uina.
— rivalis.   Val Lischanna.
— insculptus.   Val Urschai.
— alpicola.   Piz Lat.
Lina alpina.   Val Lischanna.
— Escheri.   ,,
Phædon salicinum.   ,,
Capsus nitidus.   Val Urschai.

A LIST OF CERTAIN MOLLUSCA FOUND IN THE NEIGHBOURHOOD OF TARASP.

Succinea Pfeifferi.
Helix arbustorum, subalpina.
— zonata.
— edentula.
— fulva.
— rupestris.
— strigella.
— obvia.
— ruderator.
— glabra.
— nitens.
— nitidosa.
— crystallina.

Bulimus acicula.
— lubricus.
-- montanus.
Pupa quadridens.
— inornata.
— minutissima.
— marginata.
-- avena.
— secale.
Balea fragilis.
Clausilia plicata.
— cruciata.
Limneus pereger.

LIST OF CERTAIN MINERALS IN THE ENVIRONS OF TARASP.

Rock crystal. Piz Minschun, near Aschera.
Epidote. Near Aschera, and elsewhere.
Albite.
Garnet. In the erratic blocks of Mica-schist, near Vulpera.
Cat's Eye. In the dioritic rocks, near Aschera.
Andalusite. Fluela Pass.
Asbestos. Near Vulpera, and elsewhere.
Soap-stone. Near Vulpera, and elsewhere.

Tremolite. Near Pradella, and on Piz Minschun.
Picrolite. In the serpentine.
Schillerspar.   ,,
Magnesite. Near the castle of Tarasp.
Arragonite. Near the castle of Tarasp.
Calcite. Near the castle of Tarasp.
Anhydrite. Near Steinsberg.
Chrome-iron with nickel hydrate. Near the castle of Tarasp.
Taraspite. Near Vulpera.
Lead-glance with silver. Scarl valley—old mines.

CHAPTER V.

DAVOS PLATZ.

A STUDY OF HIGH ALTITUDES FOR CONSUMPTIVE PATIENTS.

WHEN I first visited the Engadine and the Davos valley in the autumn vacation of 1869 the question of the cure of consumption by a prolonged residence in these high mountain valleys was just beginning to engage the attention of English physicians.

The wider general question whether consumptive patients generally should or should not be sent to elevated situations had already been the subject of much discussion, and the evidence in favour of an affirmative answer to the question appeared to be unusually strong. But the evidence which has been collected on this point tends to prove that immunity from consumption does not follow any particular level of elevation, as had at one time been suggested, and it would seem that the *mere amount* of elevation is not so essential as had been supposed. It is admitted that the altitude of immunity from phthisis varies in different latitudes, descending in proportion as we pass from the equator to the poles. In the tropics it is necessary to ascend to an elevation of between 8500 and 9000 feet. In the Peruvian Andes, for instance, patients are sent to mountain valleys reaching an altitude of nearly 10,000 feet. In Mexico they ascend to valleys 6500 and 7000 feet above the sea-level. On the other hand, in the Pyrenees, we are assured that at elevations varying from 1760 feet (Bagnères de Bigorre) to 4580 feet (Gavarnie), phthisis is equally rare.

In Switzerland some localities not more than 3000 feet above the sea appear as free from phthisis as others of twice that elevation. In the Black Forest and in the Harz mountains of Germany, it is stated that consumption is extremely rare at the comparatively moderate height of 1400 to 2500 feet; while Dr. Brehmer asserts that in the neighbourhood of Görbersdorf, in Silesia (1700 feet), he has never seen phthisis amongst the inhabitants.

These statements seem to point clearly to the conclusion that the freedom which any particular locality may appear to enjoy from this disease is independent of its mere elevation, and due in part to other conditions. This view is further supported by the fact that at Andermatt and at Splugen, each about 4700 feet above the sea, phthisis is known to occur, while at Klosters, which is 700 feet lower, it is unknown.

Moreover, it has long been known that a sea voyage is, in some cases, one of the most effectual means at our disposal for arresting the progress of phthisis. It would seem then that under certain conditions the *sea-level* is as curative of consumption as the highest inhabited valleys.

One of the conditions common to life during a sea voyage and to life in a high mountain valley, is the mechanical purity of the air that is breathed, and its entire freedom from organic admixture. We know that the air of large densely populated cities and towns is filled with impurities, both organic and inorganic, and doubtless in many localities this floating dust is largely composed of filthy putrescent organic matter, or infective particles, capable under certain circumstances of exciting or conveying disease. It is amongst those who have to live in the worst parts of this unwholesome town atmosphere that phthisis is most rife and fatal; and therefore to the absence of these impurities in the air of elevated regions, as well as in that of the open sea, we may reasonably attribute their beneficial influence in preventing or arresting tubercular disease.

M. Miquel's observations at the *Observatoire de Montsouris* show how comparatively free the air is from organic impurities over extensive tracts of water, as well as at great elevations; in his examinations as to the presence of bacteria in the air he found *none* at elevations of two to four thousand metres; he found on the surface of a small lake like that of Thun, in 10 cubic metres 8·0 bacteria only; on the shores of the lake (near the Hotel Bellevue) 25·0, and in the Rue de Rivoli, Paris, 55000·0!

Most of the localities which have been mentioned above as enjoying an immunity from tubercular disease of the lungs, are characterised by a pure and dry atmosphere, a dry subsoil, and a scanty population.

And it has been shown that in certain favoured localities in our own country, where these conditions of dryness of subsoil, thinness of population, and purity and dryness of the atmosphere co-exist, there also the occurrence of cases of phthisis is very rare. Too much importance must not be attached to the statements made by medical men who reside in high *and very thinly populated* districts—such, for example, as the Upper Engadine or the Davos valley—to the effect that cases of consumption are rarely observed there. These statements are no doubt perfectly true, but what is their value?

If consumption be a disease engendered by city life, by overcrowding, by breathing a damp contaminated atmosphere, we should expect it to disappear in localities where all these conditions are reversed.

It may be interesting to inquire briefly into the nature of the evidence upon which reliance is placed to support the view that elevated districts are those best suited to phthisical patients.

Before we were in possession of all the evidence that has been derived from the results of the past ten years' experience at Davos, the strongest and the

R

most unequivocal was that derived from the experience of medical practitioners resident in the large towns at the base of the Peruvian Andes, and in other similar tropical stations.

In these localities consumption is very rife, and it had long been the established mode of treatment there to remove the patients so afflicted, as early as possible, to one or other of those sheltered valleys at great elevations, which the slopes of the Andes afford in abundance. Dr. Archibald Smith, of Lima, was one of the first to call the attention of the medical profession to this method of treatment. He stated that in the Peruvian Andes immunity from phthisis was commonly observed at an elevation of between 7500 and 8500 feet. No plan of treatment could be more rational than to remove the consumptive patient from the hot, damp, reeking, unwholesome atmosphere of the densely populated town in which he had been attacked, to the pure, clear, dry, invigorating air of the adjacent mountain valleys. There is no need to marvel at the efficacy of such a process, nor does it afford any reasonable ground for assuming that, *in every part of the world,* very elevated mountain valleys are the best localities for the treatment of phthisical cases. It is very well known that, in temperate climates, some moderately elevated regions enjoy a greater immunity from tubercular disease than others of perhaps twice their altitude. Local conditions other than the single one of mere elevation determines the suitability or otherwise of each particular district. When this discussion first commenced, the advocates of Davos and the Upper Engadine as winter sanatoria for consumptive patients, who were not so numerous as they are now, laid much stress on the statement that scrofulous disease was unknown among the natives of the Upper Engadine and of the Davos valleys; and that when the inhabitants of the lower districts of Switzerland became affected with scrofula, they were restored to health on migrating to those districts. We were further told, that although fatal cases of consumption had occurred there, the disease had invariably been imported.

But we had to place against this statement another, which was also made on the authority of the medical men residing at these places, to the effect that deaths from inflammation of the lungs, from pleurisy and from catarrhal fever were common ; and we could not help associating this with another undoubted fact, namely, that consumption is a common sequel in our own country to these inflammatory affections; and we were then compelled to ask the question : Would not many of these acute attacks which prove rapidly fatal in the Engadine have merged, in a warmer climate, into those chronic conditions which are never developed there because of the fatal severity of the primary disease ?

But when we had admitted the full significance of the fact that the natives of these localities were free from tubercular disease, it was met by another fact, viz., that such was also the case in many parts of Europe, at not more than half this elevation, where the population was sparse, and the atmosphere dry and pure. Many such places are known, and a diligent search would probably discover many more. One important fact appeared to come out of the inquiry so far as it had at that time advanced, viz., that a moderate elevation of 1500 to 3000 feet was as useful in some parts of the world as an altitude of from 7000 to 10,000 feet in others.

So long ago as October 18th, 1869, I called attention in an article in the *Medical Times and Gazette* of that date, to the reputation that Davos was gaining as a winter resort for consumptive patients. After alluding to St. Moritz as a winter station, I observed: "A far greater number of persons with pulmonary complaints have wintered of late years at Davos. This is situated in a valley running parallel with the Engadine, at a somewhat lower elevation, 5105 feet above the sea. It is reached by diligence from the Landquart station, close to Ragatz, through the valley of the Prættigau, in about six hours. It is also connected by several mountain passes with the Engadine, one of which, the Fluela Pass, is traversed daily by diligence. This also is about six hours' journey—i.e., from Süs, in the Lower Engadine, to Davos Dorfli, in the Davos valley. At the time of my visit to Davos (in 1869) I was informed that there were nearly one hundred and fifty patients undergoing the 'cure' there." From other sources I learn that that number ought to have been reduced to seventy.

In the winter 1872–73 the number of winter visitors (patients) had risen to two hundred; in that of 1874–75 to four hundred; in 1875–76 to five hundred; and in that of 1878–79 to seven hundred. And the number has continued to increase so rapidly that, as we shall presently see, an outcry has lately been raised against overcrowding there, and its evil consequences.

But leaving Davos for the moment, let us inquire what has been the fate of St. Moritz as a winter sanatorium. When I first wrote on this subject in 1869, it was with regard to St. Moritz rather than to Davos that the interest of English physicians had been aroused. One of the earliest records of a winter passed at St. Moritz by an invalid is to be found in the visitors' book of the Kulm Hotel for 1868–69, from which the following is extracted :

"Any doubts we had entertained as to the possibility of keeping warm indoors in a locality where, in the open air, the temperature was often below zero Fahrenheit in the shade, were speedily dispelled. Owing to the extreme dryness of the air, we never found our sitting-room comfortable above 56 degrees Fahrenheit. The rooms are warmed by means of stoves, not open fireplaces, consequently the chief difficulty is to ventilate them properly ; to do this effectually, we left our sitting-room for five minutes every two hours, opening all the doors and windows. A pan of water kept on the stove is also indispensable to prevent the already dry air of St. Moritz becoming overdried. On an average, we were out four hours daily, walking, skating, sleighing, or sitting on the terrace reading—this latter two or three hours at a time ; twice in January we dined on the terrace, and on other days had picnics in our sledges : far from finding it cold, the heat of the sun was so intense at times that sunshades were indispensable, one of the party even skating with one. The brilliancy of the sun, the blueness of the sky, and the clearness of the atmosphere quite surprised us. The lake affords the opportunity, to those who love the art, of skating without interruption for five months. The ice has, to a certain extent, to be artificially maintained. To do this, we, with other English friends, formed a small club—First, for keeping a circle clear of snow ; second, for renewing the surface whenever it became impaired by turning a stream on to it.

"I must state, having spent part of the winter of 1867–68 at Mentone, that I derived far more benefit from that of 1868–69 spent at St. Moritz. The change from England to Mentone did me good at first, but latterly I experienced great lassitude ; whilst at St. Moritz I was far stronger at the end of the winter than at the commencement. During the whole time I had neither cold nor cough, though I was out all weathers.

" One or two days the sun-heat was remarkable, the thermometer reading 142 degrees Fahrenheit ; on the other hand, the greatest cold was −18·5 degrees Fahrenheit—viz. 50½ degrees below freezing-point— during the night.

"We left St. Moritz for Lugano and Cadenabbia at the end of March, to escape any damp the spring thaw might create ; but experience taught us that the uncertain spring weather of the plain causes far more injurious damp than the mere thawing of the snow at such a height as St. Moritz. There the disagreeable effect of a temporary and most unusual thaw in February (brought on by a warm wind and great sun-heat) appeared to be confined only to the melting of the snow on the roads and mountains, off which it ran as rapidly as it thawed, imparting no perceptible damp to the air. In the plain we had plenty of rain, snow, and mist, and on fine days it became too warm and relaxing. We thought to better ourselves by spending May on Monte Generoso, where there is a good well-managed hotel in a charming situation, about 4000 feet S.M. ; but that is too uncertain a month for this mountain. Although we have had some fine days, we have been most frequently in the midst of clouds, rains, or storms. The group of mountains, of which Monte Generoso is the most southern, seems to attract the clouds that rise from the vast Lombard plains, over which to the Apennines the view is unbroken. St. Moritz, being so much higher, is above the ordinary cloud-level, and consequently gets less rain ; most of the clouds, of which there are not a few in summer and autumn, seem to be carried rapidly over the Engadine by strong currents of wind."

Additional experiences of the winter of 1869–70 at St. Moritz appeared in *The Times* of February 21st, 1870. The writer, after describing the means adopted for providing a "beautiful surface" for skating on the St. Moritz lake, goes on to observe :

Last winter, on January 24th, the minimum thermometer (not exposed to the sky) registered −18·5 degrees Fahrenheit, i.e. 50½ degrees of frost ; this season the lowest temperature has been −13 degrees Fahrenheit on the 24th ult. The same night, at the Government Meteorological Station, a few miles distant, it was −19·3 degrees Fahrenheit. The January thermometer mean is 16·31 degrees, that of the minimum 5·56 degrees. From the 10th of October the minimum was continuously below freezing-point, on the 29th falling as low as 1 degree Fahrenheit.

These few particulars show the severity of the winter here. Of its accompanying delights I am afraid I can give your readers but a faint idea. Sleighing, which is very enjoyable, is in vogue for about six months in the year, but nothing short of an actual trial would enable any one to realise the pleasure of gliding along rapidly in a sledge, under a brilliantly blue sunny sky, with the crisp white snow beneath.

Another favourite entertainment is sliding down steep inclines on small sledges constructed for this purpose (Toboganning). The speed attainable is almost incredible. In this sport both old and young join.

The roads are quite good for walking on, since the snow is generally dry and well trodden. Any one more ambitious can easily overrun the country on snow-shoes. The snow is rarely more than two feet deep, except where it has drifted.

There is usually very little wind, and the clear frosty air is so invigorating, the sky so cloudless, and the sun's rays so penetrating, that the cold is scarcely felt. The chilly feeling associated with the winter in England is, many say, unknown here, owing to the above causes and the dryness of the climate. Of course, there is exceptional weather, and this has been, to a certain extent, an exceptional winter here as everywhere. The rain which has fallen in such abundance in Italy and elsewhere has had its counterpart here in more frequent falls of snow and cloudier skies than usual. My description refers to what an English friend of mine, who has spent five winters here, calls the " normal " Engadine winter weather, such as I have experienced fully the whole of last and part of this winter.

The following is from a private letter written from Samaden the same winter (March, 1870), and is also interesting from what may now be regarded as a historical point of view :

I have spent five months (from October 10th to February 26th) at St. Moritz, and liked the place on the whole, and the climate, I think, has done me good. During the winter we have had some splendid weather, little or no wind, very dry, and sometimes very cold. It is the *best* climate I have *ever* lived in, the air is so beautifully pure and dry. Of course we have had our share of unpleasant weather, especially at the beginning of the winter, when the snow was coming down ; but snow every day is far preferable to rain, and it is rarely accompanied by much wind.

Our party at the Kulm Hotel has consisted of four English, three Germans, three Italians, and a French lady ; in Mr. Strettell's house there have been four more English people besides himself. In this place (Samaden), which is only three-and-a-half miles from St. Moritz, but at a slightly lower elevation, there has been an English family, consisting of a lady and gentleman and five children. That, I think, is a complete list of the visitors who have passed this winter in the Upper Engadine.

The great drawback in spending the winter at St. Moritz is the want of good food. The milk and bread and butter are good; but the meat is bad, and the soup invariably requires a certain amount of " Liebig's Extract" to make it worth eating. For three months the only vegetables we had were potatoes. In fact, a person coming here for health, gains greatly as regards climate, but loses greatly from want of good food and the ordinary home comforts. As for amusements, we have sleighing, a little skating, and also billiards. There are no level walks about St. Moritz, but the hotel has a terrace; here (at Samaden) there are three level walks, and no terrace. I mention this because, when the country is all covered over to a depth of two or three feet with snow, it is of great importance to have some place to take a constitutional. In a place of this sort you are thrown very much on your own resources, and if you can sleep twelve hours out of the twenty-four, you can manage to get through the time.

Notwithstanding advocacy such as this, St. Moritz had until lately been quite unable to compete with Davos as a winter resort. While the number of winter visitors to Davos had increased from seventy in the year 1869–70 to five hundred in the year 1875–76, we find a writer in a daily paper of the latter date stating of St. Moritz:

No one has had the hardihood to pass the winter either there or at the Berninahof at Samaden since 1872 until the present season, notwithstanding the favourable testimony as to climate, comforts, and amusements of the very few who stayed through that and the three preceding winters.

And he very appropriately adds:

Several proprietors of hotels and *pensions* at St. Moritz advertise that their establishments are open the whole year, inspired partly by the success of Davos (between the Unter Engadine and the Prættigau) as a winter residence for consumptive patients, partly by the well-known inclination of some eminent doctors towards a dry-cold in preference to a damp-warm climate in certain stages of lung disease. They scarcely seem to realise, however, the extent to which provision must be made for delicate patients above the requirements of ordinary summer visitors, or even of themselves in winter. Double windows, adequate means of heating rooms and passages, carpets, arcades, and loggias for exercise in bad weather, indoor amusements, such as billiard-tables and pianos—all these the doctor recommending a novel and extreme treatment should be able to guarantee. Such provision has doubtless materially assisted the increasing success of Davos.

Writing on the 30th October, he says:

Already I have experienced the novel situation of being unable to sit out on the snow because of the heat of the sun's rays, intensified by the universal whiteness, being at times quite overpowering.

He concludes by bearing testimony that the evil reputation the Engadine once had in the matter of food is no longer merited, and that the most exacting and fastidious need no longer complain against either the quality or the cooking of the food.

Always desirous of accumulating independent testimony both with regard to the climate of the Engadine and Davos, during the winter, I obtained from a gentleman, who was himself unfortunately the subject of consumption, and who, I regret to say, is since dead, some of his personal observations both at St. Moritz and at Davos. Writing to me at the end of September, 1876, from St. Moritz, he says :

My wish has been to remain here if possible for the winter, but I think I shall shortly return to Davos nevertheless, as the winds here are decidedly trying and the cold has annoyed me somewhat. I find that, although in truth the two climates may be substantially the same thing, my bronchial susceptibility makes the Engadine too keen and searching. Then it must be admitted that all invalids would not find the winter arrangements here sufficiently comfortable for bad weather. It appears to me that Davos is considerably sheltered from wind, though of course it does not enjoy complete freedom from it. What place does?

Then in speaking of the domestic arrangements at Davos he says :

There are white porcelain stoves, from six to nine feet high ; they keep warm, if necessary, all night, but I never felt the need of much fire in my bedroom, as the rooms are very well built, not very large, but free from draughts. Double windows are used everywhere. I wore the same clothes as in winter in England in the house, but out of doors, quite a third of the time, wore only a light autumn overcoat. I think people wore perhaps rather more than in winter in England—certainly than in the South. *The feet should be protected by felt overshoes.* The snow in the road is usually hard and pleasant to walk upon ; except in spring, it is always good under foot, then it is *awful.* Sunshades are useful for ladies in winter, as are blue veils, spectacles, etc. I saw no snow blindness such as I have had myself in Colorado. Colds in the head are decidedly not nearly so common as in England, and are very quickly got rid of. Rheumatism I heard nothing of, but am inclined to think it might be brought on by exposure, but only such as would bring it on anywhere. I feared myself the depressing effects of a low temperature, but did not suffer at all from it or hear of others doing so. I am feeling similar effects here [at St. Moritz] just now, which makes me incline to Davos. I hear [at Davos] of sudden attacks of hæmorrhage, and in cases liable to it I should not think Davos good. People must not expect to find innumerable expeditions [at Davos] as here [at St. Moritz], but there are a few nice ones. Sleigh-driving is very enjoyable in the sunny, warm, cloudless winter days. There was [the preceding winter at Davos] six weeks or so excellent skating, and a few picnics on the ice, ten days' skating on the lake, the remainder in a rink. There used to be a band on the rink and many spectators. Davos appeared to be a fine place for children, as they looked well and were hardy. I saw them in winter, with no extra clothing or hat on, playing about.

The same gentleman, having passed a second winter at Davos, wrote twice again the following summer from St. Moritz.

I left Davos the 4th of June, and have been here since. Throughout the winter I did very well, hardly ever remaining indoors a whole day (not more than three or four times). I was able to walk regularly and also to skate in December and January ; my appetite was good, and though the food was not as suitable as it is here, I always managed to make good meals. I slept well ; but one thing I found trying used to be the heat of the sitting-rooms at the Hotel ——, which is a very small house. There were some twenty English all the winter (1876–77) at Davos. Among the twenty-five to thirty-five guests at the Hotel —— there were several deaths, some of them people who certainly came much too far gone, but one or two in which it seemed that the peculiar air brought on an increase of disease.

After giving the particulars of a case in which a severe and fatal illness appeared to have been induced by " the exceptional excitement, etc. of the climate," he adds :

Other deaths in the house were very sudden, and I knew of some cases of sudden attacks of hæmorrhage. There was a lady in the hotel who made great improvement, and she and I were supposed to do most credit to the place. In one or two cases where there was no disease, but weakness and want of stamina, people seemed to have gained considerably when they left. I am sure for dyspepsia the climate is a wonderful remedy. The place was very dull, of course, and rather depressing ; but as I got on well, and had been so ill on arriving, I could not complain of that. As soon as the spring began I felt less well, and since the weather has been warm have had more coughs, etc.

The temperature went very low at Davos this winter; I think some 8 or 10 degrees below zero Fahrenheit. There was an immense quantity of snow, which fell mostly after the middle of December. I stayed at Davos through the snow-melting time, as I could not think of any place better for me, and I found it no more unpleasant than much of the winter weather, and I do not think very much damper, though being warmer it was the more relaxing.

Since these letters were written, the merits of Davos as a winter station have been repeatedly before the public and the medical profession; and the number of winter residents has increased to such an extent that one of its most distinguished literary supporters and advocates, Mr. J. E. Symonds, writing in the *Pall Mall Gazette* from that place in January, 1882, says:

The Nemesis which attends sudden prosperity already hovers over Davos, and if the place increases at its present rate the ruin will be as speedy as the rise.

Four years' experience has not shaken my belief in the value of a high mountain climate for certain classes of pulmonary invalids; though I am bound to say that the hopes I entertained and publicly expressed after a few months' residence in Davos have been considerably damped by what I have since observed. At the same time this long experience convinces me that the principles upon which an Alpine cure can be expected have been steadily neglected here. When I first knew the place it was a little village, furnished with a few hotels for the reception of strangers. The life was primitively simple, the air quite pure, the houses far apart and of moderate dimensions. Since then it has rapidly expanded, and the expansion has brought the following bad consequences: 1. There is now a perceptible cloud of smoke always hanging over the valley, shifting with the wind, but not escaping, and thickening the air to a considerable extent. This smoke arises mainly, doubtless, from chimneys; but it reminds one of the breath of many hundreds of consumptive patients aggregated at close quarters. 2. The houses, which have sprung up like mushrooms, are built with so little attention to the requirements of a sanatorium that the main promenade is more than half in shadow. 3. The drainage of the whole place is infamous. One portion of the village carries its sewage down into a marsh, where it stagnates. Another portion is drained into the stream, which in winter is a shallow, open, ice-clogged ditch, exhaling a frozen vapour. To walk by the course of this river is now not only disagreeable but dangerous. The largest hotel frequented by the English has a horrible effluvium arising from the cesspool beneath its windows. In the largest hotel frequented by the Germans, a species of low fever has recently declared itself. 4. The social amusements of a watering-place have been greatly developed. Dances, concerts, theatres, bazaars, private theatricals, picnics, are multiplied. Some entertainments of the sort are no doubt not only necessary, but also beneficial. Yet it must be remembered that the peculiar severity of Alpine winter, the peculiar conditions under which consumptive people meet together here, crowded into rooms artificially heated with stoves, render all but the simplest forms of social gathering very dangerous.

The only way of averting some serious catastrophe from a health resort which has deserved popularity, and the principle of which is excellent—the only way of preventing Davos from being converted into an ill-drained, overcrowded, gas-lighted centre of cosmopolitan disease and second-rate gaiety—is to develop rival places of the same type. The valley of Davos proper, from Davos Kulm or Wolfgang down to Frauenkirch, may be said to be already exhausted for building purposes. This valley is so narrow and so much enclosed with mountains that the further development of any one of its hamlets is certain to injure the whole neighbourhood. Its torrent is too thin and hampered in its course to act as common conduit-pipe for drainage. Its boasted absence of wind causes the addition of smoke from chimneys or of exhalations from cesspools to be immediately felt in all parts of the district. If Davos is to remain what it calls itself—a Luft-Kur-Ort, or " Health resort of sun and air "—it must learn rather to contract than to expand. I have touched upon some of the obvious dangers which threaten Davos. I might have gone into more alarming problems, and have raised the question whether the accumulation of sick people in big hotels, which are really consumptive hospitals, though not subject to the precautions used in consumptive hospitals, is not attended with the gravest disadvantages. So long as the hotels remained small, and there were only a few of them in the place, the peril from this source was slight. But the tendency at Davos has been to enlarge each of the well-established *pensions*, to pack the patients together in as small a space as possible, and to build new inns at the doors of the old ones. All this is done in a climate where winter renders double windows and stove-heated buildings indispensable. All this is done for a society where the dying pass their days and nights in closest contiguity with those who have some chance of living ! Within the last few weeks two cases have come under my notice, one that of a native of Davos attached to the service of the visitors, another that of an English girl, who have both contracted lung disease in the place itself, owing, as I believe, to the conditions of life as they have recently been developed here.

Mr. Symonds' former able advocacy has done so much for Davos that this timely warning of his ought to be laid to heart by those directly interested in the prosperity of the place, and by those who, perhaps, somewhat too indiscriminately recommend its winter climate to consumptive patients; for the advocates of Davos, both in the medical profession and out of it, have become numerous and influential, and the cases which have undoubtedly derived great benefit by wintering there now amount to a considerable number. Personally, I have long been fully convinced of its importance and value as a winter sanatorium in suitable and carefully selected cases. But it would be unfair and disingenuous to conceal the fact that my personal knowledge of the experiences of winters at Davos is by no means unchequered by calamities, and some fatal occurrences there have been exceedingly sad and unexpected. The experience of different writers has varied also greatly as the seasons themselves have varied.

A writer who is most friendly to Davos * thus expresses himself on this point :

Another objectionable feature is the strong desire that exists among the local medical men, and others financially interested in the place, to suppress the number of deaths that annually occur, in order to give a false impresssion as to the marvellous powers of the climate to delay death. In fact, we do not hesitate to say that in not a few cases patients who were known to be hopelessly ill have been hurried elsewhere by order of the medical men, for no other reason than that the Davos death-rate might be kept low. We anticipate that this statement will lay us open to criticism, if it does not beget flat contradiction ; but, fortunately, we are in possession of the most reliable data to prove what we have written.

As I have already said, Davos is situated in a mountain valley in the Grisons, which runs parallel with the upper valley of the Inn, at a distance of about twenty miles north of it. Dr. Frankland makes it 5352 feet above the level of the sea, which is a

* " Davos Platz, as an Alpine Winter Station," by j. E. Muddock.

somewhat higher estimate than the one hitherto current.   It is then of just the same elevation as the village of Mürren, which is 5348 feet above the sea, and only a little lower than Samaden, which is about 5600 feet.   But it is not its particular elevation alone which gives to Davos its special suitability as a winter resort ; for, as we see, Mürren and Samaden, so far as mere elevation is concerned, ought to answer equally well.   We must seek, then, in other local conditions for the characteristic qualities of the climate of Davos. So far as purity and rarefaction of the air are concerned, it is in almost precisely the same position as the adjacent Engadine valley.   It is probably only in the greater stillness of the atmosphere, and in protection from the prevailing local winds, that Davos presents any special advantages in winter over such resorts as St. Moritz, Pontresina, or Samaden.   And now that the outcry has been heard, and heard so distinctly, of the overcrowding of Davos and its consequent evils, it becomes a question of some interest to what extent and in what cases other Alpine resorts may be equally useful as winter sanatoria.   Dr. Frankland observes ("Proceedings of Royal Society," vol. xxii. p. 317) :

The summer climate of Davos is very similar to that of Pontresina and St. Moritz in the neighbouring high valley of the Engadine—cool and rather windy ; but so soon as the Prættigau and surrounding mountains become thickly and, for the winter, permanently covered with snow, which usually happens in November, a new set of conditions come into play, and the winter climate becomes exceedingly remarkable.   The sky is, as a rule, cloudless, or nearly so ; and as the solar rays, though very powerful, are incompetent to melt the snow, they have very little effect upon the temperature either of the valley or its enclosing mountains, consequently there are no currents of heated air, and as the valley is *well sheltered from more general atmospheric movements, an almost uniform calm prevails until the snow melts in spring.*

And Mr. Symonds, speaking from long personal experience, says of the winter climate of Davos :

The position of great rocky masses to north and south is such that the most disagreeable winds, whether the keen north wind or the relaxing south, known by the dreaded name of *fohn*, are fairly excluded. Comparative stillness is a great merit of Davos; the best nights and days of winter present a cloudless sky, clear frost, and *absolutely unstirred atmosphere.* March is apt to be disturbed and stormy, and during the summer months there is a valley-wind, which rises regularly every morning and blows for several hours.

The valley is from ten to fifteen miles in length, and its direction is from north-east to north-west. It is only about half-a-mile broad, and protecting mountains rise on each side to the height of from 2000 to 5000 feet above the level of the valley. About three-quarters of a mile above Davos Platz to the north is Davos Dorfli, a sunnier spot than Davos Platz, but perhaps not so well protected from wind. Still farther north is the Davoser See, or the Lake of Davos, which affords good skating until it becomes too thickly covered with snow.

At the south-west extremity the valley is also well protected and closed in by high mountains. The Upper Engadine is, on the contrary, much exposed in this direction, and it has often been observed that storms and bad weather frequently reach this valley over the low pass of the Maloja. Then again, there are no extensive glaciers and snow-fields in the immediate vicinity of the Davos valley as there are in the neighbourhood of the Engadine, and especially of Pontresina. The smaller size of the valley, both in length and breadth, and the nature and position of its mountain barriers, with the absence in its immediate neighbourhood of great snow and ice fields, no doubt contribute to make the winter climate of the Davos valley a milder one than that of the adjacent valley of the Inn, and therefore better suited to a large class of invalids. But it is a question whether the more vigorous class of pulmonary patients, as well as those who have gained vigour and amendment at Davos, would not benefit as much, or perhaps more, in some of the Engadine resorts, and St. Moritz is perhaps the best situation in that valley for winter residence.

The winter snowfall in the Davos valley, as well as in the Engadine, usually begins early in November. An early and heavy snowfall of three or four feet is considered to promise a good winter. The snow continues to fall through November and a part of December. In the roadways it gets beaten down to a depth of three or four feet. In good seasons, fine settled weather, with absence of snowfall, sets in before the end of December. The atmosphere becomes still and calm, the air intensely cold and dry, and absolutely clear. At night the brilliant starlight, or the cold silvery moonlight streaming over the snow-mantled valley, gives it an aspect of singular beauty. The temperature at night often falls very low, frequently some degrees below zero. The days are cloudless, with an intensely blue sky, and an amount of heat from solar radiation which enables invalids to pass hours sitting in the open air; and the brilliancy of the sunshine in midwinter makes umbrellas and sunshades necessary for protection. The instant, however, the sun is withdrawn the intense coldness of the air makes itself felt, and a fall of 50 or 60 degrees Fahrenheit is common immediately after sunset. Of course all delicate invalids should be indoors before this hour. Owing however to the great dryness of the atmosphere, and the absence of wind, the extreme cold at night is by no means so much felt as might be expected. "There are no patients," says one of the local physicians, "who cannot, if they are so inclined, sleep with safety with an open window during the winter." "I was recommended," says Mr. Symonds, "to be in the open air from sunrise to sunset, to walk for two hours in the open air before going to bed, and to sleep with open windows. The invalid can take more liberties with open air at Davos than anywhere else."

Unfortunately, weather at Davos is fickle sometimes, as it is elsewhere, and a remarkably fine winter may be preceded or followed by a remarkably bad one. The winter season of 1879–80 was an exceptionally fine one, whereas the preceding winter, 1878–79, had been an unusually bad one, and had proved disastrous to many invalids there. The relaxing south wind, the *föhn*, prevailed to a great extent, in consequence of which the snow thawed at times in midwinter, and colds, which are rarely caught at Davos, were common. The following winter however many cases did remarkably well, and "wonderful recoveries" were numerous. There was almost an entire absence of wind, the air was remarkably dry and bracing, and for three months there was almost uninterrupted sunshine and clear unclouded skies. Then followed in 1880–81 another bad winter. "Davos Platz proved as capricious and fickle as our own damp and misty island. The snowfall did not set in until late, and then it was singularly light, while a high temperature and fogs and wind were the rule and not the exception. Those people who derived any benefit were in a very small minority, while the death-rate amongst the visitors rose to an alarming extent." *

During the past winter, 1884–85, and especially the latter part of it—after January—some very unfavourable weather was experienced. One of my correspondents speaks of it as " a very bad season, and up to the present (March 27th) frightful spring. . . . Davos, which is usually supposed to be free from wind, has had a north-east gale blowing for a fortnight or more." Another winter resident informs me that attacks of pleurisy and congestion were frequent. Notwithstanding the bad weather, many cases that were spending their first season at Davos did well, but those who had passed one or more previous winters there suffered greatly from the bad weather.

I have before me some records of the weather at Davos during three winter seasons. The first refers to the winter season of 1875–76. It is very brief,

* J. E. Muddock, " Davos Platz as an Alpine Winter Station."

S

and divides the days into "clear and fine," "moderately fine," and " bad."

	Clear and fine days. ·	Moderately fine.	Bad.
In November there were	12	3	15
„ December „ „	19	10	2
„ January „ „	14	12	5
„ February „ „	12	11	16
„ March „ „	10	9	12

So that out of a total of one hundred and fifty-two days, there are sixty-seven clear fine days, forty-five moderately fine, and forty bad days. The two worst months being November, when the snow begins to fall, and March, when it begins to melt. I take these figures from Mr. Holsboer's " Die Landschaft Davos," and I presume they are intended to represent a good specimen of winter weather there.

If we next take the winter of 1879–80, " perhaps one of the most perfect ever known in the Alps," and include the month of October, we find the days may be divided into " cloudless," " fine, but not cloudless," " cloudy," and " rain or snow."

	Cloudless.	Fine, but not cloudless.	Cloudy.	Rain or snow.
October had	18	4	3	6
November „	5	4	7	14
December „	14	6	6	5
January „	15	7	5	4
February „	8	7	8	6
March „	16	5	3	7

The first column includes only absolutely cloudless days, and in the second column are included days that are described as "glorious, a few white clouds," while the fourth column includes days when snow or rain, however little, fell. Out of one hundred and eighty-three, there were one hundred and nine fine days, seventy-six of which were cloudless, thirty-two days more or less overcast, and forty-two days on which rain or snow fell, fourteen of these being in November. October maintained its character for being one of the finest months in the Alps.

Let us now examine the records of the next winter, that of 1880–81 :

	Cloudless.	Fine, but not cloudless.	Cloudy.	Rain or snow.
October had	6	2	15	8
November ,,	10	6	9	5
December ,,	8	7	6	10
January ,,	10	9	10	2
February ,,	7	7	6	8*
March ,,	10	3	11	7

Mist or fog is mentioned as occurring four times during this winter—once in October, once in November, and twice in March. "No wind *in the valley*" is stated of no less than one hundred and thirty-four days, and "no upper current" on forty-one days, and a strong wind is only mentioned on five days in the whole winter. It has already been said that this was a very unfavourable specimen of a Davos winter ; there were only eighty-five fine days against one hundred and nine in the preceding winter, and fifty-one against seventy-six cloudless days. And although there were fewer days (forty to forty-two) on which snow fell, the distribution of the snowfall was less advantageous. The heavy snowfall in November of the former winter was followed by a continuation of magnificent weather, whereas the small snowfall in November of this season was followed by frequent snowfalls in December and eight consecutive days on which snow fell in February. But perhaps the most remarkable and characteristic fact which comes out of this meteorological record is the singular absence of wind in the valley. It is this peculiar stillness of the air that enables the invalid to support so well its comparatively low temperature ; so that he is not chilled and depressed by it, but, on the contrary, is braced and exhilarated.

* Snow fell on eight consecutive days.

But when we examine and compare the weather of the two last winters, 1883–84 and 1884–85, we shall find that this absence of wind cannot always be relied upon. Following the same plan, we find in 1883–84 (fairly good winter):

	Cloudless.	Fine, but not cloudless.	Cloudy.	Rain or snow.
October had	11	6	14	5
November „	11	5	14	4
December „	13	2	16	11
January „	12	3	16	11
February „	10	5	14	2
March „	13	7	11	4

in 1884–85 (bad winter):

	Cloudless.	Fine, but not cloudless.	Cloudy.	Rain or snow.
October had	1	8	22	10
November „	16	4	10	7
December „	7	7	17	9
January „	20	2	9	3
February „	8	1	19	6
March „	6	6	19	6

It is here seen that not only were there fewer fine days in last winter as compared with the preceding one, viz., eighty-six to ninety-eight, and fifty-eight cloudless days, as compared with seventy, but we cannot fail to notice the irregular distribution of the cloudless days in last winter as compared with their regular distribution in the preceding one. Out of the fifty-eight cloudless days in last winter, thirty-six occurred in two months and the remaining four months had only twenty-two between them. A fine November followed a bad October, and a fine January was followed by a bad February and March; snow or rain fell on twice as many days in these two months as during the same months of the preceding year; we observe also that there were more cloudy days during the winter than clear ones. There was rain as early as the 17th of February, whereas, in the preceding winter, no rain is noted before the 4th of April.

Next, with regard to wind in 1883–84 :

	"Strong breeze" or "High wind."	"Moderate breeze."
In October there were	4 days.	2 days.
„ November „ „		4 „
„ December „ „	2 „	5
„ January „ „	3 „	5
„ February „ „		4
„ March „ „	1 „	2

1884–85.

In October „ „	10 „	6
„ November „ „		3
„ December „ „	5 „	1
„ January „ „	2 „	2
„ February „ „	3 „	4
„ March „ „	2 „	13

This is a most instructive record, for it shows us that in this winter climate, one of the chief characteristics of which, under favourable conditions, is stillness of atmosphere, there can be a considerable number of windy days. There were fifty-one windy days last winter as compared with thirty-two the winter before. One-fourth of the days in February were windy, one-half of those of March, and more than one-half of those of October! Mist is mentioned on two days in last March.* The temperature also last winter was nearly 2 degrees Centigrade warmer than the average at Davos.

Owing to the absence of aqueous vapour in the clear dry air of this elevated region, the intensity of solar radiation on perfectly clear days is remarkable; and, according to Dr. Frankland, at Davos Dorfli, on the 21st December, 1873, at 2.50 p.m., the "mercurial thermometer with the blackened bulb *in vacuo*" recorded 113 degrees Fahrenheit, and on the same day at Greenwich the maximum reading, obtained by the same

---

* See further some interesting observations by Mr. A. W. Waters on the winds at Davos, which I have appended as a *note* to this chapter.

method, was 71·5 degrees, giving a difference in favour
of Davos of 41·5 degrees Fahrenheit. But a maximum of solar radiation amounting to 153 degrees
Fahrenheit was obtained on the 31st January, 1881,
while on the same day the maximum temperature
of the air in the shade was 42·5 degrees and the
minimum 18 degrees Fahrenheit. So that the difference between sun and shade temperature is enormous. The lowest temperature recorded during the
winters 1879–80 and 1880–81 was 16·7 degrees Fahrenheit below zero on 9th December, 1879. The mean
daily minimum for the same month was 5·5 degrees
Fahrenheit, and the mean daily maximum 23·13.
The maximum sun temperature 138 degrees Fahrenheit. This was during a month of the finest Davos
winter weather; the amount of aqueous vapour in
the air being exceedingly small, and the readings of
the hygrometer very low, as low as 3·0 degrees (!) on
one day, and never over 38·5 degrees.

The mean winter temperature in Davos Platz,
taken from twelve years' observations, is :

		Cent.
November	. . . . .	2·4
December	. . . . .	6·1
January	. . . . .	7·7
February	. . . . .	4·1 (*A. W. Waters*).

Owing to overcrowding and to defective sanitary
arrangements, chemical examination of the air in
Davos itself has shown that it has at times been by no
means so pure as it should be, or so free from admixture with organic impurities as had been imagined.
This is much to be regretted, as many invalids may
not be able to get away to breathe the purer air on
the mountain-side. The want of efficient drainage
was a year or two ago reported to be very much felt.

Of the drainage [said a writer at that time] we feel bound to
say that it is about as bad as it can be, while heaps of offal, cow-
dung from the cow châlets, and other indescribable filth are
allowed to lie exposed near the road, thereby not only offending
the sight but the smell, as well as tainting the atmosphere. It is
true that during the intensely cold winter months the low

temperature and the snow prevent ill-effects from this disgraceful sanitary neglect; but the evil is there nevertheless, and it makes itself manifest in more ways than one as soon as any appreciable rise in the thermometer takes place. After this the reader will not be surprised to learn that typhoid and other fevers, if rare, are no strangers to the valley.*

It would seem then to be incumbent on English physicians to consider whether there are not other resorts in the high mountain valleys of Switzerland to which invalids may be sent in winter, where they may escape the evils of overcrowding and overbuilding, which appear so rapidly to have overtaken Davos Platz.

Of all the places at present known, St. Moritz seems to offer the greatest attractions. It has already been tried and found to answer exceedingly well in a certain number and class of cases. No doubt it is not so well suited as Davos to the feebler classes of pulmonary invalids, who are also the victims of more advanced disease. But to many of the stronger patients, and to those in whom disease is in its earliest stage, or very limited in extent, or to those who are suffering only from general loss of tone, St. Moritz may prove as useful, or even more so, than Davos. Moreover, at the Kulm Hotel at St. Moritz invalids will now find a winter sanatorium, furnished with appliances and provided with extensive apartments and other conveniences, which it would be difficult to find in many of the hotels at Davos. Open fireplaces, so dear to the English mind, are to take the place of the close German stove, and no effort, we understand, is to be spared to make the winter life of the invalid cheerful and comfortable. The hotel possesses an excellent covered terrace for sitting out in, and a skating-rink is also close at hand. An English doctor is established in the hotel all the year

---

* "Davos Platz as an Alpine Winter Station." The writer of the above, in a later edition of this book, states that many improvements have been made in the sanitary state of Davos, and that a complete system of drainage has now been carried out.

round—another consideration of some importance to invalids who do not speak German. St. Moritz also has the advantage of being within easy reach of Davos, so that a patient who finds the climate of the Upper Engadine unsuitable can easily remove to the Davos valley.

Besides the Kulm Hotel there are now two other hotels at St. Moritz open all the year, the Hotel Caspar Budrull, near the Kulm, but on rather a lower level, with 70 beds ; and the Hotel Beau-Rivage with 40 beds, on a still lower level, on the road between the village and the Kurhaus.

Last winter nearly two hundred persons wintered at these hotels. At Samaden the Hotel Bernina also offers comfortable winter quarters, but owing to the neighbourhood of the river there are often slight fogs there in the mornings, which soon however pass away. The sun rises earlier and sets sooner at Samaden and Pontresina than at St. Moritz. At Pontresina the Hotel Enderlin is to be kept open during the winter, and this will afford another resource for winter visitors ; and, no doubt, if the number of winter guests increases rapidly, M. Müller might be induced to keep open his excellent hotel, the Julierhof at Campfer, during the winter. Its sheltered position and its elevation above the level of the river and lakes would make it an admirable winter situation.

Of the suitability of the new Kurhaus at the Maloja for residence in winter, I have no experience, and I can therefore offer no opinion. I should consider the position not well suited to pulmonary invalids ; but a hardier class of visitors might find its numerous internal resources an attraction and a help in passing the long winter evenings.

It is not usual to go out walking in the evenings, which one generally spent indoors, except when on moonlight nights tobogganing is indulged in.

On the shortest day in the Engadine the sun lasts from 10.45 a.m. to 3 p.m. There is a sudden chill when the sun goes down.

After passing a winter in the snow, one of the difficulties always has been what to do when the transitional season of spring sets in, and the snow begins to melt. "When the snows melt," one visitor writes, "winds of icy coldness blow on the snow, surpassing English east winds in their fierce bitterness, making existence barely tolerable in March and April." Some, however, boldly face the inconvenience and remain where they have wintered, and so far as we can learn, without taking any particular harm. But some springs are much more trying and disagreeable than others, and those who left the Engadine, last winter, at the end of February had reason, it would appear, to congratulate themselves that they had done so ; and, no doubt, there is a craving for a little change when spring, with its disagreeable weather, reaches the snow-covered valley. To return to England at once seems scarcely advisable, knowing what our own spring weather is like. To seek some other intermediate mountain station, of lower elevation, for a few weeks before descending to the sea-level would perhaps be the best thing to do, if such suitable stations were easily found. But there are difficulties in doing this. Many of the summer resorts between 2500 and 3500 feet above the sea are not open and available at this season, and in those that are available the accommodation is perhaps not such as invalids require. Moreover, even if a suitable intermediate station is found, it will occasionally happen that pulmonary invalids find themselves worse for the change ; and begin to think they have been ill directed in their choice ; whereas they should bear in mind that the spring is a difficult season everywhere, especially for those who suffer as they do.

Thusis, 2448 feet above the sea, is convenient and accessible, but little is known about its spring climate, and, from its situation, I should fear it would be draughty in spring. Fair accommodation can be obtained there, and it has the advantage of being on the way homeward.

Seewis, nearly 3000 feet above the sea, a village in the Prættigau, quite close to Landquart, of which I shall have more to say in the next chapter, is exceedingly conveniently situated, in a picturesque position, and, we are assured by those who have spent a whole winter there, has good, but limited accommodation.

Glion, above Montreux on the Lake of Geneva, also nearly 3000 feet above the sea, is a pleasant sunny station with very good accommodation and most picturesque and cheerful surroundings ; but it is rather out of the way for those who are returning to England.

Heiden, 2660 feet above the sea, near Rorschach on the Lake of Constance, is also conveniently accessible and in a pleasant situation, but it would probably be found dull and unprepared for spring visitors.

In conclusion, two questions must be briefly dealt with : First, what class of invalids may fairly expect to derive benefit from wintering in these high mountain valleys ? and, secondly, what are the curative agencies at work there ?

It is of the first importance to remember that these mountain climates are by no means adapted to the treatment of many well-defined forms of consumption ; that cases have to be selected with great care and discrimination ; and that regard must be had rather to the constitution and temperament of the individual than to the mere amount of local disease. Hereditary predisposition, other circumstances being favourable, offers no counter-indication to the suitability of these stations. But their remedial power is especially manifested in persons who have become accidentally the subjects of chronic lung disease, and who were the possessors of an originally sound constitution, and have obvious reserve stores of physical vigour : the constitution must have the power of healthy reaction to the exciting stimulants here applied to it. The extent to which this reaction often occurs has occasionally led to grave and even fatal indiscretion. It is the universal experience of physicians that the phthisical constitution is the most difficult of all to control; consumptive patients are for ever committing indiscretions which are perilous to themselves and in the last degree exasperating to their doctors. Cautions against over-excitement and over-exertion are therefore specially needed in climates such as we have been considering. The following summary of cases suitable to those high mountain health resorts is founded on the published testimony of a physician, whose practical experience in one of them has extended over twenty years ; and in the statements which follow my own experience is in accordance with his.

1. Where there is an obvious and well-ascertained predisposition to consumption, and when perhaps a slight hæmorrhage has occurred without the manifestation of any definite local disease ; as a *preventive* measure a residence for two or three seasons in a high mountain station is to be recommended.

2. In catarrhal forms of consumption, in the early stage, without much constitutional disturbance, the best results may be looked for. But cases with much fever from the commencement, and of nervous and excitable temperament, must not be sent to high altitudes.

3. Chronic inflammatory indurations and infiltration of limited portions of the lung, often the result of acute congestion and inflammation, are especially suitable ; not so, however, if a considerable extent of lung is the seat of tuberculous disease, or if, owing to the extent of lung involved and consequent changes in the sound lung, there is much difficulty of breathing.

4. Cases of chronic bronchial catarrh in young people ; that is to say, those cases of tendency to repeated attacks of "cold on the chest," often left behind in children after whooping-cough, measles, and other maladies. But this does *not* apply to the chronic winter coughs of persons more or less advanced in life, or to cases where there is much *permanent* shortness of breath.

5. The results, in the shape of thickenings and adhesions, of former attacks of pleurisy, to which, too often, the development of serious subsequent lung disease can be traced. The pulmonary gymnastics excited by treatment in high altitudes prove of great value in those cases.

6. Many cases of purely nervous asthma have been cured in these resorts.

7. Apart from cases of pulmonary disease, many other ailments, such as general loss of power, not dependent on organic disease, cases of nervous exhaustion, over-work, retarded convalescence in otherwise vigorous constitutions, certain forms of dyspepsia and hypochondriasis, and other less strictly definable maladies, not seldom find restoration to health and strength from prolonged residence in the pure bracing air of these Alpine stations.

Next, as to the curative agencies at work in these resorts. This question is by no means an easy one to answer decisively. When we reflect that cases of consumption are arrested in their course and apparently cured, as they certainly have been, in such a climate, for instance, as that of Arcachon, on the coast of the Atlantic, and also in such an apparently utterly different climate as that of Davos, we are led to the conclusion that we must seek for some *special relation* between the individual to be cured and the particular climate that will suit him. And it is sometimes only by actual trial that such relation can be discovered.

Purity and stillness of atmosphere are two important, and may be the most important, conditions at work. Elevation in itself, as I have already said, may also be of some importance, but it cannot be an essential ; it brings with it other conditions, however, such as dryness and purity of air, which are of great consequence. The Tartar Steppes, where the Russian physicians send their consumptive patients, and where, we are told, they are cured, are sometimes below and not above the sea-level. It is not the low temperature alone that is the cause of immunity from phthisis in these mountain valleys, for in some of the coldest parts of Russia the mortality from phthisis is more than twenty per cent. of all deaths ; but the cold, in these places, is probably associated amongst the poorer peasantry with overcrowding and other insanitary conditions of life, to which the mortality from phthisis is doubtless due.

It used to be thought that an equable temperature was of great importance in the treatment of consumption ; and within certain limits, and if associated with certain other qualities, equability of temperature is an advantage in a climate ; but unless dryness of the air is associated with it, equability of temperature is not of so much value. Indeed a too equable temperature may lead to loss of tonic property, and so diminish nutritive activity. We find, for example, that in Ceylon, which has a remarkable equable climate, consumption is exceedingly common. On the other hand, at Quito, in Ecuador, which is 10,000 feet above the sea, its immunity from phthisical disease is considered to be greatly due to its equable temperature ; the mean temperature for the year being 60 degrees Fahrenheit, and "in a large room with doors and windows open day and night the temperature varied between 57 degrees and 60 degrees only ! " But it is obvious that the climate of Quito possesses also the other conditions dependent on great elevation. It has been suggested, and with much reason, that the immunity from phthisis observed in certain places and at certain elevations may perhaps be due to the fact that the inhabitants are all agricultural or pastoral, and live out-of-door lives, and also to the relative scantiness of population.

But, as I pointed out some years ago, the chief curative agency at work in these elevated districts is probably the *antiseptic* or *aseptic* quality of the air. It has been shown that there is an almost entire absence in these localities of those *organic particles* which play such an important part in promoting putrefaction. To this fact may be added the stimulating and tonic properties of the cold pure air, promoting the desire for muscular activity, as well as increasing the power for the same, by inducing increased activity in the general forces of nutrition.   Another valuable condition is the rarefaction of the air, which necessitates greater activity of the respiratory organs.   The respirations are necessarily more frequent and more profound ; the air breathed is relatively richer in active oxygen than the air of the plains ; a more complete aeration of the blood is secured ; all the portions of the lungs which are capable of admitting air are called into full play and activity ; the air-cells are more completely dilated ; the functions of all the healthy portions of the lungs are roused and thoroughly engaged in the work of respiration.   There is less stagnation of air in the lungs, and diffusion of the gases set free at the surface of the lungs is favoured.   We are not then surprised to find that the chest expands considerably during residence in these resorts, and that portions of lung ordinarily little used in breathing (and these are the parts specially liable to be attacked in phthisis) become actively engaged, and so a compensatory activity in the sound parts makes up for the inactivity in parts which have become spoiled by disease.   The increased rapidity in the circulating functions, the more complete penetration of all the tissues of the lung by the more active blood currents, may also promote repair and recovery from the damage inflicted by disease.   The low temperature of the air inhaled also has a tendency to diminish fever.   These may not be all the influences at work in the restoration of health to the pulmonary invalids who pass their winters in these snow-covered regions, but I doubt not that they are the chief.

In winter the diligences used in summer to cross the mountain passes are replaced by sledges, each drawn by one horse, and holding two passengers.   A closed sledge has been used during the past winter in crossing the Julier.   One advantage to invalids in approaching the Engadine by the Maloja Pass during winter is that it is 1500 feet lower than the Julier.

It is best, when it is practicable, for persons in delicate health to arrive and get settled before the severe weather of winter sets in. September is usually a good month for weather. Otherwise it is important, when making the journey over the mountain passes in winter, not to undertake it in bad weather, when the exposure might prove injurious.

The following is the time-table for the post diligences to Davos during summer and winter :

FROM 15TH OF JUNE TILL 14TH OF SEPTEMBER.

Dep. Landquart ... 5. 0 a.m.   2.45 p.m. } *viâ* Kublis-Klosters
Arr. Davos Platz...12.25 p.m.   9.55 p.m. }   in 7 hours.

Dep. Coire .... 6.30 a.m.   11. 0 a.m. } *viâ* Lenz-Wiesen.
Arr. Davos Platz... 4.30 p.m.   7.30 p:m. }
Dep. Coire .... 5.20 a.m. } *viâ* Thusis-Tiefenkasten-Wiesen
Arr. Davos Platz... 4.30 p.m. }   in 11¼ hours.

FROM 15TH OF SEPTEMBER TILL 14TH OF JUNE.

Dep. Landquart ...10. 0 a.m.   1.10 p.m. } *viâ* Kublis-Klosters
Arr. Davos Platz... 5. 5 p.m.   8. 5 p.m. }   in 7½ hours.

(First dil. with one hour's stoppage at 1 o'clock in Küblis for luncheon.)

Dep. Coire ....8. 0 a.m. } *viâ* Lenz-Wiesen in 8½ hours.*
Arr. Davos Platz...4.35 p.m. }
Dep. Coire ....5.15 a.m. } *viâ* Thusis-Wiesen in 11½ hours.*
Arr. Davos Platz...4.35 p.m. }

It will be seen that there are two diligence-routes to Davos Platz. The one from Landquart is two hours the shorter. Passengers are therefore recommended to take their tickets from Bâle or Zurich to Landquart, two stations short of Coire ; they are then conveyed by diligence in seven hours (stoppages included) to Davos Platz. The hotel at Landquart is close to the station, and private carriages can be procured there if wanted. The other route from Coire is the so-called Landwasser-route, occupying nine hours to Davos Platz.

The journey in winter is performed quite as rapidly and easily as in summer, but at that season of the year invalids are strongly recommended to travel by the Landquart-route, because roomy covered sledges, furnished with hot bottles, etc., always run on it, whereas the condition of the Landwasser-route (*viâ* Coire) often makes it impossible to use closed vehicles.

Travellers desirous of visiting Ragatz, with the wonderful

---

* Neither of these routes can be depended upon in winter, on account of avalanches.

gorge of Pfæffers, must get out at Ragatz station. The hotel accommodation is very good there, and private carriages can be procured there direct to Davos Platz (the drive is one hour longer than from Landquart), or the journey may be continued by rail to Landquart or Coire, and from thence as above-mentioned.

NOTE ON THE WINDS AT DAVOS.*

"As a rule, the wind at Davos," writes Mr. A. W. Waters, "is only a valley wind—that is to say, is quite independent of the upper currents, and it will therefore be readily understood that its exact direction varies very considerably in different parts of the valley. The valley wind is really a surface wind, so that very frequently the smoke shows the direction to be from the north at the height of the roof, whereas at about double or three times this height it is blowing from the south. This occurs so often that it may almost be considered a rule when a true valley wind is blowing. Also it is often the case that in the middle of the valley the wind is from the north while at the two sides it is from the south. There must therefore on these occasions be a neutral zone, and the force must vary very greatly at different heights. This I have been able to see clearly, for on one or two days when there was a strong unpleasant wind by the ground, a vane on a pole about the height of a house showed no movement. In the two lateral valleys, Dischma and Fluela, it very often happens that the wind is blowing in opposite directions ; probably in these cases the wind may either be caused in these side valleys, or be drawn by the main valley.

"The direction of the prevailing valley wind depends upon various circumstances, and it has been a matter of scientific surprise to some people that the direction in Davos should be down the valley, and the cause has often been sought in the wrong place, but it seems to me that the explanation is a simple one, for although the direction is from N.E. to S.W. the main mountain masses lie to the south, so that, the air being more warmed over these mountain masses, the current is thus drawn to the south, although the valley falls in this direction. That this is the probable explanation receives the strongest support from the lower part of the valley. In Wiesen, which is south of the gorge of the Zuge, near the main mountain masses, the valley wind seems to have a frequent or prevailing direction from the south." In March last "the direction of the wind was mostly the reverse of that in Davos. It does not seem to some people that places within twenty miles can really be as different as they are, and this cannot be thoroughly understood without fully

* Note from "Davos Dorfli," by A. W. Waters, F.G.S.

taking into consideration the winds and their origin, but when it is seen how much they depend upon the neighbouring configuration then it can be appreciated.

"Those who know Davos are well aware that there are great complaints as to the depressing influence of the *föhn*. The *föhn* is the *favonius* of the Romans, and is now in the Engadine *favoun*, and when residing in that valley I took a good deal of trouble to find out whether the conception of the *favoun* was a definite one among the inhabitants, and found that, while most associated the direction with the warm wind, there were some who would call any warm snow-melting wind in the spring *favoun*. In consequence of the different configuration, the *föhn* affects the two places so differently that in St. Moritz we hardly ever hear any one speaking of it, whereas in Davos it is a stock subject of conversation. St. Moritz Dorf is the highest part of the valley of the Engadine, and therefore, when a *föhn* wind is blowing in the neighbourhood, it frequently deposits its moisture in St. Moritz, or if not snowing, or raining, is a damp wind. In Davos, on the other hand, the *föhn* wind has to descend from a greater height, and has passed over a range of mountains, and is in consequence of its descent a warm and dry wind, at any rate for part of its duration. The winter of 1882–83 gave some very interesting examples. On the days with strongest *föhn*—viz., November 7th, St. Moritz had, at 1 p.m., +3·2 cent., while Davos had +13 cent.; November 8th, St. Moritz had +2·1 cent., Davos +7 cent.; December 4th, St. Moritz was, at 1 p.m., −5 cent., Davos was +4 cent.; January 30th, St. Moritz was −0·8 cent., Davos was +5 cent.

"It will be seen that whereas St. Moritz has no depressing *föhn*, its absence at such times has to be paid for by a colder and damper air, together with a strong wind (which is more trying, as it is cold)."

CHAPTER VI.

OTHER MOUNTAIN HEALTH RESORTS.

BESIDES Davos and, in the Engadine, St. Moritz, Samaden, Zuz, and the Maloja, there are but few other mountain resorts in Europe which have been tried or suggested as *winter* sanatoria. *Wiesen*, 4770 feet above the sea, only two hours from Davos Platz on the Landwasser road, is one which deserves to be better known. It has frequently been visited as a spring resort by patients who have wintered at St. Moritz and Davos, and I have received most favourable reports from some of these, especially as to the food and accommodation provided at the two hotels there, reports which my own personal observation in the autumn of 1882 entirely confirms. Wiesen is in a dry and sunny situation on sloping ground 1000 feet above the Landwasser river, and surrounded by very pleasing mountain scenery. It is somewhat better sheltered from cold north winds than Davos, and, so far as I can learn, suffered less than Davos from the *föhn* wind during the past winter and spring. It gets sunshine on the shortest days from 10 a.m. to 3 p.m.

Of course the same society and amusements cannot be obtained there as at St. Moritz and Davos, but some may not think this altogether a disadvantage. A medical man, suffering from an affection of the lungs, wrote to me from Wiesen last October : " This is a capital hotel ; good table and everything very comfortable. No better place for Davos and St.

Moritz people to come to in spring;" and in a subsequent letter (November 29th) he says: "It is one
of the best places I know of for a change from either
Davos or St. Moritz, and I consider that in some
cases it is preferable to either. It is less exciting
and perhaps less bracing, but there are certainly less
draughts than either at St. Moritz or Davos, and I
would call the air at Davos more harsh, if I can use
such a term. The surroundings are delightful and
one does not tire of them. Hotel very well drained,
under a most obliging landlord. I should call this
more sheltered than any other high Alpine Health
Resort." There is a resident medical man who has
an English qualification. Wiesen lies conveniently
on the post-road between Chur and Davos and
between Thusis and Davos. It is twenty-four miles
from Chur.

Arosa, 6209 feet above the sea, is a small hamlet
at the head of a valley which opens to the west, opposite Langweis, the last village in the Schanfigg Thal,
the valley which leads from Coire to the Strela Pass,
the most direct way from that town to the Davos
valley. From Langweis (4518 feet), five hours from
Chur, there is nothing but a bridle-path to Arosa.
When I visited it in the autumn of 1882, there was
little or no accommodation for winter visitors. The
valley appeared to be well sheltered by surrounding
mountains, and if the means of access to it were improved and suitable accommodation for winter visitors
provided, it might possibly, as has been suggested,
answer as an alternative winter resort to the neighbouring Davos valley, and serve to relieve Davos of
its present excess of consumptive patients.

I am informed that *Andermatt* (4730 feet), on the
St. Gothard road, is being developed as a winter station,
but I have not been able to obtain any details as to
what measure of success has attended the enterprise.

If it should prove suitable as a winter resort it
would have the great advantage of being exceedingly
easy of access, as it is attainable by direct railway

T 2

communication both from the north and south of Europe.

In Germany there are two well-known resorts for consumptive patients during winter as well as summer. One of these, *Görbersdorf*, in Silesia, only 1700 feet above the sea-level, must be regarded as the forerunner of all these mountain winter resorts for the treatment of chest affections, which have now become so popular. Dr. Hermann Brehmer, about five-and-twenty years ago, established his sanatorium at Görbersdorf for the treatment of consumptive patients by mountain air and cold douches, combined with the strictest possible supervision of the diet, exercise, and habits of his patients. The success which attended this enterprise was mainly instrumental in calling attention to the possible value of the climate of high mountain valleys in the treatment of phthisis.

Görbersdorf, near Dittersbach Station, is situated in a long winding valley in the Riesengebirge, and has a somewhat harsh winter climate; but the admirable system of supervision in force there and the great attention paid to every detail of hygiene must, no doubt, be credited with a very considerable share in the good results obtained at Dr. Brehmer's institution.

At *Falkenstein*, in the Taunus range of mountains, at a little distance from Homburg, and about the same elevation above the sea as Görbersdorf, Dr. Dettweiler, who was formerly an assistant of Dr. Brehmer, and who is himself consumptive, has established a similar sanatorium, particulars of a visit to which will be found in the chapter on Rhenish health resorts. This establishment is open winter and summer. It is protected by mountains from north and east winds, and is surrounded by woods which afford shelter from sun and wind, and give great opportunities for out-of-door exercise.

Out of Europe there are several elevated resorts for the treatment of consumption which, though distant, are becoming well known, and are resorted to occasionally by European invalids. In the Peruvian Andes, near the city of *Lima*, there are some elevated resorts which have afforded some of the best results that have ever been obtained from the climatic treatment of phthisis. It must be remembered that in regions near the equator, even at elevations of 7000 or 8000 feet, they have a temperature in winter as high as our own summer temperature, and higher elevations than these have often to be resorted to.

The *Valley of the Jauja river*, in the Peruvian Andes, reaches an elevation of from 8000 to 10,500 feet. The towns of *Jauja* and *Huancayo* are the two chief resorts for consumptive patients from Lima. At Huancayo (or at Jauja, which is cooler) the sky is said to be always clear and sunny, the atmosphere always pure and bracing, and the temperature very equable, the annual range not exceeding 10 or 12 degrees Fahrenheit. It follows that in no European climate can invalids take so much out-door exercise or be so constantly in the open air.

Other resorts in the Andes or the Cordilleras which have been found well suited to the treatment of consumption, actual or threatening, are the following:

Santa Fé de Bogota, the capital of the United States of Columbia, about 10,000 feet above the sea, and in much repute; as also is *Quito*, at about the same elevation.

Cuzco, the ancient capital of Peru, is higher, 11,250 feet; and *La Paz*, in Bolivia, still higher, 12,200 feet.

Consumptive patients from Brazil are usually sent to high resorts in the Cordilleras; and other mountains of the Argentine Republic offer many suitable sites for sanatoria for pulmonary invalids. Of these Dr. Scrivener* has reported very favourably. He says: "the sky is pure azure, and the atmosphere bright and clear, and so very transparent that it enables you to

---

* An article on the climates of the Andine Mountains, and those of the province of Cordova, in the "Revista de Buenos Ayres."

see objects at a distance, making them appear close at hand, although it would require a journey of several days to reach them. The lightness of the atmosphere produces an exhilarating effect, and an increase of energy and activity."

These mountainous districts extend, at higher or lower elevations, from the province of Cordova to the valley of Rimac.

He recommends the mountains of Cordova to consumptive patients in preference to the Andine heights of Bolivia, as they are nearest the river Plate and contain a greater variety of objects to divert the attention and amuse. "The passage from England can be made in thirty-four days. There are several lines of merchant steamers, from London and Liverpool, as well as from Southampton and Bordeaux, which arrive at Buenos Ayres every month. From this port you can embark in a steamer for the city of Rosario, which is most beautifully situated on the banks of the river Parana, at which you arrive in about twenty-six hours. From thence you take the Argentine Central Railway and arrive at the city of Cordova on the same day." The mountainous districts commence there and extend over an area of about a thousand leagues to the valley of Rimac. He adds, "there are few spots in the world where nature has lavished such a variety of animal, vegetable, and mineral productions as the province of Cordova."

The ancient city of *Mexico*, approached by railway from Vera Cruz, lies in an extensive plateau from 6000 to 8000 feet above the sea-level, surrounded by mountain chains, and having a temperate and agreeable climate. The same may be said of the city of *Puebla*, seventy-six miles east-south-east of Mexico, which is also situated on a high plateau, 7215 feet above the sea.

The Rocky Mountains, in Colorado, United States, possess several elevated stations which are used as sanatoria for consumptive patients. *Denver* is the best known of these to Europeans. The rainfall there is said to be small, and it is also said to have three hundred and two clear days in the year, allowing therefore of much out-door exercise. Its mean annual temperature is 47 degrees Fahrenheit, but it is not such an equable climate as that found in the South American resorts just referred to. September and October are the best months for commencing residence there ; the patient then gets gradually acclimatised to the cold in winter, which is at times very severe. The days are warm and bright, but the nights are very cold, and indeed are cool all the year round. There is not much snow, and that falls mostly in early spring. On a third of the days in winter it is warm enough to dine (early) out of doors. There is a good deal of disagreeable wind during the spring months, a great drawback to the climate at this season. From the middle of September to the middle of April there is scarcely any rain. Changes of temperature are often sudden and extreme, and precautions have to be sedulously taken against chill. The city of Denver stands about fifteen miles east of the foot of the mountains, at an elevation of 5200 feet. One of the drawbacks to Denver as a sanatorium for invalids is the fact that it is a populous and rapidly growing city (30,000 inhabitants), with the amusements and excitement inseparable from such a place. It possesses good hotels and lodging-houses. The lower part of the city has a damp soil, and the drainage and water supply have been reported as imperfect. To those, however, who need not only a cure but an occupation or a career, Denver may furnish one.

About eighty miles south of Denver is *Colorado Springs.* In 1880 it had about 6000 inhabitants. Unlike Denver, this place has been selected and laid out for the purposes of a sanatorium. It is situated on a plateau five miles from the base of the mountains, at an elevation of 6023 feet above the sea. So that while Denver is about the altitude of Davos, Colorado Springs is about the altitude of St. Moritz. Its situation is a very sheltered one, and the town being built over an area of four square miles, there is plenty of ground around most of the dwelling-houses. The streets are wide and lined with shady trees. The ground has a very gentle slope from north to south. There is a top soil of two feet resting on seventy feet of sand and gravel, which is very porous, so that there is a perfect natural drainage. A supply of pure water is obtained from the mountain-side six miles off. At present (1880) there is no regular system of drainage, and as no water is taken from the soil, but all brought from a distance, the system of earth-closets is chiefly in use. They are regularly cleaned out by town scavengers. The hotels are not very good, but there are several good boarding-houses and comfortable villa residences. Farm produce is good and of moderate price, but luxuries are dear. There are plenty of horses and carriages to be hired, and the rides and drives around are numerous and interesting. There is no lack of pleasant society, or of churches, schools, and places of entertainment.

Five miles to the west of Colorado Springs, and about 200 feet higher, is the health resort of *Manitou Springs.* It lies directly at the foot of Pike's Peak, in a small valley, sheltered by hills on three sides. It is a village of only 500 inhabitants, but it has three good hotels, several boarding-houses, and a few cottages which can be hired.

In the summer months it is thronged with visitors, as it is somewhat cooler there than at Colorado Springs; it is, however, not quite so dry. It is also said to be warmer in winter, but, owing to the shadow of the hills, it has fewer hours of sunshine. Here are the springs which give the town of Colorado Springs its name. They are six in number, and some contain alkaline chlorides, others are alkaline and acidulous, one is chalybeate, and another aperient. The Indians of the Rocky Mountains used to bring their invalids to these springs of the " *Great Spirit,*" which is the meaning of " Manitou," for they looked upon their curative influence as supernatural. " The great charm

of this resort," says Denison,* "lies in its contiguity
to various places of world-famed interest. Within a
radius of a few miles one may visit the far-famed
Garden of the Gods, Glen Eyrie, Monument Park,
Cheyenne Canon, Ute Pass, and many other attrac-
tive places, or ascend to the summit of Pike's Peak.
By the Ute Pass the traveller may gain access to
South Park, and ·eventually reach Fair Play and
Leadville ; or he may turn aside, after a twenty miles'
ride, and find rest and quiet among the pines of
Manitou Park, 1200 feet higher than the spring."
There is an hotel at Manitou Park which is frequented
in the hot summer months.

The mean temperature at Manitou in winter is
about 27 degrees Fahrenheit, in spring 45 degrees,
summer 68 degrees, and autumn 48 degrees. There
are a great number of clear days, and the relative
dryness of the air is great. There are only sixty to
seventy rainy days in the year ; snow falls on about
forty of these, but does not lie so long on the ground
as in the high Swiss valleys, so that there is no disagreeable snow-melting period. Manitou is more
sheltered from winds than Colorado Springs or Denver;
in the latter place, in spring, the winds are decidedly
unpleasant. There is great variation between sun
and shade and day and night temperature ; so that
these Rocky Mountain climates are somewhat variable
as to temperature, but have a great many clear days
and great dryness of atmosphere. The most agreeable seasons are autumn and early winter ; the spring
is windy and changeable, and the summer rather too
hot; but during the hot summer months invalids
camp out or go to higher resorts.

There are many other resorts in the Rocky
Mountains, but these are, at present, the best known
and most frequented. This region presents a favourable field for those who, being threatened with consumption, feel it necessary to find an occupation while

* "Rocky Mountain Health Resorts."

they, at the same time, are seeking the arrest or cure of their disease, and who have sufficient strength, aptitude, and energy for the life of a colonist.

For those who like a sea voyage and who do not fear a long and somewhat rough land journey, the South African highlands present many attractions, with their very dry and bracing climate, hot summers, and cold winters. These high regions are usually reached by means of ox-waggons, which must be furnished with all the necessaries of life, as this mode of travelling is excessively slow, not averaging more than ten or twelve miles a day. Indeed it is a sort of modified camping-out. The start is usually made from Wynberg, a suburb of Cape Town, or Grahamstown.

Bloomfontein, 4700 feet above the sea, has a considerable reputation as a sanatorium for persons with delicate chests. It is the capital of the Orange Free State. The usual mode of getting there is to go by steamer, either direct from England or from Cape Town, to Port Elizabeth (Algoa Bay). It is usually a very rough sea voyage of two days from Cape Town to Port Elizabeth. From Port Elizabeth there is a railway to Grahamstown, a pretty town in a healthy situation, 1700 feet above the sea. It is 400 miles from Grahamstown to Bloomfontein! There is a coach which leaves every week and performs the journey in five days. Persons are said to have died on this terrible journey! The journey can also be performed in an ox-waggon in a month or five weeks; or it has been recommended as a preferable plan to buy a light spring-waggon and four horses at Grahamstown and drive oneself to Bloomfontein (where they could be advantageously sold), stopping at convenient places on the road. This would take at least a fortnight. Once having overcome the difficulty and fatigue of getting there, Bloomfontein is a civilised town, with tolerable inns and an English bishop! Its climate has been reported to be very valuable in arresting phthisis. It is said to be very dry, having a mountain range between it and the Indian Ocean to the east, and having an extensive plateau of dry, open country to the north and west. Its summers are very hot, its winters very cold, but the dryness of the air enables invalids to bear these extremes of temperature without suffering any injury.

There are many other elevated resorts in South Africa, in the Orange Free State, in Griqualand West, and in the Transvaal, but it is clear that they can only be suited to a very limited class of invalids, and, in our present somewhat unsatisfactory political relations with these countries, it is not likely that any greater facilities for travelling into the interior are likely to be forthcoming.

If there are but few European mountain stations suitable for conversion into winter sanatoria, there are a very great number adapted for resort in summer. In the remainder of this chapter I shall endeavour to mention most of these and to describe briefly the chief of them.

It may be convenient to group them roughly into those having an elevation of over 6000 feet, and possessing therefore a highly bracing climate; those with an elevation of between 5000 and 6000 feet, and having a decidedly bracing climate; those between 3000 and 5000 feet, and having a moderately bracing climate; and finally those below 3000 feet, which still possess in some respects the character of mountain climates, but whose bracingness depends much upon latitude, aspect, and surroundings, which is indeed more or less the case with all these resorts, as has been already explained.

The elevated resorts in the Engadine and the Davos valley have been already described, and some others have been briefly referred to in Chapter II.

The hotel on the *Riffel-berg*, about two-and-a-half hours from Zermatt by a bridle-track, is one of the highest of these resorts in Europe, being 8427 feet above the sea. The accommodation is fairly good, but the hotel is apt to be overcrowded in fine weather

and Alpine climbers are very early risers and make a
good deal of noise in starting on their expeditions, so
that the Riffel is only suitable as a health resort to
those who are simply desirous of recruiting their
strength in very bracing air, and who are able and
desirous of availing themselves of the excellent opportunities afforded there for excursions into the high
Alps.

The *Hospice* on the *Bernina Pass*, 7650 feet above
the sea, about three hours from Pontresina, is one of
the highest spots in Europe where any one has made
a prolonged residence solely for the sake of health.*
The food and accommodation there are both very
fair, and there is a good carriage road passes by the
hospice, providing fairly level walks for those who do
not wish to climb. The air is extremely dry and
bracing, but the nights, as the writer can testify, are
even in summer sometimes excessively cold, and one's
very bed feels as if it had been " iced."

On the *Æggisch-horn*, the *Hôtel de la Jungfrau*,
7362 feet above the sea, offers good accommodation
to those who desire to pass a few weeks in fine
bracing air, close to one of the largest glaciers—the
Aletsch—in the Alps. It is approached from Viesch,
in the Rhone Valley, by a safe and not very steep
bridle-track, in about three hours. Within ten
minutes of the hotel is one of the finest panoramic
views in the Alps—of the great peaks around Zermatt.
It is also the best starting-point for excursions on the
Aletsch glacier, and for the ascent of the higher
peaks of the Oberland.

The hotel on the *Bell-Alp*, on the opposite side of
the Aletsch glacier to the Æggisch-horn, is 7153 feet
above the sea. It is approached from Brieg, the
terminus of the Rhone Valley Railway, by a mule-
path in four or five hours. The path is steep in
points, and the ascent fatiguing to only moderate

* See " Safe Studies," by the Hon. Lionel Tollemache. Article
on " The Upper Engadine."

walkers. Its nearness to a railway station makes it, perhaps, the most accessible high mountain resort in Switzerland. The drawback, so far as invalids are concerned, is the difficulty in finding level walks. There is but one horizontal walk, and that rather a rough one, to the village of *Nessel*, high above the Rhone valley, and commanding a fine view. Excursions on the glacier involve a very steep descent in going and a very fatiguing ascent in returning. The ascent of the Sparrenhorn, a pyramidal summit (9889 feet) rising just behind the hotel, is a walk of two-and-a-half hours, and from the summit there is a magnificent near view of the great Aletsch Horn, and a glorious panorama of the peaks to the south of the Rhone valley. The food at the hotel is not always of the best, and it is difficult to get such simple necessaries as *good* bread and butter and *fresh* eggs.

A good hotel at rather a lower elevation—that on the *Reider Alp*, 6388 feet above the sea—can be reached in two-and-a-half hours from the hotel on the Æggisch-horn. It can be ascended from Mörel, in the Rhone valley, in three hours. It is better adapted for invalids who wish to make a protracted stay in a high mountain resort than its two higher neighbours.

There is a small hotel—*Hôtel Grindelwald-Rigi*—7190 feet above the sea, on the extreme north spur of the Wengern Alp, called the *Männlichen*, where there is good accommodation in the midst of magnificent scenery. It can be ascended from Grindelwald in three or four hours.

There are two hotels on Mont Pilatus, either of which might be selected for a protracted stay in fine weather. The highest, the *Hôtel Bellevue*, 6790 feet above the sea, is situated on the ridge between the two highest peaks, and is only eight minutes' walk from the summit—the *Esel*. The other, the *Hôtel Kleinsenhorn*, 5935 feet, is built on the saddle connecting two of the other peaks—the Oberhaupt with the Kleinsenhorn.

There is a bridle-path from Alpnach, on the Lake

of Lucerne, to the Hôtel Bellevue in four to eight hours, and from Hergiswyl to the Hôtel Kleinsenhorn in three-and-a-half hours.   There is the obvious objection to all these resorts on or near mountain peaks, that in bad or misty weather there are no facilities for taking exercise such as can be had in a great mountain valley like the Engadine.

The *Hôtel Alpenrose* is only a few hundred paces below the *Schynige Platte*, which is 6791 feet above the sea.   This commands one of the finest points of view in the Bernese Oberland.   It is ascended from Interlaken in four to five hours.

There is a fair Inn on the *Great Scheideck*, 6434 feet above the sea, reached in three hours from Grindelwald.

The *Hôtel Jungfrau* on the *Wengern Alp* has good accommodation ; it is 6184 feet above the sea, and can be reached in less than three hours from Lauterbrunnen.

The hotel on the *Engstlen Alp*, 6033 feet above the sea, offers a pleasant resting-place amidst beautiful surroundings and the grandest mountain scenery.   It is on the road from Meiringen to Engelberg by the Joch Pass, and is about five hours from the former place.

Arolla, 6572 feet above the sea, with a fair hotel, (Hôtel du Mont Collen), is about three-and-a-half hours from Evolena by a bridle-path, and is situated at the top of the Val d'Hérens in a fine position opposite the grand pyramid of Mont Collen.   It is surrounded by a wood of "Swiss stone-pines," here called "Arolla."

It is a fine bracing locality, fairly accessible, about eight or nine hours from Sion railway station and well suited to those who wish for exercise in mountain air with quiet surroundings.

On the southern side of the Alps, near Airolo, on the St. Gothard Railway, the *Hôtel Piora*, 6001 feet above the sea, in the valley of that name, offers a charming mountain retreat.   It is built on a hill, in a sheltered position to the left of a sequestered lake—Lake Riton—in the vicinity of pine-woods and with good points of view close at hand. It is about three hours' walk from Airolo.

There are many resorts to choose from amongst the next group at an elevation of between 5000 and 6000 feet.

Perhaps the best known of these are the several resorts on the Rigi, which vary in elevation from the *Hôtel du Rigi Külm*, about 5800 feet, to the *Hôtel du Rigi Kaltbad*, 4728 feet above the sea. At intermediate elevations are the *Rigi Scheideck*, 5400 feet, the *Rigi Staffel*, 5200 feet, and the *Rigi First*, 4750 feet. These resorts are so well known that they do not need to be described. It may be said, however, of the Kaltbad that it is an exceedingly good station for invalids requiring pure mountain air without running the risk of exposure to great cold; especially as an *after cure* from some of the German baths. It enjoys protection from the colder winds and is well exposed to the sun. The cold spring used here for baths issues from the rock at a constant temperature of 41 degrees Fahrenheit, and consists of very pure water. The hotel is admirably fitted up, the food and accommodation are excellent, and the surrounding grounds prettily laid out. Living there is more expensive than at less well known resorts.

The *Rhone Glacier Hotel*, at an elevation of 5750 feet above the sea, is nearly as high as that of the Rigi Külm. It is well managed, and affords excellent accommodation for visitors with the opportunity of breathing fine bracing glacier air. It is apt to be overcrowded in the height of summer, and is, perhaps, too much resorted to by the passing tourist to be altogether suitable for the prolonged residence of invalids.

I have already described *Santa Catarina* (p. 53), 5700 feet above the sea, and spoken of the beauty and attractiveness of its situation.

There is a small clean inn at *Guarda* (5413 feet), in a fine commanding position near Lavin, in the Lower Engadine, not far from which is another village,

Fettau, above Ardetz, also having an elevation of about 5400 feet; but neither of these offers any very special attractions as a health resort, unless it be its fine situation and the very simple and quiet life that may be led there.

The village of *Mürren*, 5348 feet above the sea, is ascended from Lauterbrunnen by mule-track in two-and-a-half hours.

It is very finely situated, opposite the precipitous western face of the Jungfrau, and is surrounded by very grand mountain scenery.

It is very popular as a bracing resort in summer, and the Grand Hotel and Kurhaus make up 250 beds. The whey and milk cure can be taken here, and baths and douches may be had.

It is not so dry and bracing a situation as the Engadine, and it is more shut in; but it is well adapted to those who find a somewhat milder mountain air better suited to their constitutions.

On the south side of the Bernardino Pass, about thirty-six miles from Thusis, we find the village and baths of *San Barnardino*, at a height of 5334 feet above the sea. It possesses a chalybeate spring, which is taken internally and used for baths. The place is much frequented by Italians during July and August. The hotel accommodation is fair, and the surrounding scenery is attractive.

There appears to be singularly little known as to its climate; it is probably warmer and moister than similar elevations on the northern slopes.

Zermatt, 5315 feet above the sea, is well known to all mountaineers. It is in the midst of some of the finest mountain scenery in Europe; but it is more suitable for the active and hardy tourist than for the valetudinarian, unless he be a muscular one, only needing rest of mind in fine invigorating air and amidst grand scenery. It is twenty-five miles from a high road, and is reached by mule track.

Quite out of the beaten track, in the Adamello district of the Italian Tyrol between Dimaro and

Pingolo, at an elevation of about 5000 feet, in a beautiful situation, is the *Hôtel* of *La Madonna di Campiglio,* close to the pilgrimage church of the same name. It is accessible by a frequented bridle-track from either Dimaro or Pingolo. Dimaro is an hour's walk from Malé, the principal village in the Val di Sole.

There are fine views and attractive excursions around, but it is too isolated and remote for invalids.

Macugnaga, 5115 feet, in the Val d'Anzasca, one of the most beautiful of the southern valleys of the Monte Rosa chain, affords excellent accommodation amidst the finest scenery. Lying on the south side of this great Alpine chain, it enjoys a milder climate than places of the same elevation on the north of it, but it should be well suited to those invalids who require a mildly bracing climate, with, however, decidedly tonic properties.

There are a very great many resorts which have to be included in the next group, viz. those between 3000 and 5000 feet above the sea. I shall mention most of these and describe the more typical ones.

Some of the highest of this group are found in the Dolomites. These are rather hot in mid-summer, and are better suited·for the earlier or later part of the season. They can scarcely be recommended to invalids from England ; but for those who are already in Italy, and who do not wish to travel north, they may serve as useful summer quarters. The accommodation is, perhaps, a little rough, and fastidious and very delicate persons had better avoid them.

Schluderbach (4820 feet), *Höhlenstein* (4750 feet), and, considerably lower, *Cortina* (4050 feet), in the Ampezzo-Thal, are all on the high road from Belluno, through the Val d'Ampezzo, to Toblach, in the Pusterthal.

In a beautiful situation, looking into the Primiero valley, is the new *Hôtel San Martino di Castrozza,* nearly 4900 feet above the sea. This is quite out of the beaten track.

U

Hospenthal and *Andermatt*, near one another, on the St. Gothard road, between 4700 and 4800 feet, more bracing than the preceding, and extremely accessible, while they offer very great facilities for interesting excursions in all directions.

The *Hôtel Alpenclub* (4790 feet), in the neighbouring and very beautiful Maderaner-Thal, which branches off from Amsteg, is in an admirable situation for a quiet health resort. It can only be approached by a bridle-path, and so is quite out of the way of the crowd of tourists, and it has fine pine woods close to it, affording both shade and shelter.

The *Baths of Lenk*, at the foot of the Gemmi, are at an elevation of over 4600 feet, and afford excellent hotel accommodation in a situation easy of access.

Morgins, 4628 feet above the sea, is situated in the Val d'Illiez, and is approached by a good carriage road from Monthez, in the Rhone valley, near St. Maurice. It is not far from Champéry, but it is more than 1000 feet higher. It is well protected from wind, and is freer from mist and fog than the somewhat lower elevations, and its air is decidedly bracing. It has a chalybeate spring, which is used in baths as well as drunk. It is rather too much enclosed to be bright and cheerful, and there is no distant view of snow mountains. It may be as well to notice *Champéry* in connection with Morgins, as it is also in the Val d'Illiez, and is reached from Monthez in three-and-a-half hours. It is the highest village in this valley (*Morgins* being in a side valley which opens to the west), and has a beautiful and cheerful situation. There are fine points of view in the neighbourhood, especially of the *Dent du Midi*. There are several good hotels and *pensions*, and these are much frequented by English people.

Champéry can be recommended as a bright, cheerful, and accessible mountain resort; but the climate is not very bracing, and it is at an elevation (3390 feet) where mists occasionally settle for a few hours at a time, especially in wet seasons.

Evolena, 4521 feet above the sea, is approached from Sion in the Rhone valley by carriage road through the Val d'Hérens. The post vehicle takes five or six hours to perform this journey of eighteen miles. It has a good hotel (Hôtel de la Dent Blanche) and is picturesquely placed in a broad, grassy valley, surrounded by pine-clad hills, beyond which are snowfields and glaciers. There are several glacier and mountain excursions to be made from Evolena, as well as more modest expeditions for those who do not feel qualified for work in the high Alps.

Comballaz, 4416 feet, three miles from Sepey and fifteen miles from Aigle in the Rhone valley, is an easily accessible mountain resort, and has good accommodation for visitors making a long stay. Beautiful excursions can be made in the neighbourhood.

Not far from Comballaz (about six miles) is *Ormond Dessus*, five hours from Aigle by diligence. It lies lower than Comballaz (3832 feet). It has a good hotel and *pension* (*des Diablerets*), and affords the most convenient head-quarters for exploring this interesting neighbourhood. It is rather hot here sometimes in the height of summer, but the air is pure and the surrounding scenery very beautiful.

In a fine situation, about three hours' walk from Brunnen, on the Lake of Lucerne, is the *Stoos Hotel and Kurhaus* at an elevation of 4342 feet. The accommodation is fairly good, the air clear and bracing, and the surroundings extremely beautiful. Lower down and about an hour nearer Brunnen is the well-known Kurhaus Axenstein (2330 feet), and a little lower the Hotel Axenfels. These command one of the finest views of the Lake of Lucerne in its grandest part. They can scarcely be called bracing resorts, but they are much more agreeable and healthful summer residences than the very hot and close town of Lucerne.

Weissenstein, 4213 feet above the sea, is a very

accessible mountain resort, being but three hours' walk from Soleure and about the same distance from the railway station of Münster in the Münster-Thal, near Bâle. There is a large Kurhaus there, surrounded by woods and pastures. It is generally full in the summer.

"No spot commands a better view of the whole Alpine chain from Tyrol to Mont Blanc."

Villars, above Ollon (4166 feet), is approached from Aigle. There is a diligence daily, which takes three to four hours to perform the journey. It commands fine views of the Rhone valley and the surrounding mountains, and has pleasant park-like grounds around, which offer many agreeable excursions. It is built on a plateau open to the south and protected by wooded hills from the north. It is somewhat exposed to mists in wet season, but is a most agreeable and bracing resort in fine ones.

A little below Villars there is good hotel accommodation at *Chesiére* (3970 feet), which commands a beautiful view.

The *Schröeken* (4134 feet) is in a wild and grand situation in the Bregenzer Wald, between Bregenz and Arlberg. It is not very accessible; it can, however, be reached in about five hours by bridle-path from Stuben in the Kloster-Thal. Numerous interesting excursions can be made into the mountains which surround it. It lies on a green hill, rising from the bottom of a vast basin which is surrounded by mountains 7000 to 8000 feet high.

In the same district is *Mittelberg* (3980 feet) the chief place in the Kleiner Walser or Mittelberger-Thal. It is reached by carriage road in four hours from Obertsdorf (2666 feet), itself a favourite summer resort amidst beautiful scenery. It is eight and three-quarter miles from the railway station of Southofen, which is five-and-a-half miles from Immenstadt.

Courmayeur (3986 feet), on the southern side of Mont Blanc, at the head of the Val d'Aosta, may be approached from Chamouni by mule-track, a journey which usually takes three days, and forms part of the *Tour du Mont Blanc.* It can also be reached by a good carriage road across the Little St. Bernard from Bourg St. Maurice in the valley of the Isère. It is somewhat inaccessible, and on that account is less frequented by invalids from the north than it deserves to be. It affords excellent hotel accommodation, it is surrounded by the grandest scenery, and its climate is mild, equable, and fairly dry. It is protected from the north by the Mont Blanc range. It is sometimes very warm in the height of summer, but the mornings and evenings are cool, and the dryness of the air renders the mid-day heat more pleasant and more wholesome. It is much frequented by Italians for its mineral waters.

I have known patients with chronic bronchial catarrh find the summer climate of Courmayeur extremely agreeable and useful to them.

Alagna (Hotel Monte Rosa), 3953 feet above the sea, is a mountain resort much frequented by the Italians, in a beautiful situation, eight hours from Macugnaga, over the Turlo Pass. It comes in the Tour of Monte Rosa from Macugnaga to Zermatt. There are also fairly good hotels at two other summer resorts which one passes in this tour, viz., *Gressonay la Trinité* (5322 feet) and *Gressonay St. Jean* (4495 feet).

Bad Gurnigel (3783 feet), a walk or drive of three-and-a-half hours from Thun; also from Berne by diligence in five hours. This is a favourite Swiss health resort, and much frequented for its sulphur baths as well as its mountain air. It is situated on a broad plateau, and is surrounded by woods, affording numerous shady walks. The picturesque country around has quite a park-like aspect. The atmosphere is not very dry, and a rather high relative humidity has been registered there; this is due, probably, to the great extent of surrounding forest. The establishment is open from the beginning of June to the end of September. There are two sulphur springs and a chalybeate one.

St. Beatenberg (3770 feet), above the Lake of Thun, is accessible by a somewhat steep path from Merlingen, on the north shore of the lake in an hour and a half, and from Interlaken by carriage road in two-and-a-half hours. Besides the Kurhaus kept by Dr. Müller, there are several good hotels and *pensions*. It lies in a sunny sheltered situation, with level walks in adjacent woods, and it is a very suitable station for delicate invalids during summer and autumn. For nervous temperaments requiring mountain air but unable to bear an exciting climate it is well adapted.

It commands a fine view of the Alps from the Finsteraarhorn to the Niesen, including the Eiger, Mönch, Jungfrau, Blumlis Alp, Doldenhorn, etc. There are also some adjacent higher points of view readily accessible to fair walkers.

The climate is mild and not very dry.

The *Kurhaus-Waldhaus-Flims* (3620 feet), about one-and-a-half miles from Flims in the Vorder Rhein Thal, traversed daily by diligences from Chur to Oberalp, is beautifully situated in the midst of pine and beech woods, which contribute to render the climate mild and windless, and impregnate the air with the resinous emanations from the pine trees. The walks through the woods are numerous and interesting. There is a small lake a few hundred feet below the Kurhaus where boating and bathing can be indulged in. This place is usually very crowded during July and August, but earlier or later in the season it offers an admirably quiet and secluded retreat. It is well adapted as an "after cure" for patients who have gone through a serious course of mineral waters. Some sixteen miles higher up the valley is *Dissentis* (3773 feet) in a somewhat exposed situation on the high road.

The village of *Gryon*, seven miles from Bex in the Rhone valley, is accessible by carriage road. It is a considerable village, being 3707 feet above the sea, and has two tolerably good *pensions*.

Château d'Oex (3498 feet), accessible from Bulle by

diligence in four-and-a-half hours—eighteen miles. It has several good hotels and *pensions*, and is a pleasant and popular summer resort, in a green open valley. It is moderately bracing. It is only seven miles from Château d'Oex to *Saanen* (Gesseney), 3556 feet above the sea, with moderate accommodation, in the centre of the Gruyère cheese manufacture.

The baths of *Lenk* (sulphur), near the village of the same name, 3527 feet, lying between Thun and Sion, in the Rhone valley. It is reached by diligence from Thun in eight hours. It is finely situated near the head of the Simmenthal, at the base of a mass of snowy mountains. The accommodation is good and the mountain excursions to be made from it numerous and interesting.

Chaumont, 3585 feet above the sea, and 2148 above the Lake of Neuchâtel. The hotel is reached in a drive of two hours from Neuchâtel. A forest of fir trees stretches down to the shores of the lake. There are good level walks around the hotel. From the " signal," 200 feet high, fine views over Switzerland can be obtained. There is good accommodation for over one hundred guests in the hotel, which is protected towards the north, but is somewhat exposed to the east and west. The weather is sometimes cold and foggy, and the north-east wind occasionally makes itself felt unpleasantly.

The following account of *Soglio,* 3569 feet above the sea, and *Promontogno,* nearly 1000 lower (2681 feet), I published, after staying at both places, in 1882 : *

About midway in the descent from the Upper Engadine to the Lake of Como, in the most beautiful part of the beautiful Val Bregaglia (Prægallia?), we come upon a few scattered houses, which compose the village of Promontogno. A low rocky promontory, jutting out here from the mountains on the left-hand side of the valley, seems, as one descends, to form an almost complete barrier across the valley ; and, indeed, the present carriage road is tunnelled through it. This part of the

---

* *Pall Mall Gazette.*

valley was called "La Porta," and formed a sort of natural
gateway between its upper and lower portions.   Coincident
with this natural geographical division of the valley there is an
almost abrupt change in the character of its vegetation, and we
pass from fir trees on one side of it to chestnuts on the other.
Fine ruins of an old castle surmount this promontory, which
must at one time have been an important fortress on the con-
fines of Switzerland and Italy.   The present political boundary
between the two countries is fixed some two or three miles lower
down the valley: The Val Bregaglia, after beginning by its
extraordinarily sudden drop at the Maloja (in going from the
Engadine), descends gradually for about sixteen miles until it
joins the road over the Splugen to Como at Chiavenna.   We
pass in this descent a series of villages, Casaccia (4790 feet above
the sea), Vico Soprano (3566 feet), Borgo-Nuovo (3471 feet),
Stampa (3379 feet), and then Promontogno (2681 feet).   Here
the vegetation changes, and we begin to feel that we are wend-
ing southwards.   Fine chestnut groves and patches of golden
corn make their appearance, the warmth of the sun and the
dryness of the air make the fir-woods fragrant and sweet beyond belief.

Besides the few houses that compose the village of Promontogno, there has sprung up there within the last four or five
years a large building, the "Kur-Hotel Bergellerhof."   This
hotel, which now forms a conspicuous object in passing through
the valley, is built just below the promontory of which we have
spoken, and close to the entrance of a remarkable lateral valley
running south of the main valley—the Val Bondasca.   It would
be difficult to convey an adequate idea of the beauty of this
valley as seen from Promontogno and from the road which
winds up therefrom to the mountain village of Soglio.   The
entrance to this valley is so narrow as to allow only of the
passage of the turbulent stream that rushes out of its deep bed
to join the Mera, which flows through the main valley, and the
path into its recess has to be reached by climbing up the mountain side.   Chestnut and fir woods clothe the mountains on each
side, and as the valley widens a fine glacier comes into view at
its head, the Bondasca glacier.   This valley is closed in by
jagged mountain summits, most striking from the rugged
grandeur of their bare precipitous granite walls and the beauty
of their colour and outline.   The road to Soglio begins to wind
up the right side of the valley almost immediately beyond the
village of Promontogno, and directly opposite the opening into
the Val Bondasca, of which, as the road ascends, it commands
fine views.

Soglio is a mountain village in a most picturesque situation on the right side of the Val Bregaglia, between Promontogno
and the next village, Castasegna.   It is built on a small plateau at
an elevation of 3569 feet above the sea, where it enjoys complete

protection from the north and the east by surrounding mountains, and is open only to the south-west as it faces down the valley towards Chiavenna and Como.

It is stated to possess an unusually mild, agreeable, and equable climate through summer and winter. But in spring— that is, in the month of May—it is exposed to storms of wind and rain ; in the month of April its climate is good. Rarely in winter does the thermometer register as much as three degrees of frost. Soglio, for a mountain village, is so accessible and so directly in the line of route which seekers after health so frequently traverse, that it is perhaps a little remarkable that it has not yet become a "health resort." It is true that it has an "Hôtel et Pension Giovanoli," which announces "Cures d'air, de lait et d'eau," and " en Septembre cures de raisins." But the " cure de raisins," the manager informed us, could not at present be carried out, as, owing to the ravages of the phylloxera, the vine cultivators in the adjacent Val Tellina would not allow of the exportation of any grapes. A good carriage road has been made from Promontogno up to Soglio, ascending in zigzags rapidly the whole way, so that it can be reached from the former place in one hour's walking or driving, while a good footpath connects it with Castasegna, running nearly the whole of the way through beautiful chestnut-woods, and passing a fine waterfall, the "Acqua di Stoll," on the road. The hotel of the place consists of a fine old palace belonging to the illustrious family of Salis-Soglio. It is an interesting old mansion, with spacious rooms, stone passages and corridors, quaint old furniture, and a cheerful garden, where the Alpine cedar is found in juxtaposition with the chestnut. There is a little library in the "Sala," in which Goethe, Schiller, Rousseau, Byron, and other modern classics are represented. Altogether it is a pleasant, well-managed place, where one might well spend a week or two of quiet repose in passing down from the Engadine or up to it from the south. Seven francs a day at present pays for everything—food, wine, lodging, and attendance ; and telephonic communication with Promontogno connects this mountain village with all the rest of the world. There are many pleasant walks in the neighbourhood, and some interesting and stiff mountain climbs over passes into the Aversthal on the north and into the Val Tellina on the south, for those who like such expeditions. The natives of this valley seem an intelligent, thriving people, of simple life and manners, and the grave Madonna-like beauty of some of the young girls is very striking.

The Italian communities of this valley are remarkable as being exclusively Protestant. On the Sunday, after Church service, they most of them engage in simple outdoor games. The males—divided into three sets according to their ages, men, youths, and boys—devote themselves to the game of bowls ; the young girls, joining hands under the shade of a chestnut or walnut tree, sing some simple accompaniment to their games; and the older women sit in the ironwork balconies calmly looking on. As evening approaches the cows come slowly in to be milked, while the goats come hurrying down the hillsides, pushing and jostling one another as if they had some important business waiting for them for which it was necessary they should be in time.

It is scarcely necessary to mention Chamounix, at an altitude of 3445 feet, which is too much overrun by tourists to render it an eligible resort for invalids. It may, however, be said that its climate is tolerably bracing, and that for those who like crowded hotels and exciting excursions it offers a congenial and healthful residence in the centre of magnificent scenery. The same remark applies to *Grindelwald* in the Oberland (3468 feet) which lies in a healthy and sheltered situation, but has the drawback of being constantly overcrowded during the summer season.

St. Cergnes (3460 feet) at the foot of the Dôle, the highest summit of this part of the Jura chain, is approached by a good carriage road from Nyon, on the Lake of Geneva. Its accessibility and its picturesque situation have rendered it a very popular mountain resort in summer, especially for the Genevese. It is also the spot from which the Dôle is ascended, the ascent taking about three hours ; the view from the summit in clear weather has been described as " wonderfully extensive and admirably diversified."

St. Cergnes is built at the bottom of a gorge exposed to the east and is surrounded with pine-woods. Dr. Lombard * of Geneva describes its climate as " essentially tonic, very useful to all those who are feeble and infirm, but too irritating for persons impressionable to cold winds. For catarrhal, asthmatic, and consumptive patients it is unsuited."

Höchenschwand (3326), in the Black Forest, is one of the highest villages in the duchy of Baden ; it has

* " Les Climats de Montagnes."

good accommodation, and commands a magnificent. view of the Alps. It has recently come into notice as a health resort. It is twelve or thirteen miles from Alb-bruck—a station on the Bâle-Waldshut railway.

Engelberg (3314 feet) has much to recommend it as a mountain health resort, having a mild but, at the same time, somewhat bracing climate, adapted especially to nervous invalids who cannot support the exciting air of higher regions for any period at a time. It is very accessible, being connected with Stanstadt on the Lake of Lucerne (forty minutes from Lucerne by steamboat) by a good coach road fourteen miles in length, which is traversed twice daily by diligence in three hours and a half.

The hotel accommodation (Sonnenberg) is excellent, and baths and douches are provided there. Engelberg lies in a bright green valley almost completely surrounded by high mountains, and therefore much protected from winds. The valley is six miles long and about a mile broad. It is often resorted to by patients who have wintered in the south, and who require a mildly bracing climate for the summer months.

The *Hôtel des Avants* (3212 feet), about six-and-a-half miles from Vevey, ascending by a good carriage road ; or from Vernez-Montreux in rather more than an hour and a half. It lies near the foot of the Col de Jaman, about 2000 feet above the lake, open to the south-west, but closed in by mountains in all other directions. It is entirely protected from the north and north-east winds. Les Avants has a pleasant mild mountain climate, but it will appear to some rather shut in. Mists, too, will sometimes hang over it for hours in the early part of the day. This is indeed the disadvantage of all such elevations in the immediate vicinity of extensive lakes.

The hotel is well conducted, and is generally full in summer. Some invalids have wintered there.

Bad Fideris and *Klosters*, both between 3000 and 4000 feet above the sea, are fairly accessible summer-resorts in the Prættigau valley, between Landquart and Davos. Nearer Landquart is

Seewis, about 3000 feet above the sea. This resort, if the accommodation were a little improved, ought to be very serviceable for residence during the snow-melting time in spring, and early summer for those who have passed the winter at Davos or in the Engadine. For the former it is very accessible, as it is only about half-an-hour's drive from the village of Pardisla, four miles before reaching Landquart from Davos.

The village has a complete southern exposure, and commands an extensive view over the lower half of the Prættigau. It is built on the southern slopes of the Vilan, which descend very steeply into the valley below Seewis, thus affording most perfect surface drainage during snow-melting or after heavy falls of rain. Pleasant walks through meadows and forests stretch up the mountain and around the village on all sides. There is a new, clean, spacious look about Seewis, which is due to the fact that more than half of it was burnt in 1863, and was then very substantially rebuilt.

It is protected from the north and east by the Scesa Plana, a mountain 10,000 feet high, and it has a mild and equable climate exceedingly free from wind.

One who resided some months there* is enthusiastic in describing its attractions.

"In June," he says, "the air at night seemed at once balmy and bracing, if such a combination be possible ; it was neither too hot nor too cold, but so soft and light as to render respiration a delight. . . . One great charm of Seewis is that, within very easy distance, there are, in every direction, beautiful walks, commanding glorious mountain views."

Seewis no doubt merits more attention as a health resort than it has hitherto received.

There are a vast number of other health resorts at

---

* S. J. Capper, " Shores and Cities of the Boden See."

lower elevations in Switzerland, in the Tyrol, in Germany, especially in the Black Forest, in Scotland, in Wales, and elsewhere, but although many of these are admirably bracing resorts, they scarcely present the special qualities of mountain climates, or only in a very modified degree.

The following list presents a choice of such resorts :

Gais, in Canton Appenzell, a pretty village, devoted in summer to the milk cure, and having an elevation above the sea of 3060 feet.

Le Prese, near the Lake of Poschiavo, on the southern side of the Bernina Pass; has sulphur baths ; is hot in summer. Elevation, 3000 feet.

Schönfels and *Felsenegg*, above Zug. Elevation, 3000 feet.

Achensee, a beautiful lake in North Tyrol, about sixty miles from Munich, with an hotel (*Achensee-hof*) much frequented in summer. Elevation, about 3000 feet.

Schluchsee, in the Black Forest, nine miles from St. Blasien ; prettily situated in the midst of a fine forest, and popular as a summer resort. It has a bath establishment for warm baths. Elevation, 2958 feet.

Baths of Weissenberg, in the Simmenthal, about fifteen miles from Thun ; diligence in three hours. Has a local reputation for treatment of chronic bronchitis and early phthisis. Sheltered pine-forests, and baths like those of Leuk. Elevation, 2940 feet.

St. Margen, in the Black Forest, not far from Freiburg, is in an attractive and healthful situation, and is much frequented in summer. Elevation, 2910 feet.

Magglingen, on the slope of a mountain, above Bienne and its lake. Elevation, 2900 feet.

Uetliberg Hotel, above Zurich, nearly always full of excursionists. Elevation, 2860 feet.

Burgenstock, with a fine view over the lake, is an hour and a half's drive from Stanstead, on the Lake of Lucerne ; good hotel and shady walks. Elevation, 2850 feet.

Seelisberg, a well-known and popular resort above the Lake of Lucerne. Elevation, 2770 feet.

Frohburg, near Olten. Elevation, 2770 feet.

Teufens, in Appenzell. Milk cure in summer. Elevation, 2740 feet.

Wildbad Kreuth, in North Tyrol, forty-five miles from Munich, is much frequented in summer, has iron and sulphur springs and salt baths. Elevation, 2720 feet.

Vorder-Todtmoos, in the Black Forest, about ten miles from St. Blasien, in a picturesque situation. It has a pilgrimage church much resorted to. Elevation, 2693 feet.

Weissbad, three-quarters of an hour from Appenzell. Elevation, 2680 feet.

Obertsdorf, nine miles from Southofen station, and about fourteen from Immenstadt, in the Bavarian Alps ; a favourite summer resort. Elevation, 2666 feet.

St. Gervais village, about half-an-hour from the Chamounix road near Sallenches. Elevation, 2657 feet. The *baths* are lower (2066 feet).

Heiden, above Rorschach, on the Lake of Constance. Milk cure in summer. Elevation, 2645 feet.

Voranen, at the foot of the Glürnisch, and about eight miles from Glarus, on a beautiful little lake, the *Klön See*, and amidst fine scenery. Elevation, 2640 feet.

Walchensee, a small village near the shore of the extremely picturesque lake of that name in the Bavarian Tyrol. It may be approached from Munich or Innsbruck. Elevation, 2630 feet.

Schliersee, on the lake of that name in the Bavarian Alps, between Innsbruck and Munich, and thirty-eight miles by rail from the latter ; a beautiful spot, frequented in summer. Elevation, 2588 feet.

Waidring, four-and-a-half miles from the St. Johann station, on the Salzburg-Tyrol railway, and on the high-road between Worgl and Reichenhall ; much frequented in summer. Elevation, 2562 feet.

Appenzell. Milk cure in summer. Elevation, 2550 feet.

St. Blasien, in the Black Forest, sixteen miles from Alb-bruck and twenty from Waldshut stations ; a popular summer resort in a protected situation, surrounded by pine-clad hills ; open also in winter ; fine views. Elevation, 2532 feet.

Zell-am-See, about fifty-two miles from Salzburg by Reichenhall and the valley of the Saale, a beautiful retired situation on the shore of the Zeller-See. Elevation, 2469 feet.

Thusis, three hours from Chur, at the entrance to the Via Mala. Elevation, 2450 feet.

Glion, the " Rigi Vaudois," connected by " funicular " railway with Territet on the Lake of Geneva. Elevation, 2400 feet.

Langenbruck Kurhaus, on the summit of the Obere Hauenstein. Diligence from Oensingen station (eight miles from Olten) twice daily in an hour and three-quarters ; a pleasant retreat. Elevation, 2356 feet.

Monnetier, on the Salève, near Geneva. Elevation, 2336 feet.

Heustrich-Bad, near Thun. Omnibus daily at 4 p.m. Has saline and sulphur baths. Elevation, 2303 feet.

Schonbrunn, a hydropathic establishment near Zug. Elevation, 2300 feet.

Mürzzuschlag, at the foot of the Semmering Pass. Elevation, 2246 feet.

Triberg lies in the heart of the Black Forest; it has a station on the railway between Offenburg and Constance. There is a fine waterfall close at hand. It is a pleasant and popular summer resort. Elevation, 2245 feet.

Obstalden, two-and-half miles from Mühlehorn station, on the Lake of Wallenstadt. Elevation, 2237 feet.

Charnex, beautifully situated above Montreux. Elevation, 2231 feet.

Hohwald, in the Vosges, nine miles by carriage-road from Barr railway station, in a sheltered and picturesque situation ; much frequented in summer. Elevation, 2198 feet.

Stachelberg (*Baths of*), about a quarter of a mile from the railway terminus in the Linththal. Elevation, 2178 feet.

The *Geissbach Hotel*, above the Lake of Brienz, and near the celebrated falls, with good accommodation, but overcrowded with tourists. Elevation, 2166 feet.

Kochel, on the Kochelsee, in the Bavarian Alps. Elevation, 1963 feet.

Starnberg, on the lake of that name (also called Würm-See) ; about seventeen miles from Munich by rail ; crowded in summer. Elevation, 1945 feet.

Rippoldsau; the most frequented of the Kniebis baths, with many pleasant walks. Elevation, 1856 feet.

Cheimsee Hotel, fifty-six miles from Munich and sixteen from Rosenheim, on the line of railway between Munich and Salzburg. Elevation, 1745 feet.

Griesbach, one of the Kniebis baths, in the Black Forest, six miles from Rippoldsau. It has a chalybeate spring and a large bath establishment. Elevation, 1626 feet.

Autogast, three miles from Oppenau station, one of the Kniebis baths, in the Black Forest, beautifully situated. Elevation, 1585 feet.

Divonne, three-and-a-half miles from Coppet, on the Lake of Geneva, a well-known hydropathic establishment. Elevation, 1543 feet.

CHAPTER VII.

MONTREUX, MERAN, AND THE GRAPE CURE.

MONTREUX, situated on the north-eastern shore of the Lake of Geneva, at its grandest and most picturesque part, has long been one of the most popular of Swiss health resorts. The district known as Montreux really comprises several villages, stretching from Clarens to Veytaux, and includes Les Bassets, Clarens, Vernex, Montreux, Torritet, and Veytaux; while Glion lies on a small mountain plateau about 1000 feet above Montreux, and Les Avants is still higher up in the mountain—about another 1000 feet.

The village of Montreux itself enjoys a more sheltered position than any of the others named. The indentation of the lake, which is here called the Bay of Montreux, is protected by the mountains around from the north and the east winds, and in some degree from the north-west wind, so that it is said to be, with the exception of Bex, the most sheltered place in Switzerland. It is also the hottest of all the Swiss stations north of the Alps except Sion, but that applies only to the summer and spring, as Montreux is warmer than Sion in autumn and winter. The "bise"—the cold north-east wind—is not nearly so much felt at Montreux as at Geneva and Morges; and it has been noticed, during the prevalence of a "bise," that it has been intensely cold at Geneva (temperature 14·3 degrees Fahrenheit) and Morges (temperature 18·0 degrees Fahrenheit), while at Montreux (temperature 23·6 degrees Fahrenheit) the air has been almost calm and not disagreeably cold. There are also less variations of temperature at Montreux—a smaller range between the maxima and minima.

The barometriç variations are not nearly so great as on the sea-coast, the maximum oscillation being 50 mm.

The climate of Montreux, though open, dry, and sunny, and with a considerable number of clear days, has a large rainfall, and must not be regarded as a warm winter climate, as the following meteorological details will show.

Mean temperature observed during seven years at the hours of :

	7 A.M.	1 P.M.	7 P.M.
October	47·0° F.	56·6° F.	49·0° F.
November	38·7	47·5	41·0
December	34·3	41·5	35·9
January	31·8	39·7	33·9
February	35·0	44·0	38·1
March	37·0	45·2	39·7
April	46·0	57·6	50·4

or

		7 A.M.	1 P.M.	7 P.M.
Spring	{ March, April, May }	46·6	56·4	49·4
Summer	{ June, July, August }	61·7	72·7	63·7
Autumn	{ September, October, November }	47·3	57·4	49·8
Winter	{ December, January, February }	33·8	41·7	35·9

	Fahr.
The mean annual temperature is	51·0°
„ winter „	36·5
„ spring „	50·8
The mean maximum for seven years (July) was	77·0
„ minimum (January) was	35·2
The absolute maximum (8th July, 1870) was	89·0
„ minimum (12th February, 1865) was	11·4

X

The number of rainy days are few—sixty in the whole year and twenty-one for the six months of winter and spring; but the rainfall is considerable, much greater than at adjacent stations. The mean annual rainfall for seven years was 50 inches, compared with 34 inches at Bex and 32 inches at Geneva.

The atmosphere is of medium humidity, and fogs from saturation are rare. Owing to its protection from winds the air is very calm and still, the number of calm days reaching 85 to 90 per cent., whereas, at Morges it only reaches 33 per cent.; and it has been noticed that the lake is often calm from Vevey to Villeneuve when it is agitated in the rest of its extent. But when the hot wind blows from the south the *föhn*, here called the *vaudaire*, blows along the Rhone valley, it makes the bay of Montreux very rough.

In an average winter a good deal of cold weather must be expected at Montreux, as its mean winter and spring temperature is some 5 degrees Fahrenheit lower than that of Ventnor and 4 degrees Fahrenheit lower than that of Torquay; but in favourable seasons, on the other hand, a good many bright, clear, sunny days may be expected and comparatively few rainy ones. In November there are often a good many cold, damp, and disagreeable days.

In spring the weather is often very variable. There are perhaps some very fine days, and then a sudden and unexpected return of cold with rain or snow; so that invalids need to take great precautions at this season. Patients often ascend to Glion at this period of the year. Few people spend the summer at Montreux, on account of the heat, but the autumn is a fine season up to the middle of October, when storms of rain frequently set in and there is occasionally a passing snowfall. It is in the autumn that the grape cure is in active progress at Montreux.

It is an advantage at Montreux to have two mountain stations of different elevations, such as Glion and Les Avants, so readily accessible; for it does happen during some seasons that there is much more sunshine to be found at the higher resorts than at the lower one, and this fact is easily ascertainable.

Although, as has been said, there are few fogs in ordinary seasons, yet seasons do occur when they are frequent, and this was notably the case in the winter of 1875–76.

The winter and spring climate of Montreux, it will be seen, is by no means a perfect one; it has, however, been pointed out, as a kind of compensation, that the hotels and *pensions* which abound here are very comfortable, and that if the weather out of doors is bad the invalid can find good shelter and protection in-doors.

The following are the various ailments which are said to be benefited by prolonged residence at Montreux or in its immediate neighbourhood :

Cases of simple chronic laryngitis, of chronic laryngo-pharyngitis, of granular pharynx—all these chronic throat affections have a good chance of cure at Montreux, especially if they are, at the same time, submitted to local treatment by inhalation, etc.

Cases of recurrent bronchial catarrh or tendency to catarrh, as well as cases of chronic bronchial catarrh when not too inveterate or severe.

Persons with hereditary predisposition to consumption, and cases of chronic phthisis and early phthisis when the general health and strength are otherwise good and there is an absence of fever.

Cases of chronic pleurisy with suspicion of the commencement of phthisis, as well as cases of chronic empyema healing slowly.

Cases of cardiac valvular disease of rheumatic origin, to ward off bronchial catarrh and fresh rheumatic attacks, also cardiac neurosis, especially if induced by excess of tobacco-smoking.

MERAN. This charmingly situated health resort is well known as one of the most favourite stations for the grape cure in the autumn, the season for which extends from September 1st to November 1st. It has also a winter season, from November to the end of February, when it is frequented by many pulmonary and other invalids; and also a spring season, for the koumiss, whey, and goats' milk cure, which extends from the 1st of April to the 15th of June.

Treatment by medicated baths, and especially treatment by *variations of air pressure in the Pneumatic Chamber*, are also carried out at Meran at all seasons. All these various methods of treatment are conducted under the watchful care of many accomplished physicians.

Meran was once the capital of the Tyrol, and is situated about eighteen miles from Botzen (with which it is now connected by a branch line) on the line of railway connecting Innsbruck and Verona; so that in approaching it from the north you travel by Munich, Innsbruck, and the Brenner; and from the south by Verona, on the direct line between Milan and Venice.

The situation of Meran is exceedingly picturesque, placed as it is on the southern slopes of the Alps of the Austrian Tyrol, at an elevation of 1050 feet above the sea, on the banks of the Passer, about half-a-mile above its confluence with the Adige. Its position is a very sheltered one, as it is surrounded on all sides by mountains except towards the south, towards the wide and extensive valley of the Adige, where it lies fully exposed to the southern sun as well as to the full fury of the south wind, which occasionally blows with considerable violence; but it is protected by lofty mountains, some rising as high as 10,000 feet, to the north, the east, and the west.

It is to this exceptionally protected situation that it owes its peculiar climatic advantages.

In the first place it is an exceptionally dry climate. It has only an average of fifty-two rainy days in the whole year, and only thirteen during the winter. Cairo is the only winter health resort that has fewer rainy days in winter, viz., nine only; Montreux has twenty-one. It may be mentioned for the sake of comparison that Torquay has seventy-nine rainy days in the same period. It has on an average seven days of snow in the winter, so that it is not a warm climate, its medium winter temperature being 41·67 degrees Fahrenheit; we find, comparing this with Montreux, that it is rather warmer, the medium winter temperature at the latter place being 40·55, but it is colder than Torquay, which has a medium winter temperature of 44·6, and much colder than Mentone, with a winter temperature of 52·28 degrees Fahrenheit. Meran then, it must be remembered, is a dry and cold winter climate; but the cold is much better borne, is more tonic and far less depressing than in a place which is damp as well as cold. The cold is also better borne because of the absence of wind, the protecting girdle of mountains keeping off all winds except that from the south. The sun-heat, owing to its exposure to the south, is very considerable, especially at mid-day, and there are an unusual number of bright sunny days in winter. Owing to the absence of wind the air is very free from dust, and mists rarely settle in the valley.

On account of the clearness and dryness of the air, and the great amount of heat due to solar radiation, the difference between sun and shade temperature at mid-day is often considerable, as much as 27 degrees Fahrenheit, so that while it is freezing in the shade, you may have a pleasant temperature of about 60 degrees Fahrenheit in the sun.

The yearly medium of humidity of the air is 67, that of the month of January 80, the highest of the year. April is said to be the driest month, with a medium of 67 and a minimum of 41.

Owing to these climatic conditions, the dryness of the air, the abundance of sun-heat, and the absence of winds, invalids are able to remain in the open air at lower temperatures than would be possible otherwise.

An invalid, who kept a record of ten years'

experience at Meran, states that on an average in the month of January he was able to "sit out of doors" during 16.8 days, that nine days more were "fair enough for walking," and that only during 5.4 days was he obliged "to remain indoors," and taking the whole months November to March (an average of ten years) he was able "to sit out of doors" 73.5 days, "walk out of doors" 45.8 more, and "remain at home" 29.5.*

There are sunny balconies to the houses for sitting in on suitable days, and there are excellent promenades with comfortable seats in the more sheltered spots, and from twelve to three a band plays in the place of principal resort.

Dr. St. Clair Thomson, writing in January last, says : " Skating has been in full swing here since early in December ; and daily invalids and ladies are seen sitting on the bank looking on, and shielding themselves from the sun with shades and fans ! Most people wear during the mid-day only a light over-coat ; and some, none at all. The morning and the evening after sunset are cold, and then feeble invalids keep within the houses, which are comfortable and well warmed. Some trees are in flower throughout the winter ; and in February the spring appears early with violets, anemones, and even the almond-tree in sheltered places. On inquiry from a physician here, I was told that in thirteen years he had never seen a case of chilblains—a testimony to the absence of dampness."

Very great pains are taken there to study the comfort of the visitors in maintaining pleasant walks and gardens and in providing seats. There is an excellent Kurhaus with a resident physician, a staff of nurses, baths of various kinds, the pneumatic chambers already alluded to, reading rooms, restaurants, theatre, etc., etc. There are numerous good

---

* Quoted by Dr. St. Clair Thomson in an article on " Meran in Midwinter."—*Medical Times, March*, 1885.

hotels and *pensions* there, as well as at the villages of *Unter-mais* and *Ober-mais* on the opposite bank of the Passer.

Ober-mais is higher (1200 feet) and cooler than Meran, and is a better place of residence during the hotter months. A great number of very interesting and beautiful excursions can be made into the picturesque and mountainous district around.

Patients often move further south about the end of February, when troublesome unsettled spring weather is apt to set in and to continue through March.

The class of invalids to whom the climate of Meran seems best suited are those suffering from pulmonary disease, who find by experience that a dry and bracing climate suits them better than a warmer moist one, and who can bear a certain amount of cold in winter without being made uncomfortable by it; e.g. certain cases of chronic bronchial catarrh, of asthma and of emphysema, especially when treatment by alteration of air pressure in the pneumatic chambers is desired, certain forms of chronic phthisis before the lung has broken down and where there is no tendency to hæmorrhage; and many other forms of chronic derangement of health, neurotic, anæmic, rheumatic, or scrofulous. The great good which many phthisical patients obtain at Meran seems to be referrible as much to the great care and constant supervision, regimenal and medicinal, which the physicians there exercise on their patients, as to the climate itself.

Dr. St. Clair Thomson, who visited Meran in January, 1885, has kindly sent me the following additional notes:

The town used formerly to be subject to inundations from the overflow of the Passer, this has now been avoided by the construction of a strong stone-work bank—the Wassermauer. This runs east and west, and on it a handsome promenade has been constructed; some of the pleasantest houses and *pensions* are built overlooking it, as they have an uninterrupted southern aspect. The upper end is laid out in a winter garden. An English church is being built at the western end, and midway between there is the Kurhaus. Here there is a resident physician, who acts under the physicians of the town ; a staff of nurses trained in massage and hydropathic treatment, and who also attend patients at their own houses; baths of various kinds, such as douche, vapour, salt and pine-leaf baths ; electrical apparatus ; and a pneumatic chamber. This latter is for the administration of compressed air in cases of asthma, emphysema, etc., the pressure in it is $1 \times \frac{1}{6}$ of the atmospheric pressure. The duration and number of the sittings of course vary with each case, the average duration is two hours, and some patients spend this time in the chamber daily for two or three months. Others have a course of twelve or fourteen, and then an interval of a fortnight or a month. The Kurhaus also contains a reading-room, with a large supply of newspapers and periodicals ; smoking-room, billiard-saloon, ladies' drawing-room; a theatre, in which performances are given twice a week, and concerts and réunions held ; and a restaurant. The band plays in front of the Kurhaus twice a day in autumn and spring, and once a day in winter. On the southern side of the Passer lies the village of Untermais. This occupies the same level as Meran, it lies more in the centre of the valley, not so close to the feet of the northern mountains. The village of Obermais, which is now but an extension of Meran, rises up the slope of the eastern mountain which guards the valley, and attains a level 46 metres higher than Meran. Here are situated the Elizabeth Gardens, and the summer garden, which are most resorted to in warm weather. It is this part which is increasing most rapidly, and from its situation has the advantage that many patients are able to spend nine months and some the whole year in it.

Except along the valley there are not many carriage-roads, but good riding-horses are to be had, and for pedestrians the country is very interesting. There is also a skating-club and a lawn-tennis club.

For a short stay it is of use to patients to break their journey in travelling either north or south, to avoid too sudden transition of climate. To the large number of invalids with whom its climate agrees it offers the advantage of allowing a more prolonged stay than most health resorts—a boon to many to whom travelling is unpleasant or impossible. With a summer visit to the neighbouring mountain villages, many spend the whole or greater part of the year here.

THE GRAPE CURE.

This account of the "Grape Cure" is translated from that by the late Professor Lebert, M.D., who, after having been Professor of Clinical Medicine at Zurich and Breslau, settled in practice near Montreux.

The nutritive value of grapes is not very great, notwithstanding the abundance of sugar and salts of potash which they contain. But their chemical composition and edible character render them not only an agreeable form of food, but also one of the few means we find at our disposal between nutritive substances on the one hand, and medicinal substances on the other.

Notwithstanding the recognised usefulness of the grape cure, its value has, however, often been exaggerated. No doubt this cure is greatly aided by the beautiful and salutary situation on the shores of Lake Leman, where it is carried out, affording as it were a combined hygienic, medicinal, and climatic cure. I had proof of this when I was Professor in Germany. In attempting to carry out the grape cure at home, in Breslau and in adjoining provinces, with a regular and fresh supply of excellent grapes from Hungary, I did not obtain nearly the same success as when I sent my patients to Meran, to the Rhine, or to Montreux.

The grape cure is, notwithstanding its refreshing and aperient effects, essentially a dietetic cure, and to be useful a moderate quantity only should be prescribed; for the excesses and the abuse of this cure which patients are addicted to, with or without the consent of their physicians, may render it injurious. The effect is aided by a good supporting diet.

In my youth I used to see eight to ten pounds of grapes per diem ordered, and all other food reduced to a minimum. As soon as I began practice I revolted against so irrational a method, especially in cases of chest disease. I soon began to regard two to four pounds a day as the maximum. Somewhat later Cruchol fixed three pounds as the medium dose in twenty-four hours.

I will add that pulmonary patients support large doses of grapes still worse than other patients, and I recommend them not to take more than two pounds a day on an average. Grapes have no special action in cases of phthisis, but they constitute a useful and wholesome article of diet. I may add that attacks of hemoptysis, which were formerly frequent during the grape cure when these large quantities were taken, have been rare at Meran and Montreux since the cure has been conducted on more rational principles.

Patients who suffer from troubles of digestion require more than one or two pounds a day. It is always advisable in dyspeptic conditions, with chronic catarrh of the stomach, to progress gently—not to exceed three pounds a day, and to carefully regulate the diet.

When gastric catarrh is accompanied with frequent attacks of pain and very laborious digestion, with some suspicion of gastric ulcer, the grape cure is counter-indicated.

Constipation, whether idiopathic or the result of some gastro-hepatic affection, is often advantageously modified by three or four pounds of grapes taken in the twenty-four hours. But cases of obstinate and aggravated constipation of long standing often resist the cure, and in such exceptional cases we may exceed the average dose up to five or six pounds, according to the effect and tolerance of the cure.

Cases of slight intestinal catarrh may be modified advantageously by grapes, but I have found them fail completely in the diarrhœa of phthisis. I have seen the same in the intestinal catarrh that accompanies Bright's disease, or amyloid degeneration of the kidneys, liver, spleen, etc.

A thoroughly laxative dose of grapes, which may reach five or six pounds a day, is very salutary in hæmorrhoidal affections without much loss of blood, also in cardiac diseases when not much advanced, and in which the venous circulation begins to be troubled, giving rise to pulmonary, renal, hepatic, and intestinal congestions.

The calculous diathesis, renal and hepatic, is sometimes modified very advantageously by grapes ; but chronic inflammatory conditions and degenerations of liver and kidneys are very rarely benefited. I have, however, seen some cases of albuminuria, with but little advanced nephrites, benefited by the grape cure.

Every year the contingent of those who come to Montreux for the grape cure is, in part, composed of a class to whom this cure is very valuable. I allude to those who are neither ill nor well, who are fatigued by a too exciting and somewhat intemperate life, or, weakened by severe illnesses, are convalescing slowly, or who, leading habitually a too sedentary and too laborious existence, find in our country, besides the grapes, which regulates the digestive functions, all those hygienic conditions which they most need.

In this way one may often prevent the passage of debility and fatigue into real disease. One must not forget that each case submitted to the grape cure requires to be carefully watched in order that it may be modified if necessary according to its effects and the way in which it is tolerated. Of two patients of the same age, and suffering from the same complaint, one will rapidly reach three or four pounds a day, whilst the other will be slow in habituating himself to the cure, taking only small quantities for one or more weeks.

I begin usually with half-a-pound of grapes in the

morning fasting, and another half pound at 5 p.m.; some patients do not bear grapes well fasting, these should take the first portion an hour or two after a very light early breakfast. After two days I order a third half-pound between 11 and 12 a.m. Little by little the dose is increased to a pound on an average. Often I do not go beyond the half-pound in cases of chest disease, and I sometimes exceed the pound in cases which bear the cure well and who experience from it a slight laxative effect. In the case of dyspepsia one may substitute for the dose in the middle of the day one or two bunches at dessert.

Ordinarily the aperient effect of grapes does not show itself for some days, and it is not unusual to observe at first some constipation, but it almost always disappears if we abstain from purgatives. In some cases, however, it persists, and it may be necessary to take an aloetic pill at bedtime.

It is not necessary to abstain from all other kinds of fruit during the cure, only they should be taken in moderation. Although the diet ought to be good and nutritious, it should not be too exclusively composed of animal food, and stimulating drinks such as wine, coffee, and tea, should not be taken strong or in large quantities.

The average duration of the grape cure is four weeks, often it may be prolonged with advantage during six weeks, and one may still continue to take grapes at dessert when the cure is finished. After each dose of grapes I recommend the mouth to be rinsed with cold water, to which a little bicarbonate of soda may be added if there is any disposition to irritation of the gums, an occurrence which is not very rare during the cure if one does not take these precautions at starting. Towards the end of the cure the quantity should be gradually diminished.

The grapes usually begin to arrive at maturity on the shores of this lake about the middle of September, but they can be obtained from Sion and from the south of France from the beginning of that month.

CHAPTER VIII.

WINTER QUARTERS.

A STUDY OF WINTER HEALTH RESORTS.*

La terre est son médicin ; chaque climat est un remède. La médecine, de plus en plus, sera une émigration, une émigration prévoyante.—*Michelet.*

AGAINST the many privileges which are said to be exclusively the lot of an Englishman must be set off the obvious disadvantage that he has to live in a climate which, for a great part of the year, is often little short of detestable. So long as he remains strong and well, he usually contents himself with grumbling, and adopts (or sometimes does not adopt) such artificial means of protection as his intelligence and experience show to be useful and necessary. But if he inherits, or acquires accidentally through disease, a feeble and ailing constitution, then too frequently he falls a victim to the rigour and inclemency of the external conditions to which he is exposed ; or, if he is both wise and rich, he follows the example of the migratory tribes of the animal kingdom, and seeks at each recurring season those conditions of climate which to him, as well as to them, are the conditions of life and comfort.

But we should be unjust and short-sighted if we failed to see any compensatory advantages in this

* A portion of this chapter appeared in the form of an article in " The Fortnightly Review."

climate of ours which we so often abuse. Many of
the best qualities of an Englishman are, to a great
extent, due to the character of the climate of the
country he inhabits. His capacity for endurance, and
for adapting himself to varying conditions and circumstances, his energy in overcoming difficulties, his
physical strength, are in some measure the outcome
of his life-long contest with unfavourable external
conditions, and of those out-door exercises and sports
to which he is driven in order to keep the blood
actively circulating through limbs which would otherwise be chilled and benumbed; or to keep his mind
free from the melancholy and depression which
inactivity under a leaden sky most surely induces.

The return of every winter necessarily brings to
many minds the consideration, where they can pass
the next five or six months with the least inconvenience and the greatest benefit to their healths. To the
too numerous victims of pulmonary consumption, as
well as to those who fear to become so, it is a question
of the greatest import ; and not to those alone, for
increasing experience of climatic conditions and
influences shows that a vast number of other chronic
maladies acquired in this climate are stayed in their
course, and not infrequently altogether arrested by
judicious change to more favourable external conditions. A few general considerations, therefore, as to
the facts and principles which should guide us in the
choice of a winter climate may not be uninteresting or
unprofitable.

It would be unwise and of no practical use within
the limits of a single chapter to hamper oneself with any
attempts at a strict classification, while an exhaustive
survey of the whole series of winter health resorts would
be, of course, impossible. It will be better to confine our
consideration to those which are tolerably accessible,
and especially to such as recent inquiries have brought
more prominently before us. I have, in another
chapter, entered fully into the question of the utility
and scope of high mountain health resorts in winter,

and especially of that portion of the question in which the public are beginning to take a great interest, the relation of those elevated regions to the cure of pulmonary consumption. It is interesting, however, to notice that in the first volume of "The Fortnightly Review," i.e. nineteen years ago, an article is to be found which foreshadows this discussion. "In cold climates," says the writer, "on the contrary, consumption is almost unknown. In Iceland it is seldom seen ; in Greenland a case is the exception, and in more northern regions it disappears altogether. In mountain ranges above 3000 feet it is an exceptional complaint. Heat is favourable to the development of the disease, especially when accompanied by moisture."*

The discussion in this country has, during the last few years, been almost limited to the examination of the merits of *one* health resort, viz. Davos in the Grisons. Fourteen years ago, when I first examined the subject, and when I first visited Davos and the Upper Engadine, it was the latter place that was chiefly in the minds of English physicians as a possible winter sanatorium for consumptives. Davos was then resorted to almost exclusively by Germans and Swiss. But the Upper Engadine did not appear at that time to maintain its position as a winter station. In "The Fortnightly Review," the Hon. L. Tollemache, writing from an intimate personal experience of the Engadine for many years, and who certainly could not be accused of any prejudice against the place, thus spoke of his own observations of the residence of consumptive patients there :

It is well known that, in the treatment of such cases, medical opinion has undergone a change, so astounding as to look like a leap in the dark, or, at least, in the dim twilight. As the remedial agent, the extreme of dry cold has suddenly replaced the extreme of moist heat ; and some patients who, only twenty years ago, would have been more or less boiled in Madeira,

---

* "Dangers of Madeira." "Fortnightly Review," vol. i. 1865.

are now frozen on Alpine heights. How far has this bold experiment succeeded? In the Engadine, certainly, the results (so far as they go) have not been encouraging. Out of the very few who, within my knowledge, have spent winters (or parts of winters) there, at least six have died—a startlingly large portion of the entire number ; whereas consumptive cases where the cure of certain disease is itself certain and certainly due to the Engadine winter, are—I will not say unknown—but exceedingly rare. But, on the other hand, there are consumptive patients whom the air seems to have kept alive, and who are, though not well, quite well enough to enjoy life.

This evidence was of great value, coming as it did from one who had, during a series of years, followed with interest the histories of the consumptive invalids he had met in the Engadine, biassed by no medical predilections, but as an earnest and honest advocate of that place as a health resort, in such cases as it had appeared to him to be of use.

It was not a little remarkable that, while the Engadine at an elevation of 6000 feet (speaking in round numbers) failed for some time to acquire a reputation as a winter station for consumptive patients, although introduced to our countrymen on high authority more than fourteen years ago, Davos, its near neighbour, not more than twenty miles distant, at an elevation of 5200 feet, grew into sudden and rapid popularity. I must, however, content myself now with merely mentioning these high mountain valleys as winter health resorts, as I have already dealt at some length with the interesting and important question of the applicability of high altitudes to consumptive patients in a former chapter, to which I must refer those who desire information on that subject.

So without dwelling longer at present on this question, I pass on to the consideration of other winter resorts ; and first with regard to the climate of Egypt, which resembles in some respects the climate of those high mountain valleys. It is dry and exhilarating, and it presents a wide range between the day and night temperatures, depending upon the powerful heating effects of the sun's rays during the day, and the great and rapid radiation of the heat absorbed during the day, after sunset, into clear cloudless space. The climate of Upper Egypt is, however, on the whole a more reliable climate than that of any high mountain valley, and less subject to variations; while the interest of the voyage up the Nile, and the diversions which it presents, render it a much more suitable resort for those who dread *ennui*, or who need occupation as well as relaxation for the mind.

The objections to Egypt are, of course, its dis-, tance, and the expense attending the journey; and, moreover, whichever route you select, it is impossible to avoid a sea voyage of at least three days. The objections to this, as well as to other sea voyages, are forcibly put by Mr. Flower : *

The principal objection to persons in delicate health undertaking a long sea voyage is the uncertainty about the influences to which he or she may be exposed ; while on land, the traveller is, to a great extent, his own master, and has power to control the surrounding conditions. He may regulate the day's journey, according to strength or inclination, he may linger in such places as have agreeable associations and environments, he may hasten over those of an opposite character; but when once embarked upon a voyage, whether he find himself crowded in a dark close cabin, with two or three uncongenial companions, lying on a narrow hard shelf, port-holes rigidly closed, and the atmosphere he breathes poisoned by noisome odours, of which the sickening smell of the oil of the engines is one of the least objectionable; the rain pouring on deck, making escape from his prison, even for a few minutes, impossible ; when he feels he would give all his worldly possessions for a breath of pure air, or a few hours' cessation from the perpetual din of the engines within and the waves without ; he is perfectly helpless, he must go through it, day after day, and night after night, until the weather changes or the voyage is ended.

There is only one period of the year when Egypt is ever visited by the European as a health resort,

* "Notes of Experience in Egypt," by W. H. Flower, F.R.S. "British Medical Journal," September and October, 1874.

and that is from the middle of November to the beginning of April, when it is considered to have the "finest climate in the world." There are several routes from England to Egypt. The shortest and most convenient is that through Italy to Brindisi, and there is now, on one day in the week, a saloon carriage attached to the train at Calais which goes through to Brindisi. From Brindisi to Alexandria is a three days' voyage by steamer. In this way the journey to Egypt is accomplished in six days. The longest but least fatiguing for those who do not mind a sea voyage, is that by P. and O. boat from Southampton to Suez, which takes thirteen or fourteen days. There is a third route *via* Paris and Marseilles by Messageries boats to Alexandria. This gives a five days' sea voyage. In returning, it is important for invalids with lung disease to bear in mind that it is not safe to return by the Southampton route, as the transition from the climate of Egypt to that of England is too abrupt. It is very important to leave Egypt before the heat becomes too great—i.e. not later than the middle of April—and it is undesirable to return to England before June. The interval may be conveniently spent in a variety of places of great interest, such as Syria, Italy, Greece, or some of the islands of the Mediterranean.

The chief characteristic of the climate of Egypt is its dryness.

In the richly wooded districts of the equatorial regions of Africa (writes Mr. Flower), where the numerous affluents of the Nile take their rise, almost continuous rains prevail; but in the deserts of Nubia and Upper Egypt, through which the great river flows in its course to the sea, sometimes years pass without a single shower. The absence of rain and absence of vegetation are obviously related to one another. The Mediterranean coast and the Delta are less dry than the upper parts of the country, and Cairo occupies an intermediate position.

We have the authority of the same writer for the statement that in an exceptionally wet season there were only eleven days out of one hundred and fifty in which rain fell, and on some of these it was scarcely more than a few drops. The days, as a general rule, are much like one another, fine, clear, bright, and sunny, and "the subject of the weather, so important to us in our island home, soon loses all interest, owing to the absence of change." Another of the characteristics of the winter climate of Egypt is the warmth or heat of the day (70 to 75 degrees Fahrenheit in the shade), as contrasted with the coldness, freshness, and heavy dews of the nights. In the night the thermometer often falls to 40 degrees or lower, seldom quite to freezing-point, so that there is a very considerable range between the day and night temperature. It has been justly observed that this is an advantage to many constitutions—that a sultry night following a hot day often induces languor and depression, and that the freshness of the Egyptian night and early morning is invigorating and bracing, and enables one better to bear the fatigues and heat of the day. Persons with delicate chests must be careful to protect themselves by appropriate clothing, and by retiring before nightfall, from the sudden change from the day to the night temperature, which they may otherwise find trying or injurious. The air of the Desert—that is, all the country above the level of the autumnal overflow of the Nile — is universally admitted to be most invigorating : a refreshing breeze, in winter at least, generally tempers even the heat and glare of the mid-day sun, and in the morning and evening it is decidedly cool. Nowhere on land is air so pure, as nowhere else is there such complete absence of all decomposing organic matters in the soil ; it has been well compared with that of the open sea.

Most of those who go to Egypt for the winter, go with the intention of making the Nile voyage ; but a winter may also be passed agreeably and advantageously at Cairo, as well as on the Nile. The thing of chief importance is to breathe as much of the Desert air as possible. It has been objected to

Cairo that the hotels and all the modern houses are built on low ground, that, until reclaimed, used to be subject to the overflow ; and that the whole of the ancient city, with its crowded population and filthy streets, is between them and the Desert; that the prevailing winds, being from the north, blow directly across the Delta. "This, and the great amount of not very clean dust which fills the air of a great city full of people and animals, form the principal drawbacks to Cairo as a residence for invalids." An alternative presents itself in a place about an hour's journey by rail to the south of Cairo, and three miles from the east bank of the Nile, named *Heluan les Bains*,* on account of the existence there of a warm sulphur spring. This station, of which I have received most satisfactory accounts from patients who have stayed there, possesses a good hotel with a medical director who speaks English ; and as it is in the open Desert, only a few miles from Cairo, and therefore very accessible, it ought to become popular with those who need to live in the winter climate of the Egyptian Desert, and yet who may not, for many reasons, be disposed to enter upon the Nile voyage. I hear during the past winter it has become a fashionable resort.

There is also an hotel at Luxor, established by

---

* The following note on Heluan is from the special correspondent of the "British Medical Journal," December 6th, 1884 : "It is to be regretted that the sulphur-springs of Heluan, a short distance from Cairo, are much neglected, and little known. Some years ago an hotel and baths were erected here by the Egyptian Government, in the Khedivate of Ismail Pacha, under the direction of Dr. W. Reil; but the baths have not been properly maintained, and are now fast going to ruin. The Heluan springs contain a much larger amount of sulphur than those of any European spa. The average winter temperature is 72 degrees Fahrenheit ; the clear atmosphere, pure desert air, and proximity of the pyramids, and places of historic interest, would seem to render this a perfect sanatorium for the valetudinarian, especially sufferers from rheumatic or cutaneous affections."

Messrs. Cook, and the surrounding district is interesting and the climate very fine, and it can be reached
by Cook's steamers from Cairo once a week. Of
the Nile voyage little need be said in the way of
description.

It is (Mr. Flower says) a perfect rest from nearly all the
little cares and troubles of the world ; the weather is almost
always fine, so that nearly the whole day may be spent on deck,
and the variety and exercise of a walk on shore can generally
be got at some time or other in the twenty-four hours ; the life
on board a dahabeeah is generally a healthy one. It is essentially an out-of-doors country life. The air, though perhaps not
·equal to that of the higher parts of the Desert, is pure and
bracing ; for, owing to the narrowness of the strip of fertile land
on the sides of the river, the air is practically that of the Desert.
On the first subsidence of the water, after the autumnal overflow, the banks are muddy and damp, so it is well not to take
to the water until December, by which time they are well dried
by the sun, though January, February, and March are the best
months. The higher the river is ascended, so the salubrity
increases. The nights are generally clear, bright, and cool, and
warm clothing is essential, as no artificial heat can be obtained
on board the boat.

Egypt as a winter resort has, then, the following
advantages : 1. It is almost rainless ; at Cairo five or
six showers would be the average in the winter. 2. It
has a generally dry and clear atmosphere ; attended,
it is true, with great changes of temperature in the
twenty-four hours, a circumstance which proves invigorating rather than otherwise, if the invalid is
careful ·to protect himself from the sudden fall of
temperature at sunset, as well as through the cold
nights. 3. Extreme cold is excessively rare. The
mean winter temperature at Cairo is about 58 degrees
Fahrenheit, and it rarely falls below freezing-point.
4. Its climate allows of constant exercise in the open
air, and exposure therefore to the tonic effect of fresh
air and sunlight.

Invalids doing the Nile voyage should be careful
not to expose themselves, on coming down the river,
to the cold north wind when it blows strongly, as it

does sometimes for days together, and as there is a
great fall of temperature at sunset they should always
retire to the saloon a little before sundown.

The climate of Egypt then is tonic and stimulating,
and it is useful in a great variety of chronic ailments,
the chief of which are the following : It is said to be
especially useful in cases of phthisis in scrofulous
persons, those cases of phthisis which have a tendency,
even in this country, to run a very protracted course ;
it is helpful, too, in most other forms of scrofulous
disease ; it is of value in gout and rheumatism, and
especially in certain important visceral changes which
gout induces ; catarrhal conditions find their relief
and cure here as well as in the cold dry air of high
altitudes, so that cases of chronic bronchial, laryngeal,
and pharyngeal catarrh get well in Egypt, as do also
some cases of catarrhal asthma. Persons suffering
from exhaustion of the nervous system from too
great excitement, worry, or undue application to
business or study, are precisely the cases for the Nile
voyage. The same may be said of those numerous
cases of intractable dyspepsia associated with hypochondriasis or hysteria.

The climate of Egypt is not limited simply to the
relief of early phthisis, but advanced cases often do
well there, though it is considered inexpedient that
they should venture on the Nile voyage or go beyond
Cairo. Cases of phthisis with a tendency to rapid
progress in irritable or highly nervous constitutions
must not, however, be sent to so tonic and exciting a
climate. The same remarks apply to cases with a
tendency to hæmorrhage.

A writer in " Blackwood's Magazine " (February,
1883) wonders that the merits of the coast of Syria as
a winter resort have not been more fully recognised.
He mentions that there are only two places on that
coast which offer the requisite accommodation—
Beyrout and *Haifa*. *Beyrout*, he says, has two excellent hotels, and nearly 300 public carriages for hire

for most picturesque drives along good roads; while other and cheaper accommodation than that at the hotels can be had; and the town possesses a street of European shops. "The valleys of Lebanon offer attractions unsurpassed by mountain scenery in any part of the world. *Haifa* is quieter than Beyrout and less civilised. There is a German colony there, and the place begins to show unusual signs of progress. It is surrounded by gardens of oranges, figs, and pomegranates, and groves of stately date-palms. The adjacent monastery of Carmel is an attraction. A German colony from Würtemberg, who decided, on religious grounds, to settle in Palestine, selected this site for their settlement. They number rather more than 300, and include Germans, German Americans, and a few Swiss. They have their own skilled physician, an architect, and an engineer."

Haifa is about ten miles from Acre, and omnibuses run between the two places in about two-and-a-half hours over the sands "along the edge of the waves." There is also a carriage road to Nazareth—twenty-two miles—and another to Cæsarea. This writer states that as a winter resort it combines comfort with economy. There is an excellent hotel kept by a German, and the Monastery affords accommodation for 100 guests; or a house can be rented and furnished, the colonists being excellent servants. Mutton, beef, pork, and chickens are plentiful, and there is a sufficiency of fruit and vegetables. It has a good port and harbour. Behind it is the vast plain of Esdraelon. "It is impossible to conceive a more agreeable climate during the winter months than it offers. From October to January the temperature is generally that of the finest summer months in England. Then it begins to get a little chilly, and a fire in the evenings is a grateful addition to the natural temperature; but this is only occasionally the case during the rainy weather. The rainiest winter month here would be considered a fine summer month in England. It does not begin to get really hot till May. The climate of Carmel is exceptionally bracing and healthy. Its most powerful attraction is the charming excursions to be made in the neighbourhood, as objects of interest within a day's ride or drive abound ; while the beauty of its situation, commanding a lovely view of the Bay of Acre and the encircling hills of Palestine, overtopped by the snow-clad Hermon, is remarkable. There is good bathing from a smooth sandy beach at all times of the year, and plenty of sport in the thickets of Carmel, including wild boar, partridges, snipe, quail, woodcock, etc. An Austrian Lloyd's steamer touches there once a fortnight, either from Beyrout or Alexandria, and letters arrive every week by land post. The Messageries and Russian boats pass it on their way from Beyrout to Jaffa. It depends on the public to remedy its present isolation, which, however, may prove rather an attraction than a drawback." Such is the account of this "New Winter Resort," given by the writer in "Blackwood," and it seems attractive enough for those who desire to get altogether beyond and away from the beaten tracks.

I purposely pass, in the next place, to the consideration of a winter climate, the characteristics of which are in striking contrast to those of Egypt and Davos. The tonic and stimulating climates of Davos and Upper Egypt on the one hand, and the soft soothing climate of *Madeira* on the other, may be regarded as at the two extremes of winter health resorts for European invalids. Madeira having been for many years greatly overrated, has, during the last few years, come to be vastly underrated. It has suffered from one of those violent oscillations of medical opinion to which all health resorts are liable ; and after such an acute disturbance we may take it for granted that it will be long before a rational equilibrium is established. It is, however, satisfactory to see that Madeira is again becoming resorted to in increasing numbers.

Situated between 32 and 34 degrees north latitude,

and 16 and 17 degrees west longitude, in the Atlantic Ocean, about 500 miles from the west coast of Africa, it is a typical representative of a warm and humid marine climate. The beauty and diversity of its landscapes, and the variety and richness of their colouring, give this island a particular charm for artistic natures.

Writing of Madeira, Dr. Lambron, of Luchon, calls it " La première résidence hivernale du monde."* The late Dr. Andrew Combe wrote : *"If I must go abroad,* I shall most likely return to Madeira, on the simple ground that, if I must forego the pleasures of home, it is better to resort at once to the *most* advantageous climate," etc. A certain Dr. Heineken, according to a writer in " The Fortnightly Review" † already quoted, was said to have lived there with a quarter of a lung for nine years ! But since the comparative want of success which attended the action of the authorities of the Brompton Hospital in the winter of 1865, the reputation of the island as a winter sanatorium for consumptives had until lately been on the decline. The Brompton Hospital sent twenty carefully-selected cases of phthisis to winter there ; of these, two only were greatly benefited ; seven improved slightly, six returned no better nor worse than when they left England, four returned worse, one died in the island. It is, I believe, sufficiently well understood now that the climate of Madeira is only suited to a very limited and carefully-selected class of cases ; but for the proper case it is a climate of the greatest utility. If we bear this fact in mind, we shall be able to reconcile the wide discrepancies which we find in authoritative and evidently unprejudiced statements about this island. Madeira is the type of what is termed an oceanic climate, i.e. a climate essentially soft and equable. It is also moist and sedative, and, no doubt, to persons with considerable constitutional vigour, it seems relaxing and depressing. But to certain persons in a state of

* " Choix d'une résidence d'hiver."
+ " Dangers of Madeira."

great debility, with much feebleness in the organs of circulation, in cases of irritative chronic bronchitis with scanty secretion, and complicated with emphysema, in some cases of advanced consumption, and particularly those complicated with repeated attacks of bronchitis, even cases that have seemed quite hopeless, a prolonged residence in the climate of Madeira has been attended often with most remarkable amelioration.   The feeble flickering lamp burns longer there than in a more stimulating and tonic air, and now and then it seems to gather renewed power and burns up again with some of its old lustre. Madeira has had a great and not altogether undeserved popularity in the past, and it will, I doubt not, have a more stable, because more discriminating, popularity in the future.   This moist sedative atmosphere allays cough in cases of irritable respiratory mucous membrane, but it often causes loss of appetite and bilious disturbance in persons predisposed to such disorders.   It would seem to be more useful in cases of chronic laryngeal and bronchial catarrh, and emphysema, than in phthisis.   It is also suitable to persons of feeble circulation, who cannot bear bracing treatment, and who enjoy a sea climate.

Funchal, the capital of the island, is the principal resort (there are also some hill stations available); it is built in the form of an amphitheatre, and looks very beautiful from the sea.   It is surrounded by luxuriant vegetation, and tropical fruits ripen there all the year round.   It is protected by mountains which rise to nearly 6000 feet, from winds from the north, north-east, and north-west.   One drawback, however, is the difficulty in getting a level walk.   Whenever one goes out there is always a steep ascent to be made, and locomotion is generally effected by means of palanquins, hammocks, or sledges drawn by horses, mules, or oxen, in which many very attractive excursions into the interior of the island may be made.

The climate of Funchal is extraordinarily equable. The mean annual temperature is about 65 degrees

Fahrenheit. The night temperature scarcely ever descends below 48 degrees Fahrenheit, and the day temperature scarcely ever rises above 86 degrees. The mean winter temperature is 61 degrees, the mean spring temperature 62 degrees, the mean summer temperature 69 degrees, and the mean autumn temperature 67 degrees, and the mean difference between the day and night temperature is only 9 degrees Fahrenheit. As might be expected, there is considerable humidity of atmosphere, the mean humidity being from 70 to 74 degrees; but this varies greatly with the changes in the air-currents. There are on an average 50 rainy days in the winter six months, 85 in the whole year. Funchal is occasionally visited by violent storms of wind, but as it is protected by mountains from the prevailing wind, the north-east, the atmosphere is generally calm from 7 to 9 a.m.; then breezes blow in from the sea till 8 or 9 p.m.; and the land-wind sets in late at night. The air, though humid, is not felt by many to be unpleasantly so, while it is pure and free from dust and rich in ozone. Mosquitoes are said to be entirely absent.

The climate then, though very equable, is not felt to be relaxing except by a few exceptional persons, and it must be remembered that there are great differences of climate to be found in different parts of the island. There are villas to be procured at various altitudes adapted for winter or summer quarters, so that there is no necessity for the invalid to leave the island during the hot summer season. The hotel keepers are always willing to make arrangements to provide invalids with such summer quarters.

The autumnal rains set in at Funchal in October and last till the end of the year. The rain is not continuous, but like our heavy April showers with intervening sunshine. Usually some heavy gales are felt about the beginning of the year from south-south-west. Then continuous fine weather sets in in February, and winter is over. It is usual to sleep with open windows at Funchal, for there is an entire absence of that chill at sunset so commonly experienced in the south, and there is no dewfall.

Besides the maladies I have already named, the climate is said to be particularly suitable to cases of chronic dysentery and malarial fever. Visitors usually live in hotels; "Miles" and the "Santa Clara" are the best. There are villas also for families. There is an English club and a good library, good horses are also to be obtained, and it is advisable, owing to the roads being paved with pebbles, to be provided with indiarubber-soled shoes. Steamers leave Plymouth and Dartmouth every Friday for the Cape and touch at Funchal, the voyage taking under four days, and the fare being nineteen guineas. There is also a weekly service from Liverpool in six days at lower fares.

From Lisbon there are steamers on the 5th and 20th of the month, performing the voyage in two days.

The *Azores*, in 38·30 degrees north latitude and 26 degrees west longitude, possess a climate resembling that of Madeira. There is even greater humidity of atmosphere, however, for "paper hangings will not adhere to the walls, and the veneering of furniture strips off." The principal island, St. Michael, has two chief places of resort, Ponta Delgada and Villa Franca. In the island of Fayal there is also Horta as a place of resort. These islands have frequently been visited by earthquakes. St. Michael possesses numerous hot springs which are renowned for the treatment of cases of rheumatism, palsy, etc.

The *Canary Islands*, in 28·15 degrees north latitude, and 14 degrees west longitude, possess a warmer and drier climate than Madeira or the Azores, and Santa Cruz and Orotava in Teneriffe offer considerable attractions to the adventurous invalid, and he has the advantage that he can move to higher elevations in the summer. The country is described as very beautiful.

Climate resorts have been classified roughly into Continental climates, i.e. those in the interior of continents—Upper Egypt may serve as an example; Oceanic climates, places situated in the open sea, as Madeira; and Littoral climates, or places on the sea-coast.* I propose to take, as my first example of a coast climate, one which has quite recently been introduced to the notice of the British public as being of extraordinary value: I allude to Mogador, on the Atlantic coast of Morocco, and nearly in the same latitude as Madeira. A French physician, Dr. Thenevin, quoted by Lombard in his "Climatologie Médicale," has resided there for many years, and it is mainly owing to his careful observations that the peculiar salubrity of Mogador has been made known to the medical profession. For the following details about Mogador and its climate I am, however, indebted to Hooker and Ball's admirable work on "Morocco and the Great Atlas."† Mogador, they say, is the last outpost of civilisation on the African coast on this side of the French settlements of Senegal. A low rocky island lies opposite the town, separated from it by a navigable channel, and affords shelter from all winds, except those of the south-west. The town is, in one respect, the most habitable in Morocco, being remarkably clean, and, in that respect, superior to many sea-ports in Europe. The narrow, but regularly-built streets contain houses mostly of two storeys, inclosing a small courtyard, which is entered by a low and narrow doorway from the street. In the Moorish town, inhabited by natives of the lower class, the houses are of one storey, and poor in appearance; but the practice of whitewashing within and without once every week makes them look clean, and, no doubt, has much to do with the remarkable immunity of the place from contagious

---

* The physical causes of the characteristics of sea and coast climates I have explained at some length in another chapter.
† Published by Macmillan and Co., 1879.

and epidemic diseases.   Its surroundings are not pre-
possessing. The low tertiary limestone rock on which
it is built is covered up to the city walls with blown
sand, driven along the shore before the south-west
wind, forming dunes that cover the whole surface;
and in most directions one may ride two or three
miles before encountering any other vegetation than a
few paltry attempts at cultivating vegetables for the
table within little inclosed plots, whose owners are
constantly disputing the ground with the intrusive
sand.   Regarded as a sanatorium for consumption,
Dr. Thenevin bears testimony to the following facts:
Phthisis is all but completely unknown among the
inhabitants of this part of Africa; while in Algeria
cases are not rare among the natives, and in Egypt
they are rather frequent.   In the course of ten years
he had met but five cases among his very numerous
native patients, and in three of these the disease had
been contracted from a distance.   He had seen,
moreover, several cases among Europeans, who had
arrived in an advanced stage of the disease, on whom
the influence of the climate had exercised a remarkable curative effect. The observations of M. Beaumier,
especially those for temperature, may help to explain
these facts, as they certainly show that Mogador
enjoys a more equable climate than any place within
the temperate zone as to which we possess accurate
information.   These are a few of the results:

		Fahr.
Mean temperature during eight years	. .	66·09°
„ „ for hottest years	. .	68·65
„ „ for coldest years	. .	65·75
„ of the annual maximum .	. .	82·05
„ „ minimum .	. .	53·00
Highest temperature observed	. .	87·08
Lowest „ „ .	. .	50·07

More striking still is the comparison between the
temperature of summer and winter. The following

results show the monthly mean temperature, derived from eight years' observations:

Summer				Winter		
	June	.	70·8°		December	61·4°
	July	.	71·1		January .	61·2
	August .		71·2		February .	61·5

Showing a difference of only 10 degrees Fahrenheit between the hottest and coldest months. It has not been possible to ascertain accurately the daily range of the thermometer, as there were no self-recording instruments employed ; but there is reason to believe that this would exhibit a still more remarkable proof of the equability of the climate. So far as observations go, they show an ordinary daily range of about 5 degrees Fahrenheit, and rarely exceeding 8 degrees. A comparison of the climate of Mogador with that of Algiers, Madeira, and Cairo, which have nearly the same mean winter temperature, will show that the mercury is occasionally liable to fall considerably below 50 degrees, and that the summer heat is greatly in excess of the limits that suit delicate constitutions, the mean of the three hottest months being about 80 degrees Fahrenheit at Algiers, about 82 degrees at Funchal, and 85 degrees at Cairo.

Rain falls at Mogador, on an average, on forty-five days in the year ; and out of one thousand observations on the state of the sky, the proportions were— clear, seven hundred and eighty-five ; clouded, one hundred and seventy-five ; foggy, forty ; the latter entry referring to days when a fog or thick haze prevails in the morning, but disappears before midday. The Desert wind is scarcely felt there. On an average it blows on about two days in each year, and then has much less effect on the thermometer than it has in Madeira, owing, no doubt, to the protective effect of the chain of the Great Atlas. These remarkable climatic conditions have been mainly attributed to the influence of the north-east trade wind, which sets along the coast and prevails throughout a great part of the year ; the average of the north and north-east winds being about two hundred and seventy-one days out of three hundred and sixty-five.   The great Atlas chain, with its branches that diverge northward towards the Mediterranean, screen the entire region from the burning winds of the Desert, and send down streams that cover the land with vegetation.   There are not half-a-dozen days in the year that may not be spent agreeably out of doors.   Some of the salubrity of the climate may possibly be due to the circumstance that the north-north-east winds come saturated with vapour, and charged with minute particles of salt from the breaking of the Atlantic waves on the reefs near the town.   There is a competent resident French physician.   The chief drawbacks at present would probably be in respect of the food supply, certain comforts necessary for invalids, and society.   It may be that Mogador is destined to succeed to the renown, as a sanatorium for consumption, once possessed by the adjacent island of Madeira.

But the best known health resort in Morocco is *Tangier*, and as it is only thirty-five miles from Gibraltar, from which place it is reached in three hours by steamer, and as it is known to possess a very fine climate, it is somewhat to be wondered at that it has not become more popular with Europeans.   But, as Messrs. Hooker and Ball observe in the work from which I have already largely quoted, Morocco, though a country close to Europe, is among the least known regions of the earth.   " Nothing is more rare," they remark, " than to find a country where neither the natives nor foreign visitors have any complaint to make against the climate, and in that respect Morocco is almost unique."   Tangier has rather a large annual rainfall, thirty inches, but it falls principally at one season, and that is October and November.   In the winter and spring the air is usually delightfully clear and bracing, and the daily temperature remarkably equable, the thermometer in the shade during the day ranging from 60 to 66 degrees Fahrenheit. Its mean winter temperature is 57 to 62 degrees. It would seem, in the case of Tangier—as, indeed, must be the case in sea voyages—that humidity of air under certain conditions is no drawback to salubrity,* for on this part of the Morocco coast " the breezes, whether they travel eastward from the Atlantic, or westward from the Mediterranean, are laden with aqueous vapours nearly to the point of saturation, and nothing dries spontaneously by mere exposure to the air."

The town of Tangier is built on rocky ground, rising steeply from the shore to the west of a shallow bay, and behind it to the westward rise undulating hills stretching up to the Djebel Kebri, or Great Mountain. From the hills to the west of the town magnificent views are obtained, including the Mediterranean, the Straits of Gibraltar, with " its double stream of vessels of every size and every nation," the coast of Spain, and the chain of mountains stretching towards Malaga. On the eastern side of the bay the shores are low and sandy, but backed by the " rugged range of the Angora Mountains, culminating in the Apes' Hill opposite Gibraltar." The city is surrounded by zigzag walls on all sides, and entered by three gates, which are closed at nightfall. " The main street is as rough and steep as the most neglected of Alpine mule-tracks, and disfigured by heaps of filth ; importunate beggars, of revolting aspect, led about by young boys, assail one at every step ; the miserable shops are mere

* Hooker and Ball make the following judicious remarks on this head : " To the human body there is nothing unpleasant in the effects of such air (at a temperature of 75 to 85 degrees) when nearly saturated with vapour, and so long as the temperature remains habitually between 70 and 80 degrees it is decidedly favourable to health." . . . . " Air at 50 degrees Fahrenheit cannot at the utmost carry more than about four and a half grains of aqueous vapour to the cubic foot ; but at that temperature it produces, when nearly saturated, that feeling on the nerves of the skin, familiar to every inhabitant of this island, which is the ordinary forerunner of colds, sore-throats, rheumatism, etc."

recesses, where, in an unglazed opening, little larger than a berth in a ship's cabin, the dealer squats, surrounded by his paltry wares." But, "unlike the towns of Southern Europe, where the main thoroughfares are cared for by the local authorities, while filth is allowed to accumulate in the byways, the dirt and offal are here let to lie under one's nose in the most public places, while the steep narrow lanes that intersect the masses of closely-packed houses are generally kept clean and bright with perpetual whitewash."

As there is no drainage in Tangier this admirable climate is marred somewhat by the drawbacks of dirt, bad smells, and the complete absence of roads, and it is only in the immediate neighbourhood of the town that Europeans can safely walk or ride without an escort.   As there are no roads and no vehicles, camels, horses, donkeys, and mules offer the only means of getting about.   So completely unlike any European resort is it, that an American gentleman said of it he "guessed he'd been taken up by the scruff of the neck and set down in the Old Testament." "We carried away from Tangier (Hooker and Ball) the impression that even on the Mediterranean shores there are few spots that combine such advantages of climate, natural beauty, and material comfort." Another writer observes that "the climate is far from unpleasantly warm ; the extraordinary prevalence of the east wind, or levante, causes it to be extremely trying to people with weak chests."

Small steamers cross from Gibraltar to Tangier in about three hours.   The landing is made in boats. There is a good hotel, kept by Mons. Bruzeaud, built on rising ground a few hundred yards outside the town, and commanding a fine view of the bay. There is excellent shooting in the country round.

Algiers will commend itself to many who are in search of winter quarters.   It will commend itself to those who have "done" Egypt, who are weary of the Riviera, and who do not dread a passage of forty or more hours across the Mediterranean. The touch of Oriental as well as modern military life there, the great interest of the town itself, as well as the variety of interesting excursions in the neighbourhood, the gaiety and vivacity of the French settlement—these are decided attractions for many of those who have to spend each recurring winter out of their own country. The journey is not a long one, and good steamers leave Marseilles on the Tuesdays and Saturdays of every week, and profess to accomplish the crossing in thirty-six hours; passing by the islands of Minorca and Majorca on the road, it is a pleasant enough voyage in fine weather. But for those who suffer much from a sea voyage—and the passage is often a rough one—there is scarcely enough in the climate of Algiers to compensate them for their sufferings. Algiers has its admirers and its detractors, which may be taken to prove that it has its bad seasons and its good seasons. There are discrepancies, too, in the different accounts of the mean annual rainfall, one giving it as twenty-eight inches, another as thirty-six; so with regard to the average number of rainy days in winter, one author making it forty-two, and another seventy-two; so also with the mean winter temperature, estimated by one observer at 62·13 degrees Fahrenheit, and by another as 57·2 degrees. There can be no doubt, however, that much rain falls during the winter months at Algiers, and many invalids have complained of having encountered very wet seasons; but authorities differ even as to which are the wettest months; one (" Encyclopædia Britannica ") says December and January, another (Murray) says November and February, and a third (Scoresby Jackson) November, December, and January. All, however, seem agreed that March and April are the best months. The winter temperature of Algiers is, on a general average, about 10 degrees Fahrenheit higher than that of the Riviera, and 18 to 20 degrees Fahrenheit higher than in England. The difference between the day and night temperature is not so

marked as on the Western Riviera; but as soon as
the sun sets the air often becomes highly charged
with moisture. The thermometer very rarely descends
to the freezing-point; one observer found it only do
so twice in twelve years. Although the winter rainfall
is so considerable, the climate is said to be the reverse
of damp and relaxing, for a rainy day in Algiers may
simply mean a heavy shower of half-an-hour or an
hour's duration, and as soon as the shower is over the
invalid can take exercise again in the open air; and
it may be worth bearing in mind that in a large city
like Algiers these occasional heavy falls of rain serve
to wash the air and keep it free from the accumulation
of impurities due to organic emanations from men
and animals. But I have heard of winters in Algiers
when the rain fell in great quantity, " nearly daily,
and often all day," in the months of November,
December, January, and February. The prevailing
wind is the north-west, a " cold and dry wind," blowing across the Mediterranean. The Sirocco blows
but seldom, perhaps for three or four hours during
four or six days in a month; but it is excessively
disagreeable while it lasts, for, coming across the
great desert of Sahara, it is laden with a fine penetrating dust, and feels hot and burning like a blast
from an oven. The climate of Algiers, less exciting
and milder, and more equable than that of the Riviera,
is not humid and relaxing like Madeira; it seems,
therefore, capable of exercising a tonic and bracing
influence in many cases of chest disease, as well as in
other chronic maladies. This kind of combination of
tonic and sedative climatic influences is peculiarly
suitable to cases of protracted recovery from pleurisy
and pneumonia, and to cases of early phthisis in somewhat feeble, lymphatic constitutions, or in cases where
the existence of nervous irritability or excitability
would counter-indicate a residence on the Riviera.

Persons, however, subject to bilious disturbance
complain that after a little time the climate of Algiers

makes them very uncomfortable, and that they are compelled to leave it.

An oft-repeated objection to Algiers is the un-healthy and foul-smelling state of the picturesque old part of the town; and the hotels in the town (of which the best is the Hôtel d'Orient) have not been considered as very safe dwelling-places. The wines are also said to be bad, and to be, together with other insanitary influences, responsible for upsetting the livers of bilious Europeans. But these objections to the town of Algiers do not apply with the same force to the suburb of Mustapha Supérieur, which consists of villa residences surrounded by their own gardens, and beautifully situated on the slopes of the hills to the south-west of the town. Family parties should arrange to take one of these villas for the winter season. They are hired from the French residents, who move into the town during the winter. Some caution is necessary in the selection of a villa. It is important that the house and garden should be exposed as much as possible to the sun all the day; for in the shade and in wet weather the atmosphere is often very damp and chilly. It is also very important to inquire carefully into the questions of drainage and water supply.

Mr. Otter* calls attention to the somewhat widely-spread error that the atmosphere of Algiers is unusually dry and rendered irritative by the pre-sence of "desert sand." He says, "the soil is, in fact, deep, rich, and damp, and there is no desert within 100 miles of the town. It is strange," he adds, "how much, even slightly, bad weather is felt in Algiers; the cold wind seemed more trying, and the rain colder and wetter, and to leave a damper feeling in the air, than in any other country which I ever visited."

The English society in Algiers is pleasant, and

* "Winters Abroad."

consists chiefly of breakfast and luncheon parties, and afternoon receptions and dances, and there are beautiful rides, drives, and walks into the surrounding country.

The following recent particulars have been kindly furnished me by one of the best resident authorities in Algiers :

Excellent hotels now exist at Mustapha Supérieur, and these are frequented by the English residents in preference to the town. It may be assumed that every foot one rises on the hill of Mustapha one gains in climate. There is, perhaps, more wind on the top of the hill, but there is more sun also, and no damp. The upper level also has an incalculable advantage ; once there, the visitor can walk for miles amidst lovely scenery, on level ground, whereas, lower down there is nothing but ascent and descent on high-roads. The distance is hardly a disadvantage as there are numerous omnibuses running. As to accommodation there—beginning at the top of the hill, there is the excellent boarding-house of Mrs. Jennings, *Villa du Palmier*, where every English comfort is obtainable, and invalids are sure of good attendance in case of sickness. Next comes the Hotel Kirsch, which is very good (landlord charges extra for table wine). The Grand Hôtel de Mustapha Supérieur, kept by M. and Madame Gebetin, is also very good.

The cost in all is very much the same—about 100 francs a week for each person. There is also the sanatorium of Dr. Landowski, who receives a limited number of invalids at a charge of £40 a month.

The same authority observes that he has seen the following plan followed, on many occasions, with great advantage : In cases of early or threatening consumption, the best chance here is to spend two winters and the intervening summer without moving ; for he has seen reason to believe that, on most occasions, the good to be obtained by wintering abroad is counteracted by the fatigue and chill inseparable from a visit to England in summer. He has found during the summer passed in a house on the highest part of the hill at Mustapha that there have been only ten days in the whole summer when the thermometer reached 80 degrees Fahrenheit. "There is no place in the Mediterranean so cool in summer, excepting Mount Lebanon and Troados, in Cyprus."

Hammam R'Irha is a health resort in Algeria which has several times within the last few years been brought to the notice of the medical profession and the health-seeking public in this country. What measure of popularity the future may have in store for it it is impossible to say ; but, for the present, it certainly suffers the reproach of having been unduly advertised, and those who became acquainted with the arrangements and the life there in the winter of 1883–84 complain that the attractions set forth in the advertisements of the place were by no means realised.

This place is said to have many natural advantages ; it is situated in a depression of the Lesser Atlas range, at an elevation of 2000 feet above the level of the sea. It possesses hot springs, adapted to the treatment of rheumatism and allied affections, which were known to and valued by the Romans; and the native Arabs now make pilgrimages to them. It is about sixty miles west-south-west of Algiers. There is first a journey by rail of four-and-a-half hours to Bon Medfa, and then a drive of eight miles to the baths. The place is surrounded and protected on nearly all sides by mountains, and there is an extensive pine-wood about a mile and a half to the north-west, and ash, oak, olive, and eucalyptus trees flourish on the hills around. Roses, violets, and geraniums flourish throughout the winter. There is, however, a defect in the mountain protection to the place in the shape of a cleft which lets in the stormiest wind in Algiers—the north-west—which blows at times so fiercely that the places which were previously warm and sunny become cold and bleak. There is a small French military hospital there for invalided soldiers, as well as the old baths, still used by the Arabs and

Jews, and near the latter is a small hotel built in 1878. Another large hotel, only partly completed and habitable (1884), is constructed 150 yards higher up the hill.

There are pleasant excursions and walks in the country around, and the winter climate is said to be suitable to rheumatic sufferers.

Looking at a table published by Dr. Saville and constructed on data furnished by the observations of Dr. G. H. Brandt made during the winter of 1883–84, it would seem that there is a fair proportion of severe winter weather encountered there; somewhat more than one looks for in a *southern* health resort.

We find that 30½ inches of rain fell between November 15th and the end of March, that 4 inches of snow fell one morning in December; that in December there was a minimum temperature of 29 degrees Fahrenheit (froze during two nights) and in March a minimum temperature of 31·0 degrees Fahrenheit; that in December there were eight rainy days and eighteen cloudy ones; that in March there were twelve rainy days and twenty cloudy ones, four days rain all day, six days cold wind, and one day Sirocco, and one day of thick fog in January.

This scarcely looks like a winter paradise for sufferers from chronic rheumatism !

The air is no doubt pure and bracing, and is said to resemble the mountain air of Scotland in the summer months, only it is drier. There is a gaseous mildly chalybeate spring within three-quarters of a mile of the bathing establishment, as well as the hot spring used for the baths, which has a temperature, at the source, just behind the hotel, of 136 degrees Fahrenheit, and contains sulphate and carbonate of lime as its chief constituents.

The baths are administered very hot, 112 degrees Fahrenheit, and can be taken privately or in swimming baths ; there are also hot and cold douches. The hot water is drunk while in the baths. The main effect of the treatment is the promotion of free action of the skin. The best results are obtained in cases of rheumatism in young and healthy persons, but even old chronic cases are benefited ; cases of chronic bronchitis and emphysema are treated successfully by inhalation of the hot vapour. Dr. Saville, who has spent some time at Hammam R'Irha, thinks it a good place to migrate to from Algiers in the spring when the latter place gets oppressively hot.

He also thinks it a suitable resort in winter for those who are recovering from protracted illness, and for those who, having broken down from worry, anxiety, and over-work, require a quiet place of rest where the air is mild, yet pure and exhilarating. It certainly does not appear to be a fit place for delicate invalids.

The accessibility of the Riviera, and the natural attractiveness of its many beautiful resorts, will always make them popular with those, whether they be invalids or not, who desire during the winter to exchange a clouded sky for a cloudless one, and the confinement of their own rooms for free sunlight and sea breezes. Not that the climate of the Riviera is by any means a perfect one. It has a heavy rainfall and a fair number of rainy days; the transition from day to night temperature is sudden and considerable ; damp and chill evenings succeeding to hot and dry days ; it can, and does, freeze there, though not often ; it is tormented by some of the fiercest and most disagreeable winds that blow, and in some parts, as e.g. Nice and Hyères, clouds of dust make life at times a burden. Yet, notwithstanding all these drawbacks, the Riviera has many recommendations, as we shall presently see. Its position with regard to the chain of the Maritime Alps gives it almost complete protection, especially its western portion, from the north; and to this and its southern exposure, and the relative warmth of the Mediterranean,* it owes its

* Dr. Marcet has shown that the temperature of the surface of the Mediterranean at Cannes is from 5 to 9 degrees Fahrenheit higher than it is on the west coast of Cornwall, and he

warm climate, for its mean temperature is from 8 to 9 degrees Fahrenheit higher than that of England. Though the rainfall is great—for instance, at Nice, nearly twice as much rain falls during the winter months as in London—yet there are fewer rainy days, for during the same period there are eighty-nine rainy days in London to thirty-six at Nice. The rain falls in heavy torrents for a short time, and then there is no more rain for days; while as to cloud, we find also in the same period, that whereas London has only twelve cloudless days Nice has ninety-seven. Then, as to relative humidity of atmosphere in winter, if we compare Cannes and London, it is (estimated roughly) as seventy-five to ninety. It must be remembered, however, that the Eastern Riviera has a much larger rainfall and more rainy days than the Western. At Genoa one-third more rain falls in the winter than at Nice, and nearly twice as much as at Hyères, and it has nearly twice as many rainy days as Hyères, and a third more than Mentone or Cannes. The greatest rainfall is in October, then November, December, and March, and the smallest is in February. There are, of course, here, as everywhere else, quite surprising variations in different years.

Owing to the general prevalence of land winds the air is much drier and more exciting than that of littoral climates generally. The north-west wind, or Mistral, is an exceedingly dry cold wind, as it parts with all its moisture in traversing Central France. It is more felt in the western towns than in those east of San Remo; it blows with greatest violence in March, and is exceedingly hurtful and unpleasant both to sick and sound; it blows, also, frequently with much violence in April, and once or twice in each of the other winter months. The Sirocco, the hot ener-

concludes "that the temperature of the air near the Mediterranean must derive a considerable accession of heat throughout the winter from that which is stored up by the water during the summer."

vating Sirocco, coming from Africa, blows chiefly in the spring and summer, and often, also, for two or three days in the winter months; it appears not to be a moist wind, as stated by some, but to be usually followed by rain. The north-east wind, or Greco, is felt most severely and frequently along the Eastern Riviera, and gives to the climate of Genoa its peculiar bleakness. "It is a biting cold wind, often accompanied with sleet, hail, or snow." The Tramontana, or north wind, owing to the northern mountain boundary which protects the Riviera, is less felt than those other winds, except in places, like Ventimiglia, exposed to winds blowing down long valleys penetrating the chain of Alps in a direction due north and south. There are also, of course, the regular daily breeze which blows from the sea on to the land, and the regular evening land breeze blowing seaward. It must not be forgotten either that in the towns along the Riviera, the invalid is exceptionally well off in point of hotels, house accommodation, the command of the best medical skill, and in many places the presence of nurses trained in the hospitals of London.

I shall attempt in another chapter to estimate the various claims and to state the different characteristics of the various important health resorts along the Western Riviera. Speaking generally, the climate of the Western Riviera is tonic, stimulating, and exciting, especially useful in cases where the vital energy is drooping and wants flogging into renewed activity. It often proves injurious to persons of a nervous and irritable temperament, and to cases which have a tendency to febrile excitement. It is on this account often ill borne by many hysterical persons and hypochondriacs.

The resorts on the Riviera di Levante—the Riviera to the east of Genoa—are not so popular as winter residences as those to the west of that famous city, and, indeed, their winter climate is not, on the whole, so suitable to invalids, especially to those who suffer from chest affections. The Riviera di Levante is colder and wetter than the Riviera di Ponente. It is not so well protected from the north; the sheltering mountains are not so high, they are usually further off, and gaps occur in the chain which admit cold currents from the north and north-east.

The rainfall is considerably greater and the relative humidity greater, though for littoral summer climates they are of only medium humidity, and would be classified rather among the dry than the humid winter stations.

Genoa and its immediate neighbourhood is peculiarly unsuited to pulmonary invalids, as its climate is very changeable and often cold, windy, and rainy; but between Genoa and Spezia there are many spots on the Italian Riviera well suited for winter residence to those who are simply looking for a milder climate and more sunshine than can be found within our own shores. A very few are sufficiently sheltered from the prevailing winds to be adapted to delicate pulmonary invalids in winter; one of these is *Nervi*.

Nervi, which may be regarded as almost a suburb of Genoa from which it is only six miles distant, lies in 44·22 degrees north latitude and 6·42 degrees west longitude. It is well sheltered towards the north and fairly so towards the east, but it is more exposed to the north-east and north-west. Owing to the steepness of its mountain boundaries, and the absence of good roads, there are but few walks and possible excursions, so that it is only suited to invalids seeking repose and quiet. Its climate is less tonic and exciting than the resorts in the Western Riviera, it is less windy, more humid, and has a greater number of rainy days.

The town of Nervi has about 5000 inhabitants, and has one or two good hotels, the chief being the Pension Anglaise, which faces due south and has fifty south windows. The Hôtel Oriental is smaller and nearer the sea, and is said to be well managed. Gardens with orange and lemon orchards occupy the

level space between the town and the sea-shore, giving
to the town a very picturesque aspect. A sunny and
sheltered promenade has been cut along the rocky
shore, and is admirably suited for invalids, as they
can enjoy there a temperature twenty or more degrees
higher than in the shaded streets in the town.

Nervi is not free from the visitations of the cold
and bitter north-west wind, which here, as elsewhere
along the whole Riviera, makes itself violently and
dangerously felt by those who incautiously expose
themselves to it. The south-east or Sirocco is of
frequent occurrence, and renders the climate, for the
time, somewhat damp and relaxing.

There are on an average about fifty-four rainy
days from December to April, and a rainfall during
the six winter months, November to April, of 25½
inches.

The temperature observations available have not
been collected with sufficient care or extended over a
sufficient length of time to be of any value for purposes of comparison. One observer gives the mean.
temperature at 8 a.m. in the months of

	November	December	January	February	March
as	56°	52°	51°	44°	49°

It would seem to be much cooler in spring than·
Mentone and the other western resorts; but in winter
it would appear, in some seasons at any rate, to be
warmer. There can be no doubt as to the mildness
of its climate, as is evidenced by the growth of
standard lemon-trees and an abundant sub-tropical
vegetation.

Nervi is at present chiefly German, and does not
offer the same social advantages which our country-
people obtain in the resorts west of Genoa.

Rapallo is another well-sheltered spot on the
coast to the east of Genoa, and, like Nervi, is
regarded as having a future before it as a winter
resort for invalids. It is about an hour and three-
quarters by train from Genoa (twenty-two miles), in

latitude 44·15. The town, which is said to have 12,000 inhabitants, is situated on the north side of a little bay, with a very narrow entrance, so that it is much shut in by a promontory to the west and another to the east. These two promontories are connected towards the north by a semicircle of hills ; this natural protection from the access of cold winds must give Rapallo a very mild climate. Olive, chestnut, and fig trees cover the lower hills, the more distant ones to the north-east rise to between two and three thousand feet above the sea.

The hotel accommodation is very limited — the Hôtel de l'Europe, an old palazzo, is to the east of the town and not far from the sea. The accommodation it affords is good and the prices moderate. There are a number of pleasant walks along the coast and inland, which give it an advantage over Nervi. It has a sandy beach, which is another advantage. Its climate is no doubt warm and rather moist. It has been said to resemble that of Pisa ; but its situation and surroundings are far more attractive. Altogether Rapallo has much to recommend it as a winter resort, the chief drawback being the limited accommodation it affords to visitors.

To those who are not so dependent on shelter from prevailing winds there are many attractive spots between Nervi and Rapallo, but they are only suited to those who have some acquaintance with Italian life and language. One of the most picturesque of these is Santa Marguerita, commanding some of the most beautiful scenery in Italy, and having a new and good hotel. The other villages that deserve to be mentioned are Bogliaseo, Sori, Recco, and Camogli ; and Quinto and Quarto, nearer Genoa.

No doubt the most attractive resort, from many points of view, on the Riviera di Levante is the beautifully situated town of Spezia.

Spezia is a town with 20,000 inhabitants and a famous arsenal, built at the north-west angle of a magnificent deep bay or gulf, formed chiefly by the projection of a rocky promontory to the west, about four miles in length. There are delightfully quiet valleys and sheltered bays to the east of the town, well suited for walks and excursions.

The western coast of the gulf is rugged and hilly, but the northern and eastern part is comparatively level for about three miles, and this is utilised for walks and drives. A charming excursion can be made daily by steamer from Spezia to the little towns of Lerici and San Terenzo, on the south and north extremities of the bay respectively, and each with a castle on a promontory projecting into the sea. Shelley's villa is the chief object of interest at Lerici. The town and gulf are open to the south and south-east, but they are protected to the north and west by steep and high mountains, whose spurs stretch down into the sea. Its climate is a mild one, but the town and bay are not protected from the prevailing wind which is from the south, nor from the Sirocco which not infrequently blows from the south-south-east. The climate is moderately warm, moderately moist, calm and tolerably equable, and free from dust. It has large and good hotels.

Spezia may be commended to the robuster sort of invalid who is not greatly dependent on protection from wind, who prefers a somewhat moist to a very dry climate, and who desires to meet with facilities for sailing, boating, riding, and a tolerably active life. Its air is said to be soothing and comforting to those who suffer from sleeplessness in the resorts of the drier and more exciting Western Riviera.

Only one other place on this coast need be here mentioned, and that is *Viareggio*, situated between Spezia and Leghorn. Having a fine sandy beach, it has hitherto been chiefly known as a summer bathing station; but its mild climate, and the immediate adjacency of large pine-woods affording both shade and shelter, tend to bring it into favour as a winter resort. It is, however, somewhat exposed to mists.

It is scarcely necessary to allude to *Pisa*, only a

few miles from the coast, except to mention its former reputation as a winter resort. This town is the dullest of places, its streets are dark and cold, and its climate damp and rainy ; the atmosphere is, however, calm, calmer than on the sea-coast, and its temperature equable. It is difficult to understand why invalids were ever recommended to winter in this very unattractive locality. There are attractions here for the passing visitor in its architectural monuments, but none to induce the invalid to make it his winter residence.

There is another group of health resorts, which, although littoral stations like those on the Riviera, have very different properties and characteristics. I allude to the comparatively sedative climate of the stations on the south-west coast of France—Arcachon, Biarritz, St. Jean de Luz—the adjacent Spanish town, San Sebastian, and with these littoral climates we may associate the neighbouring inland health resorts of Pau and Dax.

Arcachon, as an example of a sedative, yet not a relaxing climate, no doubt possesses advantages for the treatment of certain maladies. Ten miles from the Atlantic coast, from which it is separated by high sand dunes covered with pine forest, it is protected to a great extent from the fury of the west and south-west winds by the dense forest, which also offers a protection from the winds coming from the east and south-east. To the north of the town lies a great sea basin, a harbour many miles in extent, inclosed on all sides, only communicating with the Atlantic by a narrow channel running almost due south. The north and north-east winds must pass over this basin, and become thus somewhat warmed in winter and their irritating dryness diminished, while it is maintained that they also bring from the surface of this unusually salt sea water, and from the vast extent of sands exposed by the retreating tides, an appreciable amount of saline and other marine emanations, to give a special efficacy to the air in certain scrofulous conditions. It shares also in the equable temperature which belongs to littoral climates. It must be admitted that the air at Arcachon contains much moisture, owing to the west and south-west winds which blow in from the Atlantic and bring much rain and mist ; but, owing to the extreme porosity of the soil, which for miles and miles is wholly sand ("there is not a stone within twenty miles," was the testimony of a resident Englishman who knew the district well), the water is drained off from the surface as soon as it falls, so that there can never be any stagnant water on the ground. The air of the forest is also impregnated with the balsamic resinous emanations from the pine-trees, peculiarly grateful to some forms of chest affections ; and, moreover, it is found to be very remarkably rich in ozone. I have heard it somewhat carelessly remarked that the pine-trees have been ruthlessly cut down at Arcachon, and this is a fair example of the kind of superficial criticism that often passes current with respect to health resorts. It is not, of course, possible at Arcachon or elsewhere to build houses on the tops of trees ; and so long as this is the case, if houses have to be built, trees to a certain extent must be cut down to make room for them. But Arcachon contains two quite distinct parts : there is the Plage, a level tract on the south shore of the *Bassin,* which is occupied by somewhat closely packed streets and houses, and which in summer-time becomes a sort of Margate for the population of Bordeaux, from which town it is distant only an hour and a quarter by express trains ; there is also the Ville d'Hiver, separated from the former by a high sand-hill, and consisting of numerous villa residences actually built in the forest ; each house being surrounded on all sides by pine-trees. The prevailing winds, north-west, west, and south-west, being sea winds, are not cold like continental winds ; but they often blow with great violence, and were it not for the protection of the lofty pine-trees, *over* the tops of which they blow, they would form a serious drawback to the climate. They are most frequent from December to February, and they usually blow continuously, day and night, for several days in succession ; it follows that there is less sudden transition from day to night temperature here than in the Riviera. The climate of Arcachon is, in short, moderately mild and soothing, and it is especially suitable to cases of irritative bronchial or laryngeal catarrh, to cases of early phthisis with tendency to congestion or inflammatory complications, and to persons of nervous temperament. It is not suited to persons of a lymphatic and torpid habit, who do better in the tonic and stimulating air of the Western Riviera. Cases of consumption and of other chronic lung diseases have certainly been arrested at Arcachon, and dyspeptic persons, in whom the dyspepsia has been complicated with hysteria, hypochondriasis, and nervous irritability, have derived great benefit from its climate.

Dr. F. Fagge, who is in practice there, states that last winter (1883–84) they only had about two days' fog, and only five or six of frost.

Dr. Rains, however, of Charlton-cum-Hardy, who had a winter's experience at Arcachon, thus writes of it :

The villas there are most unsuitable habitations for an invalid to be located in during the winter months. They are built for other people as summer residences, and let from October to May or June to people who live by "taking in" invalids. · The villas are badly constructed for winter residence, no fireplaces with fire to cheer the drooping frame, as in England, but merely a few logs of wood burning in a square opening, around which, in December and January, you may have to sit with your overcoat on and a rug round your legs to keep warm in the evening. The place itself looks desolate and miserable, the Casino and nearly all the shops being closed till spring. The water is not good, and the drainage is bad ; but the forest, with its atmosphere impregnated with terebene, is all you could desire, provided you are prepared to wander about, talk to the trees, observe nature, and sit on the verge of the forest, overlooking the Bay of Biscay, watching a fine sunset. If a few friends go together, and take a villa on well defined terms, get an English fire-grate or two put in, and burn small coals at 40s. per ton with their logs of oak or pine, they may do pretty well, providing they are prepared to pay high prices for everything else they may require. Finally, let me say that Arcachon is pleasant for the French and Spanish in summer, but not adapted to the requirements of English invalids in winter.

I have also received complaints from winter visitors of the accommodation procurable in the "Ville d'Hiver," and some efforts should certainly be made to establish a thoroughly good and commodious hotel there, with some resources for the amusement of invalids, if it is desired to attract English patients to spend their winters at Arcachon.

At present, they justly complain that it is *triste* and unattractive.

A useful "Handbook to Arcachon" has been published by the English chaplain there, the Rev. S. Radcliffe.*

Biarritz, situated in 43 degrees north latitude, has in some respects the same qualities as Arcachon ; but as it lies exposed to all the fury of the Atlantic winds, and has no protection like the pine forests of Arcachon, it is more bracing and less mild than it, and by no means so suitable to cases of chest disease. It is, however, well suited to some forms of nervous exhaustion and irritability.

Its climate is bright and exhilarating for a great part of the year ; the relative humidity, as might be expected, is rather high, but owing to the dryness of the soil, the heavy falls of rain are rapidly absorbed, and the air is rarely felt to be damp. As a winter resort it is most suitable to hypochondriacal persons and to those who suffer from depressed states of the nervous system. Old Indians are said to find it suitable to the ailments they are prone to suffer from. In the spring it is a pleasant place for a change for those who find benefit from a sea climate with a fair amount of bright sunshine.

Some asthmatics and invalids who require no special protection from strong winds do well there.

* London : T. Lawrie, 31, Paternoster Row.

The town is built in a commanding position on bold and lofty cliffs which dominate from a considerable height the Bay of Biscay, and form a part of the eastern shores of the Atlantic. The broken rocky coast there is very picturesque, and Biarritz has many attractions both as a winter and a summer resort. The winter season, when it is chiefly occupied by English visitors, extends from November to the end of March. Those who care especially for the bracing qualities of the climate leave about the end of March, when the weather begins to be, at times, rather too warm and relaxing for such constitutions.

There are excellent hotels at Biarritz, and a Casino where various amusements are provided. There are also many interesting walks and drives, and long and short excursions around Biarritz, owing to the adjacency of the Western Pyrenees. The interesting town of Bayonne is only a quarter of an hour distant by rail, and Pau can be reached in the same way in between three and four hours.

To resume, the climate of Biarritz is a sedative one, and is in this respect contra-distinguished from the somewhat exciting climate of the Mediterranean littoral. It is also bracing as well as sedative ; it could scarcely be otherwise with the strong Atlantic winds blowing over its towering cliffs and its mighty waves dashing against their feet. It is this combination of bracing and sedative properties that makes the climate of Biarritz valuable to so many chronic invalids. It is, however, too blustering and humid a climate for many forms of chest disease, and it is on that account rarely resorted to by the consumptive. Some invalids, who find the resorts on the Riviera too exciting, and who become sleepless and feverish there, are benefited by being transferred to Biarritz.

St. Jean de Luz, situated a little to the north of the last westward spurs of the Pyrenees, as they stretch toward the Atlantic, is beautifully situated in a fine bay a few miles south of Biarritz, with the climate of which it has much in common. It is, however, more protected from winds, being surrounded by hills to the north-east and south-west, and better suited therefore to pulmonary invalids. But it would be found dull and deficient in accommodation compared with other winter stations. Owing, however, to its vicinity to the Western Pyrenees, many interesting excursions can be made from it. St. Jean de Luz, though duller than its attractive neighbour, Biarritz, has the advantage of being cheaper and of offering greater quiet and retirement for those who seek them. The mountain called La Rhune, nearly 3000 feet high, and from the summit of which a magnificent view may be obtained, is only four or five miles off. It was the scene of a bloody and victorious battle of the allied forces under Wellington, against the French army commanded by Soult.

About eight miles from St. Jean de Luz is *Hendaye*, the last French village, with a fine beach, and attempts are being made to convert it into a popular resort; one of the suggested attractions being its nearness to the old and interesting Spanish frontier town of Fontarabia, which has a Casino resembling that at Monaco, and is remarkable for the beauty of its population and for its picturesque old streets and houses.

We have little precise information about the climate of *San Sebastian*, which is only ten miles by rail from Hendaye. It certainly shares the mild sedative character of the adjacent health resorts on the south-west coast of France, while it would in all probability be found warmer and more sheltered, and therefore better suited to pulmonary visitors, while for historical interest and beauty of situation it certainly carries off the palm. It is, however, chiefly resorted to in summer for its sea-bathing, and has not yet been much frequented by English health-seekers.

The climate of *Pau* has often been the subject of much controversy, and has had its admirers and its detractors; the variability of the different seasons which is found in all European health resorts affording ample support to both parties. It appears to be generally admitted, however, that its climate is sedative, and that an unusual stillness of atmosphere has often been observed to prevail there for long periods at a time. It is well protected from the north by a series of plateaux rising behind the town, and the Pyrenean chain, at a distance of fifteen miles, affords a barrier from the enervating south winds. It is unprotected, however, to the west and east, and it is occasionally visited by severe storms. The prevailing winds are from the north and west. Its average winter temperature, from November to April, is 43·2, but little higher than that of London. Its annual rainfall is considerably greater, being forty-three inches, but it has fewer rainy days, 119 in the year. Although it has a considerable rainfall, the relative humidity of the atmosphere is not very great owing to the sandy, porous nature of the soil, which rapidly absorbs the water as it falls. Frost and snow and cold nights are not uncommon in winter. Compared with the French health resorts on the Riviera it is moister, its rainfall and number of rainy days are greater, it has less sunshine and sun-heat, and periods of cold weather are more common. On the other hand its temperature is more equable, and it is much freer from winds. The climate of Pau is sedative, not bracing, and it is most suitable to irritable nervous persons having a tendency to febrile excitement. To such persons when suffering from chronic chest affections, spasmodic asthma, emphysema, bronchial catarrh, active forms of chronic phthisis with tendency to hæmorrhage and laryngeal irritation, the sedative climate of Pau may be recommended, but is not suited to relaxed lymphatic persons, or sufferers from loss of nerve tone.

Many patients who pass their winter at Pau remove in the summer to one of the adjacent Pyrenean resorts, Bagnères de Bigorre or Eaux Bonnes, where they escape the summer heat of the town, and where they can avail themselves, if so disposed, of a course of treatment at the mineral springs.

The town of Pau is magnificently situated, facing the entire chain of Pyrenees at an elevation of 650 feet above the sea. The panorama obtained from the terrace of its old and historically interesting château of this mountain chain, and the valley of the Gave at its feet, is most remarkable and extensive ; its only defect is that the mountains are a little too far off. The town is well and regularly built, and has nearly 30,000 inhabitants. It has recently been thoroughly drained. It contains some very excellent hotels, a good English club, and many other social attractions. It is indeed a town of pleasure as well as a health resort, and even provides fox-hunting for the British sportsman. The season is from November to May. Many charming excursions can be taken from Pau as a centre. It is only a few hours to Biarritz and the shores of the Bay of Biscay, and most beautiful and interesting tours of any length can be made into the Pyrenees in the fine days of spring.

Dax, in the department of the *Landes,* is situated about midway between Pau and Arcachon, and between twenty and thirty miles from the shores of the Atlantic. It is three hours and a half from Bordeaux by express train. It was known to the Romans as " Aquæ Augustæ," this became shortened into " Aquæ," and this became modified into " Acq" or " D'Acqs," and finally " Dax " !

In and around Dax a number of hot springs are found, but it is chiefly known as possessing a large establishment for hydropathic and other general treatment where these various springs are utilised, and where patients suffering from various chronic maladies—rheumatism, paralysis, ataxy, unhealed wounds, muscular contractions, neuralgias, cutaneous affections, scrofulous disease of the bones and other organs, etc.—are systematically treated by mineral springs of various composition, by baths and douches, by a special kind of mud bath, by electricity, and by all the various means of cure which can be brought together in a well-arranged establishment of this kind. It has also been warmly advocated as an advantageous place of winter residence for consumptive patients and for sufferers from asthma and chronic bronchial catarrh, who require a soothing rather than a bracing or stimulating climate. It is said to have a higher winter temperature than Pau, and a milder air. The presence of hot springs may have something to do with raising the temperature of the air, and the surrounding pine forest probably protects it to some extent from wind. The Atlantic gales must, however, be more felt there than at Pau. Its rainfall is said to be moderate and snow rarely seen.

The thermal *Établissement* at Dax is very complete, and affords an excellent winter residence for patients who, besides protection from the cold of winter, require special treatment under the supervision of resident physicians. A uniform temperature is maintained throughout the building by heated pipes, through which the water of the sulphur springs circulates ; and patients are made independent of outdoor exercise by the presence of large well-ventilated galleries where exercise can be taken.

Passing now again further south, to the other extremity of the Iberian peninsula, we find but a very few places in the south of Spain suitable for winter quarters ; the chief of these are Malaga, Gibraltar, San Lucar, Huelva, and Seville.

These are all in the southern portion of Andalusia, and situated between 36 degrees 8 minutes (the most southern Gibraltar) and 37 degrees 26 minutes (the most northern Seville) north latitude. San Lucar and Huelva are on the Atlantic south-western coast, Gibraltar and Malaga on the Mediterranean south-eastern coast of Spain, and Seville is about forty miles inland from the former.

Malaga is the only one of these Spanish southern

stations that has ever been much frequented by English invalids.

The serious defect of Malaga as a resort for invalids, notwithstanding its admirable climate, is that it is a large, densely populated, close, and not very healthy town, without any suburban hotels or residences for the reception of winter visitors, who are consequently pent up in the close streets or confined to the single promenade along the shore. The city is built on a flat sandy plain, its streets are narrow and close, and its sanitary condition highly unsatisfactory. Its climate is, however, one of the most equable on the Continent, and it is also very dry. The rainfall is small, and there are on an average only twenty-nine rainy days in the year; so that in some seasons the drought is very serious, and the absence of water, no doubt, at times contributes greatly to its defective sanitary condition. There are great variations, however, in the rainfall in different years.

It is doubtful if this constant sunshine and absence of moisture would prove invigorating and health-giving to the majority of invalids, unless they could at the same time live in fresh country air, which appears to be impossible here. Winter, as we understand that word in England, is almost unknown at Malaga, or is very exceptional; the mean winter temperature being 56 degrees Fahrenheit; and a whole winter may pass without the thermometer sinking to 50 degrees Fahrenheit even at night.

It is sheltered completely from the north and to some extent from the east by mountains which rise behind the town to the height of 3000 feet. It lies open to the south and to the sea. The winds are occasionally trying, especially the prevalence in winter of the land winds or "Terrals," which, from passing over the snow-covered Sierras, are cold and dry to a degree which proves very irritating to some invalids.

Owing to the mildness of the climate, the sugar-cane and most tropical plants flourish there, and the eucalyptus has been introduced with success, and has, to some extent, improved the sanitary conditions of the portions of the town where it has been planted. The orange groves are particularly fine.

There is, then, much in the climate of Malaga to recommend it as suitable winter quarters for invalids who desire to find warmth and bright sunshine in winter; but the drawbacks to Malaga on other grounds are very great. These are the discomforts of the hotels, the difficulty of getting well-cooked food, the absence of objects of interest in the town, and, as I have already said, the impossibility of invalids getting any accommodation outside the town. To these must be added the trouble of getting there and the difficulty of getting away. For either there is a very long land journey—thirty-six hours from Paris to Madrid, twelve hours from Madrid to Cordova, and six-and-a-half hours from Cordova to Malaga— or there is the sea voyage of four or five days by Peninsula and Oriental steamer to Gibraltar, with the terrors of the Bay of Biscay, and another short sea trip from Gibraltar to Malaga, which is very trying in the small Spanish boats; the large boats of the Compagnie Générale Transatlantique are, however, good.

Most travelled Englishmen are acquainted with *Gibraltar,* and of late years a certain number of the robuster class of English invalids have passed some part of the winter there. It has been written of as "an uncomfortable fortress where, every way the traveller turns, he finds a hill and a different temperature at every corner of its stuffy streets." It is, however, for those who are fairly strong, and only require a little southern warmth and sunshine, a fairly healthy and pleasant place of residence from November to May, but its attractions are limited and circumscribed, and the majority of visitors begin to get tired of it after about two months, and begin to want to go elsewhere. Then it is to be avoided in seasons when the "rock fever" is prevalent, the fear of which shortens the stay of many visitors. One drawback to the climate of Gibraltar is the occasional prevalence of easterly winds, the Levanter, which proves irritating and very trying to most persons. The hotel accommodation is also very limited; the Europa is said to be the best.

Of *San Lucar*, very little need be said; it is situated at the mouth of the Guadalquiver, where this famous river pours its waters into the Atlantic. It enjoys a very hot climate, but it is a dull town of 16,000 inhabitants, devoted to the manufacture of bad sherries, and situated in a treeless, sandy, undulating country. It can be reached by river steamer from Seville.

Huelva, a few miles to the north of San Lucar, on the same coast, has much more to recommend it as a winter resort. It has a fine situation at the confluence of the rivers Odiel and Tinto, and the famous Rio Tinto mines are about twenty miles distant. It is the seat of a small English mining colony, and is also important on account of its sardine fisheries. The water supply is excellent. Seville, which is fifty-three miles off, is a four hours' railway journey, and steamers run twice a week to Cadiz. There is now a good new hotel there—the Columbia Hotel, which has been built by the Huelva Hotel Company.

Huelva has a fine winter climate, and snow has never been known there. Its mean winter temperature (October to May) by day is 67 degrees Fahrenheit and by night 56·30, so that it is warm and equable.

It is only recently that good hotel accommodation has been provided there, and more information about its winter climate will, no doubt, soon be forthcoming.

Seville is not exactly a health resort, and yet it affords an agreeable and suitable residence in winter for certain delicate persons who require more warmth and sunshine than they can find at home. Ice and snow are said to be unknown there; and although it has occasionally wet winters, the climate is usually dry, bright, and sunny. It is scarcely necessary to add that it is a city full of objects of interest and that the hotel accommodation is fair.

There are many interesting and valuable climatic resorts in the several islands of the Mediterranean, and I propose, in the next place, to allude to the chief of these.

The only really available winter health resort in the island of Corsica is the beautifully situated town of *Ajaccio*, on a fine bay facing south-west and protected from the north and the east by high mountain ranges. It is thus greatly sheltered from cold winds, but is exposed to the south-west. It has a great number of bright sunny days, but it is undoubtedly a more humid climate than the adjacent Riviera though not so much so as Madeira. Its mean humidity is 80 degrees, its average winter temperature 54 degrees Fahrenheit, and spring temperature 59 degrees. There is a moderate, but not excessive daily range of temperature.

On an average there are thirty-five rainy days during the winter and spring months. A great advantage is the absence of dust. The scenery is magnificent and the country interesting. Some who have spent a winter season there speak of it with enthusiasm, especially sufferers from asthma. A writer in the *Pall Mall Gazette* says : " Even after my arrival I was so distressed in my breathing, that I dreaded even a flight of six stairs, and now I can climb to the top of the lower hills. I have never breathed a pleasanter air. There is a new hotel, the Belle Vue, in a most favourable position." There are beautiful drives and walks along well-kept roads, the vegetation is luxurious, and flowers are abundant.

There is a steamer, belonging to the Compagnie Transatlantique, from Marseilles every Tuesday at 5 p.m., which performs the journey in fifteen hours, and from Nice every Saturday in ten hours. There are also steamers from Leghorn to Bastia—the least stormy part of the Mediterranean—every Sunday, Wednesday, and Friday in six hours.

A railway is nearly completed from Bastia to Ajaccio through Corti, and, when it is finished, Ajaccio will be only nine hours from Leghorn. Diligences run in twelve hours between Bastia and Ajaccio, conveying the traveller through the most beautiful scenery of the island.

The climate of this place is no doubt of great value in cases of protracted convalescence from acute disease; in some cases of chronic rheumatism and gout, requiring a mild and sunny winter climate and not too stimulating an atmosphere; in cases of chronic chest disease without any tendency to active symptoms, and where there is plenty of reserve force; while it would seem to be very beneficial to certain asthmatics. It must be remembered that it is a moister climate than that of the Riviera, and warmer. There are comfortable villas built round the town, where good accommodation can be obtained, and there is, besides the Hôtel Belle Vue, another new hotel—the Hôtel Continental. Ajaccio has also an English church and a public library.

The well-known gaseous, chalybeate springs of Orezza are about eighty miles distant inland. The village of Orezza is situated about 2000 feet above the sea and has an hotel and bath establishment. It is a place of summer, not winter resort.

In the island of Sicily there are three towns that have been commended as winter resorts : Palermo, Catania, and Acireale.

Catania and Acireale do not offer any great attractions to English health-seekers, and but very little need be said of them here. As a change from Palermo, when tired of that place, either of these towns, or both, might be visited.

Catania is a modern, popular, and busy town. Its mean winter temperature is 53·8, and its daily range of temperature 14·5 degrees. There is less humidity of atmosphere and less equability of temperature than at Palermo, and there is a considerable difference between the day and night temperature.

The low temperature at night is referred by Dr.
Bennet to the adjacency of the snow-fields of Mount
Etna, at the base of which this town lies. The
environs composed of lava are bare and desolate, and
the place, though warm, has appeared somewhat dull
to English winter visitors. The *Hôtel Centrale della
Carona* is said to be excellent.

Acireale is ten miles north of Catania and nearer
Messina, from which town it is about two hours by
rail. It is 560 feet above the sea, on the southern
slopes of Mount Etna. It possesses warm mineral
springs, which are much frequented in summer, and a
fine bath establishment and hotel—" Grand Albergo
dei Bagni "—which is surrounded by gardens and has
fine views. It has been found " cold and desolate " in
winter. Its climate resembles that of Catania ; but
the town is smaller, and, owing to its elevation, its
atmosphere is no doubt fresher and purer. It ought
to be a good place to migrate to in spring from
Palermo or Catania. Acireale and Catania are both
very accessible from Malta, and offer useful and available health resorts for invalids in that island.

Palermo is one of the most attractive and beautiful
of winter stations to be found in Europe. Its situation
is one of almost matchless beauty, surrounded by an
amphitheatre of mountains forming at each extremity
an arm of the beautiful bay which the town faces.
Between the town and the mountains a richly cultivated plain rises gradually for about four miles till it
reaches their base ; this, from its shell-like form, has
been named the *Concha d'Oro*. The town itself is
finely built, and beautiful public and private gardens
and luxuriant vegetation surround it. There is a
fine promenade by the sea, the Marina commanding
magnificent views, and there are varied excursions
and objects of much interest in and around the city.
There are steamers to Palermo from Marseilles every
other Friday, and from Naples daily in seventeen
hours.

Its winter climate is warm but it is also damp

and moist. It has a good many rainy days, although its rainfall is not great. Compared with Nice, Palermo has 131 rainy days to sixty in the former town, but it has only a rainfall of twenty-one inches compared with twenty-five at Nice, so that there are probably a good many moist mizzling days in winter at Palermo. It has a warmer and moister winter climate than the Western Riviera, more equable and freer from those sudden and great transitions between the day and night temperature. It is, however, a good deal exposed to winds from the north, north-east, and north-west, in which direction it is unprotected ; these, however, blow across the Mediterranean before they reach Palermo, and thus become somewhat warmed and charged with moisture. It is also subject, with the rest of the island, to occasional visitations of the African Sirocco, which, no doubt, serves to raise the average winter temperature, and is felt, while it prevails, as a most pernicious influence—dry, hot, and exhausting.

The mean winter temperature is 53·5 degrees, that of spring 59 degrees, and the mean daily range is about 10·5 degrees. The soil is damp in parts and the glare of the sun is often found very trying to the eyes.

Apart from the trouble of getting to Palermo for those who dread a long land journey to Naples and then seventeen hours of sea, or the still longer sea voyage from Marseilles, it is not well suited for invalids requiring delicate care and protection from winds, or for those whom a moist and warm climate unduly relaxes and depresses ; so also, like Algiers, it seems not to be suited to persons who are predisposed to bilious disorders. Its climate is somewhat of an intermediate one between that of Madeira and that of the Riviera. It is colder and drier than Madeira, warmer, moister, and more equable than the Riviera.

"Such a winter climate, temperate, sunny, and rather moist, may be beneficial to a certain class of patients, to highly nervous, excitable, impressionable constitutions, too much braced and stimulated by the dry tonic atmosphere of the Riviera." It may be worth trying in certain forms of "neuralgia, in spasmodic asthma, and in cases of phthisis, accompanied by much nervous irritability, or by a constant tendency to hæmorrhage."* The Hôtel Trinacria, at Palermo, is celebrated throughout Europe, but Mr. A. C. Hare speaks highly of the Hôtel de France, 47, Piazza Marina, and praises its sunny, pleasant rooms and reasonable charges.†

It has been said to be "one of the great advantages of *Malta* as a health resort that invalids can so easily get away from it!" But it is very doubtful if invalids in the strict sense of the word ought ever to go there. A low island, nowhere rising above 600 feet from the sea-level, without protection from mountains or forests, in the centre of the Mediterranean, exposed to every wind that blows, with frequent and abundant rain in the winter—such a place can scarcely be a suitable resort for delicate invalids, whatever social attractions its official society may seem to present. A medical man, writing to "The British Medical Journal" (January, 1885), thus speaks of his own experience there:

As to the desirability of Malta as a health resort, I beg to give the experience of some months' sojourn in that island. We may—for English visitors—pass over the summer, with its debilitating heat, frightful glare, indifferent water, and bad remittent fever, with diphtheria, and commence with "the season" in winter. English visitors are practically restricted to residence in the town of Valetta (where everything is "going on") or its low-lying suburb Sliema. The former is one of the most overcrowded places in the world. There are no localities or houses set apart for visitors; all the residences are on the first to third floors of houses, whose cellars and ground-floors are occupied by hordes of native families, shops, or the less unsavoury stable; so that the noise, smells, and unsanitary conditions may be better imagined than described; slumber is broken by jarring bells and most discordant street-vendors' cries.

---

* Bennet. "Winter and Spring on the Shores of the Mediterranean."
† "Cities of Southern Italy."

The rent for these abodes is exorbitant, the servants provoking, and food, when not procurable (as during quarantine) from abroad, most indifferent. The streets are so narrow, and the houses so high, that ventilation is obstructed ; the footways are so badly paved that walking is attended with the danger of a sprained ankle, etc. ; and driving on the road is, owing to the jolting, most painful to any one, to say nothing of invalids. The Government are carrying out the much needed and most necessary drainage works ; but in the meantime (as during the summer) the sewers are being opened up, the fæcal-polluted soil is exposed to the air, and disgusting poisonous emanations lower one's vitality, even if not communicating disease. Then there is no "country" to escape to. The only roads are between stone walls, and either dust-enveloped or muddy. There are no trees to speak of, no vegetation except crops, so that there is nothing pleasant to the eye, or good for the lungs. All this renders abortive the single condition favourable to health in Malta, namely, an equable climate ; and just now the severe prevalence of its endemic typhoid fever, and malignant diphtheria, has carried grief and mourning in many English families who came to Malta "for their health."

No doubt, as this gentleman states, the climate of Malta is a very equable one, the difference between the night and day temperature not being more than 4 or 5 degrees Fahrenheit in winter, and only 2 or 3 degrees (!) in summer ; and this quality may be serviceable to some who are very sensitive to changes of temperature, and who avoid tendencies to catarrhal attacks in such a climate. The relaxing effect of such an equable climate is certainly counteracted to some extent by the bracing influences of the sea winds, but in still conditions of atmosphere it must be very enervating. It should be borne in mind that seasons vary greatly at Malta, so much exposed as it is to weather, and in fine seasons it has been found a very pleasant winter residence.

Dunsford's Hotel at Valetta is well spoken of ; and the Imperial, at Sliema, a suburb of Valetta, as the best suited to invalids. As all the gardens are inclosed by high stone walls, to protect the trees and shrubs from the winds, the island has an unusually barren and treeless appearance.

A stay of two or three weeks at Malta, in connection with a short sea voyage by a P. and O. steamer to the Mediterranean and back, may be recommended in certain cases of over-work or over-strain, where temporary repose in a soothing and yet moderately bracing climate is needed.

The island of *Corfu*, in latitude 39 degrees 30 minutes, in the Ionian Sea, a few miles from the coast of Albania, has been described as " one of the loveliest spots on the face of the earth,"* but, owing to its somewhat inconstant and changeable winter climate, it is not exactly suited for the more serious class of invalids to winter in.

From the situation of the town—facing the coast mountains of Albania, which are eight or ten miles distant—it is very insufficiently sheltered from the continental winds which, in winter, blow over the snow-topped mountains opposite it. When the wind comes from the north-east, which it frequently does, it must be unpleasantly cold ; and when from the south-east, which is also a prevailing wind, it must be disagreeably relaxing. The mean temperature in January and February is said to be about 50 degrees Fahrenheit, and in April about 60 degrees Fahrenheit, so that it has a decidedly southern climate so far as average temperature is concerned.

The mean relative humidity varies between 70 degrees and 80 degrees, so that it is also rather a moist climate, and there is a considerable rainfall with about seventy-two rainy days between November and April.

It is moreover subject to fogs which often last some time ; but it is very free from dust.

Corfu is a suitable resort for the robuster class of invalids who are able to lead a fairly active life, and who require plenty of exercise and sport, if it can be got, and who would be able to avail themselves of the attractive excursions for sport and pleasure which this

---

* Bennet, " Winter and Spring on the Shores of the Medi-terranean."

charming island affords. As a change for a month in spring from some other winter resort, it is highly to be commended.

The town itself contains about 24,000 inhabitants; it lies, together with the bay, open to the full south and south-east. There is always some movement in the air, a breeze blowing in from the sea till three in the afternoon when the land breeze sets in. The streets of the town are lively and attractive, and filled with varied picturesque costumes, showing that it is a sort of link between the east and the west. The principal hotels are good, and the roads all over the island excellent.

It can be reached by steamer either from Trieste in fifty-three hours or from Brindisi in twelve hours.

The beautiful island of *Capri*, about two miles from the western shore of the magnificent Bay of Naples, is, from many points of view, one of the most attractive winter resorts for a certain class of invalids to be found in the Mediterranean.

It is situated in 40 degrees 32 minutes north latitude, and 14˙13 degrees east longitude. It is about three miles long, eleven miles in circumference, and half-a-mile broad at its narrowest part.

The western portion of the island is called Ana Capri, it is the larger part, and is about two miles long and two miles broad, and its average elevation is 800 feet; but it rises in the Mount Solaro to 1980 feet above the sea. The eastern face of this part of the island descends in steep cliffs to Capri proper, which forms the eastern and smaller part, about a mile and a half in length and half-a-mile across. This rises again at Lo Capo, the extreme east facing the mainland, to 1050 feet, where it falls away in sheer precipices to the sea.

The island is, for the most part, surrounded by precipitous cliffs affording only two landing-places, one on its north and the other on its south side. There are from 2000 to 3000 inhabitants, noted for their pleasant and courteous manners, and the beauty of their women. The town of Capri, with its white-washed, flat-topped houses, its dark-covered archways, and its palm-trees, here and there has quite an oriental aspect. Its hilly slopes are covered with vineyards and olive groves.

It is usually approached from Naples, from which it is distant about twenty miles. There are two very indifferent steamers which leave Naples every morning (weather permitting), and roll fearfully in bad weather. They touch at Sorrento, and accomplish the whole crossing in three to four hours, from Sorrento in an hour and a half, or one can cross from Sorrento by boat in two hours.

It is an ascent of about a mile from the Marina, or landing-place, to the town of Capri, and this ascent is usually accomplished on the back of a donkey.

Capri affords a delightful winter climate, and a most comfortable residence for invalids who are not very weak or delicate, but who retain a considerable amount of physical vigour, and are able to take a good amount of exercise out of doors up and down hill.

Its insular position gives it really quite a sea climate, and it has been compared to a ship in mid-ocean, and the effects of a residence there has been likened to a sea voyage, without the accidents and discomforts of the latter. It has the equable temperature of sea climates, and one is not exposed there to that unwholesome chill, when the sun goes down, which is encountered in the Riviera, and even at Naples, only twenty miles off. But Capri is also a dry climate, for much less rain falls here than on the shores of the Bay of Naples, or on the island of Ischia at the other horn of the bay. The clouds pass over the island and break on the heights behind Naples, so that there is often a great dearth of water at Capri. Owing to the amount of clear blue sky, and the reflection from the sea all round, this little island is very light and sunny. On account of its shape (it is somewhat saddle-shaped) the central depression, where the town is situated, enjoys considerable protection from east and west winds by the elevated ground at each end. It is, however, exposed to the winds from the north and south; the invalid can, however, choose either side of the island for his walks. When the north wind is blowing he can confine himself to the southern side of the island, and when the wind is from the south to the northern.

Capri has the advantage of possessing a very good hotel, the Hotel Quisisana, where excellent food and lodging can be obtained at the moderate price of eight francs a day. The landlord speaks English and understands English comforts.

There is much then to recommend Capri to the more active and robust class of invalids, who desire to escape from an English winter and get winter sunshine in combination with beautiful scenery, pure air, and comfortable entertainment. The drawbacks are its distance from home, and the difficulty and trouble of crossing from Naples in bad weather, so that one may be detained several days at Naples; also the very few roads and paths, and the absence of English doctors. It is not suited to cases of advanced consumption, but may be recommended in early or stationary cases with but a small amount of disease and plenty of physical strength.

The adjacent island of *Ischia* (twenty miles from Naples on the other side of the bay, steamers daily at 2 p.m.) is not so well suited as Capri for winter quarters; its rainfall is greater, and the habitable and most beautiful side of the island is exposed to the north, a circumstance which makes it cooler in summer but too much exposed to the northerly winds for invalids in winter. The central mountain, Mount Ipomeo, rises to 2600 feet above the sea. The town of Ischia is on the north shore, and its warm sulphur baths have long enjoyed considerable repute in the treatment of rheumatic and gouty affections and some forms of paralysis and scrofula.

Casamicciola, about a mile and a half to the west of Ischia, also has hot springs, and is most beautifully situated; unfortunately the disastrous earthquake of 1883 destroyed its fine bathing establishment, which is not likely to be rebuilt, at any rate, for some years.

Naples has such an evil reputation for unsanitariness, that whatever merits it might possess as a winter resort are far more than counterbalanced by the dangers which the invalid incurs in taking up his residence there. There are, however, localities conveniently near this beautiful city which afford suitable quarters for convalescents, as well as other visitors, who desire to spend a month or two of winter or spring in the midst of the objects of interest which abound here in such rich profusion. Castellamare and Sorrento, to the east of Naples, and Posilipo and Pozzuoli to the west, are the most popular of its suburbs. The two former have a north aspect, which makes them cooler than Naples in summer, but renders them unsuitable to delicate invalids in winter.

Pozzuoli, which has the advantage of a good *pension* (Hôtel Grande Bretagne), kept by Mrs. Dawes, has been recommended as a winter resort for invalids with chest and throat complaints, and observations have recently been made of the effect of exposing consumptive patients to the sulphuretted vapours given off by the crater of the *Solfatara* there. Professor Amaldo Cantani, and some of his colleagues at Naples, report satisfactory results from these experiments.

I have only one other winter health resort to mention in this chapter, and that is *Amélie les Bains*, in the Eastern Pyrenees. It is little thought of as a winter station for English people suffering from chest complaints, yet it is said to have very considerable recommendations. Its climate is intermediate in character between the exciting south-east and the sedative south-west coasts of France. It has proved of great value in cases of consumption and catarrhal asthma, in persons who find the climate of the Riviera too irritating and exciting, and that of Arcachon or Pau too mild and relaxing, or who do not find sea-coast climates agree with them. It is situated in the valley of Mondony, at an elevation of 550 feet above the sea. It is about eighteen miles from the Mediterranean coast and is reached from Perpignan in three hours by carriage. It possesses sulphur springs and a hydropathic establishment, and also a large military hospital, and is open all the year round. The winter temperature is reported as very equable, the rainfall moderate, and there are no fogs. The village is protected by surrounding mountains from north winds, and the Mistral is rarely felt. The mean temperature from November to January is 48·2 degrees Fahrenheit, and from February to April 52·7 degrees Fahrenheit. There are good hotels and bathing establishments, and agreeable promenades and excursions.

A very few general observations must conclude this brief survey of winter quarters. In searching for a winter health resort, what do we desire to avoid and what do we desire to find? There are three things which we desire to avoid, especially when they are found combined together, as in our own winter climate, and these three things are damp, cold, and variability. It is the combination of these three conditions which makes the climate of England so unsuitable and even dangerous to many persons. It gives rise to the distressing catarrhal conditions which are so common, and which often lead to graver disturbances of health. It is the cause of attacks of acute and chronic rheumatism, of many forms of neuralgia, and not unfrequently it is responsible for serious inflammation of internal organs. This combination of climatic conditions, necessarily associated with a clouded and sunless sky, produces a further depressing effect on the mind and spirits. It need scarcely be said that the more sensitive the organisation, the more acutely will these unfavourable conditions be felt. What we seek, then, in a winter climate is the opposite of these conditions, viz., dryness, warmth, and equability. But it is always difficult to get all we want; besides, as a matter of fact, while some invalids require a combination of warmth and moisture, others need warmth and dryness, while others do better in a combination of cold and dry air; but no one wants a combination of cold and damp, and all desire sun-heat, a clear sky, and as much of it as possible; and we shall find, as a rule, the value of a winter climate depends on the number of clear and sunny days, or the number of days and hours during which an invalid can take exercise or be in the open air. The mere absolute amount of rainfall seems of small importance, provided the nature and inclination of the soil is such that the water drains off rapidly from the ground, and that there are long or frequent intervals of clear, sunny skies. Indeed, as I have already pointed out, heavy rains often have a salubrious effect in cooling and cleansing the atmosphere. It seems quite clear, too, that diurnal variability of temperature, even within wide limits, does not render a climate unhealthy even to invalids, if it is also a dry climate and the invalid learns to protect himself from the possibility of sudden chill. Nor does humidity, when accompanied with moderate warmth, seem to be necessarily unwholesome, especially in marine climates. There are obviously many other details demanding consideration, some of which are dwelt on in other chapters. One word, however, with regard to the expense attending a change of winter quarters, which proves such an obstacle to many an invalid. Let me say to him, in the words of Dr. Johnson: "Sir, your health is worth more than it can cost;" or let me remind him, in the words of another author, whose name I cannot at this moment recall, that "if life without money is not much, money without life is nothing at all."

CHAPTER IX.

THE WESTERN RIVIERA.

A STUDY OF ITS CLIMATE AND A SURVEY OF ITS PRINCIPAL RESORTS (HYÈRES, CANNES, NICE, MONACO, MENTONE, BORDIGHERA, SAN REMO, ALASSIO, PEGLI, ETC.)

> Comment se peut-il, que . . . vous ne passiez pas vos hivers
> . . . dans un endroit quelconque où se voit le grand arbitre
> des santés humaines, Monseigneur le Soleil? Je crois que sans
> lui je serais depuis bien longtemps à quelques pieds sous terre.
> —*Lettres de Prosper Mérimée.*

THE many picturesque towns that lie scattered along the beautiful Mediterranean coast of France and Italy have long been the favourite winter resorts of the inhabitants of Northern Europe. Some of these have enjoyed a reputation as winter health resorts for a very long period, while others have quite recently grown into popularity and renown.

Passing from west to east, the health resorts of the Western Riviera may be said to begin at Hyères, a few miles from the important arsenal of Toulon, and to end at Pegli, a few miles west of Genoa. Between these, its western and eastern limits, we have the well-known French stations, Cannes, Nice, and Mentone, the principality of Monaco, with Monte Carlo; and the Italian towns of Bordighera, San Remo, and Alassio, besides many smaller and less known places on the coast between these.

In geographical strictness, Hyères is not included in the Western Riviera, the *Riviera di Ponente*, since the mountains of the Esterels, to the west of Cannes and many miles east of Hyères, form its western boundary; but as a health resort this town naturally falls into the group which I have just indicated, and with the other members of which it has much in common.

Before considering the distinguishing characteristics of each of the principal health resorts of this region, it will be convenient to consider briefly the general characters of the climate of the whole district of the Western Riviera. The Riviera is a land of sunshine and a land of winds. It is a land of intense brilliant sunshine, and of cold chilling shade. The very intensity of its sun-heat is to some extent the cause of its manifold local currents of air. The air is scarcely ever still, although, of course, some localities are much more protected from the prevailing winds than others. The climate of the Riviera, then, has conspicuous merits and conspicuous defects. But we may rest assured that a perfect climate in winter is to be found nowhere, neither on the Riviera nor elsewhere. The great thing is to know thoroughly the nature of a climate before you resort to it, so that you may obtain the full advantage of all its good qualities and guard yourself against suffering from any of its bad ones.

In examining the climate of any district the chief points to be considered are—1. Its temperature, with its variations. 2. The relative proportion of sunshine and cloud; of clear skies and of skies that are overcast. 3. The amount of rainfall and the number of rainy days. 4. The average humidity of the air, i.e. the amount of insensible aqueous vapour in the atmosphere. 5. The prevailing winds, and the amount of exposure to or protection from them afforded by local conditions.

Let us then, in the first place, consider the temperature of the Riviera.

The several health resorts of the Western Riviera lie between 43 and 45 degrees north latitude, while

London lies at 51·30 degrees; and for this reason alone the sun has naturally more power there than with us. And since the higher temperature of the Riviera is chiefly due to the greater power of the sun, to the intensity of the sun's rays, it follows that the difference between sun and shade temperatures is very considerable, and that while in a room looking south you may find brilliant sunshine and summer heat, in a room in the same house, facing north, you will encounter chill shade and winter cold. And similarly, out of doors, there is great risk of chill in passing from sun to shade. And this is a fact which it behoves delicate visitors, and indeed all visitors to the Riviera, constantly to bear in mind. If they would benefit by the great heat of the sun, they must take care to protect themselves against the corresponding chill of the shade.

It would be a mistake to regard the Riviera in the winter as a hot climate, as some persons seem to expect it should be; and, indeed, if it were altogether a hot climate it would be a far less generally useful and valuable climate than it is. Still it is a climate in which the inhabitants of Northern Europe may in the winter find, on an average, much more warmth than at home. It is found that the mean winter and spring temperature of the Riviera (and it must be borne in mind that our remarks apply principally to these parts of the year, viz. between October and May) is from 8 to 10 degrees Fahrenheit higher than that of England. I am, of course, speaking of averages, and not of coincident periods of time, for in some months of some seasons the mean temperature of parts of the Riviera will be found nearly as low as at home. For instance, at Cannes, in December, 1874, the mean excess of temperature over that of London was only 4·3 degrees Fahrenheit, and in February, 1875, still less, viz. only 3·6 degrees.

It is by no means to its southern latitude alone that the Western Riviera owes the relative mildness of its winters, for both Genoa and Florence are within the same latitude, and it is well known that they do not possess by any means the same mild winter climate. It is rather to the protection from northerly winds which is afforded it by the great mountain barrier of the Maritime Alps, which extends nearly along the whole of this coast, and at a sufficient elevation to prevent the cold winds which blow from Northern Europe, and over the snowy Alps of Switzerland and Savoy, from reaching the towns built along this part of the northern coast of the Mediterranean. Some of these towns are better and more completely protected than others from northerly blasts by reason of the relative nearness to them of this mountain wall, and by the unbroken nature in parts of the barrier it forms ; while at other parts the existence of gaps in the chain diminishes the protection it affords, and renders some of these localities quite unsuited for winter health resorts. Thus many of the towns along that part of the coast which extends from San Remo to Genoa, owing to the greater remoteness from them of the higher chain of Maritime Alps, and the comparatively low elevation of the mountains near them, are rendered much more accessible to northerly winds, and much less suited than the more western towns for the reception of invalid winter visitors.

Another cause of the mild winter temperature of the Riviera is its southern exposure along the shore of a sea the water of which is unusually warm. It has been calculated that the temperature of the Mediterranean off this coast is 20 degrees higher than that of the Atlantic at the same depth and in the same latitude; and that the temperature of the surface of the sea (off the coast of Cannes) has a mean excess of about 12 degrees Fahrenheit over the minimum temperature of the air, and a mean excess of 9 degrees over that of the sea on our own southern coast (Falmouth). Hence it follows that the atmosphere on this coast of the Mediterranean must obtain a considerable addition of heat during winter from that which has been stored up in the sea during summer,

and which is slowly diffused through the air during the colder season.

It is pretty generally known that there is a great fall of temperature on the Riviera at sunset, and that owing to this fact the time of sunset and the hour or two which follow it is a particularly dangerous part of the day to invalids and other persons, and one during which careful precautions should be observed. This fall of temperature at sunset is easily accounted for, and is always encountered whenever, owing to the absence of aqueous vapour in the air and the presence of clear cloudless skies, solar radiation is very powerful; for then when the sun is withdrawn the whole surface of the country is plunged in shade, the air no longer derives any heat from the direct solar rays, and the temperature of the whole air is a shade temperature. But this is by no means the only cooling agency that comes into operation at sunset. When the sky is free from cloud and the atmosphere clear, as soon as the sun sets the heat which has been absorbed by the surface of the earth during sunshine is rapidly lost by radiation into space, and the air in contact with or near the ground is rapidly cooled, and the moisture it contains becomes precipitated in the form of dew, and thus the lower strata of the air become damp as well as cold at and after sunset. When the sky is overspread with clouds, these prevent the radiation of heat from the earth's surface into space and reflect it back to the earth, so that the chilling of the surface at sunset is not nearly so great when the sky is cloudy as when it is clear. Hence it follows that it is especially during clear cloudless weather that invalids must be cautious of exposing themselves to the fall of temperature and deposit of dew which occur at sunset.

The temperature rises again two or three hours after sunset, and again falls to the minimum of the twenty-four hours towards sunrise, so that it is less dangerous to be out of doors three or four hours after sunset than at the time of sunset itself.

In the second place, as to the relative proportion of sunshine and cloud; the excess of sunshiny days during the winter in the Riviera over that of our winter is remarkable. If we compare Nice with London, we find that during the six winter and spring months, i.e. between October and May, there are on the average ninety-seven clear cloudless days at Nice, and only twelve in London! We are justified then in saying that the Riviera is a land of sunshine.

Next with regard to rain. It may be said, speaking generally, that it is a land of heavy rainfalls and few rainy days. But much more rain falls at the eastern end of the Riviera di Ponente, i.e. about Genoa, than at the western end, i.e. around Nice; e.g. the mean annual rainfall of Genoa being 1317 millimètres, that at Nice is 811, and that at Hyères only 746; while the rainy days from November to April, both months inclusive, number 67 at Genoa, 43·5 at Mentone, 36·2 at Nice, 45·8 at Cannes, and 37·5 at Hyères.

Compared with England, the climate of the Riviera is undoubtedly a very dry one. It is quite true that for a few days in autumn and spring there are torrents of rain, so that the total average rainfall may nearly equal that of the west coast of England; but the number of fine days "is immensely greater, both in summer and winter, than in almost any other part of Europe."

If we compare the rainfall at Nice during the five winter months, between October and April, with that of London and that of Torquay during the same months, we find that

Nice has 16·92 inches,
London   9·51   „
Torquay 12·28   „

so that nearly twice as much rain falls at Nice during the winter as in London. But now let us compare the number of rainy days during the same period, and then we have at Nice only 30·5 rainy days, while in London there are 76·5, and at Torquay 98!

On the Riviera large quantities of water fall within a few hours or days; "there are three or four successive thoroughly wet days, and perhaps nights, and then the weather clears up for some time, and the sky becomes bright and cloudless." It is very rare to encounter continuous broken weather on the Riviera, still it does occur occasionally, and the winter 1878–79 was a notable case in point. The following figures indicate the rainfall at Cannes and Mentone, and the number of rainy days during this exceptional season:

CANNES.

	Oct.	Nov.	Dec.	Jan.	Feb.	Mar.	April.
Rainfall in inches	4·34	8·35	3·79	4·46	3·94	7·73	7·50
Rainy days	10	13	12	10	15	8	16

MENTONE.

	Oct.	Nov.	Dec.	Jan.	Feb.	Mar.	April.
Rainfall in inches	10·43	6·95	2·84	2·17	3·92	4·57	8·10
Rainy days	14	14	16	9	10	9	17

· As a rule the winter rainfall is distributed in the following manner: it is common to have a heavy fall of rain in October, as many as thirteen inches will often fall in that month; the next greatest rainfall is in November, then December. Next comes March; and January and February have the lowest rainfall. In April there are heavy falls of rain again as in October.

But the Riviera, like every other locality, is subject to great variations in different seasons, and in the same months in different seasons: e.g. the rainfall in Mentone

In November,	1866, was	·27 inches.
„ „	1877, „	10·12 „
„ December,	1866, „	2·0 „
„ „	1872, „	12·94 „

The year 1877 was a very dry one on the Riviera, and if we compare the total rainfall and the number of rainy days for the whole of that year at Toulon with that of London and Torquay we get the following figures :

	Toulon.	London.	Torquay.
Rainfall in inches	12·82	26·46	42·25
Rainy days	50	172	224

Taking the average of a succession of winters, the Riviera is a very dry climate, the number of rainy and cloudy days being very few compared with the number of dry and clear days ; but it has exceptional seasons, as that of 1878–79 was, and those who were unfortunate enough to pass that winter, and that winter *only*, there, may possibly have carried away with them the conviction that the climate of the Riviera is a very wet and disagreeable one.

In the next place, if we compare the records of the humidity of the air during winter, as observed in certain stations on the Riviera, with those obtained from similar observations at certain stations at home, we get decided evidence of the superior dryness of the atmosphere of the former. Saturation being represented by 100, we get the relative humidity of the Riviera (Cannes and Mentone), as compared with London and Falmouth, represented by the following figures :

Cannes and Mentone.	London.	Falmouth.
72·4	88	84·4

It is not an easy task to write a description of the winds of the Riviera. They are legion. The Mistral, the Sirocco, the Greco, the Tramontana, the Sea wind, the Land wind, etc. Indeed, certain exposed localities on the Riviera are rendered wholly uninhabitable on account of the number and fury of these tormenting winds ; and the relative merits of its various health resorts chiefly depend on the greater or less protection afforded them against the prevailing winds by the surrounding mountains. Moreover, some difference of opinion seems to prevail amongst observers resident at different stations as to the prevalence or non-prevalence of particular winds at these stations. This is especially the case with regard to the Mistral, and the local advocates of each station rival one another in claiming a relative immunity from its visitations.

The Mistral is a wind which blows from the west and north-west. It is a very dry wind, and a wind which generally brings fine clear weather, although it is always attended with a falling barometer.* It is a wind which blows with great fury, and owing to its dryness raises clouds of dust. The air loses its humidity and becomes dry, cold, penetrating, and irritating. The dryness of this wind is accounted for by its losing all its moisture as it sweeps over Central France. It appears to be a northerly wind originally, which reaches the Western Riviera by turning the western flank of the Maritime Alps, and so gets a westerly direction given to it. It is especially the torment of the more westerly stations, such as Hyères and Nice, but it is also felt at times, usually with diminished violence, as far east as San Remo.

March is its favourite month, it then blows its fiercest, and more frequently than in any other of the winter and spring months, but it blows occasionally all through the winter.

The Sirocco is a south-east wind, it is a hot African wind, and only reaches the northern shores of the Mediterranean after having crossed this sea and so become laden with moisture; hence it is a wet wind—warm, wet, and enervating. It brings to this coast the heaviest and most prolonged rains. But these rains do not usually appear until after it has blown violently for a day or two. The spring and summer are its favourite seasons, but it may occur for two or three days in any of the winter months.

* Dr. Sparks remarks that the barometer is comparatively useless in predicting weather in the Riviera. It falls with the north-west wind, which is a dry one, and it does not fall with a south-east wind, which actually brings torrents of rain.

The east wind, which frequently blows in spring and summer, is not very often encountered in winter, and in this region it is not the formidable and dreaded wind it is with us. It is frequently followed by rain, and is most common from March to May, when it occasionally blows with great force.

A very disagreeable wind is the north-east wind, or Greco. It is bitingly cold, and not unfrequently brings with it sleet, hail, and even snow. It fortunately does not blow often. It is more felt and is more frequent and severe along the eastern portions of the Riviera di Ponente than along its western parts, and Genoa owes much of the bitterness of its climate to its exposure to this wind.

The Tramontana is the name given to the north wind. Most of the health resorts along this coast are protected from it by the chain of mountains which rises behind them, and forms a more or less complete protection from winds coming from this quarter. These northerly winds are either completely arrested by this mountain barrier, or they blow over the tops of the mountains, and are only felt at some distance from the coast. But this barrier is in some localities not so perfect and effective against these winds as at others. "Where long valleys run down in a direction due north and south, as at Ventimiglia, the north wind may have free access," and so, owing to the lower elevation of the near hills, the district east of San Remo is less protected from northerly winds than the western portion of the same coast. Nice, also, as we shall see, is but imperfectly protected from these northerly winds.

Occasionally a strong wind is felt from the south-west ;* it is a dry wind, having parted with its moisture in its course over the Spanish Sierra and the Pyrenees. A wind also often blows with considerable violence from the west ; this and the south-west wind

* Some observers state that the south-west wind is a wet wind, and it would seem that it really does bring rain to some parts of the Riviera.

are regarded by some as really "deflected Mistrals."
It is very well known that on sea-coasts generally, in
fine sunny weather, there is a breeze which blows from
the sea on to the land during the day, while at night
the reverse takes place, and a breeze is found blowing
off the land on to the sea. I have explained this
fully in former chapters. This wind from the sea is
very much felt all along the Riviera, even as far as a
mile from the sea, especially on sunny days. It begins
to blow about 11 a.m., and continues till 3 or 3.30 p.m.
It occurs then, because it is then that the surface of
the land becomes heated by the sun, the air in contact
with this surface becoming warm and rarefied ascends
towards the mountain-tops, and the colder and denser
sea air rushes in to take its place, and this goes on as
long as the heating action of the sun continues. When
the sun goes down, the air on the surface of the land
becomes rapidly cooled by radiation into space, and
therefore denser, while that in contact with the sea
remains warm, and then it is that a land breeze
springs up, blowing on to the sea.

This, then, is a brief account of the principal
winds which the visitor to the Riviera must expect to
encounter from time to time. The greatest number
of calm days occur in January and December, and
the windiest months are March, April, May, and
October. The strongest, as well as most frequent,
winds are from the east and the south-west. The
Mistral, as I have already said, is most common in
March.

Having thus briefly considered, from a general
point of view, some of the chief characters of the
climate of the Western Riviera, I now propose to pass
in review the principal resorts on this coast.

Hyères is one of the oldest health resorts on the
French Riviera. It is not actually on the coast as are
the other health resorts of this region, but it is about
three miles distant from the sea, a plain of this extent
stretching between the town and the coast. So that
the exciting influence of the sea is not experienced at

Hyères, and this is considered to be one of its advantages. It suffers, of course, much less from the sea breeze than other resorts on the coast, not only because of its distance from the sea, but also because of the protection from the sea winds afforded by the islands off the coast—the Isles d'Hyères. These "consist of three principal islands and several islets, forming a chain nearly parallel to the shore, and from two to three miles distant from it." Thus they "form a kind of wall to the south of the town."

The town itself is built along the base of a steep rocky hill, having a southern or south-eastern aspect. This hill forms part of a small and picturesque chain of mountains which bounds the valley of Hyères to the north; mountains to the east and north-east close in the plain of Hyères in that direction and project as a promontory into the sea, protecting it quite sufficiently from the north-east, but less completely from the east winds. On the opposite side of the valley, that is to the west and south-west, a series of hills rise and form a kind of screen between it and the roadstead of Toulon. The valley near the sea is swampy, but in the neighbourhood of the town it is exceedingly well cultivated, moreover, it gradually rises towards the town, which is built at an elevation of sixty or seventy feet above the level of the valley, in a situation admirably chosen for gaining all the advantages possible from the heat of the sun. The town is, therefore, much warmer and drier than the valley, which is, in parts, occasionally cold and damp.

We see, then, that the valley of Hyères is protected to the south by the Isles d'Hyères, to the north as completely by ranges of high hills, to the north-east also it is fairly well protected, but it is exposed to the north-west and west and to the south-east and east. Owing especially to its protection from sea breezes, and also from the north and north-east winds, and to the rarity or mildness of the east wind, at this distance from the coast, the atmosphere at Hyères is

sometimes exceedingly still and calm, unlike the
other health resorts on this coast, where perfect stillness of atmosphere is most rare. And the air here
is neither so dry nor so sharp as at Cannes, Nice, or
Mentone, and its climate is, therefore, less exciting
and more soothing. But these advantages are mitigated by the existence of one very serious drawback.
The valley is completely exposed, in its whole length,
to the Mistral, which from February onwards blows
with great force and great frequency. A French
physician, who had passed a winter there, says of this
wind : " It rakes the valley from end to end, with no
obstacle to stop it or turn it, it blows sometimes for six
or seven days together and nights too ! In February
and March this fearful wind prevails at least one day
in three, bringing with it much dryness and dust. I
confess I cannot understand how an invalid can
remain in such a climate without taking harm.
Favourable as I believe this station to be up to the
time that the Mistral begins to blow, equally dangerous do I believe it to be from that moment."*

The temperature records available with respect to
Hyères are not very satisfactory, but it seems to be
agreed that the climate, apart from the winds, is a
very mild one. The temperature is about ten degrees
warmer than in England. It seems also to be more
equable than at some of the other stations—not so
hot in summer and not so cold in winter—and the
daily range also appears to be less extreme. In the
depth of winter the thermometer rarely falls below
44 or 45 degrees Fahrenheit. Dr. Cazalis, however,
states that he saw snow fall three times in the same
winter, but it only rested on the ground a few hours ;
and that frost is very frequent in the valley, but very
rare in the town. There would seem to be a great
difference between the valley and the town. "The
town is warm, but the valley, at four hundred yards
from the town, is cold." Owing to the coldness and

* Dr. Cazalis, " Étude sur le Climat de Cannes."

dampness of the valley heavy dews fall at sunset, and extra precautions are required to protect oneself against the consequent chilling of the atmosphere ; and fogs are not very uncommon over the islands and the plain in spring and autumn.

There are no good and recent statistics available as to the rainfall at Hyères. It would seem, however, to be from twenty-eight to thirty inches in the year, with an average of about sixty rainy days. The winter and spring months are probably somewhat drier than at other health resorts on the Riviera. The water supply is good, and the sanitary condition of the town satisfactory, except in the old parts, which remain dirty and ill-drained. There are but few villas, invalids and visitors living for the most part in hotels, of which there are several fairly good ones, the cost of living at them being somewhat less than at the best hotels at Cannes or Nice. There are many charming and picturesque promenades, and in this respect Hyères is much better off than most other resorts on the Riviera. But life is somewhat monotonous and dull there, and the French complain bitterly of its *ennui*. As to the cases of illness to which its climate is best suited, there seems an extra-ordinary divergence of opinion amongst different authorities.   But I think it may be safely said that it is suited to persons of nervous temperament who dislike the sea, and who find Mentone and Cannes too exciting, and that they cannot sleep at these places. Some denounce it altogether for chest cases, on account of the prevalence of the Mistral during the spring months ; others think it well suited to cases with a tendency to hæmoptysis, on account of its more soothing character. It is good for nervous feeble children, and for some forms of gout and rheumatism. It is not bracing and stimulating enough for scrofulous cases, nor is it suitable to asthmatics. The natives of Hyères are said to be exceedingly healthy and long-lived. " There is no scrofula among the children,. and phthisis is scarce. At the cemetery the large number of old people buried there atttracted my attention." (Sparks.)

But the objections which have been urged against the climate of Hyères do not apply to a resort situated on the southern slopes of the hills opposite Hyères. Here there are valleys admirably situated, sheltered from the Mistral and from all but southern winds, warmed all day by the sun, and provided with many most charming and picturesque walks. "Ces situations," says Dr. Cazalis, "sont des plus belles et des plus chaudes qu'on puisse rencontrer sur nos côtes ; il n'y manque qu'une chose : ce sont des habitations : à part trois ou quatre villas, il n'y a sur cette côte aucune maison." This refers to the valley of *Costabelle*, about two miles south-west of Hyères, and destined probably to become equally famous. The principal residence there is a villa which has been built by the Duke of Grafton. There is also a good hotel there "under English management," the Hôtel de l'Ermitage, consisting now of two buildings capable of accommodating 150 guests. "The valley is embosomed in pine-woods, broken here and there by vineyards and olive orchards, and by the gardens of the few villas which have sprung up as yet. Near the shore there are large groves of olive-trees, which are finer than at Hyères itself, and indicate a warmer climate."*

The next of the health resorts on the Western Riviera which we arrive at in journeying eastward is *Cannes*. It is about three hours from Toulon by rail, and about five hours from Marseilles. The railway on leaving Toulon enters a depression, which runs at some distance from the sea, and separates the southernmost spurs of the Maritime Alps from a mass of mountains — the Montagnes des Maures, which occupies the space between this valley and the Mediterranean coast. These mountains are quite distinct from the Alps, and appear to belong to the same system of mountains as those of Corsica, of

* Sparks, "The Riviera."

which they may be regarded as the termination, jutting up, as it were, from the floor of the Mediterranean, and joining themselves on to the coast of Provence. These mountains occupy the coast-line from Hyères nearly to Fréjus. Owing to the absence of roads (there is no road even along the coast), their wild and picturesque recesses are almost inaccessible, and the beauty and grandeur of this coast-line are little known or even thought of by the hundreds of strangers who annually visit the Riviera. It is on the part of the coast formed by these mountains that the beautiful but little-known Gulf of St. Tropez is situated.

The railway again reaches the sea at Fréjus, and between Fréjus and Cannes it skirts along its shore the whole way. Most magnificent are the views of sea and mountain and coast-line which one gets, from time to time, as the train turns round the grand promontory (Cap Roux) formed by the Esterel mountains, which separate the Bay of Fréjus from the Bay of Napoule, and as it runs from tunnel to tunnel and from cutting to cutting along the eastern side of these beautiful mountains. It is impossible to be too enthusiastic about the beauty of the Esterel range, the great feature in the landscape at Cannes. "Ce qui fait," says a French writer, "son incomparable beauté, ce sont les superbes promontoires de porphyre et de grès rouge d'où l'on domine, à la fois, les deux golfes de Fréjus et de la Napoule. La contraste des eaux bleues et de la roche qui semble flamboyer au soleil à travers la verdure est d'un prodigieux effet. Le Cap Roux, dont les escarpements couleur de braise se profilent à l'angle de la montagne, est un des spectacles les plus grandioses de la Méditerranée."*

From Napoule, which is situated at the eastern base of the Esterels, and gives its name to the bay, we get a magnificent view of Cannes as it lies facing south, bathed in sunshine, stretching along the eastern

* Elisée Reclus.

end of the bay, its countless villas stretching far and
wide on the undulating rising ground which lies
between the sea and the lower hills, with the range
of snowy Alps for the distant background. For the
foreign visitor Cannes is rather the name of an extensive district than of a small coast town. Unlike any
of the other health resorts on the Riviera, it is scattered over a wide tract of land, so that its eastern and
western limits are some miles apart ; and its attractions and beauties are not limited and concentrated in
one particular spot, but are varied and widespread.
There is no sense of restraint and imprisonment in a
place like Cannes, where the landscape is wide, open,
and free. Corresponding with this great range and
variety of territory there is a corresponding range of
climate. " Grâce à la configuration de notre station,"
says Dr. Cazalis, "nous jouissons d'une gamme de
climats différents, plutôt que d'un seul ; près du
Cannet se trouvent des situations abritées des
vents, et des effluves salines ; près de la mer l'atmo-
sphère est agitée, saturé de sel, excitante. Entre
ces deux extrémités se trouve une série de positions
mixte qui conviennent à bien des genres divers de
malades."

It is customary to say that Cannes occupies two
bays, an eastern and a western ; but this seems to me
to be scarcely correct. There is one large bay, the
Golfe de la Napoule, having the Esterel mountains for
its western boundary, and the low-lying narrow point,
the Cap de la Croisette, for its eastern limit. A relatively small portion of this bay, at its eastern end, is
separated from the western part by a hill of no great
elevation, which stretches out from one of the ranges
to the north of the town and projects into the sea.
This hill, the Mont Chevalier, is surmounted by the
ruins of an old tower and the old Cathedral of
Cannes ; and along its eastern aspect the chief part of
the old town is built. This projecting rock, prolonged
by a pier, incloses a small harbour ; and this, with
that part of the shore between it and the Cap de la

Croisette, is termed the eastern bay. A fine carriage road runs along the whole length of this bay, with villas and houses on one side of it, the beach and the sea on the other. The adjacent part of the shore of the western bay is also covered, for a mile or two, with streets, and villas, and hotels, and their gardens extend in all directions for a considerable distance inland.

With regard to the climate of Cannes, in the first place, if we consider the whole district, there can be no doubt that it is less protected from winds than some of the other resorts on the Western Riviera, and that Cannes, on the whole, must be considered a rather windy place. " À Cannes, l'air est toujours en mouvement, mais ce mouvement aérien est très souvent fort minime et ne règne pas dans tous les points du territoire."*

The protecting chains of high mountains to the north are removed to some considerable distance from the coast, and scarcely offer so complete a screen from northerly currents as they do when close to the town, as at Mentone. The Esterels present a considerable barrier to the approach of the Mistral ; but the protection from this wind is not complete ; there is more or less of a gap between the hills to the west and those to the north-west, through which this wind is able at times to reach Cannes. The mountains to the east and to the north-east are not sufficiently high to afford a complete protection from winds coming from those quarters. The prevailing winds at Cannes come from the east, varying from north-east to south-east. A wind from the north is rare, and always feeble. Still more rare in winter is a wind blowing directly from the south. Dr. Cazalis gives the following as the result of several years' personal experience. Towards the last week in October the wind, sometimes from the east, sometimes from the west, becomes high. In November there is always a windy and rainy period, lasting from eight days to three weeks.

* Cazalis, " Climat de Cannes."

The wind is rarely violent, and never cold. During or after this there may be a few days of Mistral; then commences the reign of the east wind, a mild wind which lasts till February. December and January are the least windy months. West winds begin again in February, and the north-west (Mistral) may blow for two or three days. In March the winds are often violent, and bring not infrequently torrents of rain, as in November. In April the winds are very variable; but it is almost always easy at Cannes, owing to the extent and varying inclination and aspect of the ground, to find sheltered situations for exercise during the prevalence of even the strongest winds. The neighbourhood of Cannet, a village about two miles to the north of Cannes, presents many most favourable and protected sites for dwellings and for promenades; and delicate persons, especially those who suffer from chest affections, or those of sensitive nervous temperament, who find the neighbourhood of the sea too exciting, are strongly recommended to settle in the valley of Cannet.

The temperature observations taken by different persons at Cannes vary considerably, and this variation no doubt depends on the different methods followed by different observers, and the different localities in which their instruments have been placed. It may, however, be stated as an average that the mean winter temperature at Cannes is about 8 degrees Fahrenheit higher than in London; that, compared with other stations on this coast, Cannes is not so warm as Mentone or San Remo, while it is somewhat warmer than Nice. Dr. Cazalis thinks his own personal feelings may be more instructive than the varying and somewhat irreconcilable evidence of different thermometers. "When I come to Cannes," he says, "on the 15th of October, I find a temperature which reminds me of the heat of the suburbs of Paris in August; about the 25th October the temperature falls somewhat suddenly, and a light cloak is needed in the morning and evening. In November come the rains, and after that the temperature gets lower and lower till about the 15th December. Towards the end of January occasional frosts at night in cold spots may be expected, but during the daytime, in clear weather, it is warm enough for invalids to take out-of-door exercise. The temperature rises rapidly in February, but less in March, which is the worst month in the year. At the beginning of April it is often necessary to close the shutters to prevent one's apartments being overheated by the sun, and the heat in the daytime reminds one of a July temperature near Paris." There is a considerable rainfall at Cannes, about thirty-two inches during the winter-time (November to April), and about fifty-eight* rainy days. As elsewhere on the Riviera, there are heavy falls of rain, lasting often several days, in November and March; in the former month the rains have been known to last for three weeks! The winds which bring these rains are usually warm winds, coming from the south-east and the south-west. Between these periods rain is rare, and lasts but a few hours. Snow appears about once every two or three years, and never lies on the ground more than a few hours. A fog is a still greater rarity.

Occasionally a very wet or a very cold season is encountered, and then the sufferings of the visitors are very acute, much more so than at home.

The winter of 1878–79 was a most trying one on account of the prevalence of rains ; that of 1869–70 on account of the cold. Mérimée, who was devoted to Cannes, thus writes of the latter season : " L'hiver a été affreux. Il a gélé ici à six degrés, phénomène qui ne s'était produit depuis 1821. Toutes les belles fleurs qui faisaient la gloire du pays ont été détruites, beaucoup d'orangers ont gélé. Jugez de l'effet que produit sur une être nerveux comme moi la pluie, le froid, la grêle du ciel ; *on en souffre dix fois plus* ici qu'on ne ferait à Paris ! "

---

* It is difficult to come at a constant mean of rainy days from the tables of different observers. Sparks, who is ordinarily very careful, gives 58 in one place, and 45·8 in another.

In the spring of 1883 there was some very severe weather in the Riviera. The following is taken from a letter dated Cannes, 18th March, 1883 :

We have had dreadful weather. On the 8th we woke up to find the place in deep snow, from a foot and a half to two feet, down to the water's edge. Before this had time to melt more fell, with a hard frost, the thermometer sometimes at 23 degrees Fahrenheit. The evergreens were broken down by weight of the snow. Lemon and orange trees are killed by the frost; even the olive trees have been frozen. The snow lay on the ground till the 17th, when a heavy rain, following a hailstorm, made it disappear.

Another correspondent reported :

No man living on the Riviera can remember anything more disastrous than has been the winter of 1882 and the spring of 1883. There was, first of all, the fearful deluge at the end of October, which devastated the country for miles around, many people losing their lives ; and, though this visitation was followed by some lovely weather, a change soon came, and we experienced a rainfall so continuous and heavy that the whole of the flat portions of the country became a quagmire, and all work at the villas and gardens had to be for some time suspended. The winter, however, though changeable, was, on the whole, mild, and up to the 8th of March there was a prospect of an early and prosperous spring. On the 2nd instant the east wind blew up, and then for a week or ten days we had continuous falls of snow, sleet, and mizzling rain.

There was a similar visitation in the spring of 1860, and again in February, 1875 ; but the harm done this year is incalculably greater, the snowfall being the heaviest ever remembered. The tropical exotics have suffered most, the frosty mornings which even yet continue being sufficient to undo in one hour all the care and expenditure of years. The morning of the 7th presented both a sad and a pleasing spectacle—sad to think of the damage done, broken boughs of mimosas, olives, palms, and lovely plants being scattered around ; pleasing for its rarity and the grand background which the snow-clad Esterel mountains and the lower Alps formed, glinting and sparkling in the morning sun. Cannes and the neighbouring stations have suffered much more than Nice, while Mentone and its neighbours towards the Italian frontier have come off most lightly. Although the weather is now fine and the sun hot, it is not sufficient to melt the snow heaps in the streets or the layers on the hills. Out of the sun it still freezes.

The country around Cannes is exceedingly beautiful, and in the number, variety, and attractiveness
of the possible drives and excursions into the surrounding neighbourhood, it possesses advantages
over most of the other health resorts on this coast.
"There is very little ploughed or fallow land. Olive-
trees alternate with vines. The hills are verdant to
the tops with pines and forest trees, and the warm
and sheltered nooks are planted with the orange-tree,
which is here almost exclusively grown for its blossoms
used for making perfumes. Nearly every kind of shrub
and flower grows luxuriously out of doors in the gardens,
and the Cannes gardens are unrivalled in their way."

The eucalyptus tree, with its tall graceful stem
and long sickle shape, drooping leaves, now cultivated
largely all over the Riviera, is especially noticeable
in the gardens at Cannes. Dean Alford writes:
"One great advantage of Cannes over other Riviera
stations is, that you have actual forest scenery within
fifty yards of your hotel. To get such a scene at
Nice you have to walk or drive full two miles between
high walls; at Mentone, to go quite as far, and to
climb till you are worn out with fatigue; at San
Remo, to go somewhere else in a carriage." One
great drawback, however, to the pedestrian, especially
in the central part of Cannes, is the dustiness of the
roads, and the absence of cross-roads by which to pass
from one district to another. There is also a charming excursion by sea which every visitor to Cannes
makes, viz. to the Isles des Lérins, the two islands
of St. Marguerite and St. Honorat, which lie about
a mile off the coast to the south of Cape Croisette,
St. Honorat lying about half-a-mile south-south-west
beyond and behind St. Marguerite. They can be
visited in a boat from the Croisette, or by a small
steamer which goes twice a day from the harbour.
There is an interesting old monastery (des Lérins) on
St. Honorat and a fortress on St. Marguerite, celebrated as the prison in which the "man with the iron
mask" was confined, and that from which Marshal
Bazaine made his escape. There is also a very fine

view of Cannes and its surrounding mountains from these islands, extending to the northern boundary of snowy Alps.

Cannes is provided with many excellent but expensive hotels, and numerous elegant villas. There are plenty of good shops where all the necessaries and even the luxuries of life may be procured. Of society there is perhaps rather too much, if we consider the interests of the visitors only.

During the last three years numerous and notable sanitary improvements (which were greatly needed) have been effected at Cannes, but the main drainage has not yet been carried out. Differences of opinion with regard to important details have been the cause of this delay. There is now, however, strong ground for the hope that the scheme recommended by Captain Galton in his elaborate report on the sanitary requirements of the station will at an early date be acted upon, its efficient operation being still further assured by the adoption of the pneumatic system introduced some years ago by Mr. Shone, which has proved so satisfactory at Eastbourne, Warrington, and other places.

In considering what cases are best suited for the climate of Cannes, it must be remembered that Cannes is a bracing place, that its air is tonic and stimulating, and to some nervous and sensitive organisations exciting and irritating. But many who need a calmer and softer climate during the winter months are benefited by the change to the more tonic air of Cannes in the spring—the end of March and the beginning of April. It must be borne in mind also that at Cannes you can avail yourself of two somewhat different climates, according as you choose a residence in the neighbourhood of the sea-shore or inland—in the valley of Cannet for example.

All invalids, except those who suffer from scrofulous or lymphatic conditions, are advised to keep away from the shore. The extreme heat of the Boulevard de la Croisette, the fierce sunshine, the sea air, the wind, excite but do not fortify, induce fever instead of giving strength to a large class of invalids.*

Speaking very generally, it may be said that all scrofulous affections, especially in children, as well as all the milder forms of glandular affections and cases of retarded development, derive very great benefit from the climate of Cannes. These are cases in which the forces of growth, repair, and nutrition require flogging into activity, and the stimulating climate of the sea-shore—the air, the brilliant sunshine, the restless winds, are all needed to rouse the sluggish temperament into the vigour of health.

Nearly all cases of anæmia improve greatly at Cannes, especially if they lead a prudent and careful life, and take as much out-of-door exercise as possible; even cases of cerebral anæmia in the aged mend rapidly; these persons, however, must reside inland, away from the sea, and avoid too much exposure to direct sunshine. The same remark applies to cases of slow convalescence from acute disease. Of cases of chest disease, those of simple chronic bronchial catarrh do well by the sea-shore. Asthmatics, on the contrary, should avoid the sea, and live as far inland as possible. Cases of emphysema, of chronic pleurisies, and of chronic laryngitis also improve here. Cases of chronic consumption, under certain conditions, do exceedingly well at Cannes; and even in very advanced stages its climate will often help to prolong life for many years. But a number of minute details have to be carefully attended to in these cases, which it would be out of place to enumerate here; it is only necessary to say they must avoid the sea-shore. Certain forms of chronic gout and rheumatism, and of Bright's disease, are benefited by wintering at Cannes.

Hysterical and nervous maladies, and neuralgias, associated with general nervous irritability, should avoid Cannes, where their sufferings are often aggravated by the too exciting and irritating effect of the climate.

* Cazalis, "Climat de Cannes."

Attention has recently been directed to the advantages possessed by the town of *Grasse* as a winter resort.

Grasse is situated about nine miles from Cannes, nearer the mountains, on the southern slope of one of which it is built, at an elevation of about 1000 feet above the sea. It now possesses a large and well-appointed hotel (the Grand Hotel), and this fact will no doubt contribute to attract winter visitors to it. The climate of Grasse differs somewhat from that of the towns on the coast. Its mean temperature is less than that of Cannes owing to its elevation, and to its comparative nearness to the snow-covered Alps ; but it is well protected from cold winds, and its southern aspect ensures it abundance of sunshine. Its temperature is said to be more equable than that of Cannes, but until more meteorological observations have been made at Grasse, it would be better to reserve one's opinion on this point. The richness of its vegetation, however, testifies sufficiently to the mildness of the climate. The town is surrounded by immense gardens of odorous plants, the jasmine, rose, violet, orange-flower, jonquil, etc., etc.; the fabrication of perfumery being its chief industry. Its mountain air, and its distance from the sea, should render it popular with those invalids who find the sea-shore unsuited to them. It is reached from Cannes by rail in three-quarters of an hour.

The railway going east from Cannes first skirts the Golfe du Jouan, the fine bay which is bounded on the west by the Cap de la Croisette and the Isles des Lérins, and on the east by the peninsula formed by the Cap d'Antibes, which stretches far out into the sea, and forms the western limit of the wide bay, at the eastern end of which lies the town of Nice.

Antibes has a growing popularity amongst winter visitors to this coast, and many who have passed a winter season there regard it with great partiality. Life there is somewhat cheaper than at Nice or Cannes, but its climate, though an agreeable one, is said to be not so mild and equable as that of Cannes.

After passing Antibes, the line running nearly

due north along the western shore of the Bay of
Nice, the traveller finds opened out to him a very
grand and extensive view of the Mediterranean, and
of the Riviera coast as far east as Bordighera, while
to the north he obtains a fine view of the Alps behind
Nice. After crossing the river Var the train soon
reaches Nice, which is twenty miles east of Cannes,
the journey between the two towns being accomplished in about fifty minutes.

Nice, it must be honestly admitted, is rather a
pleasure resort than a health resort, and whatever
differences of opinion may be entertained with respect
to the value, in certain cases, of its winter climate, it
must also be admitted that whatever defects the
climate of the Riviera possesses, these are specially
concentrated and aggravated at Nice. Of brilliant
scorching sunshine there is during most winters an
abundance. There are constantly clear blue skies and
but little cloud ; the air is remarkably dry, bracing,
and exciting ; there are fewer rainy days perhaps than
at some of the other health resorts on the coast. But
there are occasionally heavy downpours, which may
last for days together in spring and autumn, and in
midwinter a continuous bitterly cold drizzle, with a
biting north-east wind, as disagreeable and chilling
as anything our own climate can afford, is not unknown at Nice. The frequency, keenness, and incon-
stancy of the winds at Nice are well known ; the wind
will shift three or four times a day, making it impossible to obtain any continuous shelter from it. " The
winds," says Elisée Reclus, the well-known French
geographer, "are extremely inconstant, and sometimes
of insupportable violence. At the end of winter and
the beginning of spring, when the Mistral blows with
fury, the blackish dust which it sweeps before it in a
whirlwind does not yield in intensity to the rain of
cinders showered down by a volcano." Then at Nice
you are in a large town, not in the country, and you
naturally lead a town life rather than a country one.
Of course it has the advantages as well as the disad-

vantages of a town. There is much gaiety and animation, abundance of amusements and pleasures, and a
certain amount of social and intellectual activity and
interests, so that for a certain class of invalids who dread
the *ennui* and quiet of a health resort *pur et simple,*
Nice has immense attractions. It has also considerable
dangers, for the very movement and excitement and
gaiety of a place like Nice tend to throw the invalid
off his guard, and to lead to some indiscreet exposure
or want of precaution which, in such a treacherous
climate, may have serious results. But there are
persons who must and will have amusement whether
they are ill or well, to whom life without constant
entertainment is wearisome and fatiguing, who live
in constant dread of being *ennuyé.* Let these by all
means go to Nice and take their chance there. Then
again the social elements at Nice are most heterogeneous, for it is a kind of Paris-sur-Mer, but even more
cosmopolitan. The foreign visitors include every
nationality in Europe, and many out of Europe, while
they are of every class, from princes to pickpockets.
When you descend at the railway station, especially
if you arrive by one of the evening trains which go
on to Monte Carlo, you find yourself surrounded by
faces and costumes which irresistibly recall to your
mind the Boulevard des Italiens at one o'clock in the
morning.

If Nice were the only town on the Riviera where
good accommodation and good medical skill could be
procured, as no doubt it once was, I could understand
the propriety, nay, the necessity, of sending sick
persons there in great numbers, as was once the
custom. But to select Nice, when Cannes and
Mentone and San Remo and other places are available, for the residence of the generality of invalids,
except *for some other reason* than that they are
invalids, seems to me difficult to understand. There
are, however, a select class of cases for whom the
climate and surroundings of Nice seem very suitable.
It is said to be remarkably useful to persons who,

from over-work or any other cause, have become the victims of atonic dyspepsia, with torpid livers, and a tendency to melancholia. But if such persons suffer from nervous irritability, as they often do, Nice is to be avoided.

It is also beneficial in those diseases of young children and others which depend on the scrofulous and lymphatic constitution. It is a good winter resort for many aged persons with flagging vitality and with a tendency to catarrhal attacks. Age brings caution, and invalids of this class know how to avoid the bad points and take advantage of the good points of the climate. Nice is also suitable to cases of simple anæmia uncomplicated by any nervous affection. It is also very attractive to the large class of *soi-disant* invalids. If one regards the situation and surroundings of Nice, it is easy to see why it should have so treacherous a climate. Lying at the eastern end of a wide bay, exposed to the south, and with no near hills to shelter it, it is open to winds on every side, from the land and from the sea. It is also exposed to the full force of the sun. The nearest protecting hills are to the east and south-east, where Cape Mont Boron rises to an elevation of nearly a thousand feet, and stretches out into the sea ; beyond this is the promontory of St. Jean. These would form a pretty good protection to the east were it not that there is a gap between Mont Boron and the ridge of Montalban, a little north of east, and through this gap easterly winds can reach the town. To the west and north-west it is much exposed, the low-lying Cap d'Antibes offers no protection to the west, and the hills on this side are distant and of no great elevation, so that it is most imperfectly protected from the Mistral which blows from the north-west. The mountains to the north are also seven or eight miles distant, and these are not very high, the highest, Mont Chauve, being 2824 feet, so that the Tramontana, a northerly wind, also reaches Nice. There is a gap also made in the north-east by the valley of the Paillon, whose torrent flows through the centre of the town, and along this valley the bitter north-east wind finds ready access.

The meteorology of Nice has been carefully studied by many competent observers. Its mean annual temperature is 60·3 degrees Fahrenheit, nearly the same as at Pisa and Rome. Its mean winter temperature is 49·1 degrees, and its mean spring temperature 58·1 degrees. The minimum temperature at night 26·6 degrees. The coldest months are January and February. In March there are great variations of temperature, with rough winds and clouds of dust. Falls of the barometer are almost always caused by the dry north-west wind, and rains " only cause the mercury to sink gradually and almost imperceptibly." The relative humidity for Nice is small, the annual mean at 2 p.m. being 59·6. The mean proportion of sunny, cloudy, and rainy days for twenty years is represented by the following figures : sunny 219·2, cloudy 77·3, rainy 67·4 ; and for the winter season, from October 1st to May 31st, sunny 135·8, cloudy 55·3, rainy 52·8. The mean annual rainfall is 32·43 inches, and 19·45 for the six winter months (November to April). Most rain falls, as elsewhere on the Riviera, in October and April; and the winds that bring rain are, according to some observers, the east, the south-west, and the north-east, while others maintain that the heaviest and longest rains come from the south-east. It has been calculated that for the whole year there are 83·4 days of strong wind, 258·8 of gentle wind, 22·8 of complete calm. March, April, and May are the windiest months. The east wind is the most common of the stormy winds, and blows 45 days in the year. The south-west wind is also a violent wind, especially at the time of the autumn equinoctial rains. The north-east wind sometimes brings hailstorms and snow. The Mistral blows (from north-west or west) chiefly in February and March, and is accompanied with clouds of dust. The magnificent Promenades des Anglais, one of the finest promenades in Europe, running for a mile and a half along the sea-shore, is especially exposed to the fury of the Mistral, as is also the adjacent quarter of the town; more protected from this and other winds is the Carabacel quarter, situated about a mile north of the shore, and somewhat under the protection of the northern hills; it is regarded as the most sheltered situation in Nice itself. Here many of the best hotels are built, and here invalids for whom the neighbourhood of the shore is too exciting, are advised to live. Still further north, between two and three miles from the sea, is the suburb of Cimiez; it is built on one of the spurs jutting out from the mountains to the back of Nice, and is regarded as having a much more sedative climate than Nice itself, and as much better suited to invalids with chest complaints. It is also said to be much more sheltered from the north and other winds; but it has appeared to other observers, as well as to myself, that Cimiez is in a very exposed situation, on the brow of a hill, certainly within reach of the Mistral and imperfectly sheltered from most winds, except the north. It of course escapes the sea winds and the stimulating saline emanations on the shore, and with its complete southern exposure must get all the advantage possible from the sunshine, without the reflection and glare from the sea.

The environs of Nice are exceedingly beautiful, especially the drives eastward, and the Corniche road from Nice to Mentone, a drive of about eighteen miles, is a marvel of beauty and interest. Short peeps only of this picturesque part of the coast are obtained from the railway as it passes through the cuttings and tunnels in the precipitous spurs of the Alps, which here approach close to the sea. Villefranche, Beaulieu, and Eza, all situated in the midst of magnificent scenery, are passed before reaching Monaco, half-an-hour from Nice by train. One of the most beautiful spots on the Corniche road between Nice and Monaco is the village of Turbia (Trophæa Augusti), at an elevation of nearly 2000 feet above the sea, built on a crest which unites Mont Agel to the magnificent bold promontory, the Tête de Chien, whose stupendous precipices tower above Monaco. A path practicable for mules and pedestrians descends to Monaco along the steep flanks of the Tête de Chien, but the carriage road has to make a long détour by the village of Roccabruna. But there is now another carriage road not long completed, which runs along the seashore between Nice and Monaco, and I call attention to it because about a mile or so before it reaches Monaco it passes a station which as yet (1881) possesses a name and nothing more, it is absolutely in its earliest infancy; there may be one house there but not more, and others are, or were a few years ago, only in course of construction. It is named after the village on the heights I have just spoken of, *Turbie-sur-Mer*. It is admirably situated in a small rock-bound bay, just a little to the west of the jutting promontory on which the castle and old town of Monaco are built. It is protected on all sides except towards the sea, and from its own natural advantages as well as from its nearness both to Nice and Monaco it is surely destined, when properly developed, to be a popular resort.

It is scarcely necessary that I should speak of the beauty and attractiveness of *Monaco* and *Monte Carlo*. If it were not for the dangerous seductions of the gaming-tables it would doubtless be one of the most popular health resorts of the whole Western Riviera, not simply because of its beauty but also because of its admirably protected situation, especially in parts. The old town and castle of Monaco are built on a curiously-shaped rock about one hundred and sixty feet high, which projects into the sea, first in a southerly and then in an easterly direction, so that it curves round towards the east in a manner so as to partially inclose its pretty little harbour. Between this rock and the promontory of Monte Carlo, which juts out into the sea about a mile farther east, extends the harbour and Bay of Monaco, with an excellent beach and sea-bathing establishment.

Nearly on a level with the shore, but rising gently as it extends inwards, is a new quarter named the Condamine. The north-west portion of this quarter enjoys complete protection to the west and south-west by means of this lofty rock of old Monaco, and to the north and north-west by the high mountains which here approach to within a little distance of the sea. On the opposite side of Monte Carlo, where it slopes down to that part of the coast which stretches away to Cap Martin and to Mentone, "the loveliest bit of the whole Riviera," there are admirable sites for villas and hotels exposed to the full sunshine, and protected on all sides except to the south and south-east, from which quarters the cold winds do not come.

Of Monte Carlo itself Dean Alford writes thus, and he cannot be suspected of any undue leaning towards a spot so completely identified with vice : "Situated on the platcau of Spelugues, which jutting into the sea about a mile to the west of the town (of Monaco), commands a prospect of the finest portions of this beautiful coast. The frowning height of Tête de Chien shelters it from the north wind, the little Bay of Monaco lies below, with the shadows of the steep shore and quaint old town painted on its glassy surface, and eastward olive gardens and cypress groves stretch more than halfway up the sides of the mountains, which gain in height and fantastic grandeur as the view recedes. The shelving sides of the rocks are covered with delightful gardens, the space in front of the principal building is decorated with fountains and statues; and tropical plants are planted and thrive splendidly in the brilliant sunshine." It is stated that the mean annual temperature of Monaco is two degrees higher than that of Mentone, and three degrees higher than that of Nice, and as a proof of the greater mildness and equability of its climate it is also stated that during the exceptional winter of 1870–71, when at Cannes and at Nice the frost destroyed a number of plants recently acclimatised, the same plants at Monaco did not suffer at all, although in the open air and without shelter, and that the lemon-trees, which were severely injured at Mentone, were not at all affected at Monaco. If the day should ever come when the gaming-tables at Monte Carlo are suppressed, this neighbourhood will undoubtedly become most attractive as a health resort.

Mentone is but five miles east of Monaco and fifteen (by rail) from Nice. The bay, on the shores of which the town of Mentone is built, is bounded on the west by the low-lying Cap St. Martin, covered by forests of olive-trees, and on the east by the Cap de la Murtola. From cape to cape this bay is about four miles across, and has a south-easterly aspect. As at Cannes, the old town is built on a ridge which projects into the sea and divides off a portion of this bay to the east, which forms the smaller and east bay; the western division, being much wider, forms the west bay. This division of Mentone into an east bay and a west bay represents a very essential difference in climate ; for the Mentone district is bounded, behind and on each side, by a sort of semicircle of high limestone mountains, some of them reaching an elevation of over 4000 feet, and the lowest depression or gap in which is not less than 2500 feet above the sea. The chief part of this mountain wall opposite the *western* bay is at a distance of about three miles from the town, but hills and ridges of lower elevation from 400 to 700 feet run down from it at right angles to the shore. Between these ridges, three principal valleys, with their torrents, stretch down from the higher mountains and open behind the western bay. Through these valleys currents of air descend from the north, and so produce a certain ventilation and movement of the atmosphere in this part of Mentone.

It is quite different with regard to the eastern bay. In the first place it is a much deeper indentation of the coast than the western bay, so that its curve is

almost a semicircle.   Then the hills come so close
to the shore that there is scarcely any room for the
town, which consists here of little more than a road
and a row of houses and hotels squeezed in between
the base of the mountains and the sea-shore; the
mountains, however, recede a little, farther east,
towards where the road ascends to the Italian
frontier.   Nor are there any considerable valleys
opening into the eastern bay to bring cool currents
of air down from the mountains.   It follows that the
temperature of this bay is from 2 to 3 degrees
Fahrenheit higher than that of the western bay,
owing to the reflection of the sun's rays by day
from the surface of the bare limestone rocks which
rise directly behind it, and to the gradual giving up
at night of the heat they absorb during the day.
There is also less motion in the air, and Dr. Cazalis
testifies that he has sometimes seen the atmosphere
here absolutely still, as he has also seen it at Hyères,
a very unusual thing in the Riviera.   There is also
said to be more humidity in the air of the east bay
than in that of the west.   The east bay then is very
sheltered and very picturesque, but it is found to
have a relaxing effect on some people, who also
complain of a sense of being "shut in" and
confined there, and that on bright sunny days,
and these often succeed one another with an almost
wearying monotony, the heat and glare of the sun
become really distressing.*   Then there is only one
level walk, and that is along the dusty high-road.
But for invalids whose chief care is to lounge through
the winter in a warm and comparatively still atmosphere, the east bay of Mentone is well suited ; while
the villas and houses built in the wider eastern part of
the east bay no doubt enjoy the warmest and most

---

* "The eastern bay is simply a sun trap, almost intolerable
all the noontide hours.   Often have I sought the old town and
plunged into its dark street, as into a bath, from the glare of
that faint mile of great hotels and villas."—Dean Alford's
"Riviera."

protected situation in Mentone. In the western bay it is quite different; here the higher mountains fall back, as has already been said, to some distance behind the town, and the houses not only stretch along the bay, but extend, in a more or less scattered way, over the gradually sloping territory which reaches from the bay to the foot of the lower ridges and the sides of their intervening valleys which come down to the north of the town. So that the west bay is not so much protected from winds as the east bay; it is more open to the south-west and to the west, and consequently gets more wind and is somewhat cooler and more bracing. The considerable differences of opinion which have been noticed to exist between the statements of different observers as to the climate of Mentone may, possibly, be accounted for by the circumstance that some have made their observations exclusively in the east bay, while others have made theirs in the west bay. For example, one writer, an old resident, states that Mentone enjoys "complete protection" from the Mistral; another says, "The western end is open to the Mistral, there being only the low Cape of St. Martin to ward it off;" a third writes, "The Mistral seldom blows here from the north-west, and even then is deprived of much of its violence and coldness, but it is still extremely disagreeable;" while a fourth actually gives the average of the number of days in which he had observed a Mistral blowing during the four winters he resided there, and these are said to be 4·5 in November, 3 in December, 4·25 in January, 6·7 in February, and 5·25 in March. He had observed it blow as many as nine days in February. The writer of this notice certainly found a Mistral blowing and raising clouds of dust in the west bay of Mentone on the evening of Christmas Day, 1880! Of other winds, the east wind is felt chiefly along the shore, and shelter from this wind can always be obtained in the walks and drives along the valleys behind the west bay. South-south-west and south-east winds, all coming across the sea, have free access to Mentone, but these are not, as a rule, cold winds, although they may blow at times with considerable violence. From the north wind it is completely protected.

By comparing the different means of the temperature records of different observers at Mentone, the following figures are obtained. Mean temperature for the following months :

	Oct.	Nov.	Dec.	Jan.	Feb.	March	April.	May.
East Bay	65·3	55·3	50·55	49·9	50·6	53·9	58·7	65·76
West Bay	62·2	55·6	50·69	49·12	49·46	51·1	57·64	63·1

It will be noticed that during the months in which there is least wind, December and January, there is scarcely any difference in the temperature of the two bays, but in the windy spring months the greater protection enjoyed by the east bay is shown by its higher temperature. The lowest temperature recorded during ten consecutive winters was 25·5 in March, 1877, and the highest 77 degrees in November, 1874. The mean daily range of temperature was found to be least in December, 9·2 degrees, and greatest in April, 12·5 degrees. The average rainfall from October to May inclusive is 25·61 inches, but if we omit October and May, for the remaining six months it is only 17·87 inches. The corresponding number of rainy days is 63·8 if we include October and May, 45·15 excluding them. January and February are the finest months, and have the smallest rainfall and the fewest rainy days. October is the wettest month. The average number of very fine days for the six winter months, from November to April inclusive, seems to be about 94·5, rather more than fifteen in each month.*

* The discrepancies which are observed in the figures published as "means" and "averages" in the meteorological tables of different observers are, to some extent, inevitable, and depend on the circumstance that the observations recorded extend over varying periods and different seasons. Observations which extend over only five or six years will be influenced by the occurrence of one very fine or one very bad season, and will perhaps vary somewhat from the records of another observer extending over twenty or twenty-five years.

Considered generally, the climate of Mentone may be taken as a favourable example of the Riviera climate, and it has the great advantage of possessing, as it were, two climates, suited to different classes of invalids. For those who especially desire warmth and shelter and a quiet indolent life, with plenty of sunshine and sun-heat, and who like to live close to the sea, there is the mild and sedative climate of the east bay, with its southern exposure and its almost complete protection from strong winds.* For those, on the other hand, who find an advantage from a more bracing air, who like to have the sun-heat tempered by cooling winds, who cannot feel at ease without "ample space and room enough" to wander free over hill and valley, or who are irritated by the monotonous beat of the tideless sea against the shore, or to whom the saline emanations from the sea prove exciting and discomforting, for those there is the west bay with hotels and villas, some on the sea-shore, some a little removed from it, some, and those the newest and best, far removed from the sea and high up on the hillside. The value of a climate of this kind in many forms of pulmonary affections, in certain chronic gouty and rheumatic conditions, in states of anæmia, in convalescence from many acute diseases, and in the many infirmities to which old age is exposed, is incontestible.

But existence, even in a climate like Mentone, has its troubles. Dr. Cazalis complains of the life there being "ennuyeuse et triste;" of the number of sick people one encounters in bath-chairs on the Promenade du Midi, which runs along the shore of the west bay; of the number of deaths that occur in one's hotel during the winter, "et chaque fois ce sont des scènes pénibles pour les malades qui survivent;" and, saddest of all, "le nombre prodigieux de Prussiens qui viennent s'y abattre et dont le contact est

* "There is hardly a fairer scene of languid repose to be found in all this resty land. . . . There is no edge in the breeze, no sea-air breathing from the waves."—Alford's "Riviera."

quelquefois désagréable.   Il n'y a pas là une société
d'élite propriétaire, et qui puisse faire la loi, comme
celle que nous possédons à Cannes et qui ne reçoit
pas les Prussiens " (!)

There are some very beautiful walks and drives
around Mentone, but, unfortunately for invalids, the
walks are nearly all of them steep and fatiguing, so
that, unless he is able to climb, the invalid's walks
will probably be restricted to the somewhat windy
Promenade du Midi.   But those only who can
climb up to and beyond these mountain ridges which
divide the several valleys behind Mentone will discover how exceedingly beautiful the whole district
is ; in the background a magnificent sweep of high
mountains,* remarkable for the variety and beauty of
their form and the warmth and richness of their
colour ; in front the limitless expanse of deep blue
sea, still and smooth as the surface of a mirror, or
crisped into small white crests of foam by some light
breeze ; far in the distance the snow-clad summits of
the Corsican hills, touching the azure sky, like the
ivory pinnacles of some unearthly temple ; on each
side, the exquisite coast scenery ; towards the west,
the wooded promontory of St. Martin, the picturesque
village of Roccabruna, high up on the hillside ; the
bold precipice of the Tête de Chien, and the old tower
of Turbia, above Monaco ; the rocky promontory of
Monaco itself, with its miniature bay, the glittering
towers of the Casino of Monte Carlo ; and stretching out
into the sea, far in the west, the ever beautiful range
of the Esterels.   Orange, lemon, and olive groves
are spread out at our feet ; and to the east there are
the steep, rock-bound coast of the eastern bay and
the adjacent frontier of Italy, the fine promontory
of Cap Murtola, with steep red rocks behind it, and
the carriage-road into Italy winding over it ; and extending far out into the sea, and forming the eastern

---

* " There is nowhere else on the shores of Europe so small
a locality surrounded by mountains of an equal altitude."—
Chevalier Ardoino, " Flores des Alpes Maritimes."

limit of the view, the sunny promontory of Bordighera. All this seen in the varied and gorgeous colouring of the setting sun, with its many hues of blue and purple and crimson and gold, composes a picture of almost unrivalled beauty.

" Those who need bracing," writes Dean Alford, " are apt to complain of a fevered and depressing effect at Mentone. On three separate occasions have I found this, and each time I have speedily lost it among the palms of Bordighera." *Bordighera* is the next health resort eastward from Mentone, from which it is distant about ten miles, being three miles from the Italian frontier town of Ventimiglia.* Bordighera is a conspicuous object nearly all along the Western Riviera, as it is seen glittering in the sunshine, its houses clustered together on a promontory that projects far out into the sea. It is the only health resort on this coast that occupies a position on a promontory; all the others being built round bays or depressions in the coast. It is naturally, therefore, much exposed to winds, that is to say, to all those winds that can reach it in blowing across the sea ; the east, the south-east, the south-west, and the west winds can all blow freely upon this promontory. But it is well protected by mountains to the north, north-east, and north-west, whence the coldest winds come. Moreover, it is to be remembered that all the winds that reach it must, on account of its position, come to it from the sea, and impregnated with saline emanations. And this is the sole distinguishing characteristic of the climate of Bordighera as compared with that of neighbouring stations ; the predominating influence of sea air rendering it essentially bracing and tonic. For this reason, also, its temperature is probably rather more equable—warmer in the winter and cooler in the summer—than at other places on this coast.

The old town of Bordighera.is built partly on the

* Where a tedious delay and an examination of luggage takes place at the Italian Custom House.

promontory itself, and this commands a fine view
westward of the Riviera coast, Cap Murtola, the
mountains round Mentone, the Tête de Chien above
Monaco, and even, on a clear day, the Esterels, west
of Cannes; eastward the view is not very remarkable,
the chief objects being the two capes which form the
eastern and western boundary of the Bay of San
Remo (Capo Nero and Capo Verde), and the little
bay and village of Ospedaletti. The *new* town has
been built on level ground to the west of the
promontory, on each side of the main carriage-
road. This plain, thickly covered with dense olive
groves, stretches for a distance of three miles in
the direction of Ventimiglia, and for about a
quarter of a mile inland from the shore, till it
reaches the base of the hills forming its eastern
and north-eastern boundary. The possession of
this level tract of land near the shore, and thickly
covered with vegetation, gives quite a peculiar and
attractive aspect to the western side of Bordighera.
Dean Alford, alluding to this, says : " Bordighera has
an advantage for invalids over many other resorts on
the Riviera. I .mean its level space of olive and
lemon groves between the beach and the hills.
Nowhere else can you get such delightful strolls
under the dense shade of the old olives without a
fatiguing climb. Should Bordighera ever come to
the front, as I cannot tell why it should not, as a
residence for invalids, surely this level may be made
of immense use, both for building and for laying out
in walks and drives." Bordighera is also celebrated
for its palm groves. These give a remarkably
Oriental aspect to the place. The largest groves are
to the east of the promontory, but they abound on all
sides. The church is " amidst a thicket of palms.
The promontories on either side are outlined by the
feathery tops of a hundred palms, and on looking up
the gorge the wood seems full of them. These noble
trees almost gird it round on the western and northern
sides, and grow in profusion, of all sizes, from gnarled

giants of eleven hundred years' reputed age to little suckers which may be pulled by the hand and carried to England."*

Bordighera is comparatively in its infancy as a health resort, and there is consequently an absence of long-continued meteorological observations from which we might form a decisive estimate as to the precise relative value of its climate. It would seem, however, from those that have been made† that it is amongst the mildest, most equable of the health resorts of the Western Riviera. The new town, by its position under the cape, is greatly protected from the east and south-east. It is well protected also from the north, and fairly so from the north-west, though the mountains in this direction are distant. But it is completely exposed to the west and to the south-west. At Bordighera the Mistral is a west wind, being turned completely in that direction by the mass of mountains behind Monaco, and from being forced to blow over the sea it loses somewhat of its dry and cold character. Bordighera naturally feels the local sea-breezes, which are not strong winds, more than its neighbouring resorts, and it would seem to suffer from the stronger winds in about the same proportion as these.

Its mean temperature differs very little from that of the other resorts on this coast. For the whole winter it is the same as that of Nice, a little lower than at Cannes, still lower than at Mentone. But if we look at the different months, and if we are justified in drawing an inference from the comparatively few observations that have as yet been taken, it would appear that in January and February it is warmer at Bordighera than at either Mentone or Nice, while in November, December, March, and April it is colder at Bordighera than at either of these places. Its position on a promontory jutting out into the sea would certainly tend to make it cooler

---

* Dean Alford, " The Riviera."
† " Le Climat de Bordighera," par F. F. Hamilton.

than its neighbours in the hot spring months, and would seem to point to it as a good locality for invalids to move to to escape the increased heat of this season before returning northward. Of the rainfall and number of rainy days at Bordighera it is impossible, from existing data, to speak positively; it would seem to be neither better nor worse off than its neighbours in this respect. It is especially suited to invalids who want *sea* air; to cases of scrofulous phthisis in its early stages, and other cases of early phthisis without any tendency to hæmorrhage; to cases of throat and bronchial catarrhs; to cases of chronic pleurisy; of convalescence from acute diseases, cases of anæmia, and many other conditions of constitutional feebleness. Its climate is too exciting for the very nervous and sensitive. The special facility it affords for a variety of level shady walks cannot fail to make it attractive to a large class of invalids; the number of possible carriage excursions is limited by the badness of many of the roads.

As we continue eastward from Bordighera the interest and beauty of the coast scenery diminish greatly. A drive of three or four miles along the coast brings us to the pretty little bay and village of *Ospedaletti*, shut in and protected on almost all sides by its olive-clad hills. Ospedaletti is in an advantageous situation, and recently a large hotel and several villas have been constructed there by a company which has undertaken to develop it as a winter resort. No doubt it has a considerable future before it.

Just beyond Ospedaletti we arrive at one of the most thriving of winter stations, the old Italian coast town of *San Remo*. We miss here all the fine, bold, and varied rocky scenery that we have left behind; the higher hills recede somewhat from the coast, and the nearer ones are wooded to their summits and present nothing particularly striking in form or colouring. Many beautiful spots no doubt lie concealed high up on these hillsides and amongst their

numerous valleys, but they are not accessible to the
invalid who cannot climb, unless he trust himself to
the back of a donkey, which is perhaps the best thing
to do. Dean Alford, evidently an enthusiastic lover
of the picturesque, thus writes of San Remo: "San
Remo itself is not in any sense an attractive position. The old town is indeed one of the quaintest
on the Riviera, as seen from the pier below ;
the mountain of old houses, stained and weather-
beaten, with their arched loggias and terraces,
is quite unique. And when we enter the streets the
scene is as curious—bands of masonry unite house to
house, built as safeguard against the shaking of earthquakes. But what is all this to the invalid ? There is
absolutely no scenery at San Remo, unless it be sought
by distant excursions. There is not even a level walk
commanding a view. The prospect is shut hopelessly
in by the two promontories, Capo Nero on the west and
Capo Verde on the east. There is a picturesque old
church, the Madonna della Guardia, on the eastern promontory, but this is almost the only object San
Remo can boast. If San Remo be an excellent place
for our English invalids, so far well, but other advantages it certainly has not, compared with its
beautiful rivals along this exquisite coast." And no
doubt San Remo is an "excellent place for invalids," and better suited to the wants of a very large
class of invalids than some of the more picturesque
resorts on this coast. The special recommendations
of its climate seem to be that it is less exciting than
some of the resorts further west, and on that account
better suited to nervous and sensitive organisations.
Invalids who cannot sleep at Nice and Cannes can
sleep at San Remo. Its temperature records, compared with those of the other health resorts on this
coast, show it to be as warm in winter as the warmest
of them, somewhat more equable, with less difference
between day and night temperature, and less difference between summer and winter temperature.
Owing to the greater equability of its temperature,

visitors can remain later on at San Remo without feeling the weather unpleasantly hot and relaxing, as in some other of the towns in the Riviera. The Italians use it in the summer as a sea-bathing station. It is exceedingly well protected by a triple barrier of mountains from northerly winds which blow over the town and are only felt far out at sea. The east wind is the strongest and most felt here, owing to the low elevation of Capo Verde and the absence of any other protection in this quarter. This and the south-east are the prevailing winds. The north-east blows occasionally in winter, and it is a biting cold wind. The Mistral, too, is felt here, and some observers state that it is more felt here than at Mentone. San Remo has a clay soil, and on that account it is somewhat damp after heavy rains, but this is looked upon as not altogether a disadvantage, as it tends to render the air less dry and irritating. The rainfall at San Remo and the number of rainy days during the winter season appear to be less than at almost any other resort on this coast.

The accommodation provided for visitors at San Remo is good ; the hotels are numerous and well managed ; there are plenty of shops, and there is an Italian Opera. It is not well off for drives, and the excursions into the mountains and neighbouring valleys are difficult, not so much on account of the steepness of the ascent as from the absence of paths. As to the class of invalids likely to be benefited by San Remo, it is unnecessary to recapitulate what has already been said with respect to other health resorts on the Riviera. San Remo is adapted to the same class of cases with this distinction, that its climate is rather less bracing and more soothing than some of the other stations, and therefore better adapted to nervous and sensitive constitutions. There are some invalids who have a sort of sentimental preference for Italy over France. They feel happier if they can say, " Now we are in Italy," and this has probably had something to say to the rapid growth of San Remo as a winter resort. The large German colony here may also find Italy more comfortable as a residence than France.

There remain but two other towns on the Western Riviera that can be spoken of as in any sense winter health resorts. They are Alassio and Pegli.

Dean Alford asked why *Alassio* had " never been praised as a spot of shelter for English invalids ? " Since this question was asked, Alassio has been " taken up " and its merits as a health resort have begun to be made known. It is about twenty-eight miles east of San Remo, and is best reached from the north by the line from Turin to Savona, the latter town being about twenty miles east of Alassio. Alassio is situated in a lovely bay, having a south-eastern aspect well protected between two headlands, Capo delle Male on the west, and Capo di Santa Croce on the east. It is also well protected by encircling hills to the north, at no great distance from the shore. It possesses an excellent beach of fine sand, and is popular with the Italians on that account as a summer bathing-place. The Riviera scenery again becomes very beautiful at Alassio. The fine hills behind the town are covered with olive-trees, and there are many sheltered nooks for villas as well as admirable picturesque walks and drives in the neighbourhood. Some of the views are remarkably beautiful and interesting, especially one looking west over the bay, and that of the curious island of Gallinaria just outside the bay to the east.

Meteorological observations of a thoroughly reliable kind are not yet available. It is probably not so warm as San Remo, as it is rather more open to the north-east winds, and the northern hills not being so high, the north wind (Tramontana) reaches a portion of the district close to the shore. But more trustworthy weather statistics will no doubt be soon forthcoming, and it will be time enough then to pronounce precisely on the relative merits of Alassio as a winter resort for invalids. That it is an exceedingly picturesque and beautiful winter residence may be confidently stated. Hotels are springing up rapidly, and it will no doubt soon follow the example of San Remo and develop into a popular Riviera health resort.

An asthmatic invalid, writing to me of Alassio, observes: "Alassio stands on a windy shore, and I think is the worst climate for my case I tried. I fancy the climate is too dry, and that heavy dews must fall at night, as I woke every night or early morning with violent catarrh, which did not cease till I left Alassio. I never felt well for an hour there. If some one would build an hotel up the hillside with south aspect and sheltered situation, it might prove a good resort. The hotels have not a south aspect, but both are (roughly) east and west, and they are close to the sea." This proximity to the sea is, of course, a convenience during the sea-bathing season.

Pegli is really a suburb of Genoa, from which it is distant only half-an-hour by rail. But it is very much warmer than at Genoa, as it enjoys a purely local protection from cold winds by means of hills to the north as well as to the east and west. It is a little fishing and ship-building town, situated along the sea-shore looking south, and celebrated for the presence of some very beautiful gardens, belonging to the palaces of native noblemen, especially those of the Villa Pallavicini. It differs, no doubt, considerably in its climate from the resorts at the western extremity of the Riviera di Ponente ; but there are no available meteorological tables for purposes of comparison. The humidity of the air, for one thing, is much greater, and those who have found the air of Mentone unpleasantly dry and irritating have improved much at Pegli. It has acquired a reputation for benefiting asthmatic cases, which, as a rule, do not do well (if spasmodic) on the Riviera generally.

An asthmatic patient writes to me of Pegli to the following effect, of the hotel : "Beds never properly made, nor rooms swept. Food far from good, meat and fowls often uneatably tough and tasteless. No proper supply of water in cabinets, which were therefore far from healthy. The climate is also trying, hot, and dusty in front of hotel, while the garden is as cold as a well. The hotel is very draughty. Invalids ought to beware of the trams to Genoa, which sometimes pull up for many minutes at the entrance of a tunnel in a cold wind. The road to Genoa is very dusty, and the drains very unsavoury. I began to suffer at once from catarrh, and was unwell the whole of the fortnight I was there."

In the foregoing brief sketch of the principal health resorts on the Western Riviera I have endeavoured, from the results of personal observation and from comparison of various authorities, to point out in as concise a manner as possible the chief characteristics of climate and situation of those several stations. I would repeat, by way of conclusion, that the climate of the Riviera is by no means a perfect one. But if it has cold winds and at times blinding dust, and if the air in places is exceedingly dry and irritating, it has also an immense proportion of fine days, clear skies and bright sunshine, when from nine in the morning until three in the afternoon an invalid can live in the open air. "The warm southern sun and the azure sky of the Mediterranean, far more than elevated temperature, constitute the advantages of this climate ; fine weather rather than heat is what is here sought for," and, let me say, is usually found. But if the Western Riviera has its drawbacks—and what climate has not?—it must be admitted that the number of localities which we here have to choose from gives us an opportunity of selection impossible to find elsewhere. And then it is a region of almost unrivalled beauty. A very eminent and greatly occupied London doctor was found one Christmas looking very pleased and happy on this coast. "What brings you here?" inquired a wondering friend. "I am here for three or four days' holiday."

"But is it possible that it can be worth your while to come so far for so short a time?" The doctor turned round to his friend, and, waving his hand seaward, replied: "How do I know I shall ever see such a day as this again!" And he stood on the shore of a beautiful bay, from the margin of which the blue water stretched far, far away; behind him rose range after range of encircling hills, terminating finally in the snowy summits of the distant Alps. On each side steep wooded promontories, jutting out from the mountains behind, planted their feet in the blue sea, and in a cloudless sky, shining majestically over all, the source of all its beauty, was that "grand arbitre des santés humaines, Monseigneur le Soleil!"

NOTE TO CHAPTER IX.

There are but few places in the Maritime Alps suitable, either by their elevation above the sea or by the accommodation they afford, for invalids to pass the hot months of summer. The following, however, may be mentioned as the most accessible.

St. Martin Lantosque (3120 feet above the sea and seven hours from Nice by carriage-road) affords fair accommodation in several hotels and *pensions*, and is becoming more and more resorted to during the summer, especially by the inhabitants of Nice. It is surrounded by beautiful scenery; the air is pure, fresh, and moderately bracing. There is, at times, a good deal of humidity of atmosphere; but the temperature is equable and rarely fluctuates many degrees in the twenty-four hours.

Bollène (2200 feet) is picturesquely situated about two hours' drive from St. Martin, lower down in the same valley, and it has a fairly good hotel.

Belvedere and *Berthemont* (about 2700 feet), between St. Martin and Bollène, are two villages in a cool situation surrounded by fine scenery. Berthemont has sulphur springs and a bathing establishment, and two good hotels and several villas. At Belvedere there is, or was, an English *pension*. Many interesting excursions can be made from either of these three places.

La Cascade, higher up in the valley than St. Martin, has a modest *pension* near a fine forest, and is in a cooler and fresher situation than either of the preceding.

The Baths of Valdiéri (about 4300 feet above the sea), on the Italian side of the Maritime Alps, can be reached from St.

Martin in five hours and a half, either on foot or, in fine weather, by mule-track.   There are sulphur springs here, and they are utilised in a large bath establishment, and as many as 600 visitors may be found taking the waters during the season. These baths are used for the same cases and in the same manner as the sulphur baths of the Pyrenees.   The climate is mild and equable, and the place is completely protected from winds from the north ; but it is refreshed by local currents of air which are rarely very strong.   The surrounding scenery is grand and picturesque ; the only drawback is the difficulty of access.   On the Italian side it is approached from Coni.

The Baths of *Vinadio* and those of *Certosa di Pesio* are also reached by mountain passes from St. Martin.

All the preceding resorts are approached from Nice by the Vésubie valley.   The adjacent valley of the Raya has the advantage of being traversed in its upper part by a carriage-road which connects Turin and the plains of Piedmont with the Riviera at Nice.   A conveyance runs daily from Nice to Coni in eighteen hours.

In this valley *St. Dalmas di Tenda* (3500 feet above the sea) is about fifty-three miles from Nice on the high-road between Nice and Coni.   It has an *Établissement* for hydropathy, with a fine garden, which was in the last century a Carthusian monastery, and it has also a fairly good hotel.   It is in the midst of wild and beautiful scenery.

" From Lucérain, on the above route, a military road will soon be finished along the summit of the mountain range to Fontan, a few miles below St. Dalmas.   It will give access to some high plateaux 4000 feet or more above the sea, well wooded and watered, and which promise almost faultless sites for summer stations."*

Other places of summer resort in the Maritime Alps which have been mentioned, but of which little seems to be known, are : *Bajardo*, on a mountain-top behind Mentone ; *San Romolo*, behind San Remo ; and the *Château de Thorene*, with a *pension* of forty beds, thirty miles above Grasse, close to a pine forest.

* Sparks, "The Riviera."

CHAPTER X.

MINERAL WATERS IN HEALTH RESORTS.

IN the preceding chapters, in discussing the characters and merits of various health resorts, we have been chiefly occupied in the consideration of conditions of climate, and of suitability of accommodation and environment.

In the subsequent chapters we shall have to give considerable prominence and importance to the presence and use of mineral springs in health resorts.

Fortunately, in some localities, the presence of valuable mineral springs is associated with the presence of climatic conditions which not only aid in a remarkable manner the influence and operation of the course of mineral waters, but are even of themselves active curative agents. We have seen this to be the case in those great resorts in the Engadine, St. Moritz and Tarasp.

In dwelling somewhat fully on the action of the mineral waters of these two popular watering-places, I have anticipated, in certain particulars, what I shall have again to refer to here ; and, in alluding hereafter to the modes of application and uses of the several mineral springs I shall have to notice, it may be necessary to repeat some of the reflections which will naturally find a place in this chapter.

The mineral springs in certain spas are used almost exclusively for bathing, as, for instance, at Ragatz, Teplitz, Wildbad, and, our own country, at Buxton.

In other spas, as Vichy, La Bourboule, and our own Harrogate, drinking the waters is of chief importance. But in the great majority drinking and bathing are combined.

In very many bath establishments elaborate apparatus are provided for the local application of douches and sprays of mineral water to the actual seat of disease—the skin, joints, nose, throat, air passages, etc.

In others special use is made of hot vapour chambers, as is the case at Mont Dore.

The value and extent of applicability of mineral springs depends very much on their thermality—on their degree of temperature. If they rise from the ground at a comparatively low temperature they have to be heated by artificial means before they are used for baths, as is the case at Schwalbach, St. Moritz, and in most of the chalybeate spas.

If they spring from the ground at a very high temperature they either have to be diluted with water of a lower temperature, or allowed to stand in the bath (as at Wiesbaden) until they are reduced to a suitable degree of warmth.

In some spas special baths are prepared by the addition of other ingredients to the mineral water, either of a stimulating or a soothing nature, as in the mud baths of Franzensbad, the pine-leaf baths of this and many other German spas, and the strong *mutter-lye* baths of Kreuznach and Nauheim.

The presence of a considerable amount of carbonic-acid gas in a mineral spring is another very important condition of usefulness and applicability. It not only renders waters that would be otherwise exceedingly disagreeable not unpleasant to the palate, but it also aids, in a very important extent, their digestion and assimilation. It is the absence of any considerable quantity of this gas, in the majority of mineral springs in this country, that diminishes their usefulness and prevents their attaining any great degree of popularity.

We are often reproached for not making adequate use of our own mineral springs, but it is to this

absence of carbonic-acid in them that the reproach
should be addressed; whoever will take the trouble
to compare the flat, inky tasting, chalybeate springs
of Tunbridge Wells with the clear and sparkling iron
water of Schwalbach or St. Moritz need, surely, not
be at a loss to account for the popularity of the latter
and the neglect of the former. Conditions of climate
also come into consideration in many instances, as,
for example, the sub-alpine bracing climate of Gastein
as compared with the moist relaxing air of Bath.

There can be little doubt that the good that is
accomplished in many cases by a course of mineral
waters or baths at a Continental spa is, in a great
degree, due to the concomitant excitement of travel
and change; the increased inducement to bodily
exercise, the altered diet, occupations, and habits,
and the influence of unusual climatic conditions.

It is scarcely necessary to say that the composition of a mineral water depends on the nature of the
strata through which it has passed, and that the
larger the amount of free carbonic-acid or oxygen the
water originally contains the greater will be its solvent
action on the mineral substances it encounters.

The predominating ingredient in very many
mineral waters is, as we might expect, lime in the
form of carbonate and sulphate. Other ingredients
commonly encountered are compounds of sodium,
potassium, magnesium, and iron; these are usually
found in the form of chlorides, sulphates, and carbonates. A small amount of silica is found dissolved
in many springs. The so-called sulphur waters
usually contain sodium sulphide, and some contain
free sulphuretted hydrogen.

Of less frequent occurrence in mineral sources are
compounds of aluminium, strontium, lithium, manganese, and arsenic.

Phosphates, iodides, bromides, and fluorides are
occasionally met with, but in very small quantity.

In several springs, especially in the sulphur springs,
a peculiar form of organic matter is found; in some,

this is termed *barègine*, and it is considered to be not without an important influence in the action of the water.

It is certain that at particular periods, a sort of fashion arises in favour of some particular ingredient in a spring. At one time sulphur, at another time arsenic, at another iodine, and at present *lithium* is the popular ingredient.

Since this has been the case it is remarkable how many springs have been discovered to contain minute amounts of lithium, which had hitherto been over-looked. In a recently circulated analysis of the chaly-beate springs at St. Moritz, chloride of lithium heads the list of ingredients ! !

Mineral waters have been conveniently arranged in the following classes :

1. *Simple Thermal Waters*, of which the baths of Ragatz may afford an example. These are charac-terised by their high temperature, varying from 80 to 150 degrees Fahrenheit and even higher, by their very feeble mineralisation (in some instances, as at Pfæffers, there are but two-and-a-half grains of solid constituents in 7680 grains of this water), and by their great softness.

These are sometimes termed " indifferent springs," on account of the absence of any special mineralisa-tion ; and sometimes, as they are often found in wild, romantic, wooded districts, they have been termed " wild-baths" (*wildbäder*). They are rarely used in-ternally, and if so employed they are simply intended to exert a solvent purifying influence, to wash the stomach and the tissues, and to remove effete matters.

As baths they are very largely employed, and some of the most frequented spas of Europe belong to this class, as Gastein, Wildbad, Teplitz, Schlangenbad, Ragatz, Plombières, Bath, and Buxton.

They are considered to act in the following manner : they cleanse and soften the skin, and so promote cutaneous excretion ; according to the temperature at which they are employed, they equalise or diminish the loss of heat from the body, or prevent it altogether, or even give heat to it ; they promote the circulation through the peripheral vessels, and so improve the nutrition of the skin ; they gently stimulate the organic functions and hasten tissue change ; they allay muscular and nervous irritability by their soothing effect on the peripheral nerves ; and they promote the absorption of exudations, rheumatic, gouty, and inflammatory.

They are largely employed in the treatment of chronic rheumatism, both muscular and articular, and in chronic gouty inflammation of the joints ; in sciatica and other neuralgias ; to allay hyperæsthetic states of the nervous system and certain hysterical conditions ; for the relief of painful wounds and cicatrices ; and in certain cases of loss of muscular power, when not dependent on central disease, but rather on inflammation and thickening of the nerve-sheaths.

These baths afford an essentially soothing mode of treatment, promoting elimination and change of tissue by the gentlest possible of measures and in a manner which can be tolerated by the most sensitive and delicate constitutions. Those that are situated in mildly bracing forest air, or in sub-alpine climates as Wildbad, Gastein, and Bagnères de Bigorre, are best suited to highly nervous and hyperæsthetic patients.

The more bracing the climate the higher usually the temperature at which the baths can be borne.

2. *Common Salt Waters.*—These vary greatly in their strength, from Reichenhall with 24 per cent. of chloride of sodium to Wiesbaden with only 6 per cent. It is usual to add concentrated *mutter-lye* (bittern) as at Kreuznach to the weaker springs and to dilute the stronger ones with pure water.

It must be remembered that chloride of sodium is one of the most frequent ingredients of mineral waters, but it is only when it exists in a spring in altogether preponderating amount that it belongs to this class.

These salt baths have always enjoyed a very considerable popularity. They stimulate the peripheral nerves and promote the capillary circulation. They improve the nutrition of the skin and raise its tone, and indirectly they stimulate change of tissue. When there is a considerable amount of free carbonic-acid in the water, as well as a large quantity of chloride of sodium, as at Nauheim and Rheme, the stimulating effect on the skin and on the nutritive changes is greatly augmented.

Taken internally these waters stimulate the gastric and intestinal secretions, and so promote the action of the bowels and improve the portal and general circulation.

By their stimulating influence on tissue changes and on the circulation they promote the absorption of morbid deposits.

It is necessary to bear in mind that if given in too large doses they are apt to give rise to irritation of the gastric and intestinal mucous membrane.

The presence of free carbonic-acid in these waters modifies the action of the chloride of sodium; this gas acts as a sedative to the nerves of the stomach, it promotes secretion and absorption, and augments peristaltic action. It quickens the passage of the water from the stomach into the intestine, and so augments its aperient effect.

The warmer these waters are when they are drunk the more rapidly are they absorbed, so that their local effect is made less and their constitutional effect greater. The more concentrated they are the greater is their local action.

They are employed externally in various conditions, especially in cases of hypersensitiveness of the skin ("weakness of skin") giving rise to a tendency to attacks of bronchial catarrh and to acute rheumatism. They are of value also in certain forms of retarded convalescence from acute and chronic disease, especially in scrofulous and other inflammatory enlargements of joints. Internally they are used in cases of atonic dyspepsia and in chronic gastric catarrh. In cachexias contracted in tropical climates and associated with torpidity of the portal circulation. They are valuable in cases of hæmorrhoids dependent on portal congestions ("abdominal plethora") and in congestions of the pelvic organs.

In cases of chlorosis and anæmia, when preparations of iron, by themselves, are not tolerated, much good often follows a course of these waters. They are useful in overcoming the troublesome constipation that commonly accompanies the foregoing morbid states. They are considered to agree better with anæmia and emaciated persons than the *alkaline* aperient springs, such as those of Carlsbad.

The most frequented of these baths are Kreuznach, Nauheim, Ischl, Reichenhall, Rehme, Homburg, Kissingen, and Wiesbaden ; and in this country, Droitwich, Nantwich, Woodhall, and Harrogate.

Some of the springs at these spas contain an appreciable amount of iron, which adds to their tonic effect, when taken internally.

3. *Alkaline Waters*, of which those of Vichy may be taken as the type, their chief constituent being carbonate of soda ; they also contain a larger or smaller quantity of free carbonic-acid.

The fact that some of these springs contain also a considerable amount of chloride of sodium has led to their subdivision into—

(*a*) *Simple* Alkaline Waters, like those of Neuenahr and Vichy (*hot*), and Apollinaris, Vals, and Bili (*cold*).

(*b*) *Muriated* Alkaline Waters, like those of Ems and Royat (*hot*) ; Selters and Rosbach (*cold*). Most of the commonly employed "table waters" fall under this class.

They are comparatively rarely employed as baths. In small doses they are valuable for their solvent action, and in promoting tissue changes, and for their antacid and diuretic properties. In over-doses they tend to depress the action of the heart and to cause emaciation through excessive solvent action.

They are largely employed for the relief of dyspepsia with acidity; to remove hepatic congestions from torpor of the portal circulation; in tendencies to gall-stones; in diabetes; in the uric-acid (gouty) diathesis; in renal calculi, gravel, and lithiasis.

They are also valuable in most chronic catarrhal affections of the respiratory, alimentary, or genito-urinary tracts.

The muriated alkaline waters are more tonic and stimulating than the simple alkaline ones.

4. *The Sulphated Waters.*—This class contains all the best known aperient and laxative springs, whose chief constituents are sulphate of soda or magnesia, or both together.

Some of these waters also contain considerable quantities of carbonate of soda and chloride of sodium, and this has led to the subdivision of the class into

(*a*) *Simple Sulphated Waters.* — The so-called "bitter waters," as Friedrichshall, Hunyadi Janos, and Pullna.

(*b*) *Alkaline Sulphated Waters*, as Carlsbad (hot), Marienbad, Franzensbad, Tarasp, and Elster.

These waters are greatly employed as laxatives. Those of the first subdivision are usually drunk at home as imported in bottles; those of the second subdivision are drunk also largely at their sources.

They stimulate the intestinal mucous membrane, increasing peristaltic action and causing thin fluid evacuations. If long continued or drunk in large quantity they lead to emaciation and diminish corpulence. Sulphate of soda waters are usually less irritating than those which contain sulphate of magnesia.

The alkaline sources are especially employed to remove habitual constipation with torpid portal circulation and hæmorrhoidal tendencies ("abdominal plethora"). They are most valuable in the removal of congestive enlargements of the liver and spleen;

2 F

in some forms of dyspepsia in gouty, plethoric persons; in gall-stones, especially in the stout and sedentary; in lithiasis; in diabetes. They are often prescribed for the reduction of corpulence.

5. *Iron or Chalybeate Waters*, in which iron is the chief ingredient. Some of these contain other salts, though in small amount, and often a considerable proportion of lime salts. The *purest* iron waters are those of Spa, Schwalbach, Alexisbad, Tunbridge Wells. Most of the following contain a certain proportion of other salts: Pyrmont, Orezza, St. Moritz, Rippoldsau, Bocklet, and Santa Caterina.

The most serviceable iron waters are those which contain the bicarbonate of the protoxide held in solution by free carbonic-acid, as in the St. Moritz and Schwalbach springs.

I have fully entered into the action and uses of iron waters in describing the St. Moritz springs in a former chapter.

6. *Sulphur Waters.*—This is a large and important group of mineral springs of which *Luchon* and *Les Eaux Bonnes*, in the Pyrenees, may be regarded as fairly typical examples. *Aix les Bains* and *Aix la Chapelle* also belong to this class. In nearly all the sulphur springs in the Pyrenees the sulphur occurs in combination with sodium as sodium sulphide; in others it occurs as sulphuretted hydrogen. In some it occurs combined with calcium, potassium, or magnesium.

Some of these are hot springs, as Les Eaux Bonnes, Les Eaux Chaudes, Luchon, Cauterets, and others in the Pyrenees, Aix les Bains, Aix la Chapelle, Baden in Switzerland, Baden near Vienna, Allevard, Uriage, Schinznach, Heluan near Cairo. Others are cold, as Enghien, Challes, Gurnigel, Eilsen, Neundorf, Weilbach, Harrogate, Strathpeffer, etc. Some of these sulphur springs contain a considerable amount of chloride of sodium, as is the case at Uriage, Aix la Chapelle, and Harrogate.

The action and various uses of these sulphur springs are fully discussed in the chapters on Pyrenean Health Resorts, and those of Savoy and Dauphiné, and incidentally elsewhere. They are employed largely in the form of inhalation of their vapour and sprays.

7th and last. *Earthy and Calcareous Waters.*— So named on account of the preponderance in their composition of the earthy salts of lime and magnesia, chiefly in the form of sulphates and carbonates —Leuk, Wildungen, Lippspringe, and Contrexeville may serve as examples.

In their action when used as baths, they approach in character the springs of the 1st Class, i.e. the simple thermal waters.

When drunk, as at Contrexeville, they prove soothing to the gastric and intestinal mucous membrane. When they contain large quantities of sulphates and carbonates of lime they often tend to produce constipation.

They are employed internally in some forms of irritative dyspepsia with acidity of intestinal secretions and a tendency to diarrhœa. At Contrexeville they are largely used for the relief of calculous disorders and in vesical catarrh ; courses of these waters act probably through the large quantity of water consumed, rather than in virtue of the earthy salts they contain, so that the urine is greatly diluted and the renal tubes are continuously washed out.

Externally they are often used for very prolonged baths, as at Leuk in Switzerland, in certain skin diseases, as eczema and psoriasis, where long continued soaking of the skin is considered likely to be advantageous. But it must be admitted that the precise mode of action of these earthy waters is not fully understood.

In the succeeding chapters many examples of all these varieties of mineral waters will come under consideration.

CHAPTER XI.

SOME FRENCH HEALTH RESORTS.

VICHY AND THE HEALTH RESORTS OF AUVERGNE (ROYAT, MONT DORE, LA BOURBOULE, ETC.).

VICHY.

FRENCH health resorts are certainly not so popular with our countrymen as are German and Swiss ones, or, at any rate, they have not been so hitherto. Recently, however, attention has been much directed towards the Baths of Auvergne; and in France itself, within the last few years, an active and fruitful investigation has been in progress in connection with the very numerous and important mineral springs which are found in that remarkable and interesting district of Central France.

Vichy, one of the best-known spas in Europe and the most frequented of all the French spas, is, from its geographical position as well as the character of its waters, naturally associated with the neighbouring health resorts of Auvergne, and a brief review of the medical and social aspects of the place may serve as a fitting introduction to the study of those other stations which are rapidly growing in influence and popularity.

If we compare Homburg and Vichy, it is not a little remarkable how many more English people are attracted to the former than to the latter place. Yet the waters of Vichy are, no doubt, much better suited to the treatment of many persons who go to Homburg than those of Homburg itself, and the same remark applies, but in a far more limited sense, to Carlsbad. Vichy is also quite as accessible as Homburg; it is reached in eight-and-a-half hours from Paris. The hotels are excellent. The living is cheaper, and there are quite as many amusements. But Homburg is on the road to the "regular Swiss round," it is close to the somewhat cockneyfied Rhine, and, moreover, it is the custom with German physicians when a serious course of waters has been advised, to prescribe an "after-cure" among the Swiss mountains or in the Black Forest. The French physicians do not seem to have bethought themselves yet of after-cures. But why an after-cure should be necessary to the courses at Carlsbad and Homburg, and not to that of Vichy, may, perhaps, be a little puzzling to the uninformed mind. Vichy has the misfortune, and it is one which is apparently much less thought of by the French than the English, of being removed, and yet only just removed, from agreeable and picturesque scenery. A drive of three or four miles out of the town brings you into a country sub-alpine in character; the out-jutting spurs of the mountains of Forez, valleys clothed with rich green verdure, pine-clad hills, and rippling streams. But Vichy itself, it must be admitted, is little adorned by natural beauties; it is surrounded and shut in nearly on all sides by low-lying and uninteresting hills.

Much of the land on which the new part of Vichy has been built has been reclaimed from the right bank of the river Allier, along which it lies. The river itself is shallow and straggling, with flat sandy banks, from which only a distant view can be obtained of the mountain chains of Forez and Auvergne. What nature, however, has failed to do art has done much to supply. Napoleon III., who frequently visited Vichy, like a grateful patient, did much for the place, to which he thought he owed much. He beautified and transformed it by causing handsome villas and public buildings to be erected, and by converting into a charming public pleasure-ground, beautifully laid out and planted with trees and shrubs, and furnished with flowers and fountains, all that part of Vichy lying between the town and the right bank of the river, and which is now known as the Parc Anglais.

The buildings devoted to the reception of visitors are of the most elaborate kind. The *Établissement Thermal* is furnished with every kind of appliance which " water doctors " have learnt to use—hot baths, cold baths, vapour baths, swimming baths, douches of every kind and every degree of force, direct, oblique, descending, ascending, with which your ailment can be attacked in turns ; or if it should be a particularly obstinate one you can be introduced into an ingenious cage-like arrangement, where every form of douche can be made to play upon it at the same time. It is true, however, that few are bold enough to venture into this modern instrument of torture. Inhalations of carbonic-acid are also provided, but less benefit seems to have been obtained from this mode of treatment than was formerly anticipated from it. In asthma it is now rarely used, but it is still said to yield good results in some of those troublesome chronic catarrhal conditions of the mucous membranes of the nose and throat common in scrofulous constitutions.

The Casino is an elegant and luxurious building, providing every imaginable resource to prevent the visitors at Vichy from falling victims to that most terrible of maladies, *l'ennui*. Balls, concerts, and excellent operatic and dramatic performances, in which some of the most distinguished actors and singers in France frequently take part, are daily provided. But in the Casino at Vichy, as at other French health resorts, the gaming-tables prove unfortunately one of the most powerful attractions. Crowds nightly press round the *écarté* tables from the moment play begins until the Casino closes. The game is superintended by a sort of croupier, who

makes and shuffles the cards, sees that the betting
is equal on both sides before each game begins, and
settles all disputes. Each game usually lasts but a
few minutes, the losing player immediately yielding
his place to another. In this way large sums of
money rapidly change hands, and all the worst aspects
of gambling are here represented.

Vichy and its immediate neighbourhood are extraordinarily rich in mineral springs. Wherever a shaft
is sunk within a distance of six or seven miles in the
basin surrounding the place, alkaline gaseous springs
analogous to those of Vichy are certain to be found.
Hence it is that many of the springs are private
property and bear the names of their proprietors—
such, for example, as the Source Lardy, the Source
Larbaud, and others, each proprietor naturally claiming special virtues for his own spring. Now, although
all these springs are of the same general character,
they differ, many of them, from one another in obvious
physical qualities. Some are cold, most are hot.
Some contain a considerable quantity of free carbonic-acid, some contain very little, some are clear
and sparkling, some are slightly cloudy, and at least
one throws down a large amount of an organic deposit,
which, from its resemblance to that deposited by the
waters at Barège, is called *barègine.*

Moreover, some of the Vichy springs contain iron
in appreciable quantity, while others do not, and this
fact is made use of to classify these springs into two
groups—the simple alkaline waters and the alkaline
iron waters. It is to the possession of the former,
however, that Vichy owes its great reputation. An
appreciable odour and taste of rotten eggs in some
of the springs leaves no doubt of their containing
some sulphur compound yielding sulphuretted hydrogen, although no mention is made of this gas in
the analyses published. The predominating ingredient in all the sources of Vichy is bicarbonate of
soda, and it is the chief one in the simple alkaline
springs. Of these there are three which are com-

mouly drunk.   Two are hot springs—viz. the Grande
Grille, situated in the *Établissement des Bains* itself,
and having a temperature of 41 degrees Centigrade
(106 degrees Fahrenheit), and L'Hôpital, situated in
the town, rather more than a quarter of a mile from
the Grande Grille, and having a temperature of
30·8 degrees Centigrade (81·5 degrees Fahrenheit);
the third is a cold spring, the well-known Célestins,
so called because of the existence of an ancient convent of that name close to the source.   It is situated
in a pretty and tastefully laid out inclosure, more
than half-a-mile from *L'Établissement*, at the termination of the Parc Anglais, and near the river.
The latter is a very pleasant water to drink, and has
none of the disagreeable alkaline taste which an ordinary solution of bicarbonate of soda has.   Its taste
is, moreover, quite different from that of the other
two springs, although their chemical composition is
nearly identical.   The difference of temperature and
of relative proportions of free carbonic-acid in each
may, to a certain extent, account for this.   The spring
called L'Hôpital contains by far the largest amount of
organic matter, which causes a characteristic greenish
deposit upon the sides of the basin into which it rises.
This probably gives an increased softness to the
taste of this water, and may explain why it is that
this spring is found so particularly applicable to cases
of dyspepsia with much gastric irritability.   Besides
bicarbonate of soda, these sources contain bicarbonates of potash, magnesia, strontia, and lime, chloride
of sodium, and small quantities of sulphate and phosphate of soda.   It is important to remember that
Vichy is almost the strongest soda spring known,
containing as much as five grammes of bicarbonate of
soda to a litre of water; this fact, together with the
relatively high temperature of some of its springs,
gives to these waters a very high degree of importance
medically.   It is worth noticing that these springs
contain also a minute quantity of arsenic, two to three
milligrammes of arseniate of soda in a litre.   We

shall see how this ingredient increases in quantity and assumes great importance in some of the spas of Auvergne.

The springs of Vichy which contain the largest amount of iron are the Source Lardy and the Source Mesdames. The latter arises about two miles from Vichy and is conveyed to the town in pipes. Another spring, the Hauterive, which contains a notable amount of iron, and is richest of all in free carbonic-acid, arises at a distance of three miles from Vichy, and is, on both these accounts, the one chiefly used for bottling and exportation.

It need scarcely be pointed out that serious treatment by an active agent such as the mineral waters of Vichy imperatively demands constant and experienced medical counsel and supervision. Where there are so many sources having only slight shades of difference in quality and composition, the advantage of prolonged study and experience in determining the selection of one in preference to another is quite obvious. Then there is the question of the quantity of water suitable to each case, the question of baths and of the length of time during which such baths should be taken, the propriety of using douches and the kind of douche to be employed, the length of time the cure should last, these and many other minor questions must be left to authority and experience to answer.

Speaking generally, the source prescribed for stomach affections is L'Hôpital, that for hepatic disorders the Grande Grille, and that for gout and renal maladies the Célestins. But in each case the temperament, the constitution, and the habits of the individual have to be considered, as well as the nature of the malady. L'Hôpital is said to be less exciting than any of the other springs and best suited to feeble and irritable stomachs. The Grande Grille is hotter, more stimulating, more rapidly digested, acts more quickly and energetically, and is especially indicated in cases of hepatic congestion and cases of gall-stone, with or without jaundice. It is to be preferred in lymphatic and debilitated constitutions, and is of especial value in the malarious cachexia often engendered among the French colonists in Africa, either alone or mixed with one of the ferruginous springs.

The Célestins, much preferred for its agreeable taste and sparkling quality, is said to be highly stimulating and exciting to the nervous system. The iron springs, the Source Lardy and the Source Mesdames, are especially serviceable in the case of women and children after intermittent fevers, and are also well borne by dyspeptics who require iron. The quantity of water to be taken daily of necessity varies with the malady and the individual; it is no longer the fashion to prescribe the large quantities which were at one time consumed. Indeed, very remarkable results have been obtained with quite small doses of the thermal springs ; and in cases of very feeble digestion only very small doses can at first be comfortably taken. The Vichy physicians consider the baths an important part of the cure. They are usually taken daily for an hour at a time at a temperature of 86 to 93 degrees Fahrenheit, the mineral water being mixed with an equal quantity of fresh water. This is said to be an important precaution, the neglect of which may lead to sleeplessness, headache, congestion of the brain, and many febrile and nervous phenomena. The addition of bran to the bath is a method commonly adopted for diminishing its stimulating effect. The effect of the local application of the douche in cases of gall-stone, engorgement of the liver and spleen, as well as of affections of some other organs, is highly spoken of. The cases, then, to which the course of treatment at Vichy is appropriate are those of dyspepsia, when not due to organic, malignant disease, African dysenteries, the sequels of malarial fevers, congestion of the liver and gall-stones, with the jaundice which frequently accompanies these conditions. Hepatic colic is a malady especially amenable to treatment by the thermal springs of Vichy.

Some years ago a violent dispute arose among the

doctors of Vichy as to whether gout could or could not be advantageously treated there. The dispute grew so warm that it was referred to the Minister of Agriculture and Commerce, who referred it to the Academy of Medicine, and this body, in turn, refused to give any very definite decision on so delicate a question. The conclusions of one of the greatest medical authorities in Vichy and in France on this subject may be thus summarised : Seeing that gout is a particular error or vice of nutrition, and that the maintenance in their integrity of the natural phenomena of nutrition is the chief condition that can preserve from gout, and seeing that one of the most manifest effects of the waters of Vichy, properly administered, is the regulation of the digestive and eliminating functions, and to excite in them a special activity ; it follows that the waters of Vichy tend to prevent gout or to correct the gouty constitution by maintaining nutrition intact or by re-establishing it when disturbed. It is in this general way, and not as a specific antidote, that the waters of Vichy are valuable in gout.

While it was imagined that these waters acted as a kind of chemical antidote to gout, they were often administered in excessively large and injurious quantities ; but now that a different and more rational view of their *modus operandi* is accepted, the use of small doses, the effects of which are carefully watched, is the order of the day. " Of all the diseases treated at Vichy," writes the eminent authority to which we have already alluded, "gout is the one whose treatment requires the greatest amount of precaution and watchfulness."

With respect to renal calculous affections so constantly treated with so much success at Vichy, it is only necessary to say that modern medicine no longer sees a solvent action in these waters, nor can it longer resort to that hypothesis for an explanation of this success.

In cases of diabetes the course at Vichy has been frequently of the greatest advantage; but in these cases, as, indeed, in cases of gout, the Vichy waters must not be regarded in any sense as a specific remedy, but as producing their good effects through a general amelioration of the processes of nutrition and assimilation.

Vichy is also one of the many spas to which obese persons resort for the cure of obesity, and my next neighbour at the *table d'hôte* told me he "got rid of a great many pounds of superfluous weight in a very short time here."

Cases of anæmia, pure and simple, would scarcely come to Vichy for a cure; but cases of what are termed cachectic or symptomatic anæmia—i.e. anæmia the consequence of other disease, especially of disease of the organs of digestion and assimilation—are no doubt frequently benefited to a great extent by the alkaline-iron springs which are met with here.   Improve the nutrition—improve, that is to say, the blood-making processes, and it needs no conjurer to tell us we shall make more and better blood.

The dietetic *régime* prescribed at Vichy is not a very severe one, nor is it in any way special.   A fair and moderate amount of wholesome food and wine is all that is insisted upon.   A breakfast at ten o'clock of two or three courses, of which one is usually of cooked fruit and vegetables and frequently carrots— for stewed carrots is a speciality of the breakfasts at Vichy—a little red wine and water, or tea or coffee, for a beverage; a dinner at half-past five, which differs in no respect from the dinners met with at the *tables d'hôte* of any good French hotel—these two meals form the *pension* at the hotels.   Early hours are the rule here.   At six o'clock drinking and bathing commence seriously.   Half-an-hour or an hour of absolute repose after the bath is *de rigueur.* After the ten o'clock breakfast, a lounge in the open air, or a very gentle promenade, or a ride on a donkey passes the time till two or three in the afternoon, when water-drinking begins again, and those bathe who do not like the early morning hour.   A sort of full-dress promenade in the park also begins about half-past two, and lasts during the performance of the band. On Thursdays the children are especially thought of. A cordon is drawn round a circle immediately surrounding the orchestra. There, separated from the rest of the promenaders, the children of the visitors, mostly prettily dressed, assemble in con- siderable numbers, and dance *en rond*. In the time of the late emperor, a time looked back upon with regret by many of the inhabitants of Vichy, liveried lackeys handed light refreshments to the children after or during the dance. Dinner beginning at the early hour of half-past five, is over in time to leave the visitors free either to lounge or promenade again in the park, or visit the Casino and take part in the amusements there provided.

There are many interesting excursions to be made from Vichy in carriage or on horseback, but they are most of them at some distance, and those who are seriously bent on the cure there are probably but little interested in anything else.

The climate of Vichy is temperate and is said to be just like that of Paris; owing, however, to the vicinity of the mountains of Auvergne, thunderstorms are frequent and persistent, and in autumn thick fogs are common in the mornings.

The natives of the district are a feeble-looking race, having for the most part a poor physique and an unattractive appearance. Garden flowers, and especially roses, are very abundant, and it is a feature of life at Vichy to find the doors of the hotels sur- rounded by young peasant girls, with enormous bouquets of roses and carnations, which they sell for a franc or a franc and a half apiece.

Only two miles from Vichy is another town, *Cusset*, with important mineral springs, which is striving to make itself a name side by side with its formidable neighbour. It has this advantage over Vichy, that it is two miles nearer to such picturesque scenery as the neighbourhood offers, and it possesses a spring, the Sainte Marie, much richer in iron than any of those at Vichy. In other respects the waters are of the same composition, but the waters of Cusset are cold, and any one familiar with watering places is aware how very important an element of prosperity to such a place is the possession of a thermal source. Vichy, then, has nothing to fear from the concurrence of Cusset, but it seems not improbable that the existence of the stronger iron water in the latter place and its relative nearness to pretty scenery may attract visitors to whom the Vichy course is not so well adapted.

The following Table gives the analyses of the three principal springs at Vichy—*La Grande Grille, L'Hôpital, Les Célestins*—as well as that of the spring containing the largest amount of iron in a litre* (= 1000 grammes):

	La Grande Grille. Grammes.	L'Hôpital. Grammes.	Les Célestins. Grammes.	Mesdames. Grammes.
Bicarbonate of soda .	4·883	5·029	5·103	4·016
„ potash .	0·352	0·440	0·315	0·189
magnesia .	0·303	0·200	0·328	0·425
strontia .	0·003	0·005	0·005	0·003
lime .	0·434	0·570	0·462	0·604
„ oxide of iron .	0·004	0·004	0·004	0·026
Sulphate of soda . .	0·291	0·291	0·291	0·250
Phosphate of soda .	0·130	0·046	0·091	traces
Arseniate of soda .	0·002	0·002	0·002	0·003
Chloride of sodium .	0·534	0·518	0·534	0·355
Silicic-acid . . .	0·070	0·050	0·060	0·032
	7·914	8·222	8·244	7·811
	Litre.	Litre.	Litre.	Litre.
Free carbonic-acid . .	0·908	1·067	1·049	1·908

* A litre is about 35 fluid ounces, and a gramme about 15½ grains.

ROYAT LES BAINS.

The volcanic district of Auvergne is, as might be expected, enormously rich in thermal springs, many of which rise in mountain regions, where picturesque scenery and fine bracing air contribute no unimportant addition to the effects of these mineral sources. It is said that there are no less than five hundred distinct mineralised springs in Auvergne, while the department of Puy-de-Dôme alone contains over two hundred. Most of these are but little known even in France, and, of course, still fewer are known in England. Many of them are in out-of-the-way mountain districts, at a considerable distance from railways, where the public conveyances are of the worst description and the hotel accommodation of the most primitive kind. Who has heard in England of the thermal stations of St. Nectaire, Médague, Aigueperse, Chaudesaigues, Châteauneuf, or even of Châtel-Guyon? Yet these appear as important sources in a map of the principal thermal stations of Auvergne which lies before me. Of the better known and more frequented sources, Mont Dore, La Bourboule, and Royat, the former was, perhaps, the only bath at all well known in England until lately.

Royat has already attained a popularity and an importance nearly equal to that of Mont Dore or La Bourboule; and its greater accessibility and milder climate cannot fail to make it even more popular with a considerable proportion of invalids.

It is only within the last few years that Royat has become at all well-known to the medical world in England as a health resort. Royat has been described as "in a beautiful situation not far from Clermont-Ferrand, in Puy-de-Dôme, 1380 feet above the sea-level, with an agreeable refreshing climate, and possessing several springs which in their constitution resemble those of Ems. The climate of Royat is during the months of July and August decidedly preferable to that of Ems." Since this was written the different mineral sources at Royat have been submitted to closer investigation and careful analysis, the place itself has undergone a rapid process of development, and arrangements for the fitting accommodation of visitors and invalids have been greatly extended; while it is easy to discern that the destinies of Royat are, fortunately, in the hands of able and intelligent persons, who intend to spare no effort to make it a health resort of European reputation. In this respect it contrasts favourably with its neighbour, Mont Dore, where visitors are compelled to submit to accommodation which is not only primitive—that would be a comparatively small matter—but often offensive and dirty.

In the first place, I will say a few words as to the composition and uses of the waters at Royat. It is claimed for them that they are rich in lithia, and virtues are claimed for lithia in the treatment of gout which, it must be admitted, many of the highest authorities regard as wholly hypothetical. It is necessary to keep this fact in mind, as it has been somewhat hastily maintained that lithia is the acknowledged remedy for gout; and it would be in the highest degree unsatisfactory if this statement were allowed to pass without question. If, however, it should be established by experience that very minute doses of lithia are of service in the treatment of gout, then the Royat waters may certainly claim to be efficacious in the treatment of that malady. The Source Eugénie contains fifty-nine milligrammes of bicarbonate of lithium in a litre*

---

* Since Royat has become so popular on account of the presence of lithium in its springs, it is remarkable in how many other mineral sources lithium has been discovered!

—that is to say, about a third of a grain in sixteen ounces. Arsenic in very minute quantity is also found in the Royat waters. In the St. Victor spring, according to the latest analysis, two milligrammes of arseniate of iron per litre is stated to exist.

Much importance is attached to this fact at Royat, although at Vichy no importance appears to be attached to the circumstance that arseniate of soda is found in all the springs there to the amount of from two to three milligrammes per litre. Indeed, it has been doubted by very high authorities on mineral waters whether these very minute quantities of arsenic have any appreciable effect. If it should be proved by experience that they have, this doubt also will be removed. To find arsenic in a mineral water has been looked upon lately as almost as good as finding gold in a mine. Apart, however, from the presence of lithia and arsenic in the waters of Royat, they contain other important constituents in considerable quantity. The alkaline bicarbonates predominate, as in most of the mineral springs of this district, those of soda, lime, and magnesia chiefly, and to these must be added chloride of sodium. The Source Eugénie is the most richly mineralised, and contains more than five grammes of solids in each litre—i.e. rather less than forty grains in sixteen ounces. The Sources St. Mart and St. Victor contain nearly as much; the Source César is but very feebly mineralised, and as it contains a considerable amount of free carbonic-acid, it is the best suited for use as an ordinary drinking water. It is by no means unimportant to bear in mind that three of these springs—viz. the Eugénie, the St. Victor, and the César—contain an appreciable quantity of bicarbonate of iron. There are thus four principal springs at Royat: Eugénie, St. Victor, St. Mart, and César. These are all thermal springs; the Eugénie is the hottest, and has a temperature of 35·5 degrees Centigrade (96 degrees Fahrenheit) ; the St. Victor the coolest—its temperature is 20 degrees Centigrade (68 degrees Fahrenheit).

At Royat, as in most other large spas in France, the waters are utilised in every possible manner ; the well-appointed *Établissement* provides douches of all kinds. To each bath there is a douche. There are *salles d'aspiration*, in which one sits, fitly attired, and breathes the vapour driven into the chamber from the hot springs themselves, and, after being steamed in this fashion for half-an-hour, the patient is hurried off to his hotel in a sort of sedan-chair and ordered to repose for an hour. There are also well-appointed *salles de pulvérisation*, in which jets of water are pulverised by being driven with great force against metallic discs, or the water is driven into spray by means of steam. These jets of spray and pulverised water are inhaled, and are the chief treatment employed in affections of the throat and nose. Baths and douches of carbonic-acid gas are also provided. Royat also possesses a fine *piscine*, or swimming bath, one of the largest of its kind. This is furnished with a gymnastic apparatus and is very popular with the French ladies, where they delight to disport themselves for an hour or more in the morning. The water is very buoyant, but it has not an agreeable aspect, being of a dirty, greenish, and clouded appearance. It makes the hair sticky and matted unless it is carefully washed with fresh water after the bath. There is a *professeur de natation* attached to the *Établissement*.

A hydrotherapic establishment, in which all the processes of the ordinary water-cure can be carried out, has recently been added to the other curative resources of Royat. The chief *buvette*, or drinking fountain—that of the Eugénie Source—is situated in the tastefully-arranged park, and the water is ladled out from the bubbling source itself, and handed to the numerous claimants by female attendants. Before breakfast in the morning—the serious *déjeûner* is at eleven o'clock—and again between three and four in the afternoon, are the times appropriated to drinking the waters, the dinner hour being six. A good band, often that of the artillery regiments in Clermont, plays during the afternoon drinking hour, when the park presents quite a gay and animated appearance. It is somewhat curious to observe here how completely the system of ordering large quantities of water to be drunk has given place to the other extreme of ordering very small quantities. Half-a-glass twice a day must convey a dose of lithium or arsenic small enough to gratify the most homœopathic mind. I saw a young lady, attached in an official capacity to the *Établissement,* who had, in an unguarded moment, been quenching her thirst at one of the mineral springs, and had drunk it as if it were pure water. The consequences were by no means agreeable. To use her own expression, she felt as though she could "ni la garder, ni la rendre!"—a sufficiently embarrassing state.

The next question with regard to Royat is— What are these waters good for? what diseases do they pretend to cure there? The two words most frequently on the lips of the physicians of the place are *arthritisme* and *lymphatisme.* In this we see one of the characteristic tendencies of the French mind— i.e. a tendency to rapid classification and generalisation. It is unquestionably a defect and a serious one. The English mind—I am thinking, of course, of the scientific mind—generalises slowly and with great caution. It results from this that a French physician will be dogmatic and precise when to an English one dogmatism seems dangerous and precision impossible. It must, however, be admitted that our term "gout," in its modern use, is as unscientific a generalisation as *arthritisme;* but we use the word in a conventional and popular, not a scientific sense. After all, it merely amounts to this, that certain maladies, if they arise in a constitution which has a tendency to develop gouty and rheumatic joint affections, are cured by the waters of Royat. So also certain maladies arising in constitutions which have a tendency to glandular enlargements are also cured here.

The form of gout most suitable to treatment at

Royat is that which is termed atonic, or gout asso-ciated with anæmia. The plethoric form of gout requires a far more active water, such as that of Carlsbad. The arsenic and the iron combined in the Royat waters render them decidedly tonic. Gouty dyspepsia is well suited to treatment here. I am assured also that certain forms of skin disease—indeed, nearly all forms having direct relation to the gouty constitution—are very greatly benefited by this course.

For example, what the French call *couperose*—in plain English, the gouty "red nose"—I am told, disappears under treatment here. Should this state-ment be corroborated by further experience, the valley of Royat will, doubtless, in course of time, bloom with red noses. One can imagine opposite rows of such sufferers seated at a *table d'hôte* earnestly regarding with delightful anticipation the gradually diminishing shades of colour of the noses of their opposite neighbours. Acne, too, that most intract-able and disfiguring pimply eruption on the face, is, I am informed, very amenable to the Royat waters. Gouty eczema—i.e. patches of eczema, limited and localised—belong also especially to the list of ail-ments cured here, as well as some troublesome local irritations connected with the gouty constitution. Diabetes and certain forms of Bright's disease, when associated with gout, some of the local medical men maintain, do well here.

Anæmia and chlorosis, and other diseases of women similar to those treated at Ems, are also treated at Royat. Chronic rheumatism—that oppro-brium of the medical art—is, in some instances, relieved by these waters ; but these instances are rare, and, on the whole, the subjects of chronic rheuma-tism do better elsewhere. But it is also in some forms of diseases of the chest that the waters of Royat have been lately maintained to be particularly efficacious, especially when inhaled in the form of vapour. Chronic bronchial catarrhs, catarrhal asthmas, especially when they occur in the gouty, and even cases of early consumption, are sent in considerable numbers to Royat for treatment; and its climate is considered very suitable to such conditions. It is, at any rate, premature to hazard such confident opinions as one meets with in some quarters on this point ; and especially is this the case with regard to consumptive patients, when one considers, too, the constant errors in diagnosis which occur in connection with what is called the early stage of consumption. Cases of chronic laryngitis and pharyngitis in gouty people are greatly benefited here.

I must now say a few words on the climate and situation of Royat. Royat is only a quarter of an hour's drive from Clermont-Ferrand, an important town of Central France, on the direct line of railway from Paris to Nimes, and reached by express train from Paris in between eight and nine hours. As you approach Clermont you observe that it is built on rising ground at the border of a large semicircular basin, surrounded on all sides except towards the east by those remarkable volcanic dome-shaped mountains, " Les Monts Dômes," or " Les Puys " of Auvergne. In the centre of these rises the Puy-de-Dôme, considerably the highest of these " Puys," and on the summit of which, in clear weather, the meteorological observatory erected there can be seen, as well as the magnificent ruins of a great Roman temple to Mercury, which were discovered during the excavations rendered necessary for the construction of the observatory. Royat is situated in a picturesque valley, which runs. up to a sort of ridge, which bounds a richly cultivated plateau stretching to the eastern base of the Puy-de-Dôme. It is inclosed by mountains on all sides except towards the east, where it looks down on Clermont, and thence over the whole of the rich plain of the Limagne and on to the mountains of Forez, which limit the eastern horizon.

Its climate depends partly on its adjacency to the

great central mountain chains of Auvergne, the Monts Dômes and the Monts Dore, partly on the porous volcanic soil, and partly on its own particular situation. To the first it owes its sudden storms of wind and thunder and rain, and its sudden transitions of temperature—the characteristics, in short, of a mountain climate of moderate elevation. To its volcanic soil it owes the clouds of black and reddish dust which are blown about in eddies by the sudden gusts of wind, as well as the dryness of its roads and also the dryness of its atmosphere, this dryness of atmosphere being a most important climatic condition, counteracting the ill effects of the storms of wind and rain. Then there is the peculiar situation of Royat itself. Lying as it does in the floor of a somewhat narrow valley, surrounded on all sides by mountains, and only open to the east, running, moreover, in a direction exactly east and west, and facing the east, it is particularly exposed to the direct heat of the sun. From the moment the sun rises in the east above the mountains of Forez until it sets in the west behind the gigantic mass of the Puy-de-Dôme, Royat lies exposed to its rays ; and it is therefore exceedingly difficult to find any kind of shady walks in the immediate vicinity of Royat when the sun is up and the sky is cloudless. Generally, however, when the sun goes down, the cooler upper strata of air rush down from the higher plateaux into the valley, and thus you get cool refreshing currents of air playing through the valley on the evenings of even some of the hottest days in summer.

The invalids and visitors at Royat are well provided with amusements ; the presence of two regiments of artillery at Clermont, with a permanent band, provides very good orchestral music. At the Casino there is a concert or ball or dramatic performance every night, and some of the eminent singers who are following the course of treatment at Royat often take part in these entertainments. There are reading and billiard-rooms, and the usual *écarté* tables surrounded by a crowd of gamblers. Good riding horses, donkeys, and carriages·

are supplied in profusion. The excursions are numerous, varied, and interesting ; for no more remarkable country to the geologist, the naturalist, and
the archæologist can be found than this great
mountainous district of extinct volcanoes, old mediæval
towns, historic churches, and Roman and even earlier
remains. Some of these excursions, however, are long
and fatiguing, and the real invalids will probably do
better to remain in the park or in the gardens or on
the terraces of their hotels. The ascent of the Puy-
de-Dôme is easily made from Royat; carriages of a
certain kind can be taken quite to the top. A good
walker requires about three hours to walk from Royat
to the summit. The view from the summit is very
remarkable. The surface of the country that one
looks down upon seems to have boiled up at some
remote period, like the contents of a bubbling cauldron
which has been suddenly cooled and consolidated in
the very act of boiling. The archæologist will find a
source of the greatest possible interest in examining
the very remarkable and extensive remains of a
Roman temple to Mercury, much of which is in
excellent preservation, erected here close to the very
top of the mountain, more than 4700 feet above the
sea-level. There is very little that is interesting in the
town of Royat itself. Clermont-Ferrand, however,
possesses an excellent geological museum, some curious
historical monuments, and a very fine specimen of the
Roman-Auvergnat school of architecture, the church
of Notre Dame-du-Port, built about 870; it also
possesses a celebrated incrusting spring, that of St.
Allyre.

The natives of Auvergne, especially those who
inhabit the plains and lower valleys, are a singularly
uninteresting race. They are industrious and thrifty;
but they are physically unattractive and particularly
dirty, and goître is very common. They no doubt
possess the virtue of economy, if not the vice of avarice,
but there is an entire absence in their villages
of picturesque and poetic rusticity. The plain of

the Limagne, one of the most fertile districts in France, and one of the most highly cultivated, is almost entirely cut up into small holdings, in the possession of the peasantry. An Auvergnat peasant will toil night and day himself, and make his family toil till he has amassed, say, a thousand francs. He then buys a piece of land for two thousand francs, and urges his family to renewed exertions to pay off the debt. So he goes on adding piece to piece, till he not unfrequently becomes the possessor of a portion of land of considerable value. Then his great delight is to educate one of his family to be a priest, or an *avocat*, or a doctor. Much interesting information about the " Peasant Proprietors of the Limagne " will be found in an article with this title, by Mr. Barham Zincke, in one of the numbers of the " Fortnightly Review " of 1878.

Of the dirtiness of the Auvergnat himself, as well as of his home, many curious stories are told. We were ourselves witnesses of a curious illustration. Walking one Sunday morning to the old church at Royat, where Signor Cotogni was going to sing, we were suddenly alarmed by a peasant rushing in an excited manner apparently towards a lady of our party, and flourishing a whip high in the air. We looked around and found his excitement was produced by an attempt, on the part of a pig, to get out of the window of an upper storey and take the air on the window-sill. The pig certainly was in imminent danger of falling from a considerable height into the street ; but a young and dirty damsel appeared at the window, attracted by the noise, seized the pig by the ears, and induced him to return to the society of the family circle, with which he was evidently a little bored. The only damage done was the smashing of a flower-pot. A medical gentleman, a native of the district, told me that upon an Auvergnat peasant being ordered a bath, he asked his medical man in some bewilderment what he was to do. His doctor asked him if he had never in his life taken a bath ?

"Mais, monsieur," he replied, "je n'ai jamais été malade."

On another occasion a peasant was told to go into a bath-room at one of the bathing-places and take a bath. After a short time the bath attendant heard most frightful groanings and gurglings proceeding from the apartment, and on going in to see what was the matter, found the patient on his knees before the bath making violent efforts to swallow the bath, but declaring piteously that he never should be able "to swallow all that!" Notwithstanding, then, the material prosperity of the peasants in this district, their villages form a great contrast with the cheerful picturesque villages of Switzerland, and it is impossible for any one with any delicacy to travel through Auvergne as he would through Switzerland, trusting to find a decent lodging in a country inn.

The following is the official analysis of the several springs at Royat, made by E. Willm in 1879:

	Eugénie. Grammes.	St. Mart. Grammes.	St. Victor. Grammes.	César. Grammes.
Arseniate of iron .	0·0008	0·0010	0·0021	0·0008
Carbonate of lithia	0·0322	0·0029	0·0246	0·0191
„ soda	0·7374	0·6611	0·6777	0·3371
„ potash	0·1423	0·1560	0·1564	0·0984
„ lime	0·7706	0·6172	0·7058	0·4540
„ magnesia	0·3497	0·4359	0·4519	0·2560
„ iron	0·0518	0·0141	0·0420	0·0340
Chloride of sodium	1·6728	1·5930	1·6479	0·6528
Sulphate of soda .	0·1643	0·1482	0·1612	0·0893
Silica	0·1026	0·0958	0·1450	0·0815
Free carbonic-acid	1·3955	1·5524	1·7508	1·8188
Temperature	93·2° F.	84·2° F.	69·8° F.	82·4° F.

On inquiring of Dr. Brandt concerning two other sources which have recently been employed at Royat —the *Source Goudroneuse* (Source des Médecins), which professes to be a natural tar water, and the *Source Fonteix*, he has kindly furnished me with the following information: "The *Source Goudroneuse* proceeds, like most of the other mineral springs in this region, from a sheet of mineral water, highly charged with carbonic-acid gas, and which at this particular spot is forced through a layer of bitumen, varying in thickness from one to two inches ; as the mineral water passes through this formation, a certain amount of mineral tar is dissolved, and kept in a state of solution by the carbonic-acid gas. This mineral water possesses a strong tarry odour and flavour. It keeps well in bottle, even after exportation into different climates. As regards its therapeutic action, I have found it decidedly efficacious in all catarrhal conditions of the trachea and bronchi, also in vesical catarrhs, and in one case where the urine contained pus, blood, and epithelial cells in large quantities, the use of this water during a period of six weeks produced a most decided beneficial effect. Like the other Royat waters, it is drunk pure at the spring, and mixed with wine at meals.

"The *Fonteix* spring differs little in composition from the other Royat waters ; it contains, however, a larger per-centage of carbonic-acid gas—it seems to retain the gas for a longer time than some of the other waters. Owing to the care with which the bottles are cleaned, filled, corked, and sealed, this water keeps remarkably well, and is preferred as a drinking water abroad, where a natural mineral lithia water is required. I consider these two springs a valuable addition to the Royat station."

MONT DORE LES BAINS,

At an elevation of 3400 feet above the sea, is the highest mountain health resort in Central France, and certainly the most frequented. It is necessary to go as far as the Pyrenees to find a bath at an equal altitude ; Cauterets is not so high by two or three hundred feet ; but Barèges is higher, being 4000 feet above the sea. The waters of Mont Dore are spoken of by the resident physicians, with much solemnity, as *des eaux sérieuses,* but we shall see presently that their seriousness must be in their mode of application, as it certainly is not in their chemical composition. Mont Dore is about thirteen hours from Paris, taking rail and road together. It used to be necessary to leave the railway at Clermont-Ferrand, from which it is distant, by a mountain road, about twenty-eight miles. Diligences then left Clermont-Ferrand at seven and eight a.m., and again in the evening, for Mont Dore, and were about six or seven hours on the road.* There are several routes, but the public conveyances naturally took the shortest. You left Clermont by a level road, bordered by avenues of lofty trees, and soon began to ascend gently through orchards and richly-cultivated hills, covered for the most part with vines. Les Puys, as the volcanic mountains here are termed, rise on all sides, covered with green turf up to their dome-like summits, and surrounded with chestnut-trees at their bases. The

---

* It is usual now to go on by rail from Clermont-Ferrand to La Queuille, a station on the line of rail between Clermont and Tulle. The drive from La Queuille to Mont Dore takes from two to three hours.

summit which chiefly attracts attention for some time after leaving Clermont is the one which is called Mont Rognon.   It is rather less that 2000 feet high, but it is distinguished from the adjacent Puys by having the ruins of an old castle on its top, visible for many miles around.   Like its neighbours, it appears to have been formed by volcanic action. The road also skirts the celebrated mountain plateau of Gergovia, the site of an ancient Gallic city ; it was on its slopes that the renowned Vercingetorix defeated the invading legions of Julius Cæsar.   It is related that Augustus Cæsar, many years afterwards, caused the town to be destroyed, and forced its inhabitants to migrate, thinking thereby to destroy the memory of this defeat of Cæsar's army.   Interesting remains of this ancient Gallo-Roman city have been excavated here in great abundance.   The road makes many long sweeping zigzags, and finally reaches an extensive plateau, stretching from the bases of the Monts Dôme to those of the Monts Dore.   You now see the commanding mass of the Puy-de-Dôme and the adjacent Puys from the other side to that from which you viewed them at Clermont.   There is little of interest in the road until you begin to mount a low pass which leads from this plateau into the valley of Mont Dore.   At the summit of the pass, about 4000 feet above the sea, are two very remarkable-isolated rocks, La Roche Sanadoire and La Roche Tuilière. These stand up like the pillars of a natural gateway, closing the valley.   A magnificent view over the rich plains and the extinct volcanoes of Auvergne spreads out before one, framed, as it were, into a compact picture by these lofty and gigantic rocks.

It is a singularly striking and beautiful landscape. A little distance from these rocks the road passes by a small lake, the Lac de Guéry, well furnished with fish, and incorrectly stated in some books to occupy the crater of an extinct volcano.   The road now descends rapidly, by numerous zigzags, through richly-wooded mountain slopes, till it reaches the

little town of Mont Dore. Mont Dore lies, with its
houses closely packed together, in the bed of a valley,
surrounded nearly on all sides by high mountains,
some reaching to nearly 3000 feet above it. When
you are actually in Mont Dore you seem to be shut
in on all sides, but this is not really the case. The
valley pursues a somewhat sinuous direction. Shut
in and terminated towards the south by the steep
sides of the Pic de Sancy and the adjacent mountains, the valley runs from south to north, with a little
inclination westward, and a little beyond the town of
Mont Dore it makes a somewhat sudden bend to the
north-west, and opens out more and more as it
descends, and reaches in four or five miles the
neighbouring bath-station, La Bourboule. The surrounding country is very beautiful, abounding in
richly-wooded slopes, striking mountain forms, and
pretty though small cascades, with many grand
views from easily-attained elevations. The Pic de
Sancy, over 6000 feet above the sea, the highest
mountain of Central France, is easily reached from
Mont Dore in a walk of about three hours. It can be
reached in much less time on horseback, and horses
can be taken to within ten minutes of the summit.
A very extensive view is thence obtained. In one
direction, somewhat south-east, the peaks of the
Cantal chain appear directly in front of one, and in
the opposite direction all the Puys of the Monts
Dôme stand boldly up from the plains of Auvergne.
Several small lakes come into view, and especially
remarkable is Lac Pavin, surrounded by a perfectly
circular deep rim, and obviously occupying the crater
of an extinct volcano. The imposing group of ruins
of the Château of Murols and the neighbouring lake,
Chambon, so finely described by Madame George
Sand in "Le Marquis de Villemer," form striking and
easily-recognised features in the landscape. Indeed,
Mont Dore forms the centre of numerous most
agreeable and interesting excursions of every kind of
length ; its streams and cascades, however, are poor

compared with those of Switzerland and other
mountainous countries. Horses and donkeys are
good and abundant, and cavalcading for two or three
hours daily seems to form an essential part of the
cure here. Groups of persons on horseback and
children on donkeys are constantly encountered in all
the roads round Mont Dore. The town itself has an
ancient aspect; in its centre there is a small and
somewhat sombre square, which is termed La Place
des Thermes, two sides of which are formed by buildings belonging to the bath establishment, and the
rest consisting of hotels and divers kinds of shops.

The square itself is crowded, especially in the
morning, with various species of vehicles and numerous horses and donkeys and their owners, who eagerly
urge visitors to test the merits of their respective
steeds. Young maidens and old women accost you,
regardless of an appearance indicating perfect health,
and beg you to allow them to provide you with a
costume de bain, and if you reply that you are not
here *pour suivre le traitement*, they look upon you
with astonishment, not unmingled with contempt.
Indeed, there are obvious evidences at Mont Dore
of a permanent struggle between the visitors and the
natives. The native—the grasping, greedy, dirty
Auvergnat—looks upon the invalid visitor as being
especially sent for his, the native's, benefit; the
visitor, on the contrary, not unnaturally, looks upon
the native as a person intended to provide for his, the
visitor's, comfort and accommodation. It is therefore exceedingly difficult to establish an *entente
cordiale* between persons who look upon things from
such absolutely opposite points of view. It is only
thus that this remarkable fact can be accounted for—
that, after many years of renown and prosperity,
Mont Dore does not possess an hotel really suited, not
only for the reception of the delicately nurtured invalid,
but even for the accommodation of the better class of
tourist. With the exception of the food, which, at
least in one of the hotels (Chabaury Aîné), is cer-

tainly very good, the domestic arrangements in all the hotels here call for considerable improvement.* It would seem, however, to be a deeply-rooted principle with the Auvergnat to give you the worst possible accommodation at the highest possible price. On the top of the Rigi, near the top of Monte Generoso, in the comparatively remote valley of the Engadine, and in scores of other mountain health resorts in Switzerland and Germany, excellent hotels, with admirable accommodation, are provided ; but here, within thirty miles of an important town, in the centre of a richly cultivated and prosperous country, within thirteen hours of Paris, the accommodation provided for strangers is little better than that of a country *auberge*. Such simple things as really good bread and butter it seems impossible to obtain either at Mont Dore, Royat, or any place in Auvergne. They do not know how to make either, and are probably indisposed to learn ; but a journeyman baker and a dairymaid from Paris could soon teach them. Cases of decided overreaching on the part of hotel-keepers occasionally occur here, and at the Grand Hotel an English family was charged two hundred francs as *indemnité*, because they did not like the hotel and wished to change it for another. The *Établissements des Bains* consist of two large buildings, occupying the two sides of the square I have spoken of. They are substantial, but heavy and rather dismal edifices, and the internal arrangements lack much of the elegance which may be found in other large bath establishments. One of these buildings is devoted to the baths and douches, including foot baths, which was at one time a speciality of this place, and the piscines ; it contains also a set of baths on the ground-floor for the poor and indigent. On the upper-floor (in the absence of a casino) the concert-room, reading-room, gambling-room, etc., are placed.

---

* Comfortable accommodation can now, we are assured, be obtained at the Châlet Montjoli, kept by M. Bourdassl.

The second building is termed "*L'Établissement des Vapeurs,*" and is devoted to the *salles de pulvérisation* and the *salles d'aspiration.* These are very extensive. It is a curious sight to enter one of these *salles,* filled with hot, dense, vaporous mist, through which you dimly discern the forms of the patients, some sitting, some standing, some walking, some reading, some talking. Here every morning they are shut up for half-an-hour or longer to inhale this hot mist. One comes out of it gasping for breath, into the open air. This part of the cure is no doubt sufficiently serious. Great care is taken that a patient who is being submitted to this treatment does not take cold. When he comes out of this vapour bath he is conveyed rapidly in a sedan-chair to his hotel, and is expected to return to his bed, which is heated by a *chauffeur,* and remain there for an hour. Thus it happens that, from the early hour of four a.m. at Mont Dore, the stairs and passages of the hotel are made "hideous" by the clatter of wooden shoes and the hurrying to and fro of the patients in their sedan-chairs. If you venture into those stairs or passages at these early hours, you are hustled and jostled and driven into corners till you feel that the only safe thing is to get into a sedan-chair yourself.

On looking round on the class of visitors which are met with at Mont Dore, one cannot help being struck with the fact that many of them are persons who have to use their voices a great deal. Groups of priests are encountered at every turn. The faces of well-known singers and actors and actresses are seen on all sides. At the *table d'hôte* I see Madame Christine Nilsson, with Mdlle. Baretta, of the Comédie Française, sitting opposite me, and M. Worms, of the Comédie Française, at another part of the table. In the promenade I meet Madame Marie Roze, Mdlle. Pierson, of the Gymnase, Signor Tamberlik, M. de la Pommeraye, the well-known Conférencier, and many others who come to Mont Dore to be cured of throat and laryngeal affections. This is, no doubt, its great speciality.

Mont Dore possesses eight warm springs and one cold one. The latter is not looked upon as of any importance. Of the others, two only are ordinarily drunk—viz., La Madeline and the Ramond. The Ramond contains more iron than any of the others. These two furnish the drinking fountains placed in the *salle des pas perdu* of the *Établissement.* The other sources—the Cæsar, Caroline, St. Jean, Rigny, Boyer, and Pigeon—are used for the baths. They have all very nearly the same composition, the Ramond, however, being the most ferruginous. The warm springs vary in temperature from 43 to 45 degrees Centigrade—i.e. from 109 to 114 degrees Fahrenheit. It is not a little remarkable, considering the serious claims advanced on behalf of the springs of Mont Dore, that they should be so feebly mineralised, containing only two grammes (about twenty-six grains) of mineral matter to a litre (about twenty-eight ounces) of water.

They are weak alkaline waters, their chief constituents being the bicarbonates of soda and potash and chloride of sodium. They resemble in composition the springs of Neuenahr in Germany, but contain only half the quantity of carbonate of soda. Having regard only to the chemical composition of the springs, it is difficult to understand how they produce the important results attributed to them : no doubt their thermality is an important element. Their free carbonic-acid is by no means relatively large. To say, by way of accounting for their activity, "L'électricité dynamique y est très prononcé," is to assume what it would, I take it, be exceedingly difficult to prove. But it has occurred lately to the physicians at Mont Dore to call their waters "arsenical." They do indeed contain arsenic, but in a proportion much smaller than other springs to which the term "arsenical" has never yet been applied. There is 0·00096 grammes of arseniate of soda in a litre—i.e. about one-seventieth of a grain in a quart ! But the physicians at Mont Dore construct a convenient theory to excuse,

as it were, the feeble mineralisation of their waters. It
is this: That certain salts, such as the arsenite or arseniate of soda and potash, are volatilised with greater
facility the smaller the amount of mineral matters
contained in the water; hence the waters of Mont
Dore, by their feeble mineralisation, are in an especially favourable condition for yielding to their
aqueous vapour the chief part of the arsenic which
they contain, the vapour being at the same time
condensed and under pressure. This reminds me
of the saying of a well-known French writer—that
a clever man will know how to derive advantage
even from his defects. The waters at Mont Dore
have been classified at different epochs under three
distinct heads: 1. As waters containing a mixture
of alkaline bicarbonates with iron; 2, as weak alkaline waters; and 3, at the present time, as arsenical
waters; arsenic being the remedy *en vogue* at this
moment! But it would be more in accordance
with a scientific method to be content with stating
the results of observation and experience, without
attempting to explain them by various and conflicting
hypotheses; such, for instance, as (*a*) that mineral
waters at the moment they escape from their source
are in a state of exceptional activity; (*b*) that thermal
waters possess peculiar latent chemical properties;
(*c*) that they are in a state of dynamic electrisation;
and (*d*) that their effects depend on the mode of their
administration. It will certainly be more interesting
to the public to set aside doubtful theoretical considerations and to inquire specially into the practical
methods adopted at Mont Dore, and the results that
are obtained from them. The waters are administered
internally and externally; in the form of ordinary
baths and in swimming baths; in the form of vapour
and spray; in douches, local and general; in the
form of gargles; and in certain special forms that
need not be here indicated. The waters are drunk
fasting, in the morning, from three to four glasses a
day, leaving an interval of half-an-hour between each

glass or portion of a glass ; sometimes milk, sometimes *sirop de guimauve,* sometimes *infusion de tilleul*
is added to the water to make it more digestible.

After about fifteen days one becomes, as the local
doctors say, saturated with the water ; a disgust for
the water is felt, the digestion becomes disturbed,
general fatigue is complained of, and sometimes there
is a little fever. Other signs of an overdose of the
water are said to be congestion of the face, headache,
and giddiness. When these symptoms arise the water
must be discontinued. The patients are also made to
inhale the compressed vapour of the water, as I have
already mentioned, and Mont Dore claims for itself
the credit of having first introduced this method of
utilising these waters in the year 1833. The patients
remain in the vapour from a quarter to three-quarters
of an hour. Other patients, chiefly those who suffer
from throat affections, inhale the water in a state of
pulverisation or spray. Douches of vapour applied
locally are also employed and thought of much value
in muscular rheumatism and in rheumatic inflammation of the joints, as well as in sciatica and intercostal
neuralgia. The patient is seated on a stool and an
intermittent jet of vapour is directed upon the part
affected. Douches of water, in the form of a jet or a
rose and of varying pressure, are also used. They are
applied to the spine as a stimulant to the nervous
system ; to the joints in cases of lumbago and sciatica ;
to the chest in some chest affections, and, indeed, to
any part which it is desired to influence specially.
The ordinary baths are administered either at a high
temperature, i.e. from 107 to 112 degrees Fahrenheit,
or at a moderate temperature, i.e. from 90 to 100
degrees Fahrenheit. The latter temperature is considered more suitable to feeble nervous persons, as
well as to old people and children.

Finally there are the foot baths, which are thought
of great importance here, and are believed to accele-
rate the circulation in the lower extremities and so
prevent any tendency to congestion of the head

which the rest of the course of treatment might possibly produce. The patients sit, two in each little compartment, with their feet and legs in wooden baths of hot water of the natural temperature, i.e. about 112 degrees Fahrenheit. They remain in the bath from five to nine minutes and then walk for at least half-an-hour afterwards. The duration of the treatment is from eighteen to twenty-one days.

A patient writes to me of the bathing establishment: "It requires much reform. The foot-baths are piggishly managed. The steep stone steps up to the bath-rooms try some invalids greatly, and there is too much roughness of officialism."

In the next place, what are the ailments to which the course of treatment instituted at Mont Dore is especially applicable? I have already spoken of chronic rheumatism of the joints and muscles, of certain forms of neuralgia, chiefly sciatica and intercostal neuralgia. To these must be added nearly all forms of throat disease, whether of the tonsils, of the pharynx, or of the larynx, and particularly that condition which in England goes by the name of "clergyman's sore throat ;" and all these conditions are said to be benefited at Mont Dore. Some forms of asthma and of chronic bronchial catarrh also derive great improvement here.

Of the treatment of asthma, at Mont Dore, I shall presently speak at some length, as the subject is really important.

But one of the most serious claims advanced by the physicians of this place is that they cure pulmonary consumption in the earlier stages and when it occurs in certain temperaments. And detailed accounts have been published, tending to show that, at any rate, certain cases presenting most of the general features and physical signs of consumption have recovered their health after a season, or more frequently two or three seasons, at Mont Dore. These statements, it must however be admitted, have been contested with some warmth and acrimony by the physicians of other French health resorts. For instance, a physician at the neighbouring spa of Royat, in a book on the mineral waters of Auvergne, says, "Pour ce qui est de la phthisic, que les divers médecins du Mont Dore ont la prétention de guérir, nous ne partageons point leur illusion," and the former chief physician at Eaux Bonnes, in the Pyrenees, also attacked the physicians of Mont Dore with much vigour. At Eaux Bonnes they also profess to cure consumption with their sulphur waters, and so a very lively little controversy was, at one time, in progress between sulphur on the one side and arsenic on the other. After all, this was probably simply a manifestation of the *jalousie du métier.* Experience alone can show whether these pretensions are well founded or not, and there is this much to be said in favour of making the trial—that the course at Mont Dore only requires a residence of three weeks, and that few persons suffering from so serious a malady as consumption would hesitate to spend so short a space of time as that in the summer months in an agreeable mountain station. It is difficult to believe that the good effects claimed to be produced here in such cases is chiefly due, as some have suggested, to the mountain air; for the brief period of three weeks passed at so moderate an elevation could scarcely be credited with so great an influence in arresting the course of phthisis.

Dr. Richelot, who has for many years held the honourable position of *Inspector des Eaux* at Mont Dore, thus speaks of the treatment of cases of consumption there:

In the consumptive patient, under the influence of the Mont Dore waters, nervous energy increases, the consuming rapidity of the circulation abates, the fever disappears, the inflammatory congestion of the portions of pulmonary tissue surrounding the tuberculous deposit tends to disperse, respiration becomes deeper, the circulation in the lungs less incomplete, hæmatosis is more easily accomplished, the appetite increases, nourishment is more easily digested, and without fatigue to the organs or feverish reaction, general nutrition makes rapid progress, strength revives, and the patient begins to gain flesh. If the treatment is employed when consumption is in its earliest stage, it may act, up to a certain point, as a prophylactic agent, and prevent the development of the malady by restoring the vital forces. If, however, the disease be far advanced, the treatment can, at least, check its progress and prolong the life of the patient.

The climate of Mont Dore is very variable, storms of rain and thunder coming on suddenly, and in some seasons frequently. Sudden and very localised gusts of wind are also commonly encountered. There is no doubt that the climate can be, and often is, exceedingly disagreeable. An invalid visitor told me that during one season he encountered seventeen wet days in succession; and this where the internal arrangements of the hotels are by no means satisfactory is a very serious matter. Sometimes after a heavy storm of thunder and rain, owing, I suppose, to the disturbing effect of the rain on the accumulations of filth here, all the streams, and even the river (the Dordogne, formed by the confluence of the Dore and the Dogne), and, indeed, nearly every part of the town, become most offensive from bad smells; and many visitors suffer from more or less severe attacks of diarrhœa. Owing to the situation and direction of the valley, the eastern slopes are in shade for a few hours after sunrise, but soon the sun mounts high above the eastern boundary and its rays stream down with great intensity into the deep valley, and it is often excessively hot during the whole of the day. The early mornings and the evenings are mostly fresh and bracing, and the pleasantest days at Mont Dore are those on which the sky is covered with light clouds, without rain; then the air during the whole day feels fresh and invigorating. The atmospheric pressure is considerably reduced and the average height of the barometer is 26·6. The variations of temperature during the months of June, July, and August are sometimes very considerable, the maximum being 86 degrees Fahrenheit and the minimum 37 degrees Fahrenheit. The average temperature in August is 57 degrees Fahrenheit. The hygrometric condition of the air is, no doubt, favourable to most invalids; it is decidedly a dry air compared with the air of towns—as, e.g. that of Paris—or of the sea-coast. The relative amount of watery vapour in the atmosphere of Mont Dore as compared with Paris is as 9·94 to 15·46. Fogs and mists are, however, frequent, and the season of fine weather is often extremely short. July and August are the finest months. The season scarcely commences before the beginning of July, and is soon over after the end of August.

An asthmatic patient who tried Mont Dore in June wrote to me that "Mont Dore in June was something too terrible—constant cold winds, rain, and snow (for one day). During the nineteen days I was there (4th to 23rd June) it rained, snowed, blew, and ' fogged ' to such an extent that life was a burden, and I feel sure that more than one invalid shortened his days by trying the Mont Dore cure in June. Lest invalids should be tempted to try Mont Dore in June on their way back from Italy, I mention the climate that they are likely to encounter. The place is almost deserted early in June."

The following is Lefort's analysis of a litre (1000 grammes) of the water (*Source César*):

		Grammes.
Bicarbonate of soda		0·5361
„ potash . . .		0·0212
lime		0·3209
„ magnesia . . .		0·1676
„ oxide of iron . .		0·0558
Chloride of sodium		0·3587
Sulphate of soda		0·0756
Arseniate of soda		*0·00096*
Silicic-acid		0·1552
Alumina		0·0083
Carbonic-acid (free) . . .		0·5967
Oxygen		0·9800
Nitrogen		14·2200

Traces of bicarbonate of rubidium, lithium, cæsium and manganese, borate of soda, iodide and fluoride of sodium, and bituminous organic matter.

THE TREATMENT OF ASTHMA AT MONT DORE.

Additional experience of the application of the Mont Dore cure to cases of asthma, enables me to approach this subject with more confidence than I could have done in the first edition of this work and to examine the question more completely.

Dr. Cazalis, the able and learned Deputy-Inspector at Mont Dore, has been good enough to supply me with some interesting information as to the manner in which the applicability of the waters of Mont Dore to the treatment of spasmodic asthma was first discovered. He writes to me that Dr. Michel Bertrand, the celebrated Inspector at Mont Dore, who constructed the *Établissement*, and made the reputation of the place, had repeatedly attempted to treat asthmatics there. He relied chiefly on the baths, and he succeeded in relieving certain asthmatics, who were, at the same time, sufferers from emphysema and catarrh, but he brought on violent attacks of dyspnœa in all the others, so he was led to say that "*nervous asthma was not cured at Mont Dore.*"

One day, however, an asthmatic patient begged him to take his case in hand. This he at first refused to do, but the patient insisting, he consented to commence the treatment, which he did by ordering him to take a certain number of baths, and the asthma reappeared with violence. He wished to send the patient away, but he refused to go, and came daily and passed several hours close to the Pavilion hot spring, when the asthmatic attacks began to disappear. Dr. Michel Bertrand was surprised to observe the asthma disappear under the influence of the emanations from the spring, and this observation induced him to attempt to procure a dense vapour of the water of Mont Dore in one of the douche chambers, by allowing the douche to flow for some time and break itself into spray by falling on a plank. When asthmatic patients were admitted into such a chamber and breathed this vapour, they found themselves rapidly relieved. This experience led to the construction of special chambers, into which the vapour of the Mont Dore water, obtained by special mechanical contrivance, could be forced in abundance, and in which patients could remain for some time, walking about or otherwise employing themselves.

This was the origin of those *salles d'aspiration* which have made Mont Dore celebrated. Owing, doubtless, to the nature of the mechanism by means of which these hot, dense vapours are obtained and forced into these chambers, or to the manner in which the hot springs arise, it has been found, on chemical analysis of these vapours, that they contain the same mineral and gaseous constituents as are found in the springs themselves, viz. carbonic-acid gas, arsenic, and certain alkaline salts.

It seems probable that the atmosphere of these *salles d'aspiration* contains much suspended water in the form of hot spray, and that it is not, strictly speaking, the vapour of the water that yields these matters on analysis.

Dr. Cazalis maintains that it is through the breathing of these vapours chiefly that the cure of asthma is effected ; that the cure is greatly aided by drinking the waters, and by the use of the hot douche (98 degrees Fahrenheit) along the spine ; also by means of foot-baths, which divert the " congestion " towards the inferior extremities. He protests, however, against the use of baths, and especially very hot baths, indiscriminately in the treatment of asthma.

" My daily experience," he says, " convinces me that asthmatics who take baths here find themselves much worse for them, and often they induce severe attacks. To myself and others," he adds, " it has taken the form of a principle that persons with nervous asthma who take baths at Mont Dore will not get well, while those who are restricted to the douches and the aspirations will get well. But when the asthma is of long standing, and is complicated with catarrh, with emphysema, and with abundant expectoration, the cure is more difficult. The nervous spasmodic dyspnœa will disappear, but the catarrh and the emphysema will not disappear altogether. In these cases a few baths may be given, with careful precautions; they will lessen the catarrh and the expectoration, but they should only be given at the end of the treatment after the vapours have calmed the spasmodic dyspnœic phenomena. Dry, nervous asthma is cured completely here; humid, catarrhal asthma less so; amelioration, however, is always considerable, and a complete cure not infrequent.

"Our waters are very injurious to persons with heart disease. Asthmatic children do admirably at Mont Dore, especially those in whom the attacks have appeared after the cure of scrofulous impetigo or eczema. In order to obtain a complete radical cure of asthma at Mont Dore it is necessary to pass at least two consecutive seasons there, and more commonly three."

I have quoted thus fully from Dr. Cazalis' letter because it represents the experience of a cautious and careful observer, who has had the best possible opportunities of forming a sound judgment. But it has by this time become evident to many physicians in London that better and more permanent results have been obtained in cases both of spasmodic and catarrhal asthma, by the course of treatment at Mont Dore than by any other plan of treatment hitherto available. It may be worth while, therefore, to inquire how the system of treatment adopted at Mont Dore agrees with the generally accepted views of the pathology of asthma. The routine of the treatment has been repeatedly described; I will, therefore, only indicate it as briefly as possible.

The patient is roused from his slumbers perhaps at 5 a.m., enveloped in suitable clothing, and carried off in a sedan-chair to the hot spring, where he drinks a glass of the water; he is then taken to the bath-room, where he may probably be ordered, as we have seen, to take only the foot-bath and the hot douche on the spine, or a vapour douche, or he may take a half bath, or in some cases the whole bath, the heat of which is occasionally raised even to 113 degrees Fahrenheit in refractory cases. After from ten to thirty minutes of the bath he is wrapped up in hot linen and dried, and next conveyed to the *salles d'aspiration*; he is here dressed in a loose, thick, flannel dress, made for the purpose, as everything becomes rapidly saturated with moisture in these chambers filled with hot, dense vapour, which is forced into them under pressure. He remains breathing this hot, moist atmosphere, charged with vapours of the mineralised springs, from twenty minutes to an hour. After this he is carried, warmly clothed, again to the hot spring to drink another glass of water, and thence he is conveyed to his bed, which has been well heated by a warming-pan, and he is enjoined to remain there for an hour or more. Now it is easy to see that this is treatment of a very active kind, and when applied every morning for three weeks is very likely to have some decided result.

The most important, and perhaps almost constant, effect is a very decided determination to the skin; profuse perspiration, as a rule, appearing soon after the commencement of this treatment. In obstinate cases every effort is made to produce excessive action of the skin, and in this way both circulatory and nervous energy is directed to the cutaneous surface, for in proportion to the amount of function required of an organ will be the amount of expenditure of nerve force which presides over that function. Now the cutaneous surface is always in close and sympathetic correlation with the respiratory surface, and while wandering and irregular nervous energy is thus diffused and dissipated at the surface of the body, the hyperæsthetic condition of the respiratory centre and its afferent tributaries is proportionately diminished.

An attack of spasmodic asthma may be regarded as a morbid manifestation of misdirected nervous energy, just as some forms of hysterical and epileptic convulsions may be; by the diversion of such wandering nerve energy to the skin, and its diffusion and exhaustion in connection with a natural physiological process, the nervous equilibrium is restored, and the tendency to spasm relieved. It is thus that the spasmodic nervous element in asthma is attacked.

Let us look at the processes of the Mont Dore cure from another point of view. This daily ingestion of warm alkaline fluid taken together with the profuse "sweats" that are excited, must have a very remarkable depurative influence. Retrograde tissue changes must be actively stimulated, and unstable irritating substances which may have accumulated in the organism from disordered assimilative processes, as in gouty conditions and the like, must be dissolved and eliminated, so that it also attacks vigorously those states of blood contamination which are at the root of many spasmodic and vaso-motor nervous affections, and in which category many forms of asthma may be placed.

Then again in those cases of asthma complicated with bronchial catarrhs, which form by far the greater proportion of all cases of asthma, those cases in which the morbidly increased secretions of the bronchial mucous membrane, especially when it is scanty and tenacious, act as excitants to the hyperæsthetic mucous membrane itself, just as a foreign body would; in such cases we have two obvious indications to fulfil, one to remove the catarrhal condition and free the air passages of accumulated mucus; the other to soothe and quiet the hyperæsthesia of the bronchial surface.

Both these indications are admirably fulfilled by the processes in use at Mont Dore. The use of pilocarpin in the treatment of catarrhal asthma has shown to what extent the catarrhal condition is relieved by excessive action of the skin, and here at

Mont Dore we get excessive action of the skin without pilocarpin, a remedy to my thinking far too dangerous and depressing for general use. Then there is the prolonged daily immersion, for immersion it practically is, in an atmosphere saturated with the hot vapours of the mineral water, which bathe the pulmonary mucous membrane, moisten and thin the secretions when dry and scanty, and in all cases promote their expulsion, while they must at the same time exercise a most soothing effect on the hyperæsthetic bronchial surface, and so tend directly to allay the nervous element which is, I maintain, a part of every form of asthma properly so-called.

I must not be understood as in the least desiring to undervalue the action of the mineral and gaseous constituents of the water itself on the disease ; the carbonic-acid, and the arsenic (the very small quantity which the water contains is said to be readily diffused in the vapour), and the alkaline salts may, and probably do, exercise a most important curative influence, but we can see, that even without this, the processes employed are energetic in their nature and physiological in their action. But no fact is now better established in therapeutics than the value of warm alkaline drinks in the treatment of bronchial catarrhs, and this always forms a part of the treatment at Mont Dore.

There is yet another condition, and that by no means an unimportant one, under the influence of which these processes are carried out. I allude to the altitude of the locality. Mont Dore is 3400 feet above the sea. The atmospheric pressure is considerably reduced, the average height of the barometer being 26·6. The air is much cooler and much drier than on the sea-level, and though fogs and mists are frequent in bad seasons yet the relative amount of watery vapour in the atmosphere of Mont Dore, as compared with that of Paris, is as 9·94 to 15·46. These processes then are carried out in a bracing air, in an atmosphere freer from permanent moisture than on the sea-level. It is now a well-known fact that exposure to this kind of mountain air has the effect of greatly reducing that impressionability to cold which is so common in scrofulous subjects, and which is at the root of their catarrhal tendencies; while it is especially tonic to certain hyperæsthetic states of the nervous system. A climate like that of Mont Dore tends to diminish cutaneous and respiratory sensitiveness, and is therefore admirably suited for the application of the processes there in use to the cure of asthmatic and catarrhal conditions.

I am by no means disposed to share the opinion of those who think it a matter of indifference where processes such as those in use at Mont Dore are carried out. I believe the Mont Dore cure to be the cure at Mont Dore and not elsewhere. I believe the mountain air an essential part of the cure. I believe that the processes applied at Mont Dore would prove relaxing and debilitating in the extreme to many persons, were it not for the comparatively dry and bracing climate in which they are applied. The climate itself is a nervine tonic and alterative, and I believe exercises no small influence in the good results which are obtained there. And if the gaseous emanations and the volatilised or suspended arsenical constituent of the waters are, as they well may be, important agents in the cure, it is difficult to believe that they can be reproduced at all in the same manner from a comparatively small amount of the water which has been transported to a great distance from the springs. We see then that the method of treatment pursued at Mont Dore answers many indications, and is applicable, from one point of view or another, to nearly all cases of asthma.

Is the asthma due to blood contamination from disordered assimilation or imperfect elimination, then the actively solvent and eliminative effect of the ingestion of hot alkaline fluids, and the production of profuse perspiration, answer the indications obviously pointed to in such cases.

Is the asthma purely nervous and due to a morbid

excitation of some portion of the respiratory nervous system, then we have the diffusion of nervous energy to the surface through excitation of the cutaneous functions, we have the soothing effect of the hot vapours to the bronchial membrane itself, and of the hot douches to the nervous system generally, the derivation to the extremities by means of the hot foot-baths, and finally, the well-known influence of mountain air in diminishing hyperæsthetic states of the nervous system.

Is the asthma invariably catarrhal or a complication of chronic bronchial catarrh, then we have the solvent effect of the hot alkaline vapours respired for a considerable period daily, and of the hot alkaline water drunk daily, the removal of the congestion of the bronchial mucous membrane, and the diminution of its secretion by powerful derivation to the skin, for it is in these cases that the baths at high temperatures are employed, while again we have the influence of the drier mountain air to give tone to the respiratory surface, and to diminish that impressionability to external cold which is the bane of such patients.

But in many cases the indications for treatment are complex, and we can readily see how the processes in use at Mont Dore attack the various morbid elements of these complex cases.

I say nothing of the specific influence of the small amount of arsenic in these waters—there is about one-seventieth of a grain of arseniate of soda in a quart of the water! It *may* be most influential—many of the physicians at Mont Dore believe that it is so—but its influence is hypothetical, and therefore I have not attempted to base any argument upon it.

I may add that only last season (1884) I had occasion to note the remarkable efficacy of the Mont Dore treatment in one of my own patients who had, for many years, suffered from frequent and very severe attacks of this malady. The journey was undertaken most reluctantly and despondently, but the result was extremely satisfactory and gratifying.

LA BOURBOULE,

A spa which has risen rapidly in importance and reputation, is situated only four or five miles from Mont Dore, the road following the course of the valley and the right bank of the Dordogne as it descends towards the north-west. The mineral sources there have become the exclusive property of a company, who appear to be using every effort to develop and popularise the place as a thermal station. A handsome and artistically decorated building has lately been constructed, and forms the *Établissement des Bains*, containing all the appliances of the most perfect bathing-place, arranged in the most approved and modern fashion. Many new hotels and villas have been constructed and others are in course of construction, and the whole place presents evidences of enterprise and progress, and is in striking contrast to its stagnant neighbour—Mont Dore. La Bourboule is not so high as Mont Dore, its elevation above the sea being only 2700 feet. It is not so shut in as Mont Dore, but lies in a comparatively wide open valley, surrounded by hills of moderate height and very gentle slope. It is not so well placed as Mont Dore for excursions into the interesting country around, most of the points of greatest interest being more easily reached from the latter.

There is now, as I have already said, a railway between Clermont-Ferrand and Tulle, and instead of travelling by diligence from the former place to La Bourboule, as was necessary a year or two ago, visitors to this spa descend at La Queuille station, where omnibuses and carriages are found to conduct the traveller to La Bourboule, a drive of about two hours.

The train which leaves Paris at 8.30 a.m. reaches La Queuille at 9.25 p.m., and La Bourboule (by the omnibus service) between 11 and 12 at night. The train which leaves Paris at 8 p.m. reaches La Queuille at 9 a.m., and La Bourboule (by carriage) about 11 a.m.

Besides the *Établissement des Thermes,* which is reserved for the use of the better class of patients, there are two other bath *Établissements* at Bourboule —the *Choussy,* which is more simple, but large and well-arranged, and the *Mabru,* which is chiefly devoted to the reception of patients sent by charitable institutions. The two former are very completely equipped, and contain the various kinds of douches—*salles d'aspiration* and *salles de pulvérisation,* a spacious swimming bath, *buvettes,* etc. There is also a well-appointed Casino. The two principal springs at La Bourboule —La Source Choussy and La Source Perrière—have the same composition. They are the strongest arsenical waters as yet discovered. Each litre of water contains nearly six grammes (seventy-eight grains) of mineral substances, of which nearly three grammes are bicarbonate of soda, and nearly three grammes chloride of sodium. So far it corresponds in composition with Ems water, but it is considerably stronger. Then in each litre there is contained twenty-eight milligrammes of arseniate of soda—i.e. about one-third of a grain in twenty-eight ounces. It has been seen that the Mont Dore water contains only one-eightieth of a grain of arseniate of soda in the same quantity of water. They also contain a small quantity of iron. The sources of La Bourboule have also a high temperature ; the La Perrière spring has at its surface a temperature of 50·5 Centigrade, or 123 degrees Fahrenheit. It is, however, the possession of this considerable quantity of arsenic that determines the special uses and application of the waters of La Bourboule.

Apart from the arsenic contained in them, it has been maintained that these waters resemble in composition the serum of the blood, and much stress is laid on this as showing their applicability to cases of arrested development, defective nutrition, cases of slow convalescence, and other forms of general debility. In all scrofulous affections, such as enlarged glands, scrofulous discharges from mucous membranes, diseases of the bones, etc., these waters are said to produce remarkable benefit. But it is more especially in the chronic forms of skin disease that La Bourboule claims to effect the most remarkable cures, and chiefly when they arise in connection with a rheumatic or scrofulous constitution, or as the result of simple debility. Certain forms of rheumatism and gout are also greatly benefited here. The scrofulous form of pulmonary consumption, nasal and pharyngeal catarrhs, asthma, and chronic bronchitis are all said to be improved by the use of the Bourboule waters. So that La Bourboule claims for itself a very extended sphere of usefulness, and its comparatively mild mountain climate tends to aid the curative effects of its waters. They are applied in much the same way as those of Mont Dore. It need scarcely be remarked that, containing so large a quantity of such an active agent as arsenic, their use and application needs very careful medical surveillance. It seems to be admitted on all sides that La Bourboule has a great future before it as a health resort of the first importance.

The following is the published analysis of the Bourboule springs (hypothetical composition).

In a litre = 1000 grammes :

		Grammes.
Arseniate of soda	. . . . .	0·02847
Chloride of sodium	. . . .	2·8406
,, potassium	. . .	0·1623
,, lithium	. . .	traces
,, magnesium	. . .	0·0320
Bicarbonate of soda	. . . .	2·8920
,, lime	. . . .	0·1905

						Grammes.
Sulphate of soda	.	.	.	.	.	0·2084
Peroxide of iron	.	.	.	.	.	0·0021
Oxide of manganese	.	.	.	.	.	traces
Silica	.	.	.	.	.	0·1200
Alumina	.	.	.	.	.	traces
Organic matter	.	.	.	.	.	traces
Free carbonic-acid	.	.	.	.	.	0·0518

After several consultations with one of the physicians who has had the longest and largest experience in practice at La Bourboule, we came to the conclusion that a course of the Bourboule waters might be advantageously recommended in the following affections :

1. In anæmic and chlorotic girls, and weakly children with scrofulous tendencies, or who are convalescing slowly from surgical operations, especially when undertaken in connection with scrofulous diseases of the bones, or other organs, and particularly in those cases in which the preparations of iron prove inefficacious or unsuitable.

2. In herpetic eruptions of the skin, and in eczema and in psoriasis—the first is constantly cured, the second greatly benefited, and the last frequently much improved.

3. In affections of the throat and respiratory passages, especially when found associated with a tendency to cutaneous eruptions (what the French term "herpetism"). In granular pharyngitis, chronic laryngitis, and asthma. In slow forms of chronic phthisis, associated with a tendency to glandular affections ; but it is not to be recommended in cases with a tendency to hæmoptysis, or to troublesome dyspepsia.

4. In certain forms of intermittent fever which resist the ordinary modes of treatment.

5. In chronic rheumatism, especially in those very troublesome cases known as cases of " arthritis deformans."

6. In the treatment of the gouty dyscrasia, when there is no tendency to acute attacks.

7. In diabetes, especially in emaciated subjects in whom the waters of Vichy or Vals have been ineffacious, and particularly if this disease be associated with skin eruptions.

The same authority assures me that there is no foundation for the statement that has been put forward to the effect that cases of true albuminuria are benefited at La Bourboule.

APPENDIX TO CHAPTER XI.

OTHER RESORTS IN AUVERGNE.

The following notes refer to other health resorts in Auvergne of somewhat less importance than the preceding.

Châtel-Guyon, in the department of Puy-de-Dôme, is a small town, picturesquely situated at an elevation of 1300 feet above the sea, about four miles from the railway-station of Riom, which is within half-an-hour, by train, of Clermont-Ferrand. It has several mineral springs of a temperature varying from 82 to 95 degrees Fahrenheit. It has two bath establishments belonging to different proprietors ; the *Établissement Brosson* is the most completely equipped with the necessary baths, douches, etc.

The water is rich in carbonic-acid gas, so that in the baths the skin becomes covered with bubbles of this gas, which have a stimulating effect on the surface. These baths are said to produce both a tonic and soothing effect.

Although there are only six grammes of salt to the litre, these waters are reported to have a laxative effect in small doses, and a purgative effect in large ones.

As there are scarcely any aperient waters in France, it is possible that a little too much has been made of the so-called purgative properties of the Châtel-Guyon springs.

The only aperient base in these waters is magnesia, and of this base there is only ·670 gramme in a litre. It exists in these waters as a chloride. They also contain soda, potash, and lime, in combination with carbonic, sulphuric, and hydrochloric acids.

It could only be in very large quantities of the water that there would be sufficient chloride of magnesia to act as a decided purgative. The only analysis available is published in an unsatisfactory form.

There seems to be an appreciable quantity of iron in these springs. They are reputed to be efficacious in the treatment of gastric catarrh and flatulent dyspepsia, especially when associated with constipation ; of hepatic enlargements and congestions; of jaundice; of hepatic and renal calculi; of diabetes, and some other chronic affections. Several glasses of the water are usually drunk in the morning fasting.

The season is from the 15th May to the 15th October.

Saint Nectaire is a small bath, situated about half-way between the towns of Issoire and Mont Dore, and about two-and-a-half hours by road from either place. It is said to be 2500 feet above the sea. Its situation is not a very attractive one, but it is close to some of the finest and most interesting scenery in Auvergne, the romantic and imposing ruins of the Château de Murols being only three or four miles distant.

There are four rather small bath establishments, furnished with the necessary appliances, and supplied by several springs of nearly identical composition, of a temperature varying from 64 to 115 degrees Fahrenheit. The chief solid constituents of these springs are bicarbonate of soda and chloride of sodium, rather more than two grammes of each to the litre ; they also contain a considerable amount of free carbonic-acid, and a very small proportion of iron. In moderate quantities they are diuretic, and render the urine alkaline ; in larger quantities they are said to be purgative. The baths form an· important part of the treatment at St. Nectaire. They are applied in cases of chronic rheumatism, muscular and articular ; in sciatica, and other forms of neuralgia ; in scrofulous and gouty skin diseases ; in chlorosis and chronic catarrhal uterine affections. The pulverised water has been found useful applied locally in some chronic inflammatory affections of the conjunctiva and cornea. St. Nectaire is celebrated for its petrifying springs. The season is from the 1st of June to the beginning of September.

Vic sur Cère.—A small bath establishment, near the town of this name, twelve miles from Aurillac (Cantal). It possesses a gaseous ferruginous spring, containing a considerable amount of bicarbonate of soda, and enjoys a certain local reputation.

Châteauneuf.—A bath with somewhat primitive accommodations, in an out-of-the-way part of Auvergne, fifteen miles by road from Riom station. It is in the midst of wild and picturesque mountain scenery, at an elevation of about 1250 feet above the sea.

There are as many as fifteen mineral springs scattered over this district, and varying in temperature from 59 to 100 degrees Fahrenheit; some are highly charged with carbonic-acid.

Some are gaseous and cold, and contain a notable amount of iron ; others are alkaline rather in character, containing soda, potash, magnesia, and lithia ; while others have purgative properties, from the amount of sulphate of soda and magnesia they contain. The reputation of this bath is chiefly local, and it is resorted to by two classes of invalids, first, the anæmic and dyspeptic ; and second, the rheumatic and gouty.

CHAPTER XII.

THE PYRENEES.

A SHORT STUDY OF SUMMER HEALTH RESORTS IN THE
PYRENEES.

AN interesting route to the Pyrenees, and one not
frequently followed, is that through Central France
—through Auvergne and the Cantal to Toulouse.
Clermont-Ferrand too is a good halting-place, whence
the Puy-de-Dôme may be easily ascended and a
general idea gained of the curious volcanic district
around. At Arvant, a few miles south of Clermont-
Ferrand, you leave the line of rail which goes to
Nismes, and enter upon that which passes through
the mountainous department of Cantal, and after
passing Aurillac joins the Limoges and Toulouse
line at Figeac. This is certainly one of the most
extraordinary examples of railway engineering in
Europe. The road is for a great part of the way cut
out of the steep mountain-side ; now and again it runs
through a deep cutting, spans a profound gorge by a
lofty viaduct, or leaps with a single arch over a rushing mountain torrent. Tunnels succeed one another
with astonishing frequency, and one of these, the tunnel
of Lioran, is considerably more than a mile in length,
and is, at its entrance, 3700 feet above the sea ! In
passing through it one passes from the basin of the
Loire to that of the Gironde. Ruins of ancient castles
are as numerous on the mountain-tops as they are on
the banks of the Rhine ; curious basaltic rocks, some

surrounded with numerous stages of prismatic columns, which the French liken to organ-pipes, are frequently passed. For a long time the conical summit of the Plomb du Cantal is seen as a prominent feature in this remarkable landscape,, the railway actually skirting its base. Next to the Pic de Sancy, the Plomb du Cantal is the highest mountain in Central France ; it forms, as it were, the centre of a great number of volcanic chains, which spread out from it to the north, south, east, and west.

The railway after passing Aurillac still continues to run through mountain scenery, but scenery of a richer and more pastoral kind, and the turbulent mountain streams unite here into broad swift rivers. Within twenty miles of Toulouse the country becomes flat and uninteresting. It is somewhat remarkable that this interesting mountainous region of Central France should have hitherto attracted so little attention from English tourists.

From Toulouse to *Bagnères de Luchon*, in the Pyrenees, is a railway journey of about three hours, passing by Montrejeau, where there is usually a considerable delay. Soon after leaving Montrejeau the railway crosses the Garonne and enters the department of the Hautes Pyrénées.

The valley narrows, and in parts, from time to time, some lofty peaks appear in the distance. Just beyond the confluence of the Pique with the Garonne the railway again crosses the latter, leaves the valley of the Garonne to the left, and enters the beautiful valley of Luchon.

"It is a pleasure," says M. Taine, "to be ill at Luchon." Without going quite so far as this, one may certainly say that it must be a great pleasure to be cured in such a place. Luchon is decidedly a town of pleasure. Everything is arranged to make life look agreeable. I think it is Cherbuliez who says that life is feeble in the *mise en scène*, but here, at Luchon, art and nature combine to produce a *mise en scène* which is quite irreproachable. It is as though the best bits of the Bois de Boulogne and the Boulevard des Italiens were thrown down in the midst of the grandest and fairest of mountain scenery. On each side rise immense mountains richly wooded to their very tops, and the valley is closed in by inaccessible rocky peaks, remarkable both for beauty of form and richness of colouring. Between the mountains the floor of the valley is literally an inhabited park or garden. Broad avenues of houses are partly concealed and shaded by double avenues of trees. The hotels are surrounded by gardens, illuminated at night by the electric light, for even that little bit of Paris life is not wanting. Dining-tables, with brilliantly white linen and glittering plate and glass, are laid out in the open air, and the promenaders and the diners walk and dine in presence of one another. Parisian waiters rush from table to table, and Spanish and Parisian beauties with Parisian toilettes adorn these tables. Magnificent hotels, comfortable clubs, excellent food and cooking, really fine music, in the most tastefully arranged of pleasure grounds in front of the bath establishment, gay flags and banners floating in the air—for it is rare that there is not some fête or other to excuse such a display; and on the particular occasion I speak of a glorious moonlight streaming over all—such is Luchon in fine weather and at its best. What it may be in bad weather and at its worst I cannot say. The houses around are all, or nearly all of them, elegant villas with tastefully arranged gardens. The horses which abound here—for everybody rides or drives apparently all day long—are quick and active, and the carriages are comfortable and elegant. Carriages and four are as common as butterflies; and the mounted guides (I doubt if any of them would condescend to walk) are got up in a smart costume, and have an off-hand, dashing manner, equal to anything of the kind to be seen on the stage of the Opéra Comique.

An anonymous writer in " Blackwood " observes of Luchon : " No other place in the world represents beauty and pleasure in the same degree. No valley is so delicious ; nowhere is there such an accumulation of diversions ; nowhere are there so many or such various mineral springs."

Luchon is admitted to be the queen of Pyrenean health resorts ; but it is not a place for a poor person or a solitary person. If you come to Luchon you should take Iago's advice, and "put money in your purse," for here, and indeed in most places in the Pyrenees, there is a thirst for the stranger's money as great as the most vengeful thirst for an enemy's blood. I heard a French lady remark that it was *un vol organisé*. The expression was a strong one, but possibly it was not altogether unjustifiable.

One ought not either to be solitary, for beautiful and numerous as are the excursions around Luchon, very few of them are adapted to the pedestrian. Horses or carriages and guides are nearly always necessary, and unless one belongs to a party of three or four the expense attending such excursions is very considerable. It must also be borne in mind that unlike Switzerland, where you often start at a considerable elevation, in the Pyrenees you start usually from the bottom of a hot valley, and have to toil uphill for many miles in the hottest part of the day, as nearly all the higher points of view are at a great distance from one's starting-point. The people, then, who come here for health ought to bring some of their family or friends with them, or bring introductions to families already here. It is remarked with astonishment by the physicians and hotel-keepers at Luchon that so few English come here, while many go to Aix ; yet for a sulphur bath Luchon stands pre-eminent in all respects —for the abundance and variety of its springs, the quantity of water they afford, their composition and range of temperature. It is situated at an elevation of over 2000 feet above the sea, has a mild climate, and in the months of May, June, and September, a fresh and somewhat bracing air. In July and August it can be very hot, and it is during these months that it is crowded with Parisians and Spanish ; but even in

August the mornings and evenings are often deliciously fresh.

The springs of Luchon, like those of most of the Pyrenean spas, are sulphur springs ; but it is claimed especially for Luchon that owing to the great number of its sources, the great quantity of water they yield, and the variations in their composition and temperature, it is possible here, by having recourse to one spring after another, and by blending different springs, or by the mere extent of choice, to vary and graduate and adapt the treatment to a great variety of cases and every kind of constitution and temperament.

The waters have the well-known smell and taste of sulphuretted hydrogen, but chemical analysis shows that there is but a small quantity of this gas in them. The chief amount of sulphur which they contain is in combination with sodium, as sodium sulphide. There is no doubt, however, that a considerable quantity of sulphuretted hydrogen is given off from the surface of the water in the baths, and the air over the large *piscines* contains quite one per cent. of this gas, so that breathing this air for an hour at a time, while swimming about in the water, may certainly have a decided therapeutic effect. The sulphur compounds in some of the springs readily undergo decomposition either on exposure to the air or when mixed with cold water, and they then become milky from the presence of precipitated sulphur. This also is thought a valuable property, especially in the treatment of some forms of skin disease. These waters have an alkaline reaction and a pleasant soft feeling to the skin. Besides other mineral constituents to which little importance is attached, they contain organic matter, which is found deposited often in combination with sulphur, in the reservoirs and conduits of the water : this is often termed *barègine*. There exist also some iron springs in the environs of Luchon ; the principal one, and the best and pleasantest, on account of the carbonic-acid it contains, is that at Sourrouille. Some persons drink it at table mixed with wine. The springs at

Luchon are so numerous that is has been found convenient to classify them, according to their amount
of sulphuration and their temperature, into eight
groups. The hottest are the *Bayen* (68 degrees Centigrade, or 154 Fahrenheit), *Pré*, No. 1 (60 degrees
Centigrade, or 140 Fahrenheit), *Grotte Supérieure*
(58·4 degrees Centigrade, or 137 Fahrenheit), and
Reine (57·6 degrees Centigrade, or 135 Fahrenheit).
These have, of course, to be cooled or mixed with
springs of lower temperature before they can be used
for bathing purposes. But for the purpose of what is
called *étuves sèches*, or dry vapour baths, and for the
salles d'inhalation, their high temperature is altogether
advantageous ; the atmosphere in one of these *salles*
can be raised to 46 degrees Centigrade, i.e. 117 degrees
Fahrenheit, by allowing the vapour of the mineral
waters to spread freely through it. There are no less
than twenty-two *buvettes*, or places where the waters
of the different springs can be drunk, and these are
supplied from four of the eight groups to which I
have referred. Three of the *buvettes* are in the establishment itself ; nineteen are in what is termed the
Promenoir des Buvettes, a pleasant covered walk
behind the building, and four are situated about a
hundred yards from it under an exceedingly pretty
rustic kiosque, which is called the Buvette du Pré,
placed on an elevation in the park which commands a
charming view. Here there are two *salles* appropriated to gargling, an important process at all these
sulphur baths, and it has also a little *salon au premier,*
where one can sit and enjoy the view. A number of
bottles of syrups of different kinds surround the
buvettes, and these are added by most persons to
the waters to lessen their disagreeable taste. Patients
often begin here, as elsewhere in France, by very
small doses of the waters—a third or half a glass
twice a day—and this is very slowly and cautiously
increased, for some digestions are readily disturbed
by sulphur waters ; they are taken fasting before
breakfast and during the hour or two which pre-

cede dinner.   Drinking the water is, however, but a small part of the various curative processes which are made use of at Luchon.   The *Établissement Thermal* is a handsome building, elaborately fitted up with all the appliances necessary for utilising the waters according to the most approved methods.   It has well-appointed *salles d'inhalation* and *pulvérisation*, others for vapour baths, others for *humage*, i.e. for breathing through tubes the hot vapour given off by the springs.   There are eleven different sets of baths in separate pavilions, supplied from different sources, and so arranged that the bather can take his bath with or without an atmosphere of vapour. A hundred and twenty of the baths are of marble, each furnished with a local douche.   There is also a special department for the various sorts of douches, two small *piscines* for men and the other for women, and another large *piscine de natation*.   Altogether it is one of the most complete bathing establishments in Europe.   The great central hall of the *Établissement*, the *salle des pas perdus*, is constructed of white marble, the walls handsomely decorated with frescoes, and it is conveniently furnished with chairs and lounges and tables, supplied with a number of French and other newspapers.   To this fine well-furnished hall the public are admitted free, and they can lounge there for hours reading the papers if they are so disposed.

In the next place, let us inquire what are the cases to which the treatment at Luchon is applicable. Chronic diseases of the skin, and of these especially chronic eczema, form perhaps the major part of the cases which come to Luchon for treatment.   Other chronic skin diseases often improve considerably here, but they do not yield the same satisfactory result as cases of eczema.   Chronic muscular rheumatism is often relieved here, as indeed at most thermal spas. Certain diseases of the glands, and especially of the bones, derive benefit often in a marked degree at Luchon.   The waters of Luchon are also reported to be, like those of Barèges, of great efficacy in the treatment of gunshot wounds. Several ambulances were established here after the Franco-German war, and excellent results are said to have been obtained from treatment by the sulphur waters. It is quite likely that they exercise a useful antiseptic action in such cases. Foreign bodies are stated to be extracted from the wounds by the influence of the baths. Cases of lead and mercurial poisoning are said to be cured here. Chronic affections of the throat, the ears, and the nose are largely and successfully treated by these waters, applied usually in the form of pulverisation and local douches. The source named the Pré is gaining a repute for the relief of chronic chest affections, as bronchitis, laryngitis, and all catarrhal conditions of the air passages. Some special maladies, to which I need not now allude further, are dealt with successfully at Luchon. Bagnères de Luchon then is a thermal station of much importance and attractiveness, and one which should not be overlooked by English doctors and English patients. No other European spa can compare with it in natural beauty, its elevation above the sea imparts a freshness to its atmosphere which renders the mornings and evenings always cool and agreeable ; life there is as cheerful and pleasant as external things can make it ; and it is certainly as likely to cure the ailments for which sulphur springs are appropriate as any other resort of this category.

From Luchon to Bagnères de Bigorre is a ten hours' drive by a good carriage-road through beautiful scenery. Public conveyances run during the season ; but it is better for a party of three or four to travel together, and hire a carriage and horses for themselves. The expense will be very little more, while the comfort and convenience will be infinitely greater. One can also go from Luchon to Bagnères de Bigorre by train, going round by Tarbes : this takes about five hours and a half.

Bagnères de Bigorre is a pretty little town of about 10,000 inhabitants, beautifully situated near the entrance of one of the most celebrated valleys of the Pyrenees. Its baths were certainly known and esteemed by the Romans, and in the sixteenth century it was one of the most famous rendezvous of the Southern nobility. Montaigne describes it as at that time one of the best places he had discovered *de se laver le corps tous les jours!* on account of its agreeable situation, its comfortable quarters, good food, and good company.

It presents a great contrast to *Luchon*. As a permanently inhabited town it is larger, but as a thermal station it has much less to recommend it, and it has the appearance of being much less frequented. After Luchon, it presents quite the aspect of a rural retreat; though here also there is a little bit of Paris, but it is rather a smaller bit of the Palais Royal than the grand Boulevard. Indeed, the stalls of the Promenade des Coustous, with their bad cutlery, cheap jewellery, and lottery tables, scarcely merit a comparison with anything Parisian.

The bath establishment seems small and insignificant as compared with that at Luchon, and need not be described. The waters are very feebly mineralised, of a comparatively low temperature, 33 degrees Centigrade (91·4 degrees Fahrenheit), and do not contain sulphur—a very exceptional circumstance in the Pyrenean springs. They are used chiefly as baths, and are thought to be especially valuable in cases of hyperæsthesia, in excited and feeble nervous systems; in such cases they are reported to produce remarkable calming and sedative effects. They are, in consequence, much resorted to by delicate ladies with hysterical and other disturbed conditions of the nervous system. Daily immersions for some time in tepid water have no doubt, in many cases, a very soothing effect, and this influence is no doubt aided and augmented by the calm unexciting life, the mild mountain climate, and the agreeable surrounding scenery. The town lies at an elevation of about 1800 feet above the sea, on the left bank of the Adour, in a charming situation, near the opening of the valley of Campan, and overlooking the rich plain of Tarbes. It is not *in* the mountains, like the other Pyrenean spas, but is situated just where the lesser elevations begin to rise from the plains. It is amongst the outposts of the great central chain. Its pleasant climate and attractive scenery, and its accessibility by railway, make it a much-frequented resort of the permanent winter residents in Pau; there they escape the great summer heat of the latter place. Bagnères de Bigorre is especially rich in Roman remains. "The Romans," says M. Taine, "a people as civilised and as bored as we are, came as we do to Bagnères. Rome has left her traces everywhere at Bagnères. The pleasantest of these relics of antiquity are the monuments which the patients who were cured there erected to the Nymphs, and the inscriptions upon which still exist. Reclining in their marble baths, they felt the healing virtue of the beneficent goddess spread through their limbs; their eyes half-closed, dozing in the soft embrace of the tepid spring, they heard the mysterious source, falling drop by drop, in monotonous chant, from the bosom of its mother rock; they saw the surface of the effused water glisten around them with its pale green ripples; and there passed before them, like a vision, the strange look and the magic voice of the unknown divinity who visited the light in order to bring health to unhappy mortals."

From Bagnères de Bigorre to *Barèges* is a five hours' drive through the beautiful valley of Campan and over the Col de Tourmalet—the highest carriage-road in the Pyrenees, and one of the highest in Europe, being nearly 7000 feet above the sea. It is, however, usually approached from Luz. It takes nearly an hour and a half to mount the four-and-a-half miles of gradual and continuous ascent which leads from Luz to Barèges. In traversing this short distance we pass from a region of smiling pastoral beauty, of green pastures and wooded slopes, to one of dreary barrenness and desolation. Barèges, which

is situated at an elevation of about 4084 feet above
the sea, is a little town of about a hundred or so poor-
looking houses, in a barren and dreary situation
presenting nothing that is attractive or picturesque.
"The landscape," says M. Taine, "is hideous, it
looks like a deserted quarry." Its climate is very
variable, great heat alternating frequently with sharp
cold. Cold mists from the surrounding mountains
often collect over the valley, and it is tormented by
violent winds. It is uninhabitable for five months in
the year, when it is covered with fifteen feet of snow.
"Il faut avoir beaucoup de santé pour y guerir,"
M. Taine remarks. Notwithstanding the dreariness
of its situation, its baths enjoy a very great reputation for the cure of certain maladies; especially of
wounds received in battle, of chronic articular rheumatism, of certain forms of scrofula, and of all kinds
of diseases of the bones and joints, as well as some
forms of skin disease (as at Luchon), and certain
local paralyses. Its sulphur springs are amongst the
strongest in the Pyrenees, and are rich in that peculiar nitrogenous substance to which the name of
barègine has been given. There are twelve springs,
varying in temperature from 88 to 113 degrees Fahrenheit. The waters are taken internally, and are
also used as baths and douches. The season begins
early in June, and the place becomes so crowded in
July and August that patients have sometimes to
wait their turns at Luz until they can be taken in.
We have been told that in some seasons the number
of bathers is so great that invalids have to leave their
beds soon after midnight in order to take their turn
at the baths; and that the common baths are so
crowded and badly arranged that the water and the
air become intolerable.

Owing to the repute these baths have always
maintained for the cure of injuries received in the
field of battle, a military hospital has existed there
since 1760. The place also possesses a hospital for
the reception of nuns and priests, which also, from

Sept. 1st to Oct. 15th, receives poor patients who are kept there at the expense of their *département*.

It is from Barèges that the ascent of the Pic du Midi de Bigorre is usually made, one of the finest points of view in the Pyrenees. Although about 9300 feet above the level of the sea, the summit can be reached on horseback, and in about three-and-a-half hours from Barèges. An observatory has been established there. A small hotel, where food and beds can be obtained, has been built on a *col* about 1600 feet below the summit.

A gradual and continuous descent of four-and-a-half miles brings us from Barèges to Luz, and the baths of *St. Sauveur* are only a mile from the latter place. Luz is a charming little town, situated in a well-cultivated plain—the meadows of which are remarkable for their greenness, owing to their being irrigated by numberless little mountain streams. It has been called "the most Pyrenean spot of the Pyrenees"; it is the spot from which the tourist visits the Cirque de Gavarnie, the great show-place of the Pyrenees. A drive of three hours along a road of extraordinarily wild and savage grandeur leads to the mountain village of Gavarnie ; but it is still a walk of two hours more before you reach the very depths of the magnificent cirque. The village of Gavarnie has not yet been promoted into a mountain-air cure, although it is 4400 feet above the sea. "The vale of Luz" has been said to be "the most winning" of all the Pyrenean retreats. "Its soothing calm, its welcoming tenderness, its look of friendship and of wise counsel, wind themselves round you ; and the beauty of its grassy shades, of its brakes and colour-changing hills, delights and wins you." Of the people the same writer says, "After mass, both men and women stand about and gently gossip and look good. It seems to you that you have come upon a new people, and you are more than ever sorry that they do not wash themselves!"

A good road, planted with trees, connects St.

Sauveur with Luz. *St. Sauveur* is *par excellence* the
ladies' bath of the Pyrenees, and it enjoyed for a time
the personal patronage of the Emperor Napoleon III.
and the Empress Eugénie. They erected a nice little
church there of white marble, *L'Eglise Saint Joseph*,
and a fine bridge, *Le pont Napoléon*, in grateful acknow-
ledgment of the good the baths of St. Sauveur had done
them. St. Sauveur consists of a straight row of neat
little houses, with an enclosure of trees and gardens
at one end, called the Jardin Anglais. The springs
are of comparatively feeble mineralisation ; but, like
the other Pyrenean spas, are characterised by containing sulphide of sodium, which gives them their
characteristic odour. Their temperature is also rather
low for bathing, viz., from 90 to 94 degrees F. It
has a mild mountain-climate, being at an elevation of
2360 feet above the sea, and is rather subject to
mists. It is considered sedative, and is regarded as
specially adapted to females who suffer from maladies
into which nervous irritability enters as an important
element.

To reach *Cauterets* it is necessary to drive from
Luz to Pierrefitte. This is about eight miles, and the
latter part of the road through the gorge of Pierrefitte
is very grand and striking ; the road, cut out of the
side of the rock, seems in places to almost hang over
the torrent below.

Cauterets is perhaps at the present time the most
popular of Pyrenean spas when regarded from a purely
medical point of view. It is not a resort of fashion
and pleasure like Luchon, but most of the visitors to
Cauterets come with a serious purpose. There is a
business-like look about everybody at Cauterets. The
doctors rush about with the air of busy practitioners
to whom "times is money," as they usually express a
maxim which they believe to be especially dear to
Englishmen. The patients look graver than usual,
and more bent than usual on carrying out with
business-like accuracy the details of the cure. The
bath men and women, the bath engineers, and all

the *personnel* of the various bath establishments
present the business aspect of the employés of a
Birmingham factory. "Nos eaux sont des eaux
sérieuses," is the grave utterance of all the people
interested in and engaged at Cauterets. The atmosphere, too, seemed a little heavy and business-
like, and lacked that light and exhilarating tone
which one felt in it at Luchon.

Cauterets, however, is often approached by railway, and those who avail themselves of that method
of access must needs pass by Lourdes, and then they
may encounter, as I did, on a fiercely hot day in
August, a long train packed full of pilgrims returning from a visit to that shrine. To those who
have only read of pilgrimages, the word probably
calls up in the mind a picture not without some
romantic colouring. They associate it with pious
toil and faithful sacrifice, picturesque costumes, and
venerable men and holy women! It may have been
so in the Middle Ages, but, alas! the reality of the
nineteenth century is something very different. A
modern pilgrimage is more like a crowded Crystal
Palace train on a Bank holiday: say a gathering of
Foresters, and the likeness will be more complete, for
each pilgrim, like the Forester, wears a badge. The
train pulls up, and the pilgrims rush out, hot, dusty,
and perspiring, in search of water. They are a vulgar
and noisy crowd. They have all the badge of the
sacré cœur. There are many priests among them, and
quite without prejudice it may be said, that it would
be difficult to find a less intelligent, more common-
place, and really common-looking class of men.
Indeed, I looked for some intelligent, pleasant face
amongst the crowd, and I looked in vain.

At Lourdes a short branch line runs off from that
between Pau and Toulouse to Pierrefitte, at which
place diligences meet the trains to convey passengers
to Cauterets, or to St. Sauveur and Barèges. From
Pierrefitte to Cauterets is a drive of about seven miles;
and the road bifurcates almost immediately on leaving

Pierrefitte, that to the left going to St. Sauveur, that
to the right at once commencing the ascent to Cauterets. The road ascends nearly the whole of the way,
through a picturesque valley, dominated by lofty peaks,
rugged and wild in parts. The road is here and there
cut out of the side of the rock, at a dizzy height above
the foaming torrent, the Gave of Cauterets, rushing
along beneath. As we approach Cauterets the valley
widens, and finally discloses the little town situated
at the bottom of a narrow basin, surrounded nearly
on all sides by lofty summits frowning down from
immense heights on the small town which lies crouched
between the bases of the mountains. Several mountain
valleys open into this basin, and lead to wild and
picturesque excursions into the very heart of the
Pyrenees ; none of them are carriage roads except
that leading to Pierrefitte. One of the most celebrated of these excursions is to the *Pont d'Espagne*
and the *Lac de Gunbe,* a three hours' walk from
Cauterets. Cauterets is thus quite in the mountains,
its elevation being a little over 3000 feet ; its climate,
however, is scarcely so bracing as might be expected
in a place of this elevation. It is so much shut
in on all sides by high mountains that it is capable
of becoming very hot and close in certain conditions of atmosphere. The mornings and evenings
are, however, fresh and pleasant, especially before the
end of June and after August. The basin of Cauterets
is very prone, like other places of this medium elevation, to become somewhat suddenly filled with clouds,
which may linger long, and give rather a dull and sad
aspect to the little town. The climate is also rainy,
and subject to sudden changes of temperature.

The reputation of Cauterets as a health resort is
very ancient. M. Taine tells us that Julius Cæsar is
said to have been restored to health by the spring
named after him " César," and Abarca, king of Aragon,
by the spring on that account named " du Roi." It was
here that Marguerite de Navarre, sister of François I.,
a distinguished example of the race of " superior

women," wrote the chief part of the *Heptaméron.* She came here with " her court, her poets, her musicians," interested in all subjects, reading Greek, learning Hebrew, and delighting in theological discussion ; at the same time tender and simple : " Une imagination mesurée, un cœur de femme dévoué et inépuisable en dévoûments, beaucoup de naturel, de clarté, d'aisance, l'art de conter et de sourire, la malice agréable et jamais méchante." Such is the attractive picture M. Taine gives of Marguerite de Navarre at Cauterets.

The waters of Cauterets are sulphur waters, like those of Luchon, but they are considered to be milder in their action and more sedative. They are efficacious, like those of Luchon, in diseases of the skin, in scrofulous affections, in chronic throat ailments, and especially in chronic diseases of the respiratory organs. I asked one of the leading physicians at Cauterets what they did with consumptive patients there. " We cure them," he replied, and he expressed great astonishment that we English physicians did not send more phthisical patients to be "cured" at Cauterets. But beyond this general statement I was not able to procure any definite demonstrative evidence of the cure of such cases at Cauterets. It is not easy, of course, to produce such evidence at the moment it may be asked for, and credit must be given to the medical men who practise there for intelligence and honesty ; and their testimony is so strong in favour of the great amelioration that certain cases of consumption undergo at Cauterets, that it must, I think, take rank amongst the health resorts to which persons who are afflicted with chronic forms of consumption should be sent.

There are a great number of mineral springs at Cauterets, and several bath establishments, some of which, notably the César, are most elaborately fitted with every appliance that modern science has suggested in the use of mineral springs—douches of all kinds, inhalation and pulverisation chambers, besides baths of every description.

The source, however, which is especially valued for internal administration is *La Raillère*. It is really curious to encounter the long procession of drinkers coming away from the Raillère spring, which is situated at some little distance from the town ; each, young and old, sucking a stick of " sucre d'orge à l'eau de Cauterets." It is said that ten thousand sticks of barley-sugar are sold each day during the season ! It is impossible to explain satisfactorily how the small quantity (sometimes only four or five table-spoonfuls twice a day) of this somewhat feebly mineralised sulphur spring can produce the remarkable curative effects that are claimed for it. But there seems no doubt that many chronic catarrhal conditions are greatly benefited or cured there. Moreover, the good results obtained at Cauterets are not confined to the human species. " Horses," we are told, " from the studs of Tarbes and Pau, which are afflicted with chronic, bronchial, and stomach catarrh, etc., are sent to the springs at Cauterets, and are often cured there in a week."

It would be an extremely interesting excursion for the hardy and enterprising pedestrian and balneologist, to visit the Spanish mountain health resort of *Panticosa* from Cauterets, and return by Gabas and Les Eaux Chaudes. From Cauterets to Panticosa is a rather arduous mountain-walk of eight hours, crossing a steep *col* about 8000 feet high ; a horse can be taken for about the first half of the way ; and a guide is necessary, as the first part of the descent on the Spanish side is rather difficult. We have very little information about the Spanish Pyrenean bath, Panticosa. It is said to be situated about 5000 feet above the sea, and to possess several springs, which are somewhat quaintly named, according to their uses. It is resorted to by the Spaniards in great numbers during the season, and the accommodation would seem to be very fair. To return from Panticosa to Les Eaux Chaudes, the road is easier but longer, the *col* to be crossed is lower, and one can take horses the whole of the way and a carriage a considerable part of it. The journey takes about ten hours.

Les Eaux Bonnes is another Pyrenean health resort, frequented chiefly by persons suffering from affections of the throat and respiratory organs. "Chaque siècle," says M. Taine, "la médecine fait un progrès. Par exemple, au temps de François I. les Eaux Bonnes guérissaient les blessures : elles s'appellaient *eaux d'arquebusades;* on y envoya les soldats blessés à Pavie. Aujourd'hui elles guérissent les maladies de gorge et de poitrine. Dans cent ans elles guériront, peut-être, autre chose. Les médicaments ont des modes comme les chapeaux. Un médecin célèbre disait un jour à ses élèves : 'Employez vite ce remède pendant qu'il guérit encore!'"

In following the usual carriage-road from Cauterets to Eaux Bonnes, it is necessary to descend again to Pierrefitte, and thence to the prettily situated town of Argelés. The carriage-road from Argelés to Eaux Bonnes is one of the most remarkable in the Pyrenees ; or one can take the train at Pierrefitte, and go by Pau to Eaux Bonnes, visiting the shrine at Lourdes on the way. It is now connected by rail with Pau, or nearly so ; the railway is completed as far as Laruns, and from that place to Eaux Bonnes is only a drive of twenty minutes. A little more than half-way, and just beyond the small village of Louvie-Juzon, one enters the Vallée d'Ossau, and in the distance, high above the other mountains, one sees the Pic du Midi d'Ossau, easily recognised by its curious summit of two unequal peaks. Within three or four miles of Eaux Bonnes the route bifurcates, that to the right going to Eaux Chaudes, and that to the left mounting to Eaux Bonnes. The village, about 2400 feet above the sea, is situated in a somewhat narrow gorge, stretching between the steep mountains which here bound on each side the Vallée d'Ossau. The chief part of the village consists of three rows of uniformly-built houses and hotels, forming three sides of a quadrangle, and inclosing a space planted with trees, and called the Jardin Anglais, where the inevitable band plays, and where the visitors walk, or sit and talk, or read, or work, as they may be disposed. Beyond and above the Jardin Anglais is the *Établissement Thermal,* and to the right, built in a conspicuous position on a terrace, is the handsome new church. Here also commences the remarkable and interesting carriage - road constructed through the mountains which leads from Eaux Bonnes to Argelés.

A characteristic of Eaux Bonnes is the possession of a very fine promenade, which is called the Promenade Horizontale; it begins at the Casino and is continued along the side of the mountain out of which it is cut, always on the same level, parallel to, but at a considerable elevation above, the road leading from Eaux Bonnes to Eaux Chaudes. It is planted with trees, under the shade of which many seats are placed commanding beautiful views of the Vallée d'Ossau. When it is completed, according to the original design, it will extend for more than three miles until it joins the road to Eaux Chaudes.

Eaux Bonnes, with its excellent hotel accommodation, its pleasing site, and the numerous interesting excursions into the mountains which it commands, attracts every year a considerable number of the Parisian upper classes, who find a comparatively calm and unexciting and refreshing retreat there from the gay life of Paris. Its climate, too, is especially soothing; there is exceedingly little wind there, and I was assured by an excellent authority that the air is often so still that one may pass days without seeing a leaf stir on the trees. It is, however, subject, like most other mountain stations, to thunderstorms and heavy rains.

The quantity of water yielded by the springs at Eaux Bonnes is limited, so that it has never been the custom to use them, to any extent, as baths; the *Grand Établissement,* however, contains a certain number of baths, as well as two rooms devoted to foot-baths, a chamber for gargling, another for throat douches and for *pulvérisation*. There is also another smaller *Établissement d'Orteig*, and a third *Établissement de Bains de Santé et d'Hydrothérapie*, in which the various processes of the "cold-water cure" are applied.

The principal spring, La Source Vieille, is a warm sulphur spring, in which the sulphur exists in combination with sodium, having a temperature of 33 degrees Centigrade, or about 92 degrees Fahrenheit. It is to the use of this spring that the good results obtained in so many cases of throat and chest disease are attributed.

To spend three weeks at a pleasant station like Eaux Bonnes, and by so doing to get rid of one of those exceedingly troublesome throats which are common amongst public speakers, singers, and especially clergymen (hence often called "clergyman's sore throat"), one would think ought to prove an agreeable and popular remedy. French people seem to be of this opinion, and ecclesiastics and actors and singers abound at these sulphur springs, but one meets exceedingly few English there.

It is then for the cure of chronic affections of the throat—of the pharynx and larynx—that these waters are especially renowned, as well as for the cure of chronic bronchial catarrhs. It is also claimed for them—and the claim rests on no less an authority than that of the great physician Trousseau—that they are of unmistakable efficacy in cases of consumption ; and this opinion is firmly maintained by those who have had many years of experience in treating such cases at Eaux Bonnes, as for example Drs. N. Guéneau de Mussy, Pidoux, and L. Lendet.* It is, however, in the strictly local and limited manifestations of this malady, and not in those cases in which

---

* See " Les Eaux Bonnes dans le Traitement de la Phthisie Pulmonaire," by Dr. L. Lendet.

there is obvious general constitutional infection, or in which the disease is rapidly advancing, that the cure at Eaux Bonnes is indicated. There is any amount of medical testimony forthcoming as to the efficacy of the waters of Eaux Bonnes in fitly selected cases of consumption ; while many of those chronic cases which by careful management continue to maintain a feeble but tolerable existence, by passing the winter in the south, etc., come year after year to pass some part of the summer season at Eaux Bonnes. There are many consumptive French patients, who, by the recommendation of their physicians, pass their winters on the Riviera, and their summers at Eaux Bonnes, or at one or other of the Pyrenean health resorts, and this arrangement seems to suit them well, and if their malady does not become cured, at any rate, its course is, for a time, arrested or retarded, and they obtain many years of agreeable existence which they could not insure in the north.

Les Eaux Chaudes is distant about six miles from Eaux Bonnes by a good carriage-road ; there is also a very interesting walk over the mountains between the two villages, commanding magnificent views of the grand surrounding mountain scenery. The carriage-road for the first three miles is the same as that traversed in coming from Pau. It will be remembered that this road bifurcates within about three miles of Eaux Bonnes, the branch to the right going to Eaux Chaudes. When we reach this bifurcation we enter a narrow defile, sombre but picturesque, bounded on each side by enormous mountain walls, with a blue band of sky overhead. The road keeps to the left side of the gorge, often at a great height above the river—the Gave—which 500 feet beneath roars and foams along its steep and stony bed.

We come somewhat suddenly upon the village of Eaux Chaudes, a simple village of a few houses and hotels, most charmingly situated in the very bosom of the mountains. As a health resort, as a bathing station, Eaux Chaudes seems as it were to

have fallen asleep; it presents none of the life and
activity and outward signs of material prosperity
encountered in most of the other Pyrenean spas.
Its *Établissement des Bains* is almost deserted;
there are baths, but no bathers; there are springs,
but no one comes to drink them. The bath attendants
invite you imploringly to avail yourself of their
services; the rows of empty unused glasses mutely
appeal to you to "come and drink." And yet Eaux
Chaudes is one of the most picturesque spots in the
Western Pyrenees; but it is not in vogue at present.
The tourist comes and admires the beauty of the
site, drives to the village of Gabas, most grandly
situated amidst wild mountain scenery, the magnificent Pic du Midi closing in the horizon, and then
probably returns to his hotel at Eaux Bonnes. But
the patients at Eaux Chaudes are few. When the
Route Internationale between Eaux Chaudes and the
famous Spanish mountain health resort Panticosa is
completed—and it has been recently determined that
it shall be completed—passing by Gabas and the
Col de Portalet, Eaux Chaudes may possibly recover
some of its ancient reputation. The springs are
sulphur springs, as elsewhere in the Pyrenees, but
they are rather weaker and of lower temperature,
notwithstanding the name of Eaux Chaudes.

Eaux Chaudes enjoyed at one time some reputation as a "ladies' bath," but St. Sauveur is now the
resort *par excellence* of the French ladies.

It is common, after a "cure" at one of the Pyrenean
health resorts, to consolidate the results by a visit for
a few weeks to one of the adjacent watering-places on
the coast of the Atlantic; and Biarritz or San Sebastian
or St. Jean de Luz or Arcachon are all conveniently
situated for that purpose.

It has been said of the Pyrenees that "their
function is not to be high but to be pleasant; to
behave as a summer district of amusement and health;
to conduct themselves properly as rivals of Vichy and
Trouville. . . . In the popular imagination they are

composed of carriages-and-four, of *capulets* and *bérets*,
of mineral waters, rocky gorges, Luchon, admirable
roads, bright green valleys, two hundred and thirty
hotels, and the Cirque of Gavarnie.   A few peaks and
a little snow are thrown in here and there by Nature,
like seaweed on a beach, or weeds in a pond, just to
give local colour."   The Pyrenean chain of mountains
has been compared to a fern branch, the stalk of
which corresponds to the line of the crests, while the
leaflets and the spaces between them represent the
diverging ribs and ravines.   It is so steep that not
more than fifty passes have been discovered over it
between France and Spain, and of these only four are
practicable for carriages and thirty for mules.

Another peculiarity of the Pyrenees are the *Cirques*,
strange amphitheatres of crags, many of them thousands
of feet high, which are found in the central part of the
chain—probably the most stupendous monuments of
rock-building the earth can show.   There are very
few lakes, and these are quite small and lie very high;
the lowest is 4000 feet above the sea.   Hot springs
swell up everywhere, from close to the Bay of Biscay,
at Dax, nearly to the shores of the Mediterranean, at
Amélie les Bains.   There are already 600 known.

Of the charms of this region, the most striking are
the "amazing greenness of the slopes, their delicious
freshness, their flowers, their grass, their bushes, and
their trees.   Their ravines are all beauty and delight."
They have none of the grandeur and sublimity of the
Alps, but "their immense freshness is full of charm,
and brightness, and colour; they are graceful and well-
mannered, and it is impossible to be bored by them."

The inhabitants are often very picturesque; Cata-
lonian and Aragonese costumes are to be seen in
every village, and the men commonly are seen with
scarlet sashes round their waists.

The preceding is a brief sketch of a rapid personal
survey, during an autumn vacation, of a few of the most
important health resorts in the Pyrenees.   It is some-

what remarkable that so few English health-seekers
find their way to these thermal stations; yet they
possess many attractions; most of them are in situations of great natural beauty, and surrounded by
scenery possessing every kind of interest. The hotels
are for the most part good, and the natives civil.
But there are also certain drawbacks which weigh
heavily on the English mind. Even in some of the
largest and best hotels in the Pyrenees—and I would
instance the large and handsome new hotel at Cauterets
—there is an inattention to decency and cleanliness in
some important domestic arrangements which would
not be encountered in a third-class hotel in Switzerland. Then the Pyrenees are expensive, and if the
natives are civil and polite they require to be paid
handsomely for their civility. Moreover, you can
scarcely move without a guide. There is hardly an
indication anywhere, as in other districts visited much
by tourists, by which a pedestrian can find his way;
indeed, it seems pretty obvious that it is intended that
he should not find his way about alone; so pedes-
trianism is not in vogue in the Pyrenees, horses or
carriages and mounted guides being the order of the
day. And no doubt in some of these hot valleys it is
better to ride or drive than to attempt to walk. Then,
again, you do not encounter in the Pyrenees any of
those excellent hotels built in attractive spots at considerable elevations, such as exist in great numbers in
Switzerland, and from which the higher mountains
can be conveniently explored. But for invalids who
come with a serious object, the life at such resorts
as Luchon or Eaux Bonnes presents much that is
attractive and agreeable.

There are a few other less-frequented Pyrenean
baths which may be briefly noticed here.

Aulus (Ariège) by rail from Paris to Toulouse and
Toulouse to St. Girons, then eighteen miles by carriage;
it is situated in a valley about 2400 feet above the

sea, surrounded by high mountains. There are many charming excursions and mountain ascensions to be made here. The waters are warm (temperature 68 degrees Fahrenheit) and contain sulphate of lime chiefly, a smaller proportion of sulphates of magnesia and soda, small quantities of iron, manganese, and a trace of arsenic. It is drunk and used in baths and douches. It is also largely exported. These waters are said to be useful in debility of the stomach and intestines, in vesical catarrh, anæmia, and chlorosis, and to exercise a specific effect in cases of inveterate syphilis.

Ax (Ariège) about 2300 feet above the sea, is beautifully situated in the midst of charming scenery, seventeen miles by carriage-road from the railway station of Tarascon. It is no doubt of great antiquity as its name implies (from *Aqua*). It has four bath establishments and a great number of warm sulphur springs, resembling those of Luchon in their composition and the variety of their strength and temperature. Only forty of the springs are utilised.

These springs have necessarily much the same action and uses as the sulphur springs of Luchon and Cauterets.

Cambo, only eleven miles from Bayonne, on the river Nive, no doubt owes its growing popularity as much to its accessibility, especially from Biarritz, as to its pleasing situation and mineral springs. Those who have passed the greater part of the winter at Biarritz find it an agreeable change, when spring comes, to migrate for a few weeks to Cambo and exchange the Atlantic winds, sea air, and coast scenery of the former place, for the soothing and sedative climate of this sub-Pyrenean station. The river Nive divides Cambo into two parts about half-a-mile distant from one another, *Le haut Cambo* and *Le bas Cambo*. The former, where the hotels and *pensions* are situated, is about 200 feet above the sea on a steep terrace. It commands a charming landscape. At the foot of the hill the clear stream of the Nive forms a graceful curve, and from its right bank stretches away a vast extent of fertile fields and wooded country. On the other side rise the lower spurs of the Pyrenees, mountains of no great height but giving variety to the landscape, which, although not grand, yet presents a pleasing, cheerful calm and rusticity, full of freshness and delight.

Cambo possesses a warm sulphur spring of a temperature of 75 degrees Fahrenheit, analogous in composition to the other sulphur springs of this district ; it has also a cold iron spring.

The bath establishment has the usual appliances for the administration of these waters internally and externally. Very many charming walks and excursions can be made into the surrounding district.

Capvern (Hautes Pyrénées).—The baths are about half-an-hour's drive from the little town and station, which is a few miles from Tarbes on the line between that town and Toulouse. The popularity of this station has increased rapidly of late years. There are two springs and two bath establishments. The springs are warm and have a temperature of 77 degrees Fahrenheit.

Their mineral constituents consist chiefly of sulphate and carbonate of lime, and sulphates and chlorides of sodium and magnesium. They are said to powerfully stimulate digestion, to increase the renal hepatic and alvine secretions, and to be of especial value in gravel and catarrh of the bladder, in hepatic engorgements, and in gall-stones, and in many uterine affections.

Encausse (Haute Garonne), in a picturesque and pleasant situation a few miles from St. Gaudens, on the line between Toulouse and Tarbes, possesses warm sulphate of lime springs which are credited with almost as many healing virtues as the shrine at Lourdes. They are said to act most efficaciously in the cure of intermittent fevers, and to be useful in the same cases as those treated at Capvern.

Olette (Pyrénées Orientales), reached by rail from

Perpignan to Prades, thence by a carriage-road of a few miles. There is a good bath establishment, 1900 feet above the sea-level, with all the appliances usually found in such institutions. The springs, of which there are many, are warm, and vary in temperature from 80 to 170 degrees Fahrenheit. They contain sulphide of sodium, and resemble in their properties and uses the other numerous sulphur springs of this region. Interesting walks and excursions abound.

Saint Christan (Basse Pyrénées), a few miles from Oleron, is approached from Lacq (three or four hours of carriage-drive), a station on the line between Bayonne and Pau. It is a small bath, distinguished by the presence of sulphate of copper in its springs. These formerly had a reputation for the cure of leprosy. The springs are cold ; they also contain iron as well as copper. They are employed in obstinate scrofulous and syphilitic skin affections, and to heal chronic ulcerations. They are said to be of especial value in chronic ophthalmia, keratitis, and some other affections of the eyes.

Siradan, Sainte Marie, a short distance from Saléchan, on the line of railway from Toulouse to Luchon, is about twelve miles from the latter. It has four springs, two of which are ferruginous and cold and two warm ; the latter resemble greatly in composition, character, and uses, those of Capvern. They enjoy considerable local popularity.

Ussat (Ariège), situated in a picturesque valley, through which the river Ariège flows, within half-an-hour's drive of Tarascon station, is a popular "ladies'" bath, and is annually resorted to by a great number of invalids suffering from uterine and allied nervous and hysterical affections. There is a fine *Établissement Thermal*, situated in a spacious park, and the bathing arrangements are exceedingly good. There is a small hospital for the reception and gratuitous treatment of poor patients. The temperature of the springs is about 100 degrees Fahrenheit. Their mineralisation is feeble, and consists chiefly in small amounts of the carbonates and sulphates of lime, magnesia, and soda.

These baths, when employed at a moderate temperature, are said to be exceedingly soothing to the nervous system, and of great value in the treatment of chronic metritis, dysmenorrhœa, and other affections of a kindred nature.

It would be tedious to enumerate more of these smaller Pyrenean spas, which resemble one another greatly and have chiefly a local reputation. They are most of them too remote and inaccessible to attract many English visitors.

CHAPTER XIII.

THERE is an important group of Health Resorts
in Savoy, of which *Aix les Bains* is the chief, com-
prising also such well-known stations as St. Gervais,
Evian, Amphion, Challes, Marlioz, Brides, and
Salins. Together with these, we may conveniently
consider in this chapter the adjacent resorts in
Dauphiné—Uriage, Allevard, and La Motte.

AIX LES BAINS

May justly be considered as the most important
bathing resort in France. Its reputation is universal;
and situated as it is on one of the great European
highways between the north and the south, between
Italy and France, it is most easy of access to that
great stream of travellers who annually migrate, in
search of health or of pleasure, from the north to the
south of Europe, or even to Africa and the far East.
Thus it is that Aix is far more cosmopolitan than
any other bath in France, and with our own country-
people it is especially popular, partly, no doubt, from
its great repute in the treatment of those very English
maladies, rheumatism and gout. But Aix has also
attractions for the sound as well as for the sick, and
it is by no means, like so many other bathing stations,
given over entirely to invalids. Its attractive site and

picturesque neighbourhood, its excellent hotels, its large and gay (perhaps somewhat too gay) Casino, and the brilliant society which may frequently be found there combine to render it a town of pleasure, as well as a resort for health.

The train service between Paris and Aix is exceedingly good, as the express trains to Turin, Brindisi, and Rome stop there. There is one which leaves Paris at 7 p.m. and arrives at Aix at 5.33 a.m., and another which leaves Paris at the convenient hour of 8.55 a.m. and gets to Aix at 8.26 p.m.

Many have written of the pleasantness of Aix as a place of residence, and of the interest and beauty to be found in the numerous excursions into the surrounding country. The town itself is situated about a mile from the small but very picturesque *Lac du Bourget*, 823 feet above the level of the sea and 90 feet above the lake. It lies in a wide open valley surrounded to some extent, but by no means shut in, by the adjacent Alps of Savoy. The town is well built, with wide streets and shady avenues of trees, and gardens surrounding many of the hotels and private houses.

Hotels and boarding-houses and furnished apartments abound, the accommodation being, as usual, proportionate to the price paid for it. During the height of the season, from June to the end of September, living is somewhat expensive there, and good accommodation, including board and lodging and wine, cannot be obtained for much less than eighteen francs a day. Recently some hotels and villas have been built on the high ground overlooking the town and lake, and these have the advantage, during the hot season—and it can be very hot at Aix—of a fresher and cooler atmosphere. Steamers daily, and rowing-boats, take visitors to the different objects of interest on the lake; and by means of a canal connecting the lake with the Rhone, steamers are able to make the voyage from Aix to Lyons in a day, returning to Aix the day following.

Numerous interesting Roman remains are exhibited at Aix, of which local guide-books afford full information. There are numerous walks and drives, short and long, all full of interest and attractiveness, and many excursions by railway which will delight those who can spare the time to take them. The strong and healthy companions of invalids, who often find the time hang so heavily on their hands in other bathing stations, are at no loss for active occupation and amusement at Aix. The long but interesting excursion to *La Grande Chartreuse* is usually taken in two days, the night being spent either in the monastery or an adjacent inn. The very picturesque old town of Annecy and its lake are distant only an hour and twenty minutes by rail from Aix.

"After a long experience of travel, seldom have I seen any place combine so much to charm and interest as Annecy. The dignity of a Prefecture has led to the formation of delightful gardens on the shores of the lake, which is about twelve miles long by three miles wide, and is surrounded on three sides by Alpine snow-covered mountains. Thus Annecy adds to the loveliness of Como the grandeur of Lake Leman." *

These few introductory observations on the beauty of the neighbourhood of Aix, which might be easily extended to a length, which the objects of this work would scarcely justify, are however due to the almost unique attractions of the locality. I must now pass on to considerations more exclusively practical, and first with regard to the *climate* of Aix. There is really little precise information available on this head. It is said to enjoy a mild and equable climate, and to be protected from winds by the surrounding mountains, but I should doubt if the immediate surrounding hills are high enough or near enough to afford any very complete screen from prevailing winds. That it is often very hot in summer there can be no doubt ; whether its climate

* Lord Lamington in "Nineteenth Century," August, 1883.

is trustworthy as a good transition station in spring for those invalids who are returning North, as the South becomes too warm and relaxing, we have not yet sufficient detailed information to enable us to form a quite reliable judgment. It seems probable that the spring at Aix is as uncertain and very variable a season as it is for the most part elsewhere. Nervous subjects are said to find the climate soothing and productive of sleep, while those who find it too warm and relaxing in' the town can resort to the cooler and more bracing atmosphere of the surrounding hills.

It is calculated that between 24,000 and 25,000 persons went through the course of treatment at Aix in the season 1883, just double the number of the season 1873, showing that the popularity of this bath had doubled in ten years, a sufficiently forcible testimony to the esteem in which the treatment at Aix is held by the medical profession and the public.

The sources of Aix are sulphurous, and their characteristic ingredient is sulphuretted hydrogen. Their temperature ranges from 113 to 115 degrees Fahrenheit, and the quantity yielded daily is enormous—as much as 1,320,000 gallons and over.

With regard to the analysis of the water, I must here express my great regret that the published analyses of some of the most important springs in France present very great discrepancies. This is the case at Aix as well as elsewhere. It may be that these sources vary considerably in the amount of their solid constituents from time to time, but in that case it would be wise to say so, and to give the *average* composition.

I will here give the analysis of the Aix springs as contained in Dr. Brachet's excellent monograph, and also that in the work of Dr. Vintras, for the sake of comparison.

I may mention that the so-called "Alum" source is badly named, as it contains no "alum," and only an extremely minute quantity of aluminic sulphate, even less than in the other spring, the *Source de Soufre.*

	Source de Soufre.		Source d'Alun ou de St. Paul.	
	Vintras.	Brachet.	Vintras.	Brachet.
Calcic carbonate .	0·14850	0·1894	0·18100	0·1623
Magnesic ,, .	0·02587	0·0105	0·01980	0·0196
Ferric ,, .	0·00886	0·0010	0·00936	0·0008
Calcic sulphate .	0·01600	0·0928	0·01500	0·0810
Magnesic ,, .	0·03527	0·0735	0·03100	0·0493
Sodic ,, .	0·09602	0·0327	0·09240	0·0545
Aluminic ,, .	0·05480	0·0081	0·06200	0·0003
Sodic chloride .	0·00792	0·0300	0·01400	0·0274

Silica, calcic phosphate in minute quantities, and traces of lithium and iodine ; also some organic matter (*barègine*).

They also contain free gaseous sulphuretted hydrogen, nitrogen and carbonic-acid, and some sulphur in the form of hyposulphite.

The precise chemical composition of any mineral water is probably of far less importance than its mode of application, its temperature, and the particular mode in which its gaseous constituents are associated with it.

These hot mineral springs are utilised in an *Établissement Thermal* which is one of the most complete in Europe. It has four swimming baths, two of which are furnished with cold douches ; two " family " swimming baths, also with douches ; forty-one single baths ; twenty-five large douches with two *doucheurs* or *doucheuses* to each ; twenty douches with a single *doucheur* or *doucheuse;* two douches *en cercle*, three *à colonne;* six vapour baths (Berthollet) ; two inhaling-rooms; three for spraying; five vapouria; six *bouillons* (steam baths) ; four ascending douches; and four foot baths. With all these various bathing, douching, and other resources, we need not be surprised to hear that a maximum of from 2000 to 3000 patients have been treated daily.

The waters are administered internally as well as applied externally, but it is not usual to give them in large quantities, or to enforce them on persons with delicate digestions, who may find difficulty in tolerating them. Two to four glasses a day form the average quantity, and when stronger sulphur waters are indicated it is usual to prescribe the strong sulphur water of Challes. But it is to the mode of application, externally, of the hot sulphur springs, especially to the combination of douching and shampooing, that Aix chiefly owes its reputation. The patient sits on a wooden chair in a light and spacious room, his feet in hot water, and two *doucheurs* (or *doucheuses*), most of whom are extremely skilful, shampoo and apply *massage* to different parts of the body, over which jets of hot water, varying in temperature, are propelled.

"It is astonishing," says Dr. Grainger Stewart, who was himself submitted to the process, "with what skill, what patience, tenderness, and firmness the shampooing and passive movements are performed. When every joint has been moved to the utmost extent possible, the patient is made to stand, while from a distance a powerful stream of water is propelled upon the different limbs, especially about the articulations chiefly affected. When the bath is over, the patient is rapidly dried, wrapped in flannel sheets and blankets and is carried back to his hotel in the curious sedan-chair. Having reached his apartment he is lifted into bed, still swathed like a mummy, is covered up with additional blankets and a quilt, and left to perspire for a longer or shorter period. After twenty minutes or half-an-hour, he is carefully rubbed down by an attendant who had accompanied him to the bath."

Sometimes a *steam bath*, called the Berthollet, is applied. The patient "enters an apartment which contains a curious wooden box, with a round hole in its movable lid. After undressing, he steps into the wooden box and finds that he is shut in all except the head, the round hole being occupied by his neck. Immediately a valve on the level of the floor is opened, the hot vapour rises about him and he soon begins to perspire freely. The perspiration running from his brow, trickles down his face. Presently he feels the streams flowing down his sides and his legs, and very speedily a feeling of oppression and debility comes on, and after ten or twenty minutes the bath is opened up, the patient is carefully dried and removed to his hotel."

Then there is the *local vapour bath.* " By ingenious contrivances the bath-man is enabled to steam one arm or one leg. Speedily the limb begins to perspire and the parts become soft and comparatively flexible. Perspiration occurs all over the body, especially in those who have been undergoing other forms of treatment, and so great care requires to be taken to prevent a chill. When the parts have been thoroughly softened, manipulation, shampooing, and passive movement of joints are carefully carried out, just as after the douche, but only confined to one limb.

" On certain days the patient is sent to the spacious and comfortable *swimming baths,* and there he is allowed to disport himself for a longer or shorter time, practising, amid the somewhat warm water, active movements of the limbs. When his swim is ended he may have a cold douche or not, according to the direction of the doctor. He is rapidly dried, and, if well enough, is directed to walk about smartly in the gardens, which are close to the establishment."

Ordinary cold water is used when necessary to lower the temperature of the mineral springs; or Challes water is added if a very strong sulphur bath is desired.

Rooms are specially devoted to inhaling the sulphurous vapour of the water (*humage*), and also to the inhalation of the atomised thermal water (spray), which can be directed upon any part of the body, and is especially applied in affections of the nose, throat, face, and eyes.*

* For much of the information contained in this chapter I am indebted to the excellent monograph on " Aix les Bains," by Dr. Brachet.

Any time between the beginning of May and the end of September the treatment at Aix may be undertaken. July and August are often very hot months, but they are not found too hot for many rheumatic patients.

It is necessary, in the next place, to consider the action of these waters and the kind of diseases to the treatment and cure of which they are applicable.

It is easy to see that the object of such a method of treatment as has been just described must be to stimulate powerfully the action of the skin as an agent of elimination. It is as it were a purgation of the skin ! Whatever excrementitious matters are retained in the blood, or too slowly eliminated, that can be got rid of through the skin, must be discharged from the system by such active stimulation of the cutaneous surface.

We see also how, by the attraction of the blood to the surface of the body and its retention there in the dilated vessels, congestions of deeper parts must be relieved and a general stimulus given to the circulation of the nutritive fluids of the body ; so that indirectly a great stimulus is given to the healthy nutrition and normal tissue change. It is necessary, however, to bear in mind that usually the existence of a certain latent vigour and power of reaction is assumed in the application of such stimulating measures.

Indeed it is this point that must frequently exercise and test the judgment and discrimination of the physicians. If from age, or general exhaustion, or a naturally feeble constitution, this power of reaction is absent, then these modes of treatment only excite and exhaust, and a state of feverish debility is produced, and the patient is left worse, instead of better, for the treatment. So that it is occasionally necessary to resist the not unnatural desire of such patients for active treatment of this kind.

It is not uncommon for the temperature of the body to be raised two or three degrees during the douches, and the pulse thirty or forty beats in the minute; moreover, a slight degree of feverishness is often induced (thermal fever) which needs careful management, especially with respect to diet and exercise.

The period of treatment ordinarily lasts about twenty-five days, with a few days intermission for rest; but it may be, and often is, necessary to suspend the treatment for a time, so that invalids should allow themselves five or six weeks, and if everything goes well the last week may be advantageously spent at one or other of the more accessible mountain resorts in the neighbourhood. The Hôtel Mont Blanc, at St. Gervais, which has an altitude of 2600 feet, has been recommended, but this involves a carriage drive of seven hours, as well as a railway journey of an hour and a half; Glion (2400 feet) or Les Avants (3200 feet) are more easily reached, in a few hours, from Aix. Exercise in the open air is insisted upon; a moderate, careful diet, free from any excess, is enjoined; as are also early hours, especially in the damp evenings of the spring and autumn, and aperients are often required by gouty and rheumatic patients.

Among the cases best suited to treatment at Aix, chronic rheumatic and gouty affections take the foremost place. It is maintained by the local authorities, that by the absorption of sulphuretted hydrogen (internally or externally), the peculiar poison of rheumatism is eliminated by the skin and the kidneys; but this is a very hypothetical statement. The personal explanation of Dr. Grainger Stewart on this head deserves to be quoted:

The treatment of Aix is of extraordinary value in various rheumatic conditions. *First,* it is of great service in the way of removing the thickness and stiffness which so often remains after attacks of acute rheumatism—a stiffness due partly to changes within the joint, but mainly to thickening of the fibrous tissues around the articulation. *Second,* in cases of chronic rheumatism, where a slow, inflammatory action is going on in and around the joints, it suffices both to remove inflammatory products, and to diminish the tendency to rheumatic inflammation. *Third,* in rheumatic affections of the muscles, fascia, and nerve sheaths, it affords in many cases the most decided and speedy relief. *Fourth*, in the wasting of muscles, which so often occurs in connection with rheumatic processes, the manipulation and shampooing, along with the electrical stimu- lation which the doctors superadd, generally prove distinctly serviceable ; and *fifth*, on the occurrence of slight rheumatic threatenings, it appears that the use of the Berthollet, or vapour bath, often suffices to prevent the further development of the disease.

In that troublesome affection of the joints, which is called by some "rheumatic gout," by others "chronic rheumatic arthritis," and by others "arthritis deformans," Dr. Brachet maintains that the local application, twice daily, of the sulphurous vapours, together with two glasses of Challes water, to which iodide of potassium or lithia is added, proves of great service.

As to the treatment of gout at Aix it is admitted that, at the beginning of the course, an acute attack is sometimes induced, necessitating a suspension of the treatment until the severe pain has passed away ; but it is maintained that the ultimate result is to diminish the frequency and intensity of the attacks and to eliminate the gouty poison from the system. It is usual, in many of these gouty affections of the joints, to combine with the thermal treatment the application of the continuous electric current to the wasted muscles. Skin diseases of gouty origin, psoriasis and eczema, are especially adapted to the treatment at Aix.

Rheumatic forms of neuralgia and especially sciatica, whether of rheumatic gouty or syphilitic origin, are benefited by the combinations of bathing, douching, and shampooing as practised here singly, alternatingly or combined, as may be thought most appropriate to the particular case.

Chronic affections of the nose and throat are also sent to Aix for amelioration or cure. Chronic in- flammation of the mucous membrane of the nose with offensive odour and discharge (ozœna) are treated by nasal douches, the inhalation of the natural vapour of the hot sulphur springs, and by local applications of the aqueous spray; swimming baths are also enjoined as well as the internal use of Challes water.

Precisely the same method of treatment is applied with advantage to cases of chronic inflammation of the pharynx, granular pharyngitis, etc., often associated with a gouty and rheumatic as well as a scrofulous diathesis.

Chronic laryngeal catarrh, with hoarseness and loss of voice, and often some irritative cough, induced by over-fatigue of the larynx in public speaking and singing, or by excessive smoking, or by alcoholic drinks, is benefited by the inhalations of the atomised sulphur water as applied at Aix and in the *salles d'inhalation* at Marlioz. The climate of Aix and Marlioz is also believed to aid the treatment of these cases greatly.

Some forms of chronic bronchial catarrh are reputed to have been relieved by treatment by the inhalations, etc., at Aix and Marlioz, and the same is said with respect to hay-fever and certain types of phthisis.

As an aid to the effect of specific remedies in the treatment of constitutional syphilis, Aix les Bains has always enjoyed a considerable reputation, though Aix la Chapelle surpasses it in this respect; as to the power and influence of this treatment in *revealing latent* syphilis the greatest divergence of opinion exists, and the latest tendency is to a denial of this property. Scrofulous affections of the bones, joints, and glandular system, scrofulous ophthalmia, and even lupus are reported to be greatly benefited and even cured by the course at Aix.

Cases of chronic skin disease which travel all over Europe, from bath to bath, seeking relief, and, it must be added, often finding none or but very little, naturally visit Aix les Bains in considerable numbers. One of the most troublesome and inveterate of these is *psoriasis*, which is often of a gouty and rheumatic nature ; this disease is rarely completely cured, but they claim at Aix that under the influence of treatment there, the patches become fainter, the scales are shed, and the absorption and action of internal remedies are promoted.

In cases of eczema the first effect of the treatment is often to exaggerate somewhat the symptoms, but if the course is prolonged sufficiently the manifestations disappear.

In the relief of acne the swimming baths, together with the local application of the sulphur spring, prove very efficacious.

In hysteria, and in certain chronic uterine affections, in anæmia, and chlorosis, and even in cases of myxædema, treatment at Aix proves beneficial. In some of these cases removal from home influences, which are not infrequently injurious, change of air, scene, and food, the regular occupation of bathing and shampooing, and the enforced exercise in the open air, combined with judicious medical supervision, have probably more remedial influence than the *sulphur* in the water.

In certain cases of paralysis, or loss of motor power, the combined stimulating influence of bathing, shampooing, and the application of the electric current produces a beneficial effect.

Certain cases of traumatic disease of the bones and joints are usefully submitted to the treatment at Aix.

I have dealt thus fully with the subject of the cure at Aix les Bains, because it is one of great and increasing importance.

MARLIOZ

is but a quarter of an hour's walk from Aix, and may practically be regarded as belonging to it. Its springs, three in number, are *cold* sulphur springs, and they are used chiefly in the form of inhalation and pulverisation, i.e. in vapour and spray. At

Marlioz there are special arrangements in the shape of well-arranged *salles d'inhalation* for the application of the vapour and spray of these sulphur springs to the treatment of chronic affections of the respiratory passages, such as chronic coryza, chronic lanyngitis, and pharyngitis, and chronic bronchial catarrh. Patients following this treatment can either reside at Aix, or, if they prefer a quieter life, they can find comfortable apartments in the *Château* and *Villa* attached to the *Établissement* at Marlioz, which is situated in pleasant and extensive park-like grounds.

In the bath establishment at Marlioz the usual bath and douches are provided, and every appliance for hydrotherapeutic treatment either with the sulphur springs or with ordinary water. There are some little discrepancies in the published analysis of the Marlioz water. It, however, appears to contain sulphur in combination with sodium as sulphide of sodium, just like the Pyrenean springs, according to some authorities, while others have found sulphuretted hydrogen in it. Besides carbonate and chloride of sodium it is said to contain iodide and bromide of sodium.

CHALLES.

This important source is of comparatively recent discovery, and rises in a very picturesque district about three miles from Chambéry, and within an hour's journey of Aix les Bains. The Challes water is said to be the strongest sulphur water known in Europe, and it also contains iodides and bromides of sodium and potassium. It is on this account largely prescribed by the physicians at Aix as an important adjunct to the treatment pursued there ; it is also added to the baths there when it is thought desirable to increase the amount of sulphur in them. The Challes water is very rich in sulphide of calcium, and contains, it is said, as much as 0˙513 grammes of this compound (about 7 grains) in a litre, as well as a gramme of bicarbonate of soda. It is a cold spring, its temperature being 53 degrees Fahrenheit.

The amount of water yielded by the spring is limited to 3000 to 4000 litres a day, so that its bathing establishment is necessarily restricted; it is, however, very well fitted up, and contains inhalation and pulverisation rooms in addition to the ordinary baths and douches.

This water is especially valuable in scrofulous affections of the skin and other organs; in goitre, in chronic glandular enlargements, in chronic ulcers, in scrofulous disease of bones, in constitutional syphilis, and in chronic inflamation of the nose (ozœna) and throat, in tuberculosis of scrofulous persons, also in the treatment of chronic rheumatism.

For all these purposes the Challes water can be taken alone or in combination with other treatment at Aix; but those who prefer to reside at Challes can obtain good accommodation in the old Chateau there, which has been converted into a hotel. It has a picturesque situation and commands a fine view of the Alps of Dauphiné and the surrounding country; many interesting excursions can be made from it.

BRIDÈS-LES-BAINS AND SALINS MOUTIERS

May be considered together, for they are only two miles apart. They are situated in the Tarantaise, a somewhat remote part of Savoy, exceedingly picturesque and Swiss-like in character. These baths are rather difficult of access, or, no doubt, they would be much better known. Albertville, the nearest railway station, is about an hour and a half or two hours from Chambéry, and is a branch off the main line. From Albertville to Moutiers is a beautiful drive of about eighteen miles, and two miles from Moutiers is Brides-les-Bains. It is situated in a charming valley, 1870 feet above the sea, surrounded by mountains and glaciers, and with a fresh though mild climate. Many most interesting excursions can be made into the surrounding country.

Brides possesses a comfortable and well-managed *Établissement des Bains,* fitted up with the various descriptions of baths and douches now in general use in such resorts.

The season begins on the 15th of May and lasts till the 15th of October. June is the pleasantest month, as the air is fresh though mild, and one avoids the greater heat of July and August. There are several good hotels and *pensions.* The life is simple and quiet ; but a small casino, with musical and dramatic performances, contributes to the amusement of the visitors.

The mineralisation of the water is moderately strong ; in a litre there are 5·7200 grammes of solid constituents.

		Grammes.
Chloride of magnesium	. . . .	0·3071
„ sodium	. . . .	1·3601
„ potassium	. . . .	0·0670
„ lithium	. . . .	traces
Sulphate of soda	. . . .	1·6113
„ magnesia	. . . .	0·1941
„ lime	. . . .	1·8200
Bicarbonate of lime	. . . .	0·4380
„ protoxide of iron	. .	0·0112
Carbonic-acid gas	. . . .	0·0837

Its temperature is 95 degrees Fahrenheit. The average dose is from two to six glasses daily, according to the effect required. In small doses it is said to be tonic and to improve the digestion, and in larger doses it is gently aperient. It is also diuretic. This water is recommended in cases of sluggish liver and biliary obstructions, in abdominal plethora, and hemorrhoids ; in cases of dyspepsia, with obstinate constipation ; in malarial affections of the liver, spleen, and gastro-intestinal canal, e.g. chronic diarrhœa and dysentery, induced by residence in hot climates; in obesity, which it is said to remove without weakening the system; in renal calculi and catarrh of the bladder, especially in the gouty and rheumatic. Chronic uterine diseases are also treated advantageously here and at Salins.

Salins Moutiers lies between Brides and Moutiers, about two miles from the former. There is a regular service of omnibuses between the two villages, for the majority of visitors prefer living at Brides, where there is more society and amusement, where the valley is wider and more open and 300 feet higher. There are, however, two hotels and several *pensions* at Salins.

The source at Salins is a strong thermal salt spring impregnated with carbonic-acid, resembling the springs at Kreuznach and Nauheim. It has a temperature of 95 degrees Fahrenheit, and its flow is very abundant.

The following analysis shows that it is very richly mineralised.

In a litre = 1000 grammes:

	Grammes.
Chloride of sodium	10·22
„ magnesium	0·30
Sulphate of lime	2·40
„ soda	0·98
„ magnesia	0·52
Carbonate of iron	0·15
„ lime	0·75
Carbonic-acid	0·68

Traces of bromium, arsenic, iodine, and lithium.

The Salins water is chiefly used for bathing, but it is also given internally in small doses. The small bath establishment contains special appliances for diseases of women, and separate rooms for douches, etc.

It is particularly efficacious in the treatment of scrofulous affections of the glands, joints, and other organs. Feeble, sickly, scrofulous children are said to rapidly gain flesh and a rosy colour under this treatment. It is also specially applicable to diseases of the uterus and maladies peculiar to women. It is claimed for these waters that, while they strengthen and restore, they do not excite the nervous system.

ST. GERVAIS LES BAINS

Is situated amidst the grand and picturesque scenery in the immediate vicinity of Mont Blanc. It is usually approached from Sallenches, on the highway between Geneva and Chamouni, by a good *char* road of about six miles. It is picturesquely situated at the bottom of a wild rocky gorge (the gorge of Bon-nant), and the spacious *Établissement Thermal* fills up the whole width of the valley. The bath establishment is most commodious and complete ; it contains 100 chambers, a large library, and salons for conversation, dancing, and music. It is at an elevation of 2067 feet above the sea.

As might be supposed, in an attractive mountainous district like this the number of interesting excursions amongst the surrounding valleys and mountains are numerous, and of all varieties of length and difficulty. It is only a walk of four-and-a-half hours across the mountains to Chamouni. The ascension of Mont Joli is one of the most popular and attractive of the easy mountain excursions.

The village of St. Gervais is reached by either of three most agreeable paths, in about half-an-hour, from the baths. It is a more suitable place of residence during the heat of mid-summer than the baths, as it is somewhat higher (2680 feet above the sea) and more open ; and, moreover, very good hotel accommodation can be obtained there, especially at the Hôtel Mont Joli. Patients from Aix les Bains who are recommended to finish their cure by a week or ten days of mild mountain air at St. Gervais, should choose the village to reside in rather than the baths.

The waters at St. Gervais les Bains are both hot and sulphurous and cold and chalybeate. There are four principal hot springs, varying in temperature from 95 to 108 degrees Fahrenheit. They give off a considerable quantity of sulphuretted hydrogen and carbonic-acid gases.

They are employed internally and also externally in the various forms of baths and douches now usually found at all such establishments. They are adapted to the treatment of cases of chronic rheumatism, certain chronic skin diseases, and some forms of abdominal congestion.

The baths at St. Gervais are especially indicated in those cases in which it is desired to combine with bath treatment the tonic and soothing influences of a mild mountain climate.

EVIAN AND AMPHION.

The other two health resorts in Savoy which it is proposed to consider in this chapter are situated close together on the southern shore of the Lake of Geneva just opposite the port of Lausanne. The Evian water enjoyed a considerable reputation as a pleasant and wholesome table water, in the immediate neighbourhood, long before it obtained its present vogue as a remedial means.

The agreeable situation of the little town, just in the centre of the attractions of Lake Leman, its mild and genial climate, and its ready accessibility, by rail and steamboat, both from the North and from the South, have no doubt contributed greatly to its popularity.

Connected by means of a new railway with Aix les Bains, it is considered by some physicians an excellent plan to supplement the course at Aix with a course of twelve or fifteen days at Evian. Certainly, in the months of July and August, the borders of the Lake of Geneva must be fresher and cooler than the valley of Aix; and the pure, pleasant water of Evian an agreeable change from the sulphurous beverage of the former place.

It is said that the air of Evian and its vicinity is peculiarly favourable to those delicate, anæmic, hypersensitive young children who do not prosper at the sea-side. It is assuredly not uncommon to meet with young feeble children, who, if sent to the sea-side for a tonic change, become bilious, languid, and irritable, lose their appetite, and become weaker instead of stronger. There is something in the air of the sea-coast which seems to irritate instead of brace them, and it would seem that such children obtain great benefit from the more mildly tonic air of Evian and recover there their strength, their vivacity, and their colour!

Evian possesses five springs of nearly identical composition, the two chief are La Source Cachet and La Source Bonne-vie. They give their names to two distinct bath establishments.

These springs are very feebly mineralised, less so than ordinary drinking water; they contain, however, some amount of oxygen, nitrogen, and carbonic-acid gases. It has been suggested that these springs act much in the same manner as distilled water would, and by their great purity and the absence of solid mineral substances, their solvent and purifying properties are greatly increased; while, by their fresh and pleasant taste, they are rendered much more digestible and more readily absorbed than distilled water would be. Very large quantities of the Evian water can be consumed and absorbed daily, if some little care is taken to accustom the stomach to its use, "from four to twenty-five glasses a day"! When taken as a solvent of so insoluble a substance as uric-acid, it is considered that the possibilities of absorbing such large quantities of this very feebly mineralised water is altogether an advantage, because it possesses the important property of saturating itself with the salts encountered in its passage through the organism.

This water may be regarded, then, as washing the blood and the tissues, and removing from them deleterious excrementitious substances difficult of solution; hence its value in the presence of the uric-acid or gouty diathesis.

The temperature of the Evian springs is about 53 degrees Fahrenheit, the water is soft and agreeable to the skin in the baths.

The published analyses of the Evian sources vary somewhat, but the following may be regarded as a fairly accurate approximation. In a litre (1000 grammes):

	Grammes.
Bicarbonate of potash	0·0037
„ soda	0·0134
„ magnesia	0·1227
„ lime	0·2787
Chloride of sodium	0·0024
Sulphate of magnesia	0·0028
Phosphate of soda	traces
Free carbonic-acid	0·03672

It will be seen by the above that it is a slightly alkaline and very feebly mineralised water. With regard to its properties, it is said to be remarkably diuretic, especially before the function of the skin has been stimulated by the employment of baths and douches. In the next place it promotes appetite, and in large doses it excites the processes of retrograde metamorphosis to such a degree that, notwithstanding the consumption of an increased quantity of food, many persons become much thinner.

One of its chief properties, as a curative agent, is that of augmenting the urinary excretions, thus rendering it a valuable remedy in cases of gravel (uric-acid deposits), renal calculi, etc. "It cleanses the urinary passages throughout their whole course without fatiguing them." Its use is also credited with producing a calming influence in cases associated with nervous and hypersensitive constitutions.

The quantity taken varies from two to twenty-five glasses a day, according to the case and the object in view. It is usual to begin with small doses, and to take at least two-thirds of the total quantity between getting up and half-an-hour before the *déjeûner à la fourchette*. The dose of water must not be incautiously increased, but it must be first observed if it passes away freely by the kidneys, and if it is absorbed without any trouble or inconvenience. If it either purges or constipates some medical treatment should be had recourse to, in order to modify this anomaly.

The baths generally have a soothing, sedative effect, relieving the pains of chronic cystitis and nephritis, and restoring sleep to irritable nervous subjects. They, however, not infrequently excite a return of sub-acute gouty attacks in those who are prone to them, and in some highly neurotic excitable patients they occasionally irritate and exhaust, instead of calm and strengthen. The alkalinity of the baths is of use in some other affections, especially the lichenous and pruritic cases.

It is usual to drink one or two glasses of the water while in the baths.

Injections of the water while in the baths are useful in some uterine affections.

With respect to the maladies especially suited to treatment at Evian, the various forms of chronic dyspepsia must first be mentioned; and the cases of nervous and gouty dyspeptics, with tendency to great discomfort from flatulent distension, are treated in a special manner here by means of what is called the "dyspeptic douche." This douche is applied over the abdomen and stomach, and is accompanied by skilful massage. This method is said to yield remarkably good results, even in very inveterate cases. The douche is applied twice a day, together with massage, and from five to twelve glasses of the water are drunk.

In the next place must be mentioned renal and bladder affections, and amongst these certain forms of chronic albuminuria are said to be greatly benefited by treatment at Evian, especially the form due to catarrhal desquamative nephritis, and also the cases of parenchymatous nephritis. Speaking of the use of Evian water in this latter affection, Professor Noël Gueneau de Massy says: "This is the most easily digested water known, and it is voided easily on account of its eminently diuretic properties. It carries with it, without fatiguing the kidneys, all the epithelial and other débris which in case of inflammation encumbers the renal filter ; I generally find it successful."

The cure at Evian is also recommended in cases of chronic pyelitis, especially when it has been induced by lithiasis and preceded by the passage of renal calculi.

But it is in cases of uric-acid deposits and excess of urates in the urine, betraying the existence of a gouty condition, that the water of Evian is considered especially beneficial, and it is more particularly indicated in those individuals who are the subjects of asthenic gout, and who require gentle methods of treatment. It is desirable to begin with small doses of water, three to four glasses a day, which may be increased slowly up to twenty.

Cases of renal colic, due to the presence of urinary concretions (gravel and calculi), are fit subjects for the Evian course, and are often remarkably benefited thereby.

Cases of vesical catarrh not due to any organic cause but the result of chill, or of some temporary local or constitutional morbid condition, are said to be cured at Evian.

It is further claimed for the course of treatment and the régime pursued at Evian, that it is very beneficial and calming to persons of hypersensitive nervous organisation, who are suffering from over-fatigue of the nervous system, sleeplessness, loss of appetite, etc.

The ordinary hydro-therapeutic treatment can be carried out in well-appointed establishments for the purpose at Evian. The life at Evian is, during the season, gay and cheerful, and a variety of amusements, balls, comedies, operettas, and concerts, are provided daily at the Casino. There is a great variety of attractive excursions to be taken on the lake or into the surrounding country.

AMPHION

Is within a twenty minutes' walk of Evian. Its situation on the border of the lake is very pictu-resque, far more so than that of Evian. The bath establishment is situated in its own grounds, which extend down to the lake, and it has a cheerful and yet a quiet and retired aspect. It is not one large building, but consists of three detached resi-dences, and is specially suited to persons who need repose and prefer retirement. There is a landing-place and jetty for steamers and boats close at hand. The hotel and restaurant arrangements are excellent, and are under the management of the proprietor of the *Hôtel Beau-Rivage* at Cannes. It lies at the foot of a hill which protects it, in the heat of summer, from the southern sun, and as the neighbourhood is well wooded the air is usually fresher and cooler than at Evian.

There are good arrangements for baths and douches, but on a smaller scale than at Evian.

Amphion possesses three mildly alkaline springs almost identical in composition with those at Evian, and therefore it is appropriate to the treatment of the same cases as are sent to the larger spa. Amphion, however, boasts of the possession of a spring contain-ing iron, and the use of this alkaline-chalybeate spring is recommended in cases in which anæmia and chlorosis play a predominant *rôle*.

For invalids requiring a quiet, picturesque, and cheerful summer residence in mildly bracing air, away from the excitement and gaiety of a fashion-able spa, and yet with such a resort and its resources within ready access, Amphion presents exceptional attractions.

Having passed in review the chief bathing resorts in Savoy, it only remains for me to notice, in this chapter, the principal baths in the adjacent province of Dauphiné; the most renowned of these is

URIAGE.

The present popularity of Uriage is of comparatively recent date, although from the results of excavations and the monuments thus brought to light, it is certain that its springs were known to the Romans and utilised by them.

The little town of Uriage, 1358 feet above the sea, is situated between seven and eight miles from Grenoble (station *Giers-Uriage*, and then forty minutes' drive by omnibus), in a green, fresh, and pleasant valley surrounded by wooded slopes, amidst grand and picturesque mountain scenery at the eastern extremity of the Alps of Dauphiné. The town is composed chiefly of handsome and convenient buildings for the accommodation of the visitors.

The *Cercle*, or Subscription Room, is especially remarkable for the elegance and comfort of the accommodation it affords.

The *Établissement des Bains* is one of the finest and most complete in Europe : it is built in the form of a square, and within the square are the chief hotels of the place. There is also a châlet belonging to the *Établissement*, where private apartments can be had. There are also numerous furnished villas and *pensions*. Daily concerts, theatrical performances, balls twice a week, etc., are provided for the amusement of the guests.

Most interesting excursions of various lengths can be made into the mountainous country around. The summit of the Montagne des Quatre Seigneurs (3094 feet) can be reached in an hour and a half, and commands a fine view of the surrounding country. Another favourite excursion is to the

Cascade de l'Oursière, nearly four hours from Uriage, and another is to the ruins of the Chartreuse de Prémiol, which takes about two hours.

The ascent of the Belle-donne is a more ambitious undertaking, but the view from the summit is said to be remarkably fine.

The Château d'Uriage possesses a good gallery of pictures and a museum of antiquities, and there is a beautiful view from its terrace.

The adjacent hamlet of *Vaulnaveys* contains several inns and lodging-houses, where the poorer class of patients, who come to Uriage for treatment, are usually to be found.

The bathing establishment contains numerous bath rooms, several apartments for the various kinds of douches and vapour baths, as well as inhaling rooms for breathing the vapour and spray of the water.

The drinking fountain is under a covered gallery, where the drinkers can take exercise in bad weather.

The spring, to the possession of which Uriage owes its reputation, is a combined sulphur and salt spring, a rare and valuable combination, blending, as it were, the properties of a sulphur bath with those of a sea bath.

There is another spring containing iron, which is regarded as of quite secondary importance ; still it is no doubt an advantage to possess also a chalybeate water for cases which require the administration of iron.

The saline and sulphur spring has a natural temperature of 80 degrees Fahrenheit. It is perfectly clear at its source, but becomes turbid, from deposited sulphur, when exposed to the air. It has a sulphurous odour and taste.

Besides a notable quantity of sulphuretted hydrogen, each litre (1000 grammes) contains from six to seven grammes of chloride of sodium (common salt).

The following is the analysis of this spring given by Lefort, in 1000 grammes:

	Grammes.
Chloride of sodium	6·0569
„ potassium	0·4008
„ lithium	0·0078
Sulphate of soda	1·1875
„ magnesia	0·6048
„ lime	1·5205
Bicarbonate of soda	0·0555
Arseniate of soda	0·0021
Silica	0·0790
Sulphuretted hydrogen	0·0113
Free carbonic-acid	0·0062

Taken internally in moderate doses, two or three glasses a day, it generally has a gentle aperient action; in larger doses it is purgative. In small doses it is said to be alterative, and to stimulate the functions of digestion. The baths are tonic and restorative. If they induce at first some nervous excitement, this usually passes off after a few days.

The system of combining douching and massage, as practised at Aix, is also in vogue at Uriage.

In the spray (pulverisation) and inhalation rooms, the local influences of sulphuretted hydrogen gas and the pulverised sulphur water on the air passages and the skin is especially aimed at.

With regard to the cases suitable for treatment at Uriage, all forms of scrofulous affections must first be mentioned. Scrofulous diseases of the eyelids, of the throat (pharyngitis granulosa), scrofulous eruptions of the face and head (eczema, acne, etc.), are treated by the vapour and spray applied locally.

Scrofulous affections of the glands, bones, and joints are greatly benefited here.

It is said that delicate infants with scrofulous taint do exceedingly well at Uriage.

Most varieties of chronic skin disease are considered appropriate for treatment here, but those of an eczematous character derive the greatest benefit.

The baths and waters are useful in certain forms of chronic rheumatism; they are useful also in menstrual irregularities, and as an active auxiliary to mercury in the treatment of constitutional (" tertiary ") syphilis.

The waters of Uriage possess, therefore, a wide range of application and usefulness; and their important properties, combined with the mild but tonic climate, and the picturesque and interesting country around, cannot fail to increase the popularity of the place as it becomes more widely known.

ALLEVARD.

The baths of Allevard lie about midway between Chambéry and Grenoble, being twenty-three miles from the former and twenty-five from the latter town. It is about an hour and a half's drive from Goncelin, a station on the line of railway connecting these two towns.

The village of Allevard is built on both banks of the river Breda, at an elevation of 1550 feet above the sea. It owes much of its popularity, no doubt, to its beautiful scenery and mild climate, and to the many admirable excursions and points of view in its neighbourhood. One drawback, however, is the prevalence of *goître* and cretinism amongst the native inhabitants. It is necessary to remember that there is often a rapid fall of temperature immediately after sunset.

The bath establishment is surrounded by a large park, and is provided with all the usual appliances for the utilisation, in the form of baths, douches, vapour, and springs, of the sulphur spring for which it is celebrated. The natural temperature of the water, 75 degrees Fahrenheit, is not high enough for the baths, and it has to be raised artificially by a method which avoids any alteration in its chemical properties.

As at some of the Pyrenean spas, gargling is an important process at Allevard, and a new hall is specially appropriated to this operation.

The water is rich in sulphuretted hydrogen, and contains also some chloride of sodium, sulphates of soda, magnesia and lime, as well as carbonates of lime and magnesia ; but its solid constituents are small in amount, and its chief ingredient the gaseous sulphuretted hydrogen.

Like other sulphur springs it is chiefly employed in diseases of the skin and respiratory organs, and especially in the pulverised form in chronic affections of the pharynx and larynx, loss of voice, etc., it is also used in cases of bronchial catarrh and asthma, and it has been said to arrest the progress of phthisis. Externally, in baths and douches, it is used in rheumatic affections.

In conclusion I must allude very briefly to

LA MOTTE,

A small bath situated about twenty miles from Grenoble, and possessing a saline spring, containing chlorides and bromides, and said to have a composition almost identical with that of sea-water.

The bath establishment is contained in the old château, and the price of each room and apartment is affixed inside. It is situated at an elevation of about 1800 feet above the sea in a valley enclosed by mountains. There is a dense shady wood in the immediate neighbourhood, with numerous paths affording admirable protection to promenaders during the heat of the summer. The spring from which the water is obtained rises at some distance below the château, and the water has therefore to be pumped up some 200 yards, and as it loses heat in the process (its temperature at the source is 144 degrees Fahrenheit), it is again heated before it is employed in the baths.

The *Établissement* is provided with the usual baths and douches, with a vaporarium, *salles d'aspiration*, etc.

The cases treated at La Motte are all forms of

scrofula, chronic rheumatism, old dislocations and fractures, spinal-caries, and chronic inflammation of liver and stomach.

Speaking of La Motte, Dr. Leudet observes (*Compte Rendu de la Session*, 1883–84, *de la Société d'Hydrologie Médicale de Paris*): "Grave nervous phenomena, dependent on spinal-sclerosis, due to the combined action of syphilis and arthritism, are arrested in their development and obtain from this bromo-chloride treatment, alternating with the exhibition of iodides, a real and lasting benefit, which they could not obtain from the use of specifics alone."

CHAPTER XIV.

THERE are many French health resorts, some of them scarcely less important than those that have taken a prominent place in the preceding chapters, which do not admit of being treated of in definite groups; these will now be briefly reviewed without making any attempt at an artificial classification.

Contrexeville, a small and pretty village in the Department of the Vosges, pleasantly situated in a valley open to the north and south, at an elevation of about 1000 feet above the sea, is in the heart of the Vosges mountains. It is one of the most important of French spas, and is frequented by a great number of patients during the summer months, from May to the end of September. Its climate is somewhat variable, and the temperature in the mornings and evenings is often considerably lower than during the day, so that delicate visitors require the protection of warm clothing.

The Contrexeville waters have long been celebrated for the treatment especially of diseases of the urinary passages, of uric acid and other forms of gravel, of the gouty constitution, and of some forms of hepatic disease.

It has five chief springs, four of these, the *Pavillon, Prince, Quai,* and *Souveraine,* are used on the spot, and the water of the fifth is used for exportation. They are all cold. The Pavillon spring may be taken as a type of all of these. Their chief constituents are sulphate and carbonate of lime, and, in smaller proportions, of sulphates of soda and magnesia. A small quantity of lithium also appears in the analysis of this spring, which is as follows. In a litre (1000 grammes) :

		Grammes.
Sulphate of lime .	. . . .	1·165
„ soda	. . . .	0·236
„ magnesia	. . .	0·030
Bicarbonate of lime .	. . . .	0·402
„ magnesia	. . .	0·035
iron	. . . .	0·007
„ lithium	. . .	0·004
Silica	. . . .	0·015
Chloride of potassium	. . .	0·006
„ sodium	. . . .	0·004

Traces of fluoride of calcium and of arsenic.

The *Source du Prince* contains more iron and arsenic, and is, therefore, more tonic. The *Source du Souveraine* contains more magnesia and is more laxative. There is enough free carbonic-acid in the water to make it sparkling and not disagreeable to drink. The effect of this water when drunk in considerable quantity is to largely increase the flow of urine ; it is also said to act as a gentle laxative, to increase the flow of bile, to improve the appetite, to favour perspiration, and to quicken the circulation.

It is reported to cause an increase in the urinary secretion to an extent far beyond that of the amount of fluid consumed. This statement probably requires qualification, and is rather an instance of *excess of zeal* on the part of the advocates of Contrexeville; for it would hardly be an advantageous thing to give, for a long time, a remedy which abstracted water from the blood in great excess of the quantity supplied to that fluid !

These waters dissolve the catarrhal mucus which accumulates in the urinary passages in chronic cystitis and allied affections, and carry it away together with the small solid urinary concretions that are apt to be found associated with these cases. If the urine shows an excess of acid it rapidly removes this excess, and if, owing to decomposing changes undergone by the urine before it is voided, it happens to be alkaline, these waters, by bringing the urinary mucous membrane into a healthier state, and so checking ammoniacal decomposition, restore to the urine its normal acid reaction.

There appears to be a general agreement of authorities as to the value and efficacy of these waters in the treatment of calculous affections, especially of renal colic, of chronic cystitis and irritable bladder with enlarged prostate. They are also said to have a tonic effect.

The waters are used externally as well as internally, in the form of baths and douches ; the application of the douche to the region of the kidneys being regarded as of special value.

The Contrexeville springs are considered to afford a very efficacious and *gentle* mode of treating atonic forms of gout ; cases in which the use of stronger and more highly mineralised waters is inadmissible.

Cases of congestion and torpor of the liver with a tendency to inspissation of the biliary secretions and the formation of biliary calculi, are reported as deriving great benefit from treatment there. Troublesome cases of nocturnal incontinence of urine have been cured by these waters. Gouty *diabetic* patients derive much benefit from treatment at Contrexeville, as indeed they do at Vichy, Neuenahr, and many other alkaline spas.

The cure at Contrexeville is no doubt aided by the calm and quiet life amidst pleasing mountain scenery, where many agreeable and interesting walks or drives can be made.

The hotels and *pensions* are excellent, and the *Établissement des Bains* is fitted with all necessary appliances. There is a pretty park for promenades, and a Casino with the amusements and resources usually provided at French watering-places.

Plombières is a neat little town situated in a picturesque, deep, narrow valley, 1321 feet above the sea, in the Vosges mountains; it is reached by a branch line from Port d'Atelier near Vesoul on the Eastern of France Railway. The springs are hot springs, and vary in temperature from 59 to 160 degrees Fahrenheit. It has been called the French Teplitz, as it is the custom here, as at Teplitz, to give the baths at a high temperature and for long periods. The mineralisation of these springs is absolutely insignificant; indeed, they are almost like rain water. In 1000 grammes of the water there is about ¼ gramme of solid constituents, the chief of which are sulphate of soda and carbonate of lime.

Notwithstanding its extremely feeble mineralisation, it is said to effect most wonderful cures, both from its internal as well as its external application. Probably much is due to the manner of employing the waters, to the mode of life insisted upon, and to the healthful surroundings of the place.

Plombières possesses no less than six hydropathic establishments, where its springs are utilised in the form of baths and douches; and hot-air baths and *massage* are also employed. There are also four *buvettes* for drinking at. There is also a hospital for the reception of patients, both civil and military. Moreover, it possesses an iron spring which is only used for drinking.

These waters are said to be "slightly stimulating, diuretic, aperient, and tonic!" They are said, when taken internally, to cure obstinate forms of gastrointestinal catarrh, especially when associated with chronic diarrhœa; but the hot baths have probably the chief share in these good results.

Plombières is very largely resorted to as a "ladies' bath," and claims to cure sterility when dependent on functional atony of the generative organs. The baths are prescribed also for nervous headaches, and various forms of neuralgia, as well as for spinal irritation and feebleness. Certain forms of chronic rheumatism and gout are benefited by treatment here.

Plombières is largely resorted to in summer, and is a pleasant and attractive place of residence, somewhat hot in the daytime, but fresh and cool in the mornings and evenings. There are many charming promenades and excursions in the neighbourhood.

Vittel in the Vosges is only three miles from Contrexeville, and its renown as a health resort is of quite recent growth. It is situated in a pleasant wooded valley 1100 feet above the sea.

The mineral springs are at some distance from the village, in a beautiful situation surrounded by parklike grounds and commanding a fine view of the adjacent portion of the Vosges chain of mountains.

It has four principal springs of a temperature of about 52 degrees Fahrenheit; the *Grande Source* resembles in composition the *Pavillon* at Contrexeville, its chief constituents being the sulphates of lime, magnesia, and soda; it is, however, somewhat richer in the laxative sulphates of magnesia and soda; the *Source Marie* is stronger in laxative salts, and the *Source Salée* is even still stronger, and therefore more aperient. The fourth spring, the *Source des Demoiselles*, contains iron and gives traces of iodine. This is a tonic, and is used in cases of anæmia, chlorosis, and functional uterine disorders.

The other springs are employed in the same cases as the Contrexeville waters, viz. cases of uric-acid diathesis, gravel, chronic cystitis, and vesical catarrh; also, on account of their greater tendency to purge, in cases of congestion of the liver and gastro-intestinal catarrh, associated with constipation, or in persons who have become too corpulent.

The bath establishment is elegantly fitted up, and has a covered gallery for exercise in bad weather.

The life at Vittel is simple but cheerful, and the excursions around are numerous and interesting.

Bourbonne les Bains (Haute Morne) is situated near the Vosges mountains, of which it commands a fine view, and is approached by a branch line from Vichy on the Eastern of France Railway. The little town is built on a hill between 800 and 900 feet above the sea.

It possesses several hot springs, which vary in temperature from 120 to 150 degrees Fahrenheit. It has a large military as well as a civil hospital, maintained by the State. It has been termed the French Wiesbaden, as its waters resemble those of that spa in their uses, composition, and temperature. They are weak, common salt springs, and contain free carbonic-acid, which constantly bubbles up from the surface of the hot springs.

The *Établissement Thermal* consists of two distinct buildings, one for first and the other for second-class baths. The waters are taken internally and used externally, in the form of baths, douches, fomentations, injections, and for vapour baths.

The following is the analysis of these common salt springs, which, though common in Germany, are rare in France.

In 1000 grammes (1 litre) :

	Grammes.
Chloride of sodium	5·800
„ magnesium	0·400
Carbonate of lime	0·100
Sulphate of lime	0·880
„ potash	0·300
Bromide of sodium	0·065
Silicate of soda	0·120
Alumina	0·130
Peroxide of iron	0·003
Manganous manganic oxide	0·002
Traces of iodine and arsenic.	

The baths and douches (in combination with electrical treatment) have been found very efficacious in the treatment of chronic rheumatism and some forms of paralysis. They are also valuable in the various manifestations of scrofula, diseases of the bones, joints, and glandular system, and are especially useful in gun-shot fractures and other slowly-healing wounds. They have been found useful in the cutaneous affections of the gouty and the syphilitic. Some forms of dyspepsia are also treated here.

The Casino provides dancing and other amusements, and the picturesque neighbourhood affords many pleasant excursions.

Martigny les Bains, in the Vosges mountains, is situated in a valley between Contrexeville and Bourbonne les Bains, surrounded by wooded and vine-clad hills. It has a station on the Eastern of France Railway.

Its reputation as a bath is of quite recent growth ; it has cold mineral springs, of which the chief ingredient is sulphate of lime, thus resembling those of Contrexeville and Vittel. But its claim to distinction is that its waters possess a considerable amount of lithium—three centigrammes per litre of chloride of lithium—and that they are, therefore, of special value in the treatment of the uric-acid diathesis and the gouty affections dependent thereon. Besides cases of uric-acid gravel, they are used advantageously in chronic cystitis, diabetes, enlargement of the prostate, hepatic congestions, chronic gastric catarrh, and some forms of rheumatism and neuralgia. They are chiefly employed for internal uses, but baths and douches are also given.

Bains les Bains is also in the Vosges, and is situated between Contrexeville and Plombières. It has a station on the Eastern of France Railway, and is not far from Epinal. It has the advantage of being adjacent to the fine forest of Tremonsey, and is pleasantly situated in a valley, 1000 feet above the sea, through which the River Sémouse flows.

It has several springs and two bath establishments. In the public baths the two sexes, appropriately clothed, bathe together. The temperature of these springs varies from 84 to 122 degrees Fahrenheit.

They belong to the class of indifferent thermal baths, their mineralisation being very feeble. Two of the springs are used for drinking, but the chief use of these waters is for hot baths. Like other baths of this kind they are used and found beneficial in cases of chronic rheumatism, muscular and arthritic, in chronic uterine affections, and in cases of want of tone and exhaustion with enfeebled digestion. Residence in agreeable mountain scenery and in fresh, pure air contributes to the good results obtained there.

Luxeuil is situated in a plain 1300 feet above the sea, at the foot of the Vosges mountains, and has a station on the Eastern of France Railway. Its hot springs, eighteen in number, resemble those of Plombières in their composition and uses. Luxeuil has a special reputation as a ladies' bath, and its *Établissement* is on that account fitted up with great elegance and comfort.

As many as 300 persons can bathe at the same time in its three large public baths or *piscines*, in which the water is continuously renewed.

Arrangements are provided for the application of the various forms of douches now in use, and for local and general vapour baths, inhalations, etc.

The mineralisation of these springs is very feeble. The chief solid constituent is chloride of sodium, of which there are about 10 grains to a litre of the water.

Luxeuil also possesses several chalybeate springs, some of which contain a notable proportion of peroxide of iron.

These waters are employed in cases of functional disturbance of the female organs of generation; in sterility, uterine congestions, and displacements; in nervous exhaustion and hysteria; in anæmia and chlorosis; and in some forms of rheumatism and neuralgia. The dyspepsia, which is a common associate of these disorders, is also relieved by drinking one or other of these springs.

The surrounding neighbourhood abounds in picturesque walks and excursions, and the Casino provides a variety of amusements for the visitors.

Néris is a well-known hot bath resembling in its uses the baths of Teplitz and Baden-Baden, in Germany. It is situated in Central France, in a rather hot valley, about 800 feet above the sea, not far from Montlucon, in the department of Allier, 207 miles from Paris. It is reached by a drive of three miles from the Chamblet station of the Orleans Railway.

It has two bath establishments, the larger of the two being one of the finest and most complete in France.

Besides baths and douches of all kinds, special provision is made for *massage*, and for inhalations, and all the various resources of hydropathic treatment. There is also a hospital for the reception of poor patients. The temperature of the springs varies from 115 to 126 degrees Fahrenheit. Their mineralisation is feeble, but they contain a considerable amount of free nitrogen and carbonic-acid gases.

From the following analysis it will be seen that the chief solid constituents of these springs are the bicarbonates of soda and lime, sulphate of soda, and chloride of sodium.

In a litre (1000 grammes) :

		Grammes.
Bicarbonate of soda	0·4169
,, potash	0·0129
magnesia	0·0067
,, lime	0·1455
,, iron	0·0042
Sulphate of soda	0·3896
Chloride of sodium	0·1788
Silica	0·1121

These waters are chiefly employed in baths, and are found efficacious in some diseases of the nervous system, especially certain paralytic and convulsive affections, chorea, tremblings, convulsions, neuralgias,

sciatica, hysteria, and insomnia. Cases of chronic,
gouty, and rheumatic arthritis ; of chronic uterine inflammation in nervous subjects ; and some cutaneous
affections (eczema, acne, lichen, etc.) are treated with
advantage at Néris. Life at Néris is calm, quiet,
soothing, and economical. The hotels and *pensions*
are good, and there are pleasant walks in the surrounding country.

Bourbon l'Archambaut, 210 miles from Paris, is
about ten miles by carriage from Souvigny station, on
the Orleans Railway. It is situated in a valley, in
the department of the Allier, between four steep hills,
at an elevation of eight or nine hundred feet above
the sea. The Bourbon waters have a very ancient
reputation, and were known and employed at the time
of the Roman occupation. Louis XIV. often came
to these baths, and Madame de Montespan passed
the last twelve years of her life at Bourbon, and died
there.

The most important spring is the *Source Chaude,*
which has a temperature of 126 degrees Fahrenheit.
The other spring, the *Source Jonas,* is not so hot.

The Bath Establishment is of a somewhat modest
appearance compared with some similar establishments elsewhere. It contains some small *piscines,*
each furnished with a douche, rooms for private baths
and douches, and especially *eye* douches, as well as
for hot-air baths and for pulverisation. Patients are
conveyed, as at Mont Dore, from their beds to the
baths in a sort of sedan-chair, and back again.

There is a civil and military hospital for poor
patients, and it is said that many who arrive on
crutches are soon enabled to depart without them.

The chief constituent of these hot springs is
chloride of sodium, of which there are about thirty-
eight grains in a litre of water ; bicarbonate of lime,
magnesia, and soda are next in amount. The
presence of bromine and iodine in these waters has
also been established. Taken internally and as

baths they act as diuretics and diaphoretics, promoting elimination by the skin and kidneys. The baths are given at a temperature of 82 to 88 degrees Fahrenheit; at higher temperatures they prove too exciting. They are found to be of great efficacy in many complaints, especially when associated with scrofulous or lymphatic constitutions; in inflammation of the joints and glands, in torpid ulcers and fistulas, in scrofulous disease of the bones, in scrofulous ulcerative ophthalmia; in the latter the water is applied by a special apparatus for instillation, as well as in douches.

Chronic rheumatism is especially benefited by the baths, and so also are those forms of paralysis which are dependent on rheumatism or hysteria. It is rather a dull place, but the Casino affords the usual amusements.

Bourbon-Lancy, in Central France, 226 miles from Paris (Lyons terminus), is reached from Gilly station in an hour and a half's carriage drive. It is built on the eastern slope of a hill, with steep granite rocks rising above it to the north and west. The hot springs which arise at the base of these rocks have been known for many centuries.

Fine views of the mountainous country around can be obtained from the summits of the adjacent hills.

There are two hospitals there and a fine *Établissement Thermal*, with admirable arrangements for baths, douches, etc.

The springs, of which there are seven, are chloride of sodium springs; six are hot and one is cold. They contain a considerable amount of pure carbonic-acid and a small amount of iron. Traces of iodine and of arsenic have also been found in them. Their mineralisation is feeble. There is about eighteen grains of common salt to a litre of the water. The temperature of the principal spring is nearly 130 degrees Fahrenheit.

Taken internally these waters increase the secretion of the mucous membrane, and in large doses are laxative; the iron they contain gives them, however, a tonic property.

They are employed in chlorosis and anæmia, with disordered uterine functions and in sterility. They are useful in scrofulous affections of the bones and joints, and in promoting the healing of torpid ulcers and wounds received in battle.

They are also said to be of value in cases of flatulent dyspepsia, especially when associated with uterine irritation. In chronic rheumatic thickenings and stiffness of joints, in some paralytic states, and in certain forms of neuralgia, the hot baths appear to be of much efficacy.

There are many interesting excursions to be made in the neighbourhood.

Saint Honoré is remarkable as the only hot sulphur bath in central France. It is about 190 miles from Paris, and four miles by carriage-road from the station of Vaudenesse (Lyons terminus).

It is situated at the foot of the mountains of Morvan, at an elevation of 920 feet above the sea, between two wooded hills, on one of which stands the *Château de la Montagne*. The springs, of which there are five, are considered to resemble those of Eaux Bonnes in their composition and action, but they are less exciting.

The mineralisation is extremely feeble; they contain however free sulphuretted hydrogen and carbonic-acid gases, and a very small amount of alkaline sulphides. They have a distinct odour of sulphuretted hydrogen.

These baths were known to the Romans. The present *Établissement Thermal* is very complete in all its appointments, including a large public swimming bath, in which the water is continuously renewed, and rooms for pulverisation and inhalation. In the bath the skin becomes covered with little beads of gas,.

which, as we have pointed out in other chapters, produces a decidedly stimulating effect on that organ.

As at Eaux Bonnes the chief therapeutic use made of these waters is in the treatment of affections of the respiratory organs, especially chronic pharyngeal, laryngeal, and bronchial catarrhs. Bronchial asthma is said also to be benefited at St. Honoré.

The earlier stages of pulmonary consumption are benefited, it is said, by the inhalations here. Scrofulous and lymphatic affections are also treated here, and functional uterine disturbances in neurotic subjects are much benefited. Moist skin eruptions are relieved by this treatment.

The surrounding country is interesting and presents many objects for agreeable excursions, and the life at St. Honoré itself is cheerful and pleasant.

Pougues les Eaux is about five hours from Paris on the Lyons Railway, near the banks of the Loire, and only a few miles from the interesting town of Nevers. *Pougues* is recovering the popularity it once possessed some three centuries ago, when it was much resorted to by some of the chief persons in France. In the reading-room of the Casino the walls are covered with portraits of some of its ancient patrons.

There are two springs, both of which are cold— the *Source Saint Leger* and the *Source Bert*. They are used chiefly for drinking. The *Établissement* has however twenty-six baths, and all the appliances for a complete hydrotherapic course.

These waters are slightly alkaline, and contain the bicarbonates of lime, magnesia, potash, soda, and iron in small quantities. They are considered to be especially suitable to cases in which the waters of Vichy prove too powerful. In dyspepsia in delicate nervous persons, in the gouty diathesis, with uric-acid deposits in the urine, in diabetes, and in some forms of chronic vesical catarrh and renal inflammation. The presence of an appreciable amount of iron in

these waters makes them valuable in the treatment of
the chlorotic anæmic states associated with menstrual
irregularities.

Although Pougues is a small station, the life there
is bright and cheerful.

Vals in *Ardèche*, between Lyons and Marseilles,
an hour's drive from Aubenas station, is one of the
most important spas in France, and were it not for its
distance from the capital (434 miles), it would be as
much frequented by foreigners as Vichy is. Its waters,
as exported, are of world-wide reputation, and its
popularity as a watering-place has grown greatly of
late years. The little town is picturesquely situated
on the right bank of the river Volane, and is surrounded by an amphitheatre of volcanic mountains.
It consists of little else than a long street of hotels
(twenty or more in number), cafés, and *maisons
meublées.* Vals is surrounded by most interesting
country, rich in picturesque sites of wild and savage
beauty.

The springs, of which there are fourteen principal
ones, are cold, and contain large but varying proportions of bicarbonate of soda and large quantities of
free carbonic-acid.

Vals possesses two bath establishments. The
larger of the two possesses the following sources :
Rigolette, Précieuse, Madeleine, Désirée, Saint Jean,
and *Dominique;* the smaller one has the sources
*Marquise, Souveraine, Pauline, Chloé, des Convalescents,
Saint Louis,* and *Constantine.* All these contain considerable but varying quantities of bicarbonate of
soda, except the *Dominique,* in the larger establishment, and the *Saint Louis,* in the smaller one ; these
two are ferruginous and arsenical springs.

Owing to the large amount of carbonic-acid in
these springs, even the strongest of them are not unpleasant to drink.

We may take the *Source Constantine* as a representative of the stronger springs, containing, as it

does, over 7 grammes of bicarbonate of soda per litre; and the *Source Pauline* as a representative of the weaker springs, as it contains only about 1½ grammes of bicarbonate of soda per litre; and we have, between these, several degrees of mineralisation, as, for instance, the *Source Désirée*, with 6 grammes per litre; the *Source Rigolette*, with rather more than 5 grammes; the *Source Souveraine*, with 2½ grammes; and the *Source Chloé*, with 2 grammes.

The following is the detailed analysis of the Sources Constantine and Pauline (per litre = 1000 grammes):

	Constantine.	Pauline.
Bicarbonate of soda . . .	7·05300	1·6117
„ potash . .	0·07100	traces
„ lithia . . .	0·01075	traces
„ lime . . .	0·43700	0·0288
„ magnesia . .	traces	0·0083
Carbonate of iron . . .	0·00670	0·0090
Silicate of alumina, potash and soda	0·15900	0·1824
Chloride of sodium . . .	0·28000	0·0414
Sulphate of soda . . .	0·20400	0·1696
Phosphate of soda . . .	traces	traces
Organic matter . . .	traces	traces
Free carbonic-acid gas . .	2·10000	1·0820

The Vals waters are richer in bicarbonate of soda than any known mineral waters, and the graduation in quantity presented by the different sources is of great convenience in their therapeutic application.

The following is the analysis of the *St. Louis* ferruginous and arsenical spring; the *Dominique* is somewhat stronger.

Free sulphuric-acid	0·09960
Arseniates	0·00100
Silicate of iron	0·00810
„ alumina	0·04540
„ manganese	traces
„ lime	0·01780
„ soda	0·01850

Sulphate of protoxide of iron . .	0·02822
„ sesquioxide . . .	0·01165
„ lime	0·03200
„ potash	0·04797
„ soda	0·11250
„ magnesia	traces
Chloride of calcium	traces
Phosphate of lime	traces
Iodine	traces
Organic matter	traces

The Vals waters have a large range of applicability. They are useful in the various forms of acid dyspepsia and gastric catarrh, in hepatic congestion, jaundice, and gall-stones ; in diabetes, in gravel and renal and vesical calculi, and in prostatic enlargement and irritability of the bladder ; in some forms of chronic gout and rheumatism, in splenic enlargements as a sequel of malarial fevers, and in certain anæmic and chlorotic conditions associated with functional uterine and gastric disturbances.

If it were not for the long journey, Vals would, as I have already said, be attractive to invalids on account of the pleasant life there and the beautiful drives and walks which the surrounding country affords.

La Malou (Hérault), a watering-place somewhat remote and inconvenient of access, which has recently been advocated on account of the utility of its waters in certain chronic affections of the nervous system, especially in locomotor ataxy and other diseases of the spinal cord. It is in the southernmost part of France, and is reached *viâ* Tarascon, Montpellier, Fangères, and Bedarieux (nineteen and a half hours' railway journey from Paris) ; there is then about an hour's drive to La Malou. It is agreeably situated in a valley about 600 feet above the sea, surrounded by mountains, the sides of which are covered by vineyards and chestnut-trees.

The several springs, which vary in temperature from 61 to 97 degrees Fahrenheit, are of pretty nearly

the same composition, and contain alkaline carbonates
and iron, together with much free carbonic-acid, as
the following analysis shows. In a litre :

		Grammes.
Bicarbonate of soda	. . . .	0·7711
,, potash	. . .	0·1242
Carbonate of lime	. . . . .	0·4528
,, manganese	. . .	0·1863
Peroxide of iron	. . . . .	0·0251
Chloride of sodium	. . . .	0·0187
Silica	. . . .	0·0638
Alumina	. . . .	0·0302

There are three well-appointed bath establishments.
These waters are taken internally as well as employed
in the form of baths, douches, and inhalations.
Besides those affections of the spinal cord mentioned
above, cases of chronic rheumatism, especially cases
of *arthritis deformans*, are benefited by treatment
there ; also the various forms of neuralgia and neurotic
hyperæsthesia. Cases of anæmia and chlorosis associated with chronic disturbance of the uterine functions,
and cases of sexual inability, are considered suitable
subjects for treatment at La Malou.

The baths are considered to be remarkably soothing
as well as strengthening.

Enghien, only thirty minutes by train north of
Paris, has cold sulphur springs in a somewhat damp
situation. It is a prettily built suburb, having a
small lake and tastefully arranged pleasure-grounds.
It is a more highly sulphuretted water than some of
those found in the Pyrenees, but it has the disadvantage of being cold, and it contains no *barègine*.

A good bath establishment provides the usual
conveniences for baths, douches, inhalations, etc.
The springs which supply them are numerous, and
have all nearly the same composition. Sulphate
and carbonate of lime are the chief solid constituents,
and they contain free carbonic and sulphuretted
hydrogen gases The Enghien waters are employed

chiefly in cases of skin disease and in affections of
the respiratory organs ; of the former, herpetic affections associated with rheumatic tendencies are the
most benefited ; and of the latter, cases of chronic
granular pharyngitis, laryngitis, bronchitis, and asthma
in the rheumatic and herpetic. Cases of early phthisis
in scrofulous subjects are said to be benefited at
Enghien.

As may be imagined, Enghien being a suburb of
Paris, life there is characterised by much gaiety and
brightness ; concerts and balls in the Casino, regattas
on the lake, etc., etc.

Pierrefonds, eight and a half miles' drive from
Compiègne station and sixty from Paris (Northern
line), lies at the foot of a hill surmounted by a castle
and near a small lake. Its mineral springs, which are of
recent date, resemble those of Enghien in composition,
and contain free sulphuretted hydrogen. These
waters are used in the same cases as other sulphur
springs. Treatment by the application of pulverised
water (introduced here by Dr. Sales-Girons) is
especially popular ; and a special room in the
well-appointed bath establishment is devoted to
pulverisation!

Pierrefonds possesses also an iron spring in which
the iron exists in combination with arsenic. The
cases treated there are those of chronic pharyngitis,
laryngitis, bronchitis, asthma, and phthisis ; also
herpetic skin affections, rheumatism, and functional
uterine disorders associated with anæmia and chlorosis.

The neighbourhood of Pierrefonds offers many
attractive walks and drives, and it has the advantage
of the adjacency of the magnificent forest of Compiègne. Life is pleasant and agreeable there, as a
variety of amusements are constantly provided,
theatrical, musical, etc.

Saint Amand, celebrated for its mud baths, is a
few miles from Valenciennes and about 160 from

Paris, with a railway station (Gare du Nord). It has the advantage of being near a large forest. It has a well-organised *Établissement Thermal*, about a mile from St. Amand, accommodating about 100 patients; in one part of the building the "mud" baths are given, while the other part is devoted to private baths and douches. The springs, of which there are four, are moderately warm (70 degrees), and have a sulphurous smell and deposit a considerable quantity of *glairine*. They contain small amounts of the alkaline and earthy sulphates and carbonates. They are drunk and taken in the form of baths and douches.

The mineral mud, which is collected on the spot, is saturated with sulphurous water, and has a temperature of 77 degrees Fahrenheit.

Patients are immersed in this black mud from two to five hours at a time in compartments so arranged that they can read, write, and feed in the bath. After the mud bath the patient is conveyed to the ordinary baths and washed.

The cases benefited by treatment at St. Amand are those of chronic rheumatism of the muscles and joints; some forms of paralysis; chronic diseases of the bones and joints; sprains, and gunshot wounds, some forms of uterine disease (hypertrophy and ulceration); obstinate sciatica and dry cutaneous eruptions, as psoriasis, lichen, and lepra.

Life is calm and serious at St. Amand, as the patients are mostly confirmed invalids.

CHAPTER XV.

SOME RHENISH HEALTH RESORTS.

VACATION STUDIES OF SOME OF THE PRINCIPAL HEALTH RESORTS IN THE RHINE DISTRICT.

" Monsieur! Monsieur ! à six heures du matin vous prendrez à la Pauline trois verres ! Trois verres à la Pauline !! A dix heures vous prendrez un bain, en sortant du bain vous prendrez . . . . encore deux verres, et à cinq heures du soir, Monsieur, vous prendrez . . . . encore trois verres ! Monsieur ! ces eaux vous feront beaucoup de bien."—*Bubbles from the Brunnen of Nassau.*

BETWEEN Bonn and Mainz, on both sides of the Rhine, lies a district richer in important mineral springs than probably any other part of Europe of similar extent. The names of some of these are very familiar to the ears of English doctors and English patients, and it may seem that there can be little that is new to be said of any of them. But medical science and, therefore, medical practice undergo perpetual changes, and the remedy of one generation is no longer a remedy in the next, or, if it still holds its place, it is most likely as a cure fór some quite different malady. Cures come into fashion and fall out of fashion, and the utility and applicability of various health resorts vary in popular and, indeed, in professional estimation, as knowledge and experience of their effects grow and widen. It will not therefore be unprofitable labour to take a brief survey of even some of the best known and most familiar health resorts, and note in what respects they have changed in reputation and in what respects they remain as they were. Many considerations, too, of great public interest apply with especial force now to the more popular and frequented of health resorts. I allude especially to their sanitary condition and to the provisions which obtain in them for the prevention of the origin and spread of epidemic and contagious diseases. The methods also still in vogue in the use and application of the stronger, and indeed of all mineral waters, are probably in many respects faulty, too much a matter of mere routine and too much the result of uninterrupted precedent. A little disinterested scrutiny and friendly criticism may not be without good fruit. On the other hand, some quite novel and useful modes of application of mineral springs and other curative agencies in connection with some of the Rhenish health resorts cannot fail to be of interest to the many seekers after health which our ailing humanity always includes. It is now many years ago since Sir Francis Head, coming to Schwalbach for the benefit of his own health, wrote those charming sketches of life at some of the Rhenish spas, which he published under the title of " Bubbles from the Brunnen of Nassau." The taste for long digressions and interpolated legends which form so large a part of that volume has probably completely passed away, still many will not fail to find much that is attractive in the pleasant and graceful style of those "literary bubbles." In the sketches which I have attempted in this chapter of some of the same resorts as those described by Sir Francis Head, the reader will, I fear, miss the literary charm of that gifted writer, but, on the other hand, I have endeavoured to provide a detailed professional estimate of their respective merits.

NEUENAHR.

In journeying up the Rhine the first health resort of any note that we come to is Neuenahr. It is situated in the valley of the Ahr, a small river, which joins the Rhine at Sinzig, a little above Remagen, and about midway between Bonn and Coblenz. It used to be necessary to descend from the train or disembark from the steamer at Remagen in order to visit Neuenahr. Remagen is an hour and a half's railway journey from Cologne, and now a branch line extends to Neuenahr.

If the weather should be clear no one will regret having to spend an evening at Remagen, and witnessing a sunset, from the adjacent Apollinarisberg, over this exquisite piece of Rhine scenery. The Apollinarisberg is crowned by an elegant Gothic church, with four towers, the Apollinaris Kirche, and it is from the platform on which the church is built that a view is obtained over the most exquisite bits of river scenery that this beautiful stream affords. Seen by the setting sun as the shadows of evening creep over river and mountain, the silvery surface of the stream and the contrasting deep, sombre tints which cover the mountains give a majestic grandeur and beauty to the scene which is almost inconceivable to those who have only seen the river from the deck of a Rhine steamer. It is true that this is one of the finest sunset views on the Rhine. It takes in the banks of the river from Hönningen to Königswinter. The bend of the river incloses in its curve a fertile tract, which forms the foreground, while mountains close in the landscape on all sides, and in one direction the Siebengebirge, with the Drachenfels, forms a prominent and striking feature in the landscape. No one who has ever witnessed a sunset from the Apollinaris Kirche at Remagen will ever again speak slightingly of Rhine scenery.

The Ahr valley from Remagen to Neuenahr is uninteresting. The valley is wide and bounded by low and gently sloping hills ; only one considerable hill, the Landskron, is remarkable, as it stands alone, a truncated cone of basaltic rock about 900 feet high, and forms a very prominent object. Near this rock we pass the Apollinaris Brunnen, the source of the Apollinaris water so familiar now to the British palate. The thousands of glass and stone bottles piled up over a great extent of ground indicate a commercial activity in connection with this spring altogether unparalleled. The scenery improves as we approach Neuenahr, which is built on both sides of the river at the foot of a high wooded hill, rising for nearly 1000 feet above the village, and commanding a magnificent view. But the really picturesque part of the Ahrthal, and that which forms the principal excursion for the visitors at the baths, commences about two miles from Neuenahr, at the ancient little town of Ahrweiler, completely surrounded by its old walls, in excellent preservation, and with a fine gate tower at each end of the town. From this tower to the ancient village of Altenahr, where the beauty of the scenery culminates, the valley for nearly seven miles is picturesque in the extreme ; the river winds with many a sinuous curve along its tortuous bed, while precipitous rocks hem it in on each side, save where here and there the valley widens and the swift stream spreads out into a broad and shallow river. The steep black slate rocks on the left bank, with their southern exposure, are covered from base to summit with terraces, over which extend, all along the valley, vineyards kept with the greatest care, the vines yielding a generous red wine which is much valued.

At Altenahr the scenery is extremely wild and beautiful. Enormous masses of black jagged rocks, most picturesque in form and situation, stand boldly up from the floor of the valley. On one of these stand the ruins of the Castle of Altenahr—an ancient fortress dating from the tenth century—dominating this narrow part of the Ahrthal. A visit to the Castle of Altenahr enters into the programme of every visitor to Neuenahr. It is within an hour and a half's drive; there is a good hotel at the end, and the scenery quite repays the trouble. But the serious business at Neuenahr is to bathe in and drink the waters and get rid of whatever ailment brings one there; and it is a serious object which brings most people to Neuenahr, as there is little or nothing else there to attract the idle and aimless. There is a comfortable Kurhaus in pleasant and extensive grounds, with shady walks and agreeable spots to lounge in. The food is good and the life simple and placid. The baths are in the Kurhaus, and are well organised, and furnished with the usual array of douches of all kinds to stir and stimulate our sluggish organs. The springs at Neuenahr belong to the group of mild alkaline waters. They are warm and contain a considerable quantity of carbonic-acid. This gas is allowed to escape freely into the air from the principal spring, which is thus seen to seethe and boil and foam as it pours from its source into a basin, around which a well-shaped inclosure is built. Here the carbonic-acid can be seen to form a dense and dangerous atmosphere over and around the spring—a dangerous one indeed, for on one occasion a few years ago a little child while at play descended into the inclosure to get a ball it had lost, as it did not return another followed it, and then a third. They were all three removed in a state of insensibility, and one, the first who descended, did not recover. There are probably great quantities of this gas stored up, a little below the surface, in this part of the valley, and it is noticed in consequence that mice are never found in many of the houses. There are four springs in use—the chief of these are the Victoria and the Grosser Sprudel. The temperature of the latter is 104 degrees Fahrenheit; the others are not so hot. A pint of the water contains about eight grains of bicarbonate of soda and two-and-a-half of magnesia, besides lime and small quantities of common salt and sulphate of soda and an appreciable amount of iron—appreciable I mean to the palate, although there is only about one-twenty-fifth of a grain to a pint.

The following is the analysis of this water:

Bicarbonate of soda.	. . .	8·20 grains.
„ magnesia.	. .	2·50 „
„ lime.	. .	2·40 „
„ protoxide of iron	.	0.04 „
Chloride of sodium .	. . .	0·70 „
Carbonic-acid .	. . . .	17 cubic inches.

The springs contain rather less carbonate of soda than those of Ems, and only a quarter part of those of Vichy, while they contain practically no chloride of sodium, whereas there are seven grains in Ems water and four grains in Vichy.

The uses of these springs are various, depending on their mildly solvent alkaline action, so that they are especially applicable to cases of biliary and urinary concretions and their consequences; to cases of atonic gout, and especially those cases where it is important to avoid the stronger alkaline and saline springs of Homburg, Tarasp, Carlsbad, and Vichy. Cases of gout which require very gentle treatment and which support badly the depletory effects of stronger mineral waters—cases, for example, associated with a weak heart and feeble circulation—these are well suited to treatment at Neuenahr.

Cases of chronic rheumatism are treated here, and notably that which produces great deformity and crippling of the joints (*arthritis deformans*). The excellent resident physician, Dr. Richard Schmitz, assures me that he has obtained remarkably good results from the use of the Grosser Sprudel followed by shampooing, in some of these distressing deformities. Of chest diseases, cases of chronic bronchial catarrh are much benefited, so are certain cases of chronic Bright's disease. Forms of dyspepsia with obesity, or fatty liver, and associated with a feeble heart, are better treated here than by the stronger and more depressing courses of Marienbad or Carlsbad. But it is, above all things, to the successful treatment of diabetes that Neuenahr owes its great and increasing reputation. The physician whose name I have just quoted has made a careful and special study of this disease at Neuenahr, and the results he has published of the effects of treatment there are very remarkable and encouraging. Some of these diabetic patients begin by drinking enormous quantities of the water, which they gladly do to quench the intolerable thirst from which they suffer. A patient beginning with nine glasses of the Grosser Sprudel daily gradually increased it to five glasses before breakfast in the morning, six glasses at ten a.m., and five more at five p.m. This did not, however, quench his thirst, for he drank during the night nearly a quart of cold Sprudel! In thirty-eight days this patient had entirely lost the characteristic sign of diabetes.

Dr. Schmitz is careful to insist on a strict observance of a correct dietary, although he by no means shares the prejudices commonly encountered in German spas as to certain articles of diet. He makes some of his patients eat salad three times a-day; and he laughs at the prohibition of butter, which, as he says, could with the alkaline water only make a little soap, and is far less objectionable than the greasy *ragoûts* one is so constantly presented with at the *tables d'hôte.* I was glad to find that this experienced physician was strongly impressed with the folly of making feeble and delicate invalids rise at very early hours in the morning to drink quantities of cold or even warm mineral water. He had observed, as most careful and unprejudiced physicians must have noticed, how badly these early draughts of water, as well as bodily exercise at these early hours, are borne by many feeble and delicate persons, and he had known many instances of great impairment of health produced by adherence to this routine.

Upon inquiry into the sanitary state of Neuenahr, I was assured that it was very satisfactory. Dr. Schmitz had only known of one case of typhoid fever during a residence of eighteen years. Yet there is no drainage, properly so called ; the sewage matters are allowed to accumulate in cesspools and these are emptied between the hours of eleven p.m. and four a.m., and their contents buried in the earth. There is no doubt that many of the wells from which the poorer inhabitants draw their supplies of drinking water are largely contaminated with sewage matter, and the absence of typhoid proves that sewage matter of itself is not capable of generating typhoid unless it contains the germs of typhoid derived from some infected organism.

Finally, with regard to the climate of Neuenahr, it is said to be very dry and healthy, and to have a very even temperature ; but from my own observation I should say that, in common with much of the Rhine district, it is subject at times to heavy morning mists. There is generally a current of air blowing along the valley, "but from the position of the town, with protection on the north by vine-covered hills and on the south by wooded heights, it is usually warm. The tortuous rocky defile below Altenahr closes the valley on the west, and there are no side valleys by which eddies and cold currents may be produced. During the spring and summer the wind is usually from the south-west or south, and coming over great woodland districts is pure and refreshing, but never cold." I am told that wild boars abound in the adjacent forests and afford good sport, and that not long ago three of these animals bounded through the garden of the Kurhaus, much to the astonishment and alarm of the visitors.

EMS.

The next considerable spa in proceeding up the Rhine is the well-known Bad Ems, twelve miles from Coblenz, in the beautiful valley of the Lahn. Ems has been called the "pearl of German baths," and there are many reasons for calling it so. Its natural situation is very beautiful, and art has been liberally applied to aid nature in its embellishment. The society assembled there is often of the most select description. Princes and ambassadors, kings and emperors, walk about among humbler people whom the common infirmities of humanity bring, if not to the same level, at any rate to the same Source. There is a considerable solemnity, however, about the very gaiety of Ems, which indicates a consciousness that imperial and princely shadows have long and often rested on its pathways; none of the loud, noisy gaiety such as you find, for example, at a bath like Luchon, in the Pyrenees.

Ems extends for a considerable distance along the right bank of the river Lahn. Along this bank, at the lower part, is the old Dorf Ems, where most of the poorer inhabitants dwell, while the upper part consists of handsome hotels and shops, and the Kursaal and the Kurhaus and the springs are all on that side of the river, together with some prettily laid-out, shady, park-like walks. On the left side of the river there are also bath-houses, and many fine villas and hotels, a Russian and English church, etc. for the most part protected from the sun by the shade of the surrounding woods among which they are built, and many shady walks and drives extend up the wooded moun-

tain side on this left bank of the river. On both sides
of the river rise high hills; those on the left are
clothed with trees to their summits, and afford many
a pleasant shady walk or drive; those on the right
are steeper, barer, and more rugged, but with many
wooded and vine-clad patches.

The pedestrian can find many attractive walks,
with fine points of view, varying from half-an-hour to
an hour and a half's distance from his hotel. One of
these, and the most distant, about three miles from
Ems, mounting by a zigzag road through shady
woods, is called by the attractive name of the Schöne
Aussicht; but it is very disappointing, and presents
a much less pleasing view than many of the less
distant heights. It is the highest point (1300 feet) in
the neighbourhood, and for that reason there is little
that is picturesque in the view, which is, however,
very extensive, stretching to a great distance over a
high tract of land of somewhat monotonous and
nearly uniform elevation. It is true a small and
distant bend of the Rhine can be seen in one direction and the Taunus hills in another; but, although
extensive, the view is exceedingly tame. It is one of
the most striking instances of the loss of the picturesqueness of surrounding scenery by looking down
on it from a considerable elevation. Above the upper
part of Ems, on the right bank of the river, rises
somewhat abruptly a high, sharp-crested, jagged hill,
on the top of which there is a tower, the Concordia
Thurm, which commands a fine view over the town
and the river. The interesting old town of Nassau is
also a beautiful drive of four miles up the valley. But
there is no lack of good excursions from Ems, either
for riding, driving, or walking, while on both sides of
the river there are an abundance of shady walks and
lounges. A fine covered walk in the centre of the
public gardens, near the Kursaal, affording shelter
both from sun and rain, was erected in 1874, at the
special request of the Emperor William, and is a real
boon to the place. An historical monument, of which

Ems is proud, is a small white stone let into the ground near where the band plays in the morning, with the simple inscription: "15 Juli, 1870, 9 Uhr Morgens," and which marks the spot where King William stood when he caused his memorable answer to be given to the French ambassador, Benedetti. The people of Ems are very fond of their old emperor, and many stories are current there of the friendliness and simplicity of his manners. On one occasion he invited all the doctors of the place, twelve in number, to meet him. He received them sitting on the edge of a table, and talked familiarly with them for some time, contriving to show that he knew something of each of them individually, and on bidding them adieu he said: "Now, gentlemen, you need not know me when we meet again, for I am a poor man and cannot afford a new hat frequently." Ems used also to be a favourite resort of his nephew the late czar, who was in the habit of coming here in great state, bringing with him a long retinue of servants and courtiers, and forty horses, while his imperial uncle was content with four. Corruption, the natives assert, was rampant in the Russian train. Diamond rings and snuff-boxes were always distributed largely at the end of the czar's stay, and were bought back again at reduced prices to do similar service elsewhere.

Ems has the reputation of being hot and relaxing, and no doubt it can be very hot at Ems during the height of summer; but the early mornings and the evenings are cool, especially in the months of May and September, which are probably far the pleasantest months of the season, although by no means the most popular. Great pains have been taken, by cultivating trees in the public gardens, to afford as much shade as possible, and to provide cool retreats from the midday heat, while the abundance of water available enables the roads and the trees to be watered three times daily.

Ems is abundantly supplied with mineral springs; there are five principal ones, all having nearly the same chemical composition and differing only· in temperature. The hottest is the Kessel Brunnen, 120 degrees Fahrenheit; then the Fuersten Brunnen, 102 degrees; the Augusta Quelle, 101 degrees; the Kraenchen, 90 degrees; and the coolest, the Victoria Quelle, 80 degrees. Those of the higher temperature are best suited to a certain class of cases, those of the lower temperature to others. The Ems springs belong to the medium alkaline group, containing ten grains of bicarbonate of soda to a pint, while they also contain a moderate quantity of common salt— seven grains to the pint. They also contain carbonic-acid in moderate amount, and they are warm. Of all the German waters containing a considerable amount of both carbonate of soda and common salt, Ems is the only one that is warm. It is this that characterises its springs, and to a great extent determines their application and utility. They also contain lime and magnesia, and a small amount of lithia.

Ems is a ladies' bath *par excellence*, and is indicated when soothing rather than bracing treatment is required. Of the other uses of its waters, the chief are chronic catarrhs of the throat and air passages, and for the treatment of those affections they are drunk, used in the form of inhalations, and as gargles, and also they are bathed in. It is applied in certain forms of gout, especially those that require gentle rather than active treatment, or that are associated with bronchial catarrh or urinary calculi. In some forms of dyspepsia, with congestion of the liver, or inflammation of the bile ducts or chronic diarrhœa, the Ems waters often do good. Those chronic affections of the joints resulting from attacks of rheumatic fever are said to do well here. It used to be the fashion to send consumptive patients to Ems, but *nous avons changé tout cela* with a vengeance. The warm, relaxing air and the hot springs of Ems are now regarded as absolutely prejudicial in case of phthisis, and it is certain that attacks of hæmoptysis have frequently been induced by drinking the hot springs here. The climate of Ems is mild and relaxing, and from the beginning of July to the middle of August the midday heat is often very great. The town is, however, reported to be very healthy. It is abundantly supplied with water and well drained, the outflow of sewage being into the river some distance below the town.

Each house, however, is "furnished with well-cemented and air-tight cesspools," the upper part of which is connected by a pipe with the general system of drainage, but the solid sewage is allowed to accumulate, and is removed during the winter and used as manure. One of the great recommendations of Ems is its exceeding accessibility and the good hotel accommodation provided there.

Tradition still governs the method of drinking the waters at Ems, and the invalids, young and old, active or feeble, begin to consume their daily allowance between six and eight a.m. The dose is repeated in the afternoon between four and six. The early morning hour is, no doubt, a good time for taking warm solvent water in many cases, but it does not admit of doubt that many feeble and delicate constitutions would do better to remain in their beds until eight or nine o'clock, take a light breakfast on rising, and drink their first dose of water at eleven.

HOMBURG.

The manifold virtues of common salt give to the group of health resorts, which includes Homburg, Nauheim, Soden, etc., their reputation, and to their mineral sources their curative effects. The springs contain many other constituents besides chloride of sodium, but it is the predominating and characteristic constituent in all of them. For example, in the strongest of the springs at Homburg—the Élizabethen Brunnen—out of the one hundred and two grains of solid constituents which are contained· in a pound of the water there are seventy-five grains of common salt ; and in the weakest—the Louisen Brunnen—out of thirty-two grains there are twenty-three of common salt. The strongest of the springs at Nauheim contains, out of three hundred and twelve grains of solid constituents, two hundred and sixty-five grains of common salt ; and the weakest one hundred and nine grains out of one hundred and thirty-seven. At Soden, also, the chief constituent in all the springs is common salt in varying proportions. These springs, then, are all salt springs ; their uses depend very much on the relative proportions of common salt in each of them.

But the possession of springs containing common salt is scarcely sufficient to account for the extraordinary popularity of Homburg, for Nauheim has springs which contain much more. Nor is it exactly accurate to say, as one of the local authorities does, that "art and nature have combined to make of Homburg and its environs an earthly paradise." If, with the touch of a magician's wand, we could remove everything from Homburg that art has placed there—the cultivated trees and shrubs, the ornamental park-like grounds, the pretty villas, the magnificent Kursaal and its surroundings—the natural situation of Homburg could scarcely be called picturesque. The grounds to the east of the town have been very prettily laid out, trees have been freely planted, and shady avenues abound. The lawn-tennis ground especially is very prettily placed ; and the visitors are received into lodging-houses of an attractive and superior kind. But nearly all this is the work of art rather than nature. An exceedingly ugly, uniformly flat stretch of country lies between Homburg and Frankfort. This plain rises gradually till it reaches the foot of the Taunus mountains, and Homburg lies in the upper part of this plain, at an elevation of 350 feet above Frankfort.

It is true that on the east side of the town the ground is undulating ; the wells are situated in a depression, and some low hills covered with woods afford many shady walks and drives close to the town, while less than an hour's walk brings one upon the beautifully wooded slopes of the Taunus mountains; but the site of Homburg itself is not picturesque, and if a visitor never went beyond its ugly main street, with its rough pavement, he could not fail to carry away a most unfavourable impression of it. But almost everything that art could do has been done to make the surrounding of the springs attractive ; and pleasant shady avenues and covered walks, palm-houses and parterres of flowers, and other attractions give to the neighbourhood of the mineral springs a very pleasing aspect. Certainly if one is doomed to drink physic every morning for three or four weeks, it is an excellent idea to get it bubbling fresh and clear and cool out of the ground in a pretty park, handed to you by a comely maiden, in the company of well-dressed people in the fresh air and early hours of a bright summer morning, to the strains of sweet music with the enlivenment of cheerful conversation, and, perhaps, not without a little flirtation to fill up the intervals between the doses of one's medicinal restorative. A local advocate of the virtues and attractions of Homburg approaches this part of the subject very boldly and writes : "Are there not hopeful sons and blooming daughters to be married and provided for, and have not many of them found eligible partners in the watering-places ?" And he adds : "The writer himself knows several such couples who were struck by Cupid's arrow at the Stahl and Louisen Brunnen, and Hymen afterwards had to take pity on them !" There can be no manner of doubt that the attractiveness and the popularity of Homburg are quite as much dependent upon social considerations as upon medicinal ones. Something must be said in favour of the climate of Homburg. It is unusually dry and bracing for a place of such moderate elevation —650 feet above the sea. This is due to the absence of streams and rivers and to the absorbent nature of the soil, and to the position of Homburg on a raised portion of a wide plain, the neighbouring mountains being sufficiently distant to keep the mists and clouds which frequently settle over them from influencing its atmosphere.

Owing to its somewhat exposed situation it suffers from the heating effect of the direct rays of the sun, but, as a compensation, cooling currents of air blow down from the distant forest-clad hills, and the mornings and evenings are fresh and exhilarating. There is a very general consent among those who have frequented the place for many seasons that " Homburg is bracing."

Homburg has several mineral springs ; the chief of these, and the one that has obtained a world-wide reputation, is the Elizabethen Brunnen, and it is around this source that the gaily-dressed crowd gathers during the early morning hours, from six to eight, when mineral waters are presumed to have the greatest efficacy. The Ludwigs Brunnen has also its admirers, but it has an afternoon popularity rather than a morning one. Then there are the Kaiser Brunnen, the Louisen Brunnen, and the Stahl Brunnen. A few words about each of these and their uses will not be

2 P

uninteresting to those who have to drink either of them. They all contain salt, common salt, but in very different amounts. The Elizabeth spring contains seventy-five grains in a pint, the Ludwig thirty-nine, the Kaiser fifty-five, the Louisen twenty-three, and the Stahl about forty-five, so that the Ludwig is about half and the Louisen about a third of the strength of the Elizabeth spring. Then they all contain iron in varying proportions; the Stahl Brunnen, as its name implies, contains much more than the others—three times as much as the Elizabeth and nearly twice as much as the Louisen, which is the next strongest in iron, while the Ludwig contains scarcely any. Perhaps the next most important ingredient in these springs is carbonic-acid gas. Here, again, there are considerable differences. The Kaiser contains the most—one-third as much again as the Elizabeth or the Louisen, and the Louisen contains the least. The presence of a small amount of sulphuretted hydrogen gas in the Kaiser Brunnen makes it less pleasant and less acceptable as a drinking spring, and the existence of an appreciable quantity of this gas in the Louisen and Stahl Brunnens undoubtedly detracts from their merits as chalybeate springs. With regard to the other constituents, there is wonderful uniformity in all these sources, the only difference being one of quantity. Besides the chloride of sodium (common salt) they contain chlorides of magnesium, calcium, ammonium, potassium, and lithium. The first two of these amount to ten to twelve grains in a pint of the Elizabeth, and are of some importance.

Is it easy to determine what are the special uses of the Homburg springs from a consideration of their composition? It would seem not; there appears to be abundant room for differences of opinion. It is easy to dispose of the iron at once by admitting that it gives a tonic property *pro tanto* to all the springs into which it enters as an ingredient, and that, according to the case or the constitution, its tonic effects may be accelerated or hindered according to the nature of the other constituents with which it is combined. It may also be said of the carbonic-acid that its presence usually (not with all persons, however) aids the digestion of the water, and exercises a beneficial stimulating effect on the stomach and also on the skin when the waters are taken in the form of baths. It must be borne in mind that it is next to impossible to decide altogether *à priori*—i.e. from a mere consideration of the component parts of a mineral spring—what will be its effect in all cases or in any particular case.

In the final appeal experience must be the test, especially for an appreciation of delicate details of applicability. For it has been well observed that mineral waters are very composite remedies, and we cannot regard the combined action of a great number of substances merely as the sum of their separate actions, since they may partly aid and partly hinder each other in their effect upon the organism. In these very springs, for instance, we find the aperient chloride of magnesium counteracted by the astringent carbonate of lime. It is often said by those whose opinions should carry great weight, and on *à priori* grounds, that drinking the Homburg waters is useless in gout. It is an opinion which merits very careful consideration, but if this opinion is met by the universal cry, "Whereas I was lame, now (after drinking at Homburg) I walk," a general consent of this kind, founded on experience, would outweigh all theoretical statements. But if we meet with a number of persons who, on the contrary, say, "Lame I came, and lame I depart," then it may be as well to reconsider whether Homburg is the best place to send gouty patients to indiscriminately. To this question a negative answer would be given by many authorities. If Homburg, it is said, in addition to its Elizabethen Brunnen, had a mild, warm, alkaline spring, like the springs of Neuenahr or Ems, it would be worth all its other springs put together in the treatment of many forms of gout—forms of gout which are unmistakably made worse by drinking these sometimes exciting common salt waters. The best test that a mineral water is doing one good is the very practical one that one daily feels better and stronger. The appetite is better, exercise is less fatiguing, sleep is sounder and more refreshing, and there is a consciousness of returning and increasing energy, both intellectual and physical. He is a bold physician who, in the absence of any of these signs, relentlessly urges the unfortunate patient to persevere in irritating his stomach and his nerves with the promise of some far-off advantage.

The Homburg course is, no doubt, of great value in some cases, especially in certain forms of dyspepsia, where the organs of digestion require vigorous stimulation, and rousing out of their sluggish inactivity. Many such cases have gouty and rheumatic tendencies, and then a combination of drinking and bathing in these salt and carbonic-acid waters is of great service. But it would seem that many rheumatic and gouty cases require a less highly mineralised, less exciting, and more solvent alkaline water. This is the point to be borne in mind, and it must not be overlooked that common salt waters, unless very weak, are irritating to certain constitutions, and quite capable of exciting gastric catarrh, which in others they may cure.

Much advantage no doubt attends the routine of rising early every morning and promenading in cheerful society for an hour or two in the fresh morning air; but some are induced to do this who had better remain in their beds. No inconsiderable benefit is doubtless derived from the simple ingestion of a quantity of water, independently of and sometimes in spite of its mineral contents. The action of the chloride of sodium in these springs is believed to depend on its influence in promoting those changes of the tissues of the body necessary to healthy nutrition. It helps us to get rid of the "old man," the worn-out or half worn-out parts of us that linger about our bodies and clog and impede their mechanism, and interfere with our vital chemistry; and it helps us to build up a "new" and better man. It helps us to get rid of our somewhat ill-defined aches and pains and infirmities, which an artificial existence and a too busy or too careless life induces; and it would do this, perhaps, far more effectually if we did not find here too much of that very artificial and conventional life which brings these evils in its train.

Scrofulous diseases of the glands, of the bones, of the skin—"torpid scrofula," as it is called—are treated advantageously with these salt waters, and especially with the salt baths; although, as we shall see, Kreuznach claims especially the cure of these maladies. Many sufferers from chronic muscular rheumatism and chronic gout find great advantage from combining the use of these Homburg waters with that of the pine-leaf baths which are prepared here. An extract of pine leaves is added to the heated salt spring, and a very grateful aromatic bath is thus produced. Mud baths, such as are given at Marienbad, can also be obtained at Homburg, but they are costly.

A few words as to the sanitary condition of Homburg. Some excitement was caused, not unnaturally, by the circumstance that several cases of typhoid fever occurred here in the season of 1879, and another and a fatal case the next year. It seems, however, that all these cases were distinctly traced to the consumption of water from one particular well in one particular hotel. This well has been sealed up, and strict police regulations have been carried out to prevent the possibility of so serious an accident again occurring. All the houses are compelled to draw their supply of water from that which is brought direct from the distant hills, and every effort is made to keep the drains well flushed and to avoid the possibility of any insanitary conditions again arising. The outbreak of typhoid having been thus traced to its source and fully accounted for, every precaution has since been taken to maintain the sanitary state of the town.

NAUHEIM.

Nauheim, although comparatively little known in England, is one of the most important salt baths in Germany.   Unlike its more popular neighbour, Homburg, it is altogether a "serious cure," and not a resort of pleasure.   The visitors to Nauheim are chiefly, if not exclusively, invalids.   The magnificent Kursaal bears witness of the time when gaming-tables existed there, and when it shared with Homburg the patronage of gamblers.   Now a few invalids only are seen wandering about the spacious apartments and the adjacent shady and pleasant grounds.

Nauheim is situated at the foot of a fine-wooded hill, the Johannisberg, one of the outlying spurs of the Taunus hills, at a distance of about twenty miles from Frankfort, an hour by rail, and twelve or thirteen miles from Homburg, a pleasant drive of two or three hours.   Its springs are sufficiently rich in salt to make its extraction a profitable industry. Of its springs, five in number, three are used exclusively for bathing and two for drinking.   The three bathing springs are hot, and issue from the ground at a temperature varying from 82 to 96 degrees Fahrenheit.   They also contain an abundance of carbonic-acid, and one of them shoots out of the ground with great force and with much bubbling and foaming, and sometimes rises in a jet to a height of about forty-four feet.   This is the Frederick-William Sprudel, a very important spring, on account of the amount of common salt it contains—two hundred and sixty-five grains in a pint—on account also of its warmth (96·4 degrees), and of its richness in carbonic-acid,

48,000 cubic feet of this gas escaping in twenty-four hours. The combination of these three properties— high temperature, richness in common salt, and carbonic-acid—renders it, according to the learned resident physician, Professor Beneke, unique among European mineral springs. The two drinking springs, the Kur Brunnen and the Carls Brunnen, are weaker, though these are considered by Professor Beneke too strong to be drunk pure in most cases, and he has been the first to introduce the wise practice of diluting these waters with another gaseous spring, containing but a small quantity of chloride of sodium and an appreciable amount of iron; and by this means he obtains a water scarcely distinguishable in its taste and its effects and of nearly the same composition as the celebrated Ragoczy water of Kissingen, the chief difference being that the Ragoczy contains four-and-a-half grains of sulphate of magnesia in a pint, which the Kur Brunnen does not contain; while it contains, when diluted, five grains of chloride of calcium in a pint, which does not exist in the Ragoczy. Professor Beneke had often observed that the undiluted salt springs did not agree with the digestive organs, and this led him to the practice of dilution, an innovation which deserves imitation. The bathing springs are used of their natural temperature and composition; very rarely they are made stronger, as at Kreuznach, by the addition of "mother lye."

The duration of these strong salt baths varies from five minutes to an hour, according to the effect desired to be produced. Sometimes they are used in the form of steam baths, fresh brine running into the baths at one end and a similar quantity running out at the other during the whole course of the baths. This kind of bath has to be employed with caution, for by constantly surrounding the body with fresh brine and fresh carbonic-acid, a highly exciting effect is produced on the nerves of the skin. The salt water is also used as douches and for local fomentations. There is also a *salle d'inhalation*, where the brine is driven into spray by suitable apparatus. In some cases it is found necessary to dilute the mineral water before using it in the baths, on account of the irritating effect it produces on the skin ; while in other cases, where it is thought necessary to produce very active stimulation of the skin, "mother lye" is added to make it stronger. The abundance of carbonic-acid gas set free from these springs has led to the use of a gas bath at this spa. The patient sits enveloped up to the neck in an atmosphere of carbonic-acid gas, and in some cases of gout it has proved serviceable.

As Nauheim may be taken to be the type of a strong salt spa—a "sool" bath, as it is called in Germany—it will be interesting to inquire what are its real uses and value, and in what respects it differs from hot and cold sea-water baths. First come the various forms of scrofulous diseases, and of these, scrofulous eczema seems to have been specially benefited, as well as cases of eczema in which there was no observable taint of scrofula. This cure of eczema often occurs as an "after-effect." Professor Beneke writes : "Some of my patients, who left without any improvement, reported to me, about eight or ten weeks afterwards, their perfect recovery."

In scrofulous affections of the glands and of the joints, the nightly application of the salt water or the "mother lye" to the parts affected has been found highly useful, but in many cases of scrofula, especially when associated with much debility, and when there is danger of over-stimulating a feeble and irritable nervous system, Professor Beneke believes sea air and sea baths to be far more applicable. It is in certain forms of gout and rheumatism that these salt springs are said to be specially efficacious. To use rather technical language, "retarded metamorphosis of nitrogenous compounds" is regarded as lying at the root of these complaints. The nitrogenous compounds, entering into the composition of our food, or formed in the processes of nutrition, are checked in their process of metamorphosis; and it is the property of these salt springs, as I have already hinted, to promote and hasten these necessary changes. For such patients simultaneous regulation of the diet is of extreme importance, and Professor Beneke would have the hotels and the *tables d'hôte* at all such places placed under the strictest medical supervision; and from this point of view he ridicules the moue of living pursued at Homburg.

One very interesting and important result has attended the careful scientific observations of this excellent physician at Nauheim : it is that the heart affections resulting from attacks of rheumatic fever, so far from counter-indicating the use of these salt baths, as was formerly supposed, are especially benefited by a careful application of them. He has observed the valvular defects in some instances removed, and in most cases the natural effort at compensation greatly aided; while the pulse, instead of being accelerated, is usually diminished from six to ten beats in the minute.

Paralysis dependent upon undoubted changes in the nervous centres themselves, Professor Beneke does not find benefited by the Nauheim baths; but if the paralysis depends on rheumatic affections of the coverings of the nervous centres, then improvement often follows.

One great advantage of these salt and carbonic-acid baths is that, owing to the combined stimulating effect of the salt and carbonic-acid on the skin, persons can remain much longer in these baths at a lower temperature than in other baths. Sea baths lack the presence of carbonic-acid, and cold sea baths cannot be supported for any length of time. On the other hand, the tonic and alterative effects of sea air present advantages which, in combination with warm sea baths, gives them a value which inland salt baths do not possess.

SODEN

Is another of the common salt spas which we find on the southern slope of the Taunus mountains. It is connected with the Taunus railway by a branch at Höchst, and so is brought within half-an-hour of Frankfort. It is about nine miles from Homburg, and is prettily situate in a valley bounded to the north by wooded hills, which form, as it were, the base of the two highest peaks of the chain, the Alt König and the Feldberg. It lies open to the south, but is protected also by hills of gentle elevation to the east and the west—hills which are covered with rich verdure and with vineyards and fruit-trees. Soden is also celebrated for its roses, so that its pleasant gardens, green meadows, and shade-giving trees make it a most agreeable retreat; for it is much more of a retreat than German watering-places usually are. The houses are, for the most part, scattered about among the trees, and surrounded with gardens. The Kurhaus is in a sort of thickly wooded park, which almost conceals it from sight. From its protected situation the climate of Soden is essentially a mild one, the air is balmy and soft and still, though the close vicinity of the mountain-chain often causes a freshness in the evening air which is grateful and invigorating. But mildness of climate and equability are what characterise Soden, so that it is chiefly resorted to by those who need a soft and soothing air. It is possible there for invalids to spend much time in the open air with advantage, and this they often do, extended in hammocks suspended from the branches of trees. It is found that persons with chronic catarrhal condition of the throat and air-tubes, and cases of consumption that require soothing rather than bracing treatment, do well at Soden.

But it must always be borne in mind that it is a soothing climate. Cases, for instance, which require bracing are not suited to it. Cases of nervous asthma cannot breathe there, while the catarrhal cases do well. Soden has a great number of mineral springs, twenty-three altogether, and, although they vary scarcely at all in the nature of their ingredients, they vary greatly in their quantity and in their temperature. There are springs as strong as well as stronger than those of Homburg and Kissingen. There are others which are as pleasantly mild as seltzer-water. One of these milder springs has been found of great use in cases of chronic bronchial catarrh. Soden possesses a conveniently arranged bath-house, where douches and all other varieties of baths can be given. The judgment pronounced by the celebrated French physician, Trousseau, many years ago, respecting the relative claims of Soden and Homburg are as true now as they were then. "Soden," he says, "would compete with its more fortunate rival if its medicinal virtues alone were considered. It is neither less useful nor less favoured by nature; its waters contain the same ingredients and render exactly the same services as those of Homburg; and those who dread the artificial life of watering-places, and who need calm, must prefer the gentle and peaceful life of Soden to the more tumultuous pleasures of Homburg or Wiesbaden." But persons who need bracing treatment must not go to Soden.

THE CURANSTALT FALKENSTEIN.

Within three miles of Soden are the beautifully situate ruins of Königstein and Falkenstein. The Castle of Königstein is particularly fine—perhaps one of the finest ruins in Europe, especially when approached from the hill which separates the little town of Königstein from the village of Falkenstein. Here the grand and extensive old ruins are seen crowning an elevation to the west of the little town, and surrounded by the forest-clad hills of the Taunus.

In a beautiful situation on the southern aspect of the wooded hill to the north-east of Königstein, and just below the ruins of Burg Falkenstein, commanding a most extensive view in the direction of Frankfort, has been established a health resort of a somewhat novel kind, and of considerable interest and importance. It is the Curanstalt Falkenstein. It is an establishment intended chiefly for the treatment of consumptive patients, but cases of anæmia and of convalescence from acute diseases are also received. The medical director, Dr. Dettweiler, himself a sufferer from consumption, formerly assisted Dr. Brehmer, who was the first to establish a sanatorium of this kind at Görbersdorf, in Silesia. The principles of treatment are very simple, and so far they have proved successful in a fair proportion of cases. Dr. Dettweiler does not consider that the moderate elevation, about 1700 feet above the sea, has much to do with the results obtained, but that carefulness in small things is of chief importance. "Im kleinen gross," "great in little things," is his motto. It is by the careful systematic regulation of everything that belongs to the daily life of the invalid—his food, his exercise, his repose, his occupations ; by restraining his morbid caprices, his over-sanguine tendencies ; by keeping a constant, firm, but kindly supervision over him, that Dr. Dettweiler obtains his good results. A judicious application of hydrotherapy, chiefly in the form of douches, and of electricity, and of all the resources, medical and general, which modern researches have proved to be useful, and all under the constant personal supervision of the resident physician —such is the system upon which the treatment of patients at Falkenstein is founded. Living for many hours each day in the open air is a primary consideration, and hammocks are swung from the branches of trees in the surrounding forest, in which patients can rest for several hours in the open air, and they assemble also in covered terraces with a southern exposure, where they sit in the open air whenever the weather does not forbid it. The situation of the establishment is admirably chosen for this purpose, the Taunus hills behind protecting it from the north, and gentler elevations shielding it from the east, and, in a lesser degree, from the west.

WIESBADEN.

Wiesbaden is one of the oldest as well as one of the most popular spas in Germany. Unlike most of the other spas, it goes on all the year round, having a winter season as well as a summer season. So Wiesbaden has grown to be a considerable town, and presents on that account many advantages to the invalid visitor. It has excellent hotels, and any number of lodging-houses of all kinds and of all prices, many of them handsome villas in pretty gardens. Education is good and not expensive, and the same may be said of its amusements. Wiesbaden is very proud of the excellence and cheapness of its amusements; it has a theatre, which is one of the four or five in Germany which receive a subvention from the State, and it provides in consequence dramatic and operatic entertainments of a high order all the year round. An orchestra stall costs three shillings, and for eighteen shillings a year one has unlimited access to all the resources of the Kursaal—library and reading-room, restaurant, concerts, weekly balls, etc. ; and one may drink any quantity of the Koch Brunnen into the bargain. Wiesbaden does not rely merely on its mineral springs to attract invalids into residing there; it aims at a much wider sphere of usefulness; it aims at providing nearly everything that all classes of invalids may require, so it possesses special resources for all kinds of special maladies. There are excellent establishments for the application of the water-cure, with douches, fine swimming-baths, etc. and for the application also of medical electricity. Milk and whey cures are also provided, cows and goats being fed in a special manner for the purpose; and the grape cure is also introduced in the autumn. This is particularly recommended as an " after-cure," after a course of the waters. In the year 1878 as many as eighteen thousand pounds of grapes were imported for this purpose. In one of the hydropathic establishments there are also compressed air baths. Wiesbaden is also well supplied with special medical skill in the shape of oculists, aurists, dentists, etc., so that it forms a sort of invalids' compendium; while it lays claim to virtues as a winter climate, which, it must be admitted, can scarcely be granted it as a summer residence. Surrounded on all sides by hills, except to the south, to which it lies completely exposed, it has the disadvantage of being very hot in summer during the daytime; the early mornings and the evening are, however, cool, for cold currents of air blow down into the valley from the Taunus mountains after the sun goes down. In many of the hotels, too, in summer the presence of mosquitoes becomes a serious plague.

In winter the climate of Wiesbaden is very healthy, and though it is really colder than in England—that is to say, the average temperature is lower—yet it differs from our cold inasmuch as it is accompanied by a dry, bright, and clear atmosphere. There is far more sunshine, and, from the protection afforded by the surrounding hills, much more stillness in the air, and although the temperature is lower, it is more equable, and the cold is, therefore, not nearly so much felt as in England and in other parts of Germany, and in certain parts of the town the abundance of the subterranean hot springs makes a distinct difference in the temperature of nearly 4 degrees Fahrenheit. Wiesbaden thus possesses many attractions as a winter residence for invalids and their families, not only on account of its climate and the medical resources to be found there, but also because of the agreeable life, the cheapness and excellence of its amusements, and the educational facilities it affords. With regard to the springs at Wiesbaden, their

virtues are widely known.   Their chief characteristic
is their high temperature, the Koch Brunnen having
a temperature of 156 degrees Fahrenheit; and their
chief constituent is, like so many of the neighbouring spas, common salt.   Out of sixty-three grains
of solid ingredients in a pint of the water fifty-two
consist of common salt.   There is nothing characteristic in the other ingredients, which are for the
most part those generally found combined with common salt in other salt springs.   The complaints which
are said to be especially benefited by the Wiesbaden
springs are those, in the first place, which fall under
the category of chronic rheumatism, chronic gout and
neuralgia, old rheumatic and gouty deposits, and
thickenings about the joints, as well as muscular rheumatism, and of neuralgic disorders, sciatica in particular.   If these can be cured at all, Wiesbaden, it is
said, will cure them.   Cases of paralysis due to chronic
inflammation of the coverings of the spinal cord are
benefited by the hot baths; so are diseases of the
bones, and especially those resulting from gunshot
injuries.   Chronic ulcers of the skin and some forms
of skin diseases, such as eczema and scrofulous enlargement of glands, are appropriate to treatment at this
spa as well as at other salt spas.   It is generally
believed that the high temperature of the baths is the
chief special influence which is operative in the Wiesbaden cure, but most of the patients are expected to
drink the water as well as bathe in it, and regarded as
a drinking spring, other ailments must be added to the
list of those already named, chiefly catarrhal conditions of the mucous membrane—e.g. chronic gastric
catarrh, chronic intestinal catarrh (diarrhœa), and
chronic bronchial catarrh; all these are said to be
amenable to cure, or, at any rate, to great amelioration
by drinking the Wiesbaden waters.

The routine of drinking begins at six a.m., and
from that hour till eight the young women at the
Koch Brunnen are busily engaged in supplying the
crowd of applicants for glasses of the steaming hot

beverage, too hot to be drunk at a draught, so it has to be slowly sipped or allowed to cool a little before it can be swallowed. I suppose one must not object to the now universally accepted belief that the water of the Koch Brunnen tastes like chicken-broth. I must, however, protest that the comparison would not have occurred to my unaided faculties. It has been recommended, and apparently with good reason, instead of drinking several glasses of the water at this early hour to take, say a glass and a half then, the same dose between eleven and twelve, and a third at five p.m.

The baths are taken either early in the morning or about an hour after breakfast, and patients are required, according to the case, to remain in the bath from twenty minutes to an hour, and a period of complete repose after the bath is earnestly enforced. The water is much too hot to be used as a bath as it issues from the springs, and it is therefore allowed to cool during the night, either in the baths themselves or in reservoirs connected with the bath-houses. The baths are all given in certain hotels and bath-houses, of which there are thirty altogether, containing eight hundred and fifty baths. These are for the most part in the neighbourhood of the hot springs. It is clear that little strict supervision can on this account be maintained over the bathing arrangements, so that this system lacks in some points what it obviously gains in convenience. It is undoubtedly an immense convenience, especially to those who are crippled by their maladies, to be able to get their baths in the hotels they live in. In some of these bath-houses, where the water is allowed to cool during the night in the baths themselves, and as the partitions between the baths are not complete, but only extend up for a certain height, the air of the bath-house becomes exceedingly hot and stuffy, unwholesome, and necessarily depressing. In one or two of the hotels better arrangements obtain for the ventilation of the bath-houses. The baths are taken in the form of "full

baths" and "douches." A "rain douche" of moderately cold water must indeed be an excellent method
of refreshing the patient after half-an-hour in these
hot baths.

The period of the cure must not be circumscribed
by a hard and fast line. The typical twenty-one
baths must often be extended to perhaps twice that
number. The environs of Wiesbaden are exceedingly
agreeable, and many beautiful walks and drives may
be taken through the forests which cover the surrounding hills.

SCHWALBACH.

It is a pleasant drive of three hours from Wiesbaden to Schwalbach, over gently undulating ground and through pine forests covering the lower hills of the Taunus. It is a most refreshing change from the hot streets of Wiesbaden to the cool fresh air of these forest roads. On first reaching Schwalbach you see nothing but the tail of a long straggling village. This is the older part of Long (Langen) Schwalbach.

After passing through this you reach the new part of Schwalbach, the Curviertel, or modern *quartier,* which has quite a different aspect. This part is built on the slopes of two converging valleys, which meet at the Kursaal, and from this point the older village stretches away like a half-disjointed tail. The situation of the modern portion is certainly very pretty ; many handsome villas crowd the wooded slopes of these northern declivities of the Taunus. Agreeable shady walks ascend through the woods in various directions, and the surrounding country affords a great number of most picturesque carriage excursions of almost any required distance. Although Schwalbach is nearly 1000 feet above the sea, it is much hotter than might be expected in a place of that elevation, owing to its being so much protected by the hills around, so that it becomes very hot in the floor of the valley during the middle of the day in the months of July and August. As the visitors to Schwalbach are for the most part anæmic ladies, the life there is, as may be imagined, very quiet, and devoted chiefly to drinking the waters and bathing in them. Until recently Schwalbach did not possess a Kursaal, but since the year 1881 a handsome new building has been opened, which now affords its visitors all the resources found at neighbouring spas—a good dining-room and restaurant, reading and conversation rooms, billiard and smoking rooms, ball and concert rooms, etc. The hotels are good, and situated close to the springs and the bath-house, but many persons live in lodging-houses, which are numerous and good, and in some cases extremely moderate in price, so that living at Schwalbach is much cheaper than at most German spas. Schwalbach is especially an iron cure. It is *the* iron cure of Germany. It belongs to the class of simple iron springs in which the iron exists as the chief constituent, unassociated with any ingredients which can complicate or interfere with its action. It is one of the strongest and purest iron waters in Europe. It also has the advantage of possessing a very large proportion of free carbonic-acid, which makes it sparkling and pleasant to drink, increases its digestibility, and renders it valuable as a medium for bathing in. There are only two springs that are used for drinking—the Wein Brunnen and the Stahl Brunnen. The Stahl Brunnen contains about half as much again of iron as the Wein Brunnen and rather more carbonic-acid. There are many other springs, the water supplied by which is used for the baths, and others which are not utilised at all. Sir Francis Head, in his " Bubbles from the Brunnen of Nassau," mentions that the Pauline, which is now never used for drinking, was the " fashionable Brunnen " in his time, and adds : " But as the cunning Jews all go to the Stahl Brunnen," by way of getting the most for their money, " I strongly suspect that they have some good reason for this departure from the fashion." The Jews were in the right; nobody drinks at the Pauline now.

The bathing arrangements at Schwalbach are good. The chief benefit of the bath being believed to consist in the action of the carbonic-acid which the water contains on the skin, it is so contrived that as little as possible of this gas shall escape in the conveyance of the water to the bath, while the bath itself is of brass, and is provided with a double bottom. Between the two bottoms is a chamber, into which steam is conveyed for the purpose of heating the water to the required temperature. As the water in the bath becomes heated it gives off myriads of bubbles of carbonic-acid gas, and the contact of this gas with the skin is considered to have a beneficial effect upon the superficial vessels and nerves. There can be no doubt that the gas does exercise a distinct influence on the skin, which becomes red and experiences a diffused tingling sensation.

The course at Schwalbach is considered especially applicable to cases of bloodlessness, arising as a consequence of hæmorrhages, or of any exhausting disease, or in retarded convalescence from acute maladies; also in anæmia, so often associated with disturbances of the nervous system—hysteria, etc.— in these latter cases, the use of the tonic water, the soothing baths, and the calm, but, at the same time, cheerful surroundings of the place should exercise an undoubtedly curative effect.

Schwalbach is usually approached from Eltville, a railway and steamboat station on the Rhine, from which it is distant a pleasant drive of only nine miles, so that it is very accessible.

SCHLANGENBAD.

About midway between Eltville and Schwalbach is the beautifully situated bath of Schlangenbad. It is difficult to imagine a more picturesque spot for a watering-place than Schlangenbad—a winding valley, turning upon itself with a sharp bend, so that one end of the village is brought nearly on a line with the other, surrounded by high, richly-wooded hills, their lower slopes covered with scattered villas surrounded by flower-gardens and bright green lawns—such is Schlangenbad. A pretty, quiet, peaceful retreat. It has a sort of miniature Kursaal, with shady walks and seats around, two or three nice clean hotels, and three bath-houses, one quite modern, where the patients can live in apartments some of which are even elegantly furnished, at a fixed price, which is printed over the door of each room. On the ground floor are the baths ; these are extremely well arranged, and more comfortable, airy, and pleasant than those of any other German spa with which I am acquainted. Those in the modern bath-house are really luxurious. Reclining in one of those luxurious baths, the water, with its delicious softness and pleasant temperature, seems to envelop the whole body with a sort of diffused caress; while, from some peculiar property in the water, it gives a singular lustrous beauty to the skin, which seems to be suddenly endowed with a remarkable softness and brilliancy. It certainly tends to put one upon the best possible terms with oneself, and one can readily understand the calming influence which these baths are said to exert over irritable and disturbed states of the nervous system.

The reputation of the Schlangenbad water as a

cosmetic is widely spread, and large quantities are exported in bottles to Paris and St. Petersburg especially for toilet purposes.

Schlangenbad, 900 feet above the sea, belongs to the group of so-called "indifferent, earthy baths," and thus resembles Gastein, Pfeffers, and Wildbad ; the natural temperature of the water ranges from 81 to 86 degrees Fahrenheit; it is raised in the baths to from 87 to 92 degrees. It is less stimulating than the same kind of water at higher temperatures, as at Teplitz and Gastein, and it is, therefore, more suitable to those sensitive organisations whose nervous systems above all things need a soothing treatment, and its comparatively slight elevation above the sea accords with this indication, for its climate is mild, though fresh and equable. The surrounding woods afford every opportunity for open-air exercise and lounging, and the quiet, yet pleasing life there makes this place the type of cheerful repose. A great authority on bathing-places says of Schlangenbad : "We know of no thermal bathing resort which produces such a calming and at the same time refreshing effect upon the invalid requiring gentle management. It has effected (in some cases) all that we should otherwise have expected from the remote and often rainy Gastein." Delicate ladies who suffer from hysterical and other forms of nervous excitability and exhaustion, and who need repose, are especially suited for this spa. Painful forms of spinal disease, with loss of muscular power, such as are unable to bear the stimulating treatment of the thermal salt baths, are often soothed and benefited by treatment here. It is a little remarkable that there are not more cases of gout in irritable nervous subjects to be found here. Cases that do well at Buxton may fairly be expected to benefit at Schlangenbad, and it has the advantage over Buxton of greater warmth and a complete change of living ; moreover, at Buxton it rains every other day! It is very accessible, being only five miles from the station of Eltville, on the Rhine.

KREUZNACH.

The last of those Rhenish health resorts that I have to notice is Kreuznach, in the valley of the Nahe, about ten miles from Bingen, on the left bank of the Rhine. It enjoys a pre-eminent reputation among salt baths for the treatment of all forms of scrofulous disease. Whatever diseases can be traced to a scrofulous tendency, or can in any way be identified or associated with scrofula, are regarded as suitable for treatment and likely to be ameliorated or cured at Kreuznach. The springs have been termed " bromioduretted," and they contain a certain very small amount of compounds of iodine and bromine ; but those who are most familiar with their use and application place but little reliance on the presence of these compounds, and regard them rather as strong salt springs, which they fortify in a special manner and apply also in a special fashion.

The only spring used for drinking, the Elizabeth spring, arises in Kreuznach itself, quite close to the Kursaal ; but the other springs used for bathing are found at some distance from Kreuznach, especially at Theodorshalle, a mile off, and at Münster-am-Stein, two-and-a-half miles distant. The Elizabeth spring contains ninety-four grains of solid ingredients in a pint, and of these seventy-three grains are chloride of sodium ; of bromide of magnesium there is rather more than a quarter of a grain, and of iodide of magnesium rather more than three-hundredths of a grain. Of chloride of calcium there is rather a large proportion —thirteen grains. This is the only spring used for drinking, and it is drunk first in small quantities, and gradually increased to about a pint daily. It is not sparkling, as it contains no carbonic-acid ; but it is not very unpleasant to drink. It is chiefly, however, to the use of the baths that the physicians at Kreuznach trust for producing the good effects which they claim from the use of their springs ; and these baths are administered in a special manner. Most of the baths at Kreuznach are fortified by the addition of what is called "mother lye." This is prepared at Theodorshalle and Münster-am-Stein, in connection with the salt works at these places. Immense hedges of rough twigs and brambles are built up, and the water from the springs is allowed to flow over these, and by thus exposing a great extent of surface to the air, it becomes concentrated to a certain degree ; it is then collected and boiled in large pans, and after boiling it is kept at a high temperature for several days ; in this process the chloride of sodium for the most part is separated by crystallisation, and the liquor left behind, after further concentration, forms the "mother lye." This differs much in composition from the water of the springs from which it is derived. It is a yellowish brown, oily-looking liquid, containing but a relatively small quantity of common salt ; while the other chlorides, especially the chloride of calcium, are in larger amounts. It also contains an appreciable quantity of bromide of potassium. In one thousand parts there are three hundred and thirty-three of chloride of calcium, nearly seven of bromide of potassium, and fourteen of chloride of lithium. It is usual to add to a single bath about two litres of "mother lye," and two-and-a-half kilos of common salt ; but this quantity of "mother lye" is largely increased in certain cases. In a bath of this kind the patients are retained for a long time, often for an hour, and a long period of repose is also needed after a prolonged bath of this sort. It produces much drowsiness and a feeling of exhaustion, and a considerable rest after it is essential.

Kreuznach is celebrated especially for the treatment of scrofulous, glandular, and other enlargements, certain diseases of women, certain forms of skin disease, and certain forms of gout and rheumatism. The system pursued there is regarded as the typical mode of applying strong salt springs, internally and externally, for the relief of chronic maladies.

It has the advantage of possessing an abundance of springs, many of which belong to the salt works at Carlshalle, Theodorshalle, and Münster-am-Stein ; and the large supply of "mother-lye" and the salts extracted therefrom, always at hand, enables the physicians to use baths and local applications of any degree of strength they may desire and for any length of time. Such local applications prove most efficacious in promoting the absorption of scrofulous and other hypertrophies and deposits.

The immediate surroundings of Kreuznach can scarcely be called picturesque. It is itself a dull uninteresting town ; but there are no doubt many interesting excursions in the neighbourhood, most of them, however, too distant for invalids. It cannot bear comparison for attractiveness with such baths as Homburg or Ems. Münster-am-Stein, two-and-a-half miles higher up the river Nahe, also a bathing station, where the same mode of treatment is carried out as at Kreuznach, is much more picturesquely situated. It lies at the foot of precipitous red porphyry cliffs—one, the Rheingrafenstein, rising to a considerable height almost perpendicularly from the river, forms a grand object ; and another, the Gans, an indented ridge of porphyry, rises some two hundred feet higher, and commands a most extensive view of the valley of the Nahe. On the former there are some fine ruins of a castle which belonged to the Rhenish Counts, built in the eleventh century ; while opposite are the ruins of Ebernburg, once the residence of the celebrated Franz von Sickengen.

I may conclude these sketches of some of the principal watering-places in the Rhine district by a few general considerations more or less applicable to all of them. It should be remembered that all these valleys, in which for the most part the springs I have been speaking of are situated, are generally during the months of July and August intensely hot, and exercise during the middle of the day is almost impossible ; while, on the other hand, the months of May, June, and September have the advantage of being much cooler, while the hotels are not so crowded, and the difficulties attending life at a popular spa are reduced to the minimum. It is much wiser, then, for the English invalid to visit a place like Ems or Kreuznach in May or September than in July or August. There is another point upon which the spa doctors make complaint against our country-people, and I think justly. They say that an English patient will often come to them and say : " I have come here to be cured, and I must be cured in three weeks. I have no more time to spare, and my doctor in England tells me that that is long enough." Now, it should be borne in mind, that English patients who are sent to German spas are commonly the subjects of chronic diseases, maladies of long standing, and the object aimed at is often neither more nor less than to induce a change in the whole constitution, getting one's system to throw off, as it were, an inveterate habit, and it is not to be supposed that this can be accomplished in three or four weeks.

If those who have had a life-long experience of a particular process of treatment tell you that little or no good can be obtained from less than a stay of six weeks, cheerfully accept your fate and stay, or give it up altogether.

Another point that cannot be too strongly insisted upon is that the patient should make some inquiries into the sanitary state of the place he or she proposes to stay at, more especially if it is intended that any of the younger members of their families are to accompany them. Germany is too poor, as one of the physicians I encountered frankly told me, to be able to put all her bathing resorts into perfect sanitary condition ; and if by

chance disease of a contagious nature is introduced
from without, there often exist the precise conditions
most favourable to its spread.   The outbreak of typhoid
at Homburg was traced to one particular hotel, and to
the drinking of water from one particular well there,
and now this town is probably in a better sanitary
state than many of its fellows.   In one of the places I
visited one of the local authorities told me there had
been sixty cases of diphtheria, with ten deaths, in the
schools there during that season.   It is always easy to
send a few plain questions to one of the physicians of
a Continental spa and ask for clear and direct answers
to them.   It should scarcely be necessary, yet I fear
it still is so, to insist upon the impropriety of drinking
the water, especially that put in the bedrooms, in
these as well as in other resorts.   Whatever care
may be taken with regard to the water put on the
table, one can never be sure what water a careless
housemaid will give you to drink in your bedroom.
So many ladies drink nothing but water that it is better
to be provided with the means of boiling water and
then mixing it with a little fruit syrup, or making it
into toast and water, and so escaping the fearful
penalty that occasionally follows the drinking of
impure water.

One must be prepared also to find life at a fashionable German spa somewhat expensive, particularly if
one lives in a large hotel ; but if one feels independent
of hotel society and contemplates a prolonged stay, it
is much less expensive to take an apartment in one of
the furnished houses which abound at these places, and
get one's dinner at the Kursaal or elsewhere.   Many
small comforts can be provided in a lodging-house at
very little expense, which are charged very highly for
at the best hotels.   In all the principal German spas
life is made very easy, although it may be somewhat
monotonous.   At the Kursaal and elsewhere good food
and refreshments can always be obtained at moderate
prices.   There is no lack of amusements of a certain
kind, and those who are content to lounge about in

gorgeously decorated and luxuriously furnished apartments, or in pleasant gardens and parks, will find that
a subscription of a few shillings will obtain for them the
entire range of such luxurious establishments as one
finds, for instance, at Ems, or Wiesbaden, or Homburg. Then it is no small thing to many that good
music is to be heard at most of these spas twice, sometimes three times in the day; and there are few places
where good lessons cannot be obtained, at quite
moderate prices, in music and languages.

CHAPTER XVI.

THE PRINCIPAL BOHEMIAN HEALTH RESORTS.

CARLSBAD, MARIENBAD, FRANZENSBAD, AND TEPLITZ; TO-
GETHER WITH KISSINGEN, IN BAVARIA, BOCKLET, AND
BRÜCKENAU.

KISSINGEN.

I SHALL begin this chapter with a short account of Kissingen, for it is situate intermediate between the group
of Rhenish health resorts I have spoken of in the last
chapter, and those important health resorts in Bohemia
whose names are at the head of this page, and whose
strong claims to our careful consideration I shall next
endeavour to do justice to.

Kissingen is one of the most popular spas in
Europe, and justly so. It is not exactly fashionable,
like Homburg, but it is better than that, it is useful
and health-giving. Not that Kissingen is a dull
place, or without its due share of celebrities, of whom
Prince Bismarck is the chief, but, being about five
hours further from England than Homburg is, London
"society," unless under special medical orders, rarely
gets beyond the latter place. So one's mind at
Kissingen is free from any other absorbing thought
except to cure one's dyspepsia, or one's gout, or one's
rheumatism, or whatever other ailment has brought
one there. Kissingen is 122 miles from Frankfort,
viâ Schweinfurt, and the journey takes about five

hours and a half by fast train. It is situated in a pleasant open valley, through which flows the Franconian Saale. This valley is bounded on each side by picturesque wooded hills, and is 640 feet above the sea-level. It was known as a health resort as long ago as the sixteenth century ; but its great popularity is of recent growth, and the quiet village of former times has of late years developed into a handsome well-built town of 10,000 inhabitants.

It possesses a fine spacious promenade or Kurgarten, between the Kurhaus and the Kursaal, which presents an animated appearance between the hours of 6 and 8 a.m. and 6 and 8 p.m., when the band plays and the Kur-guests take their waters and gently exercise themselves. The two principal springs—the Ragoczy and the Pandur—are on the south side of this promenade ; the Maxbrunnen, the milder spring, much like seltzer, is on the north side.

The "Actien Badhaus," where the celebrated "mud baths" are given, is situated opposite the garden on the right bank of the river; it has two wings, the *left* containing baths for ladies, and the *right* for gentlemen. The new Casino, with restaurant and reading-room, is adjacent to it. There are many agreeable walks and longer excursions to be made amongst the surrounding hills and the adjacent country.

The usual daily life at Kissingen is to drink the waters from 6 to 8 a.m., breakfast in one's own rooms from 8 to 9, after that those who bathe do so ; one o'clock is the dinner hour, the dinner being, as a rule, plain, and strictly governed by medical orders ; after dinner coffee is generally taken in the open air, and the time between this and 6 p.m. is devoted to exercise and amusement. From 6 to 8 the waters are again taken, and there is a general promenade with music in the Kurgarten. Then supper and bed. There is a small theatre for those who require evening distraction.

Kissingen is the type of a moderately strong, cold, common-salt spring, with abundance of free carbonic-acid. Compared with Homburg it contains but little more than half the quantity of common salt. The Ragoczy at Kissingen, the spring usually drunk there, contains 44 grains of common salt to the pint, whereas the Elizabeth spring at Homburg, which is the one usually drunk at that place, contains 75 grains to the pint.

The Pandur spring scarcely differs in composition from the Ragoczy; it contains a little less common salt and a little more free carbonic-acid. The Maxbrunnen is much weaker—it is a very weak gaseous common-salt spring.

The following analysis of the Ragoczy and the Maxbrunnen gives the composition of the stronger and weaker springs. In sixteen ounces:

	Ragoczy.	Maxbrunnen.
Chloride of sodium (common salt) .	44.71	17.520
„ potassium . . .	2.20	1.140
„ lithium	0.15	0.004
„ magnesium . . .	2.33	0.510
Sulphate of magnesium . . .	4.50	1.820
„ lime	2.99	1.060
Carbonate of lime	8.14	4.620
„ protoxide of iron .	0.24	—
Free carbonic-acid	41 cubic inches	

It will here be seen that the Ragoczy contains, besides common salt and carbonic-acid, an appreciable quantity of the aperient chloride and sulphate of magnesia, as well as a moderate amount of the tonic carbonate of iron, so that it may be said to be stimulating, aperient, and tonic.

The Kissingen waters have proved beyond all others especially valuable in certain forms of atonic dyspepsia, in nervous as well as in gouty persons. In chronic gastric catarrh (i.e. an excessive secretion of mucus from the mucous membrane of the stomach) and the digestive troubles it involves, a course of Kissingen waters often proves more effectual than

any other remedy. I have known many instances of
this form of dyspepsia, which have resisted all forms
of dieting and medication in London, recover completely at Kissingen. Much may, of course, be due
to the wholesome life and careful diet which are
enforced during the cure ; but these, though important auxiliaries, would not of themselves effect the
good that is done here. These weaker salt waters are
certainly more useful to and better borne by the class
of nervous, irritable dyspeptics than the stronger ones,
which sometimes only aggravate their sufferings.
The moderate quantity of salt, and the considerable
amount of carbonic-acid contained in the Ragoczy
springs, are sufficiently stimulative to the gastric
mucous membrane to rouse it into greater activity ;
while the appreciable amount of iron in it gives tone
to the debilitated and exhausted constitution, and
the aperient ingredients promote the abdominal circulation, and tend to remove congestions of the liver
and improve the functions of that too often erring
organ. These springs, therefore, act as aperients and
hepatic stimulants.

By their aperient action and their tendency to
increase tissue changes, both destructive and repro-
ductive, they are useful in diminishing corpulency,
and are better borne by feeble persons who have this
object in view, than the stronger springs which are
often prescribed for this purpose.

Gouty, rheumatic, and neuralgic conditions, when
they are obviously associated with digestive troubles,
are suitable cases for treatment at Kissingen, and in
the removal of these conditions the excellent baths
available there afford important aid.

The warm mud baths prove exceedingly soothing
and ameliorating to many cases of chronic muscular
pains, chronic joint pains, and chronic neuralgias,
especially when they are of rheumatic or gouty origin.

What are called "sool-spray baths" are also
procurable here. These are a kind of vapour bath in
which the atmosphere is kept at a temperature of

from 78 to 86 degrees Fahrenheit, saturated with vapour, and in which particles of salt are held suspended by means of mechanical pulverisation of the water. This is an excellent stimulant to the surface, and is not only useful in some forms of rheumatic and gouty pains, but is of value also in those chronic cases of slight bronchial catarrh which are met with in gouty, over-fed persons, somewhat beyond middle age. For these the moderately active aperient waters are also serviceable, as they tend to relieve the "abdominal plethora" from which they also suffer.

Persons affected with malarial cachexia acquired in India, Algiers, or elsewhere, are benefited by the Kissingen course.

About a mile, or a mile and a half from Kissingen to the north, on the Saale, leading to which there are walks on both sides of the river, is a strong salt spring containing large quantities of carbonic-acid ; here there are salt works as at Nauheim, and the strong gaseous salt water is used for baths and douches of various kinds, and especially in the form of a *wave* bath—"a broad radiating sheet of water cuts the water of the bath at the point where it is brought in contact with it, lifting the surface into waves resembling those of the sea." This bath is employed at a lower temperature than most saline baths, and is said to be very refreshing and invigorating ; although it used to be objected to this mode of employment of the strongly gaseous salt spring, that, by the movement of the water, the escape of carbonic-acid is facilitated, and that this gas accumulates to a deleterious extent in the atmosphere around the bather.

There is also an arrangement here for collecting the carbonic-acid gas as it rushes from the spring, and using it in various forms of gas baths in cases that are thought appropriate.

BOCKLET

Is a prettily situated village, with chalybeate springs, about five miles up the valley from Kissingen, where mud baths are also prepared. It has a pleasant, mild climate, and thé accommodation there is simpler and living cheaper than at the popular neighbouring spa. It is completely súrrounded by a ridge of hills, except towards Kissingen, and these afford numerous agreeable walks for the visitors.

The following analysis gives the composition of the chalybeate spring in sixteen ounces :

Bicarbonate of protoxide of iron.	. .	0·67
„ lime.	. . . .	0·54
„ magnesia.	. . .	3·60
Chloride of magnesium	. . .	4·43
„ sodium	. . . .	6·55
Sulphate of sodium	. . . .	2·54
„ magnesium	. . .	3·23
Free carbonic-acid	. . . 39 cubic inches	

It will be seen that it is a strong gaseous chalybeate water, with some aperient constituents, viz., the aperient sulphates and chlorides of magnesium and sodium.

It is admirably adapted for the treatment of cases of anæmia, associated with constipation, where a calm quiet life is desirable, and in such cases it has an advantage over those chalybeate springs in which the iron in them is associated with a considerable quantity of the astringent carbonate of lime.

BRÜCKENAU

Is another of the Bavarian spas in the neighbourhood of Kissingen, from which place it is a four or five hours' drive by diligence. The nearest railway station is Jossa, about eleven miles off, on the line between Elm and Gemünden. Its situation is very picturesque, lying in the valley of the Sinn, which abounds in beautiful scenery, at an elevation of 915 feet above the sea. The village is built on the western declivity of the Rhôn mountains, but the springs, with the Kurhaus and handsome Kursaal, are two miles from the village.

The climate is mild and agreeable, the average summer temperature not exceeding 63 degrees Fahrenheit.

The springs and the Kursaal, and the hotels and houses for the accommodation of the visitors (all of which belong to the King of Bavaria) are situated in beautiful gardens and pleasure grounds, surrounded by hills covered with beech forests; through which there are picturesque walks and excursions of various distances. One of the finest excursions is to the summit of the Kreuzberg (2750 feet), from which a very extensive view of the surrounding country is obtained; this takes about four hours, and a guide is said to be necessary for the last hour and a half.

The living here is cheap ; the prices of the rooms are all fixed and marked on them, and visitors can live here for four or five marks a day. The rooms are, however, very poorly furnished, and the attendance and food leave much to be desired.

The waters are very mild and contain iron only in very small quantity ; but they contain abundance of carbonic-acid, which makes them very pleasant to drink. There are three springs, the chief and strongest of which is the Stahlquelle, but they contain only 3·4 grains of solid constituents to the pint, and of these only 0·09 is bicarbonate of iron. If we remember that the spa water contains 0·37 and the Schwalbach water 0·64 of this salt of iron, we shall see that this Brückenau spring is an exceedingly weak chalybeate—too weak, indeed, for cases where a serious course of iron is required. It has been questioned whether it should be called an iron spring, and the other constituents are of absolutely no importance. These waters are said to be useful as an after-course to Kissingen, and if, after three or four weeks of water-drinking at the latter place, one still has time and inclination for more, this inclination might, doubtless, be harmlessly and agreeably gratified by a week or two amongst the beautiful scenery around Brückenau, and the consumption of a certain amount of its sparkling and refreshing springs.

The medical authorities of the place declare, however, that their waters are tonic and blood-restoring, and useful in many diseases associated with debility; that, combined with warm milk, they are beneficial in pulmonary affections, in chronic bronchitis, in dyspepsia, and in scrofulous diseases of children. The baths, no doubt, are stimulating to the skin, on account of the amount of carbonic-acid they contain, and they are reported to be very useful in some forms of female maladies.

CARLSBAD.

Of all the spas of Europe, Carlsbad may be regarded as, perhaps, the most important, if we have regard only to the activity of its mineral springs and the gravity and seriousness of many of the maladies for which it is prescribed.

It is one of the oldest as well as one of the most frequented of German spas, and a vast concourse of invalids from every part of Europe resort to it yearly.

Situated in the north-western corner of Bohemia, a few miles from the town of Eger, it is rather a long journey from this country, and since it is an especial injunction of the physicians of the place that invalids should not travel there rapidly I may as well mention the following route as perhaps the easiest and best one to follow. From London or Dover to Cologne; if this long journey is taken in one day the invalid should sleep at and start from Dover; and if fatigued should rest a day at Cologne. The next stage is also a long one, from Cologne to Nuremberg, which takes ten-and-a-half hours. At Nuremberg one must stay the night. It is six hours from Nuremberg to Eger,

and about one and three-quarter hours from Eger to
Carlsbad.

Carlsbad is situated in the valley of the Tepl, at
an elevation of 1200 feet above the sea.   This river is
a very small, poor stream, and is said, in hot and dry
seasons, to be little better than a sewer, and, as the
valley is narrow and often stuffy, it is therefore desirable to procure lodgings on the hill, in what is
called the English quarter on the Schlossberg, an
eminence situated just above the Schlossbrunnen.
Here one can live free from smells.   The best of the
lodging-houses on this hill are the " Victoria " and
the " König von England."   There is no *table d'hôte*
at the hotels or lodging-houses, and the meals are
usually taken at the restaurants, which are many.
The food at all these is strictly subject to medical
regulations and supervision, and this is considered to
form an important part of the cure.   The dinner-hour
is from twelve to three, the supper-hour from seven to
eight.   Living is rather expensive here during the
height of the season.   As at most other resorts of the
kind, there is a free concert every evening at the
Kurhaus, a daily theatrical performance, and a dance
once a week.   Of course there is an English church!
As the Jewish race tends fatally to obesity, and as the
Carlsbad waters possess the property of making fat
people thinner, there are always great crowds of fat
Israelites here from Germany, Hungary, and Poland.
The town of Carlsbad is to a great extent built on the
crust of a vast common reservoir of hot mineral water,
the "Sprudel-Kessel." It is built on the lid of the kettle.
The stream of this subterranean cauldron escapes
through artificial apertures made in the rock, to
prevent this natural boiler from bursting! and notwithstanding these artificial vents the water has been
known to force new passages for itself.   It is recorded
as a curious fact that during the Lisbon earthquake
of 1755, the Sprudel ceased to flow for three days!

The narrow valley in which these springs are
found is surrounded by pine-clad slopes, through

which there are paths in all directions, and besides these agreeable, shady promenades in the vicinity of the springs, there are many pleasant excursions, of various distances, into the surrounding country. One of the favourite walks in Carlsbad is the "Alte Weise," where there are some good shops.

The springs at Carlsbad, nineteen or more in number, all contain the same constituents, and differ only in their temperature, which ranges from 48 to 166 degrees Fahrenheit.

The well-known Sprudel-brunnen, situated on the right bank of the river, is the hottest; it rises to a height of about three feet from the ground, and every few minutes will suddenly leap to a height of twenty to twenty-five feet. The Sprudel, with some adjacent springs, is inclosed in a "Colonnade;" and the "Muhl-brunnen Promenade" incloses this and several other springs. The remaining springs are in various parts of the town.

There are six bath-houses capable of giving 230 baths; in them you can procure either mineral water baths, mud baths, vapour or gas baths.

The band plays at the Sprudel and the Muhl-brunnen from 6 to 8 a.m., and drinking begins at a very early hour. In the height of the season, the crowd of patients is so great that they have to wait their turn *en queue* for a quarter of an hour at a time. To avoid this nuisance, it is better to begin drinking a little later than the crowd, say 7.30 or 8 o'clock, and breakfast a little later.

I have said that the composition of all the springs is very nearly identical, and the selection of the spring suitable to particular cases is determined by its temperature and the amount of free carbonic-acid it contains. The Sprudel, being the hottest, contains the least carbonic-acid. The two chief constituents of the Carlsbad springs are the aperient sulphate of soda and the alkaline carbonate of soda, and the next in importance is chloride of sodium (common salt).

The springs in chief repute at Carlsbad are the

Sprudel, the Marktbrunnen, the Mühlbrunnen, and the Schlossbrunnen. The following table gives the analyses and temperatures of three of these:

ANALYSIS OF THE THREE CHIEF SPRINGS AT CARLSBAD.
(In 16 ounces.)

	Sprudel.	Mühlbrunnen.	Schlossbrunnen.
Sulphate of soda . . .	18·21	17·96	17·24 gr.
„ potash . . .	1·26	1·71	1·46
Chloride of sodium . . .	7·91	7·89	7·52
Carbonate of soda . . .	10·45	10·86	9·66
„ lime . . .	2·28	2·02	3·06
„ magnesia . .	0·95	0·26	0·38
„ protoxide of iron	0·02	0·02	0·01
Carbonic-acid (cubic inches) .	11·80	14·80	20·60
Temperature . . .	164·2°F.	125·6°F.	124·7°F.

The Carlsbad cure is indicated in many serious maladies, and great benefit often results from passing one or more seasons there.

It is especially useful in certain derangements to which gouty persons and free livers are prone; in cases of what the German doctors call "abdominal plethora"—cases, that is, of passive engorgement of the liver and of the intestinal vessels, with a consequent tendency to gastric and intestinal catarrh. Such persons often suffer from a severe form of dyspepsia, with much stomach pain and flatulency, occasional vomiting in the morning, and obstinate constipation; or there may be a tendency to frequent incomplete actions of the bowels. Such cases derive much benefit from a carefully directed course of the Carlsbad waters.

The Carlsbad cure is notably adapted to cases of jaundice dependent on the presence of gall-stones, and it is also valuable in cases where there is a tendency to the formation and passage of gall-stones, or to the formation of thick, inspissated bile, even if there be no jaundice.

The Carlsbad course greatly diminishes the frequency and violence of such attacks, even if it does not altogether arrest them. Gall-stones are often passed during the course. Enlargements of the liver, due to passive engorgement, from over-feeding and insufficient exercise, a condition frequently associated with hæmorrhoids, are suitably treated there.

Enlargements of the liver and spleen, induced by exposure to malaria in hot climates, and associated with constipation, find relief at Carlsbad.

Cases of gravel (uric-acid deposits) and renal calculi, if connected with hepatic congestion and constipation and accompanied with catarrhal conditions of the bladder, are benefited by treatment there, but cases of uric-acid deposit and renal calculi pure and simple, without any hepatic or gastric disorder, are better suited to Vichy.

No doubt the gouty condition itself, apart from the particular modes in which it may express itself, is ameliorated by the Carlsbad course.

The Carlsbad waters are great and powerful purifiers of the body, for if bathing be associated with the internal consumption of the waters we submit the organisation to a threefold purifying influence, for while the hot mineral baths stimulate the excretory functions of the skin, the internal use of the waters greatly promotes the discharge of effete substances through the evacuations of the intestinal canal and the kidneys; in this manner the blood and the tissues of the body become purified of retained effete and excrementitious substances.

Carlsbad, on account of this effect of its waters, enjoys the reputation of reducing corpulence, and it does, no doubt, lead to a moderate diminution of fat; but unless a very strict diet be followed after as well as during the course, the fat readily returns.

Carlsbad has a special reputation for the treatment of diabetes, and there is no doubt whatever that it leads to a very great improvement in many cases, and to a temporary or permanent cure in others.

The quantity of water necessary to be drunk will naturally vary with the nature of the malady and the constitution of the patient. Very large quantities of water such as were at one time taken are no longer prescribed. From two to six glasses a day are sufficient for all but quite exceptional cases ; and any excessive aperient action of the waters should at once be taken as an indication for lowering the dose or suspending the course for a day or two.

As much exercise as possible in the open air is usually prescribed by the Carlsbad doctors, and this, together with the strict regulation of the diet there, contribute considerably to the good usually obtained.

The springs at Carlsbad may be resorted to at any period of the year, and it is said the number of winter visitors increases yearly ; the season, however, may be regarded as extending from April to October. Diabetic patients who should take the waters twice a year are advised to visit Carlsbad in April or May and again in October. It is suggested that the more robust and vigorous patients should come in the cooler months of April, May, September, and October, and the feebler sort in June, July, and August.

The length of the course varies from three to six weeks according to the nature of the disease, the age and strength of the patient, and the observed effect of the waters.

It is considered important that those who intend submitting themselves to the Carlsbad cure should adopt a careful and rational diet some time *before*, as well as after the course. It is also suggested that a few bottles of Carlsbad water or a few doses of the Carlsbad salts should be taken, at home, for a week or ten days before setting out for the spa ; and rest, mental and physical, is also recommended during that period.

Although it is the rule at Carlsbad to drink the waters early in the morning, before breakfast and only then, it is quite permissible and even advisable for persons who cannot digest the waters well on an empty stomach, to take a cup of tea, or coffee, or thin cocoa, or beef-tea half-an-hour before they commence drinking the waters. The warmer the spring the more slowly should it be drunk, and an interval of a quarter to half-an-hour should be allowed between each glass. After the last glass of water an hour's walk before breakfast is recommended ; only, of course, where the patient is strong enough.

In those exceptional cases which require unusually large doses of the water, it is best to drink them at separate times of the day, at mid-day and from 4 to 6 p.m., as well as in the early morning.

Patients should clothe themselves warmly for the early dose of water, so that the action of the skin, which the warm drink promotes, should not be checked. The baths are best taken in the forenoon, about an hour and a half or two hours after breakfast. Repose before and after the bath is desirable.

The following is the diet usually recommended during the course :

Breakfast : Coffee, tea, or weak cocoa, and two or three milk or water rolls, not more, to which may be added one or two soft-boiled eggs if required.

Dinner : Soups, one or two light dishes of meat, or fish, or poultry, or game, fresh vegetables in small quantity, mashed potatoes or stewed fruit. Only the lightest kind of puddings, in small quantities, are allowable. One glass of claret, or one glass of good ale.

It is usual at Carlsbad to drink freely of a very excellent cold, gaseous spring in the neighbourhood, which is bottled for table use and for exportation, the Giesshübler-Sauerbrunn.

A cup of coffee, with or without a roll, is allowed in the afternoon.

Supper : Soup, or two soft-boiled eggs, or a small quantity of freshly-roasted meat.

It is by no means unimportant that the patient should, after the Carlsbad cure, continue the *régime* which he has followed there, more or less closely, for a few weeks, and it is often highly advantageous to pass a week or two at some sub-alpine health resort, in Switzerland or the Tyrol, or the Black Forest, where a quiet out-of-door life in pure air can be enjoyed.

MARIENBAD,

Like Carlsbad, is situated a few miles (eighteen and a half) from the town of Eger, in Bohemia, and it is therefore reached in the same way as Carlsbad. It lies, however, in a different direction, being on the line of railway which runs from Eger to Pilsen and Vienna.

Marienbad is a favourite modern spa, having much in common with Carlsbad, but differing, at the same time, in some not unimportant particulars. It is situated in a beautiful, broad, open valley, with pine-clad hills on three sides, and is at an elevation of 1912 feet above the sea. This comparatively high situation gives it a fresh and somewhat bracing climate in summer.

The wooded hills around are intersected in every direction with footpaths, which afford delightful walks.

There are many fine points of view that are easily accessible ; one especially, the Mecséry Temple, about twenty minutes' walk, affords a beautiful view of the basin in which Marienbad lies and the Bohemian mountains in the distance. Another fine point of view, an hour and a half's walk, is the summit of the Podhorn (2750 feet), accessible to carriages. This commands a magnificent and extensive view of the Erzgebirge, Fichtelgebirge, and the Bohemian forests.

Another object of interest in the neighbourhood is the abbey of Tepl, to which the springs belong. It is situated about nine miles to the east of Marienbad, and, besides a fine library, possesses much to interest the visitor.

The lower portion of the valley, where the springs are situated, inclosed in "temples," is laid out in pleasure grounds, and these, with a fine Kursaal, colonnades for shops, a theatre, and excellent lodging-houses, render Marienbad a very agreeable, though quiet, place of residence. It is not so gay a place as Carlsbad or Teplitz, but its pretty situation and quiet life will seem more attractive to many ; and living is more moderate than at some other spas.

The waters resemble those of Carlsbad, only they are *cold*, and they contain more of the aperient sulphate of soda (Glauber's salt) and more carbonic-acid. Some of the springs also contain a notable quantity of iron. The chief drinking springs are the Kreuzbrunnen, the Ferdinandsbrunnen, and the Wald-quelle. The two former are brought by pipes, from about a mile distant, to the Promenaden Platz.

The Marien-quelle and the chalybeate waters of the Ambrosius-brunnen and the Carolinien-brunnen are used chiefly for bathing. There is also a pure acidulated chalybeate spring, the Kron-prinz Rudolf's Quelle, which is a decided iron tonic. The following table gives the analysis of this as well as the two principal saline-aperient springs :

ANALYSIS OF SPRINGS.

	Kreuz-brunnen.	Ferdinands-brunnen.	(*Chalybeate*) Kron-prinz Rudolf's Quelle.
Sulphate of soda . . .	38·04	38·76	0·81
„ potash . .	0·40	0·50	0·17
Chloride of sodium . .	13·06	15·39	0·45
Carbonate of soda . .	9·02	9·89	1·06
„ lime . . .	3·99	4·30	8·57
„ magnesia . .	3·33	4·20	5·14
„ protoxide of iron	0·27	0·47	0·32
Free carbonic-acid (cubic in.)	15	22	16
Temperature	48·2° F.	48·2° F.	48·2° F.

Marienbad is celebrated for its baths as well as its drinking springs; and you can obtain these moor or mud baths, pine-cone baths, gas baths, and ordinary alkaline and chalybeate mineral baths. The waters are taken in the early morning fasting, and again, if necessary, between six and seven in the

2 T

evening; the band plays in the Kreuzbrunn Promenade from 6 to 7.30 a.m., and from 6 to 7 p.m.

The waters are taken in smaller quantities than at Carlsbad as they are stronger (the Kreuzbrunnen and the Ferdinandsbrunnen), and their aperient action more marked. They are adapted to the treatment of the same cases as Carlsbad (except that diabetes is not treated here), but are better suited to those which require a more decidedly aperient effect. Marienbad has acquired a great reputation for the reduction of corpulency and the treatment of cases of plethora and abdominal congestion. The waters of Marienbad, like those of Carlsbad, possess the property of diminishing the amount of fat accumulated in the body, without injuriously affecting the digestion or the processes of blood formation; so that while the fat is disappearing there is no loss of muscle and the general health and nutrition are maintained. This property is due chiefly to the sulphate of soda in these springs; and Marienbad is the chief representative of these cold, gaseous, sulphate of soda waters. The fat, however, soon reforms if great care in diet is not observed *after* as well as during the course.

In cases of corpulency it is often thought desirable to allow some of the gas in the water to escape, lest it should be too exciting, and this is managed by pouring the water from one glass to another before drinking it. The presence of an appreciable amount of iron in these waters may impart to them a certain tonic effect.

Like the Carlsbad waters they are also employed in cases of chronic enlargement of liver and jaundice and in various gouty states.

Moor or *mud* baths form an important part of the treatment at Marienbad, as they also do at

FRANZENSBAD.

Franzensbad is only three miles from Eger, in a fresh and airy situation, but with little of the picturesqueness and beauty of Marienbad. It is 1300 feet above the sea, the climate is healthy, and the bathing arrangements particularly good, but visitors complain of the costliness of living there. It has a special reputation as a ladies' bath, although some doubts have been expressed whether this is anything more than a matter of fashion, or altogether justified by the results obtained.

It will be seen by the following analysis that the springs are very like those of Carlsbad and Marienbad in composition, but much more is made here of the small amount of iron they contain:

ANALYSIS OF SPRINGS.

	Salzquelle.	Wiesen-quelle.	Sprudel.	Franzens-quelle.
Sulphate of soda . .	18·000	25·00	27·0	25·00
Chloride of sodium .	9·000	9·00	8·0	9·00
Bicarbonate of soda .	9·000	6·00	7·0	8·00
„ lime .	2·300	1·60	2·3	2·30
Carbonate of protoxide of iron . . .	0·016	0·37	0·2	0·06
Carbonic acid (cubic in.)	27·000	31·00	39·0	40·00
Temperature . . .	50° F.	51° F.	51° F.	50·3° F.

They are cold and contain a considerable quantity of sulphate of soda and of carbonic-acid, and a small quantity of iron.

It is customary at Franzensbad to pursue a very gentle mode of treatment, giving these strongly aperient waters in small doses, and it is therefore well suited to delicate persons requiring gentle management, to cases of anæmia and hysteria, and to thin, ill-nourished hypochondriacs.

The good effects obtained here are not thought to be referrible directly to the iron in the water, but rather to the small doses of active substances promoting nutritive tissue changes in weak and excitable constitutions. They promote nutrition and cell formation by stimulating digestion and change of substance. They are better, therefore, for weak, thin, hæmorrhoidal subjects than the active treatment pursued at Carlsbad. They are valuable also in those anæmic cases in which the simple chalybeate springs are not well borne on account of their producing constipation. They are useful in various forms of debility in females, and in cases of spinal irritation, and in these cases the co-operation of *mud* or *moor* baths is much insisted upon.

The town stands on a drained peat-bog, and the black peat earth is ten feet thick in places. It is from this the baths are made. It is carefully sifted and treated for years with simple or mineral water, and the mixture thus formed (which is, chemically speaking, a very complex one) is mixed with the mineral water till it forms a black thickish fluid of a specific gravity of 1·2 to 1·3. This is heated by passing steam through it, till it has a temperature of 80 degrees Fahrenheit or more.

These moor baths, although they produce, on account of their moist heat, an effect analogous to that of warm water baths, are thought to be more soothing and less exciting to some persons ; they are suited therefore to the treatment of rheumatic and gouty exudations in those delicate people who cannot support more vigorous thermal treatment, but they are not so well suited to cases of muscular rheumatism. They are suited also to cases of hyperæsthesia associated with paralysis, in hysterical spine, and in tabes dolorosa. They are useful in cases of paralysis with muscular contractions; in such cases it occasionally happens that they allay the irritation and clonic spasm and restore power of locomotion. The more soothing effect of the moor bath has been thought by some to be referrible to the fact, that the temperature throughout the bath is much less uniform than in the water bath.

TEPLITZ.

This is one of the oldest as well as one of the most popular of German baths. It is usually approached from Dresden by Bodenbach and Aussig. It is on a branch of the line between Dresden and Prague, and it takes about four hours and a quarter, inclusive of considerable delays, to reach it from Dresden. It is very celebrated for its hot springs, of which it possesses many, ranging between a temperature of 76 and 120 degrees Fahrenheit.

It is situated in a somewhat open valley, 700 feet above the sea, between the Erzgebirge and the Mittelgebirge, with a somewhat variable climate. The park and gardens of the Schloss, belonging to Prince Clary, afford agreeable and shady walks and lounges, and contain a "Gartensaal," with restaurant, theatre, ball-room, reading-rooms, etc., etc.

The village of Schönan is now continuous with Teplitz, a row of new buildings, chiefly lodging-houses, connecting them, and it is the more fashionable resort of the two. There is a hill, the "Mont de Ligue," between Schönan and Teplitz, commanding a fine view; and to the east of Schönan there is another admirable point of view, the Schlossberg, 1280 feet above the sea, easily ascended in half-an-hour. Teplitz is surrounded by pleasing country, into which many attractive excursions of various lengths may be made.

There are excellent hotels and lodging-houses here, some of the best being in the road which connects Teplitz with Schönan.

The springs are alkalo-saline, and contain but a small amount of solid ingredients, the chief of which is carbonate of soda.

The following is an analysis of the principal source,

the Hauptquelle or Ursprung, which has a temperature of 120 degrees Fahrenheit.

In a pint (sixteen ounces) :

		Grains.
Sulphate of potash	. . . . .	0·098
„ soda	. . . . .	0·200
Carbonate of soda	. . . . .	2·635
Phosphate of soda	. . . . .	0·014
Fluoride of silicon	. . . . .	0·351
Chloride of sodium	. . . . .	0·433
Carbonate of lime	. . . . .	0·330
„ strontia	. . . .	0·027
„ magnesia	. . .	0·088
„ protoxide of iron	. .	0·019
„ protoxide of manganese	.	0·021
Sulphate of alumina	. . . .	0·020
Silica	. . . .	0·443
Crenic-acid	. . . . .	0·034

4·803

There are several bath-houses in Teplitz and
Schönan supplied by water from eleven or more
different springs. In one there are public baths for
the gratuitous use of the poor, and there are also here
military hospitals for invalid soldiers. As many as
4000 baths daily can be given here, and yet in the
height of the season (July and August) it is necessary
to be careful to bespeak one's bath beforehand, as
they are occupied from four in the morning till late in
the evening. Besides the mineral water baths,
Teplitz is also celebrated for its moor or peat baths.

Teplitz is the warmest of the "indifferent" thermal
baths, and they have been accustomed for many
years to give *very hot* baths here of 106 to 110 degrees
Fahrenheit, usually followed by one or two hours'
gentle perspiration in bed. It is now usual to give
the baths at a somewhat lower temperature in many
cases, as the strongly stimulating effect they produced
was found not to be without risk in irritable, nervous
persons requiring gentle treatment, there was also the
danger of weakening the skin and giving rise to
tendencies to take cold.

This active thermal treatment was intended, by

its powerful stimulative action on the skin, to promote the absorption of morbid exudations in persons suffering from chronic gout, rheumatism, and certain forms of paralysis; and it was found to be best suited to persons of sluggish and lymphatic temperaments. The course is also considered useful in chronic diseases of bones and in the after treatment of gun-shot injuries. It is believed to promote nutrition by increasing the activity of tissue changes.

The modified courses recently adopted are better suited to sensitive and delicate persons who require baths of a somewhat lower temperature, and such patients are allowed to take the baths later in the day, as the early morning hours proved exhausting and hurtful to many.

The debility of the skin induced by these very hot baths renders it often necessary to supplement them by a course of sea-bathing. The springs are used almost exclusively for bathing, and the bathing arrangements are very complete and excellent.

CHAPTER XVII.

MINERAL WATERS AT HOME.

IT must be frankly admitted that Great Britain offers but comparatively poor resources for courses of natural baths and mineral waters.

No amount of patriotic advocacy of our own mineral springs will alter the facts that we have no sparkling gaseous chalybeate springs like those of Schwalbach; no hot sulphur springs like those of Aix; no acidulated alkaline springs like those of Vichy or Vals; no gaseous salt waters like those of Homburg and Kissingen; no hot alkaline aperient springs like Carlsbad; no simple gaseous, slightly saline, acidulated springs like those of Selters and Apollinaris; no mud or pine-leaf baths as at Marienbad and Franzensbad; and even the common non-gaseous aperient "bitter" waters we are obliged to import from abroad, as is evidenced by the large consumption of Friedrichshall and Hunyadi, of Pullna and Æsculap waters.

A very short list will comprise all the mineral springs well known or much resorted to in this country :

Bath, Buxton, Cheltenham, Dinsdale, Droitwich, Harrogate, Leamington, Llandrindod, Matlock, Moffat, Tunbridge Wells, Strathpeffer, Woodhall Spa ; and of these, three only, viz., Bath, Buxton, and Harrogate, are really much resorted to.

Bath.—There are some signs that the former

popularity of the Bath waters is likely, in some measure, to be regained.

These springs resemble the indifferent earthy thermal springs which are so numerous on the Continent, the chief solid constituent of which is sulphate of lime, as those of Gastein, Teplitz, and Leuk. There are four hot springs at Bath, varying in temperature from 104 to 120 degrees Fahrenheit, and the supply of water is most abundant. The bathing arrangements are good and complete.

The chief use of these baths is in the cure and alleviation of chronic rheumatism, muscular and articular, and in some forms of lumbago and sciatica, as well as in paralytic affections, when not due to serious changes in the nerve-centres. In the earlier states of gouty arthritis these baths are also of use; and they have been found valuable in some forms of skin disease, as in chronic eczema and lepra—the prolonged soaking of the skin being probably the chief advantage thus gained. Their internal use has been commended in dyspepsia, especially in gouty persons, in gastralgia, and in vesical irritability.

They have been advocated internally and in baths in cases of sterility and in functional uterine disorders. The baths are said to prove soothing and restorative in cases of nerve exhaustion or irritability from over-work or anxiety. As compared with Buxton, these springs have the advantage of being much hotter.

The baths are usually taken on alternate days, for twenty minutes at a time, and the treatment is commonly prolonged from two to three months, a much longer period than it is the custom to spend at Continental spas.

The amount of water recommended to be drunk daily is from a pint to a pint and a half, in divided doses. They probably have much the same cleansing effect as the drinking of ordinary hot water.

One advantage of the cure at Bath is that it can be followed in the winter, when it would be impossible to visit more distant resorts, indeed, some consider that to be the most suitable season.

Bath is said to be from 3 to 5 degrees Fahrenheit warmer than London in winter. The hills around protect it to some extent from northerly and easterly wind. The lower part of the town is certainly felt to be relaxing in summer.

Buxton and its waters have enjoyed a well-merited celebrity for many centuries, and it is now probably the most popular bath in this country. It has a large hospital, the Devonshire Hospital, where the efficacy and applicability of the waters have for many years been studied.

Buxton is a simple thermal spring like Ragatz or Plombières. The temperature of the water is 82 degrees, and its peculiarity is the large amount of nitrogen gas it contains, 63 cubic inches to the pint. The supply of water is very abundant. It is doubtful if it can rightly be said to have any active properties when taken internally, and its efficacy must be solely due to its external action. That it should have been found useful when drunk in certain affections of the stomach and bladder must be regarded as due simply to its solvent and diluent action, such as would follow the drinking of any warm pure water. There is a weak chalybeate spring sometimes drunk there, and another spring especially used for bathing the eyes!

But Buxton is particularly celebrated for the cure or alleviation of chronic gout and rheumatism ; its baths have also proved very useful in cases of hysterical paralysis and joint affections, in some forms of paralysis not dependent on central lesions, in sciatica, and in the removal of inflammatory thickenings of joint.

Massage in the baths no doubt contributes to the cure of many forms of chronic joint affections and neuralgias. The average duration of the bath at Buxton is ten minutes.

Buxton being at an elevation of 1000 feet above the sea, and in the midst of some of the finest scenery of Derbyshire, is a pleasant and cheerful place of resort, and has fine bracing air, which is not without its influence in the successes obtained there. It has the great drawback of being a very rainy place.

Harrogate has long been celebrated for its sulphur springs, of which a great number arise in the neighbourhood. They are all cold, and vary considerably in their strength. They contain sodium sulphide, and free sulphuretted hydrogen, as well as considerable but varying proportions of common salt.

Some of the sulphur springs are very strong and others are very weak, while others are of intermediate strength, as at some of the Pyrenean spas, so that they present a sort of natural graduation adapted to the requirements of various cases. The strongest sulphur springs at Harrogate contain a very large proportion of common salt. Harrogate is also renowned for its chalybeate springs, some of which are strong and some mild. One of the milder chalybeate springs is artificially charged with carbonic-acid, which makes it much more agreeable and refreshing to drink. The stronger springs have an exceedingly unpleasant taste. Taken in large quantities they are purgative, like the common-salt waters of Homburg and Kissingen, and in smaller quantities they are said to be "alterative." Their use as aperients is not attended by any debilitating effect.

They are recommended for the cure of chronic dyspepsias, constipation, congestion of the liver, and conditions of abdominal plethora, in short, in the same cases as the waters of Homburg and Kissingen. They are given internally and in the form of baths in cases of chronic gout and rheumatism. In certain forms of skin disease they lay claim to peculiar efficacy, as in chronic eczema, in psoriasis, lichen, acne, and troublesome syphilitic eruptions.

These waters are useful in the treatment of lead

and mercurial poisoning. The chalybeate springs are administered in cases of anæmia and chlorosis, and are found of much use in functional uterine disorders, when taken in association with the baths.

The arrangements for drinking and bathing at Harrogate are good and complete; and the fine bracing air of the surrounding country no doubt contributes, in great measure, to the good effects obtained from treatment there. Harrogate is in an open, upland country, about 430 feet above the sea.

Cheltenham.—The waters of Cheltenham were at one time largely resorted to, but now they are almost entirely neglected. The town is handsome and well-built, and lies at the foot of the Cotswold range, in Gloucestershire. It enjoys a mild climate, and has a rainfall of about thirty-two inches. It has good houses and fine streets, with rows of trees, parks, and public rooms and drinking-halls for the waters. The springs are numerous. The chief one of them contains, in the pint, sulphate of magnesia, seventeen grains; sulphate of soda, fourteen grains; and fifty-two grains of common salt. "Such waters are very distinctly aperient, and are applicable when action on the alimentary canal is required. They have proved useful in dyspepsia, in hepatic inactivity, and sluggishness of the portal system and of the bowels. This explains how the springs were for a long time considered appropriate in liver affections, especially of those who had been in the tropics." These waters are now little used unless in the form of the Cheltenham salts, extracted from the spring-water. The explanation of this is readily found. Cold waters of this composition are rarely drunk at the spas where they are found. These so-called "bitter waters" are very unpleasant to drink in quantity, and other waters answer the purpose required better. When they are found of considerable strength at a spa, they are usually exported for home use in great quantity; but the Cheltenham waters are not strong enough for this purpose, and so we employ a number of imported waters of much the same composition, only very much stronger.

Dinsdale-on-Tees is a small but rising sulphur spa, which has recently been remodelled and reorganised, and is certainly growing into popularity.

This spring takes its rise near the banks of the Tees, at one of its most picturesque spots, five miles from Darlington, about three miles from Croft, and only a mile from another railway station, possessing the singular name of " Fighting Cocks," on the line from Darlington to Middlesborough. Not far from the spring, and connected with it by a pleasant walk, is the prettily-situated village of Middleton-One-Row, where many of the visitors to the spa reside while they take the waters. The country round is exceedingly beautiful, and the air is fresh and bracing. The spring issues from a rock behind the bath-house, on the margin of the river, and, it is said, yields twelve gallons a minute. It is used for drinking, and in the form of cold and hot baths. The temperature of the water at the source is 52 degrees Fahrenheit, and in order to furnish the warm baths it has to be heated by a furnace. The accommodation has of late been greatly improved and augmented. The spring contains a considerable amount of alkaline and earthy sulphates, some chloride of sodium, and free sulphuretted hydrogen.

The sea-bathing station of Saltburn, one of the most recently developed and most picturesque of the watering-places on the Yorkshire coast, is only about twenty miles from Dinsdale, and Redcar and Coatham are still nearer, so that the baths of Dinsdale are most conveniently situated for combining a course of hot sulphur baths with a course of sea baths, while the advantage of a short stay in a bracing sea-side health resort after a course at a sulphur spa is very generally recognised.

Droitwich.—The strong salt springs of Droitwich, in Worcestershire (three-and-a-half hours from London), have recently been developed in a degree which will doubtless be rewarded by a greatly increased popularity.

The brine of Droitwich appears to be as strong as that of any of the celebrated brine baths of Germany. A large swimming bath has been constructed there containing brine at a temperature of 80 degrees Fahrenheit. These hot brine baths are found very useful in some forms of chronic gout and rheumatism, and in the removal of chronic inflammatory exudations and thickenings around the articulations. Like the baths at Kreuznach they are useful in many forms of scrofulous disease and in chronic uterine affections.

An account of the physiological action of these strong salt springs will be found in the chapter on Rhenish Health Resorts, in the remarks on the salt springs of Nauheim.

Droitwich is a somewhat dull old town, but the surrounding country is interesting. There is a good hotel there (the Royal Brine Baths Hotel), and tolerable lodgings can be had in the town. Droitwich brine baths can also be obtained at the Imperial Hotel, Great Malvern, where the brine is conveyed in tanks by rail.

Leamington, near Warwick, is a cheerful town, with fine streets and parks and an Arboretum, and with good drinking-halls for its waters. It has four saline springs of nearly the same composition. They are cold, and contain a certain amount of carbonic-acid in solution.

Their chief constituents are sulphate of soda and the chlorides of sodium, calcium, and magnesium. They all contain a trace of sulphuretted hydrogen and a small proportion of iron, traces also of iodine and bromine.

A German author observes (Braun) : " We can see no reason why Leamington should not again become a much-frequented and useful watering-place, the more so as in many instances it may be resorted to in early autumn, or even winter, on account of its great mildness of climate compared with Continental spas."

Dr. Eardley Wilmot, of Leamington, is disposed to lay great stress on the presence of a considerable amount of chloride of calcium in the Leamington springs (twenty grains to a pint). He observes: " The action of calcium chloride upon the glandular system, as an absorbent of morbid products and deposits, is well known, and the mineral waters of Leamington have always enjoyed a certain reputation for this action. They are, indeed, in constant use at the present day in this town, and with the best results, for the purpose above indicated. Bronchocele; tabes mesenterica; strumous, lymphatic, or other glandular enlargements, in any situation, external or internal; chronically enlarged tonsils; scrofulous diseases of joints, have all been severally benefited, in many well-noted cases, by the use of the Leamington spa waters."

Certainly the composition of these springs is of sufficient importance to make it desirable to reinvestigate their uses.

Llandrindod is the chief of the mineral springs in Central Wales. It was brought into notice about a hundred years ago, and after a lengthened period of success fell into decay. It has been revived of late years and has become a very considerable place of resort. It stands on an open, elevated, breezy plateau, at a height of about 800 feet. It has not in itself any very great beauty, but there are objects of interest within driving distance. There are very fair hotels, and lodgings, and shops, and there is some talk of building a new hotel on a large scale.

There are two establishments which supply the mineral waters, which contain sulphur and iron and a little chloride of sodium. The chief one has recently had some new baths added to it. They are, in fact, rather weak Harrogate waters; but, just as some of the weaker springs at Harrogate are in certain cases more useful than the stronger, so these waters are often efficacious. What between the fine air and the waters very excellent effects are produced, mainly in dyspepsia, gout, and chronic rheumatism. Patients certainly have arrived here in advanced stages of disease and have received very marked, at least temporary, benefit.

For some years the waters have been aërated and exported.

Matlock.—The springs at Matlock are of a temperature of 68 degrees, and contain no important constituents. Carbonate of lime is the chief solid ingredient in them. Its hydropathic establishment is, however, very well known, and the treatment of chronic rheumatism, of dyspepsia, and some glandular affections has been attended there with considerable success. It is situated in a beautiful part of Derbyshire, amidst very fine scenery, and in a healthy and invigorating climate.

Moffat, in Dumfriesshire, in the upper part of Annandale, about 400 feet above the sea, has a cold sulphur spring, containing sulphuretted hydrogen, with chloride of sodium and some sulphate of soda. It is resorted to chiefly, however, for its good air, and for the hydropathic treatment pursued at the establishment there, which is widely known.

Tunbridge Wells was greatly visited two centuries ago for its chalybeate waters. Its spring contains about the same amount of iron as the Spa water. It would be a waste of time to dwell now on the virtues of the Tunbridge waters. It has been suggested that the waters of the spring should be well aërated in order to make them popular; but the cases which are sent to Schwalbach or to St. Moritz, to drink the chalybeate waters at either of these places, are sent there, not on account of the waters alone, but for the attendant advantage of change of air and scene, and, in many instances, for the baths, rich in carbonic-acid, which they obtain at these places.

Tunbridge Wells is now chiefly resorted to as a healthy, almost suburban residence, as it is only thirty-two miles from London ; and at *Bishop's Down Spa* accommodation is provided for invalid visitors who require a course of baths or douches, or electrical or other treatment.

The surrounding scenery is pleasing and rural ; the absorbent sandstone of which the soil is composed renders the atmosphere comparatively dry and bracing; the water supply is abundant, and the drainage good. *Tunbridge Wells* can therefore be recommended as a healthful and agreeable retreat in summer, most conveniently near the metropolis, and especially suitable to convalescents with whom the sea-side does not agree. It has never, it has seemed to the writer, been able to shake off a certain appearance of solemnity and rather oppressive respectability, not altogether exhilarating to feeble and depressed invalids.

Strathpeffer is at the head of the sulphur wells of Scotland ; it is situated in the remote north of Ross-shire. The climate is good if the season happens not to be rainy. It lies in a picturesque valley, a few miles from Dingwall. It has strong sulphuretted wells—not so powerful as the stronger Harrogate waters, however—a bathing and a drinking establishment. The latter is far the most important. Within the last few years a magnificent hotel has been added to the resources of the place.

Strathpeffer is largely resorted to, chiefly by the natives of the northern counties, and the clerical element is strong here.

The waters are useful in dyspepsia, in chronic rheumatism and gout, and in some cutaneous affections.

An effervescing chalybeate spring has recently been reported to have been discovered at Strathpeffer.

Woodhall Spa, near Horncastle, Lincolnshire, has acquired reputation recently on account of its so-called "Bromo-iodine spring." Woodhall is situated in an elevated part of Lincolnshire, about eighteen miles from the sea-coast. Its subsoil is dry sand, and the climate is said to be dry and invigorating. The country around is wooded and agreeable. It is in a part of England which has the smallest annual rainfall—not much more than twenty inches.

There is good hotel accommodation (the Victoria Hotel) as well as comfortable lodging-houses near the baths.

Dr. Frankland, in his analysis of this water, states that it contains "unusually large proportions of iodine and bromine." In a gallon he found 4·396 grains of bromine, and ·616 grains of iodine. It also contains a small quantity of arsenic.

From the amount of chlorine combinations it contains it has been classed amongst the sool-waters. It has been calculated that in a pint of the Woodhall spring there are 120 grains of chloride of sodium and 21 grains of the chlorides of calcium and magnesium, so that the Woodhall spa is a moderately strong common-salt water with an unusually large proportion of iodine and bromine compounds.

The cases especially suitable for treatment at Woodhall spa are scrofulous glandular tumours, and scrofulous diseases of the bones, joints, and skin, cases of atonic gout and rheumatism, and tertiary forms of syphilis. Some forms of chronic uterine disease ought to derive benefit from treatment here as they do at Kreuznach.

The Woodhall spring is no doubt an important one, and as it becomes more widely known to the profession and the public, it will doubtless attract to itself an increasing amount of popularity.

Of other mineral springs still made use of in this country it may be mentioned that there are aperient waters at *Scarborough* and at *Purton Spa,* near Swindon.

There are numerous iron springs to be found in various parts of Great Britain, but they none of them have achieved anything more than a limited local reputation. There are also a number of small sulphur wells, such as at *Gilsland, Shapfell, Nottington,* near Weymouth, *Askern,* near Doncaster, and *Croft,* near Darlington.

There is a salt spring at *Builth,* near the banks of the Wye ; and also in Scotland, at the *Bridge of Allan,* near Stirling. The latter is visited on account of its protected situation and mild climate.

Lisdunvarna, in County Clare, Ireland, has sulphur and chalybeate springs, and enjoys a considerable local reputation.

INDEX.

Aix la Chapelle :

Aix la Chapelle, a very popular sulphur bath in Rhenish Prussia, not far from Cologne, 534 feet above the sea. The springs are hot (129 degrees Fahrenheit), and contain sulphuretted hydrogen, sodium sulphide and sulphate, a fair amount of common salt, and a small quantity of carbonate of soda. These waters are drunk, and used as baths of thirty to forty-five minutes' duration, and also in the form of vapour baths and douches.

The cases treated here are those of chronic muscular and articular rheumatism, chronic gouty eXudations, some forms of paralysis, of metallic poisoning, and especially all forms of constitutional syphilis.

According to Professor Justus von Liebig, this water

contains, in 10,000 grammes, or 10 litres :

Chloride sodium	..	26·161
Bromide ,,	..	0·036
Iodide ,,	..	0·005
Sulphide ,,	..	0·136
Sulphate soda	..	2·836
,, potash	..	1·527
Carbon. soda	..	6·449
,, lithion	..	0·029
,, magnesia	..	0·506
,, lime	..	1·579
,, strontia	..	0·002
,, iron oXide	..	0·095
Silicic-acid	..	0·662
Organic matter	..	0·769
Grammes	..	40·791

Alexisbad :

Alexisbad, in a valley of the Hartz Mountains, 1352 feet above the sea, has a chalybeate spring resembling in most respects the *Pouhon* at Spa, but with less carbonic-acid. It is resorted to also for its fresh forest air.

Altwasser :

Altwasser, near Liegnitz in Silesia, between Salzbrunn

PAG

Baden :

Baden, near Vienna, is situated in a most picturesque,
beautifully wooded,and mountainous spot, and is a fashionable resort of the Viennese.
Its sulphur springs resemble
greatly those of *Aix la Chapelle,* but are of somewhat
lower temperature; they are
used in the treatment of the
same affections. It is a cheerful and agreeable resort, and
affords abundance of distractions.

Baden :

Baden, near Zurich, in
Switzerland, is generally regarded as a hot sulphur bath
(120 degrees Fahrenheit), but
though smelling of sulphuretted hydrogen there is no
appreciable quantity to be
found on analysis. Its chief
constituents are salts of lime
and also some *barègine.* It
also contains a small amount
of chloride of sodium and free
carbonic - acid and nitrogen
gases.

It may be regarded as an
"indifferent thermal bath,"
and it is employed in those
cases which usually derive
benefit from baths of a high
temperature and of long duration, as at Teplitz, etc. It
has a mild climate at an elevation of 1180 feet above the
sea, and the life there is
simple and economical.

Baden-Baden :

Baden-Baden, in a beautiful
situation on the borders of the
Black Forest, 616 feet above
the sea, is one of the most
complete and magnificent
bathing stations in Europe.
Its springs are hot, their temperature varying from 114
to 154 degrees Fahrenheit.
These waters are not strongly
mineralised; they contain
about sixteen grains of common salt to the pint, and
but a small quantity of free
carbonic-acid. For external
use they must be classified
amongst the "indifferent thermal springs." These weak salt
waters are, however, well
adapted for internal use in
the cases treated here, in con-
junction with the employment
of baths. These cases are
atonic forms of chronic gout
and rheumatism, the slighter

Cannstatt, near Stuttgart, 600 feet above the sea, with a

mild climate, in a beautiful valley, has, besides the resources afforded by its mineral springs, two special establishments, one for the treatment of skin diseases and the other for spinal affections.

Its springs are scarcely warm —64 to 70 degrees Fahrenheit; they contain only a small proportion of chloride of sodium —about 20 grains to the pint— 6 or 7 grains of aperient sulphates, and a certain amount of free carbonic-acid. They are indifferent baths, but with the moderately stimulating effect of carbonic-acid. The presence of this gas promotes the digestion of these waters when drunk. These springs are useful in atonic dyspepsia and chronic gastric catarrh, and in some forms of anæmia and scrofula.

Cudowa, Silesia, close to Bohemian frontier, a few hours from the railway stations of Frankenstein and Josephstadt. High table-land, 1235 feet above the sea, with a weak gaseous alkaline chalybeate spring. Mild climate.

Eilsen :

Eilsen, 273 feet above the sea, in an agreeable wooded valley, near Bückeburg, not far from Hanover in North Germany. It is a little frequented cold sulphur spring; the water contains a considerable quantity of sulphuretted hydrogen gas which is used for inhalations. The chief solid constituent is sulphate of lime; it also contains some sulphate of magnesia and a considerable amount of iron, which is, however, probably converted into an insoluble sulphide within the body. The water is used for drinking and bathing and also for inhalation. There are also "mud" baths and "moor" baths there. The same cases are treated there as at other sulphur springs.

Elster :

Elster, a Saxon bath, 1460 feet above the sea, on the Bohemian frontier, between Plauen and Franzensbad, with a railway station on the line from Altenburg to Eger, and having a mild but fresh climate. Its springs, which are cold, resemble those of Franzensbad in composition, and contain sulphate and carbonate of soda, common salt, and a notable proportion of iron (which latter fact distinguishes them from those of Franzensbad). They also contain free carbonic-acid. The same cases which go to Franzensbad are suitable for treatment at Elster, but its waters must be regarded as more tonic on account of the greater quantity of iron they contain. As at Franzensbad, moor baths form a prominent part of the cure.

Elster has also a spring, the Salzquelle, rich in sulphate and carbonate of soda and common salt, but without iron—in short, a cold alkaline aperient; and it also possesses a simple acidulous table water, with only a small amount of sulphate of soda.

One of these springs has been reported to contain a relatively large proportion of carbonate of lithium.

Fachingen :

Fachingen, in the Lahn Valley, Nassau. Like Vichy and Bilin it is one of the strongest soda springs. It is only exported. It contains 28 grains of bicarbonate of soda in 16 ounces, and in its other constituents, as well as in its effects and uses, it resembles precisely the springs at Bilin.

Flinsberg :

Flinsberg, Silesia, close to the Bohemian frontier; two hours from Griefenburg station; beautifully situated, with a fresh climate, 1530 feet above the sea. Life there simple and cheap, and pure but weak chalybeate springs.

Gastein :

Gastein, or Wildbad-Gastein, 3430 feet above the sea, twenty-one miles from land, a station on the Salzburg and Tyrol railway (diligences three times a day from land in four hours), possesses several indifferent thermal springs, varying from 77 to 120 degrees Fahrenheit in temperature, without taste or smell, and containing an extremely small proportion of mineral ingredients. These springs have been known since the seventh century. They are now very popular, the number of visitors, which are chiefly of the upper classes, reaching 3000 annually. The season lasts from 15th of May to the end of September; the usual course consists of seventeen baths. It is a very rainy place, and there are as many as sixty per cent. of rainy days in June, July, and August. This fact has led to the construction of a long, covered, glass gallery, which serves as a promenade in bad weather. The rain, however, does not usually last long, only a few hours. The life there in the midst of grand mountain scenery, is quiet, sedate, and soothing.

The baths, which resemble those of Wildbad and Teplitz, are given at rather a high temperature, 95 to 99 degrees Fahrenheit, and last, according to the nature of the case, from ten minutes to an hour. Gastein, on account of its elevation, is usually selected, in preference to Teplitz or Wildbad, for persons of feeble and irritable constitutions who might be over-excited by baths of this high temperature at lower elevations, the soothing, and at the same time fortifying Alpine climate enabling them to support this active treatment.

These baths are considered useful for certain forms of chronic rheumatism, and for the absorption of gouty exudations. In cases of paralysis, tabes dolorosa and spinal debility, hysteria and hypochondriasis in weakened and debilitated persons. There is good accommodation in hotels

Lippspringe :

Lippspringe, 442 feet above the sea, in Westphalia, two hours from Paderborn station, is a resort for phthisical patients. Inhalation of an atmosphere rich in nitrogen is a part of the treatment. A considerable quantity of the warm mineral water is also drunk fasting. Caseous deposits are said to be expectorated, and the cavities thus formed to heal there. The spring contains small amounts of sulphate of soda and magnesia, carbonate and sulphate of lime, and chloride of sodium, and 1·4 cubic inches of nitrogen in 16 ounces. The temperature is 70 degrees Fahrenheit. The climate is equable and of high relative humidity, cool and moist west winds prevail. It is a quiet, somewhat dull place, with agreeable promenades.

Luhatschowitz :

Luhatschowitz, in a pleasant valley of the Carpathian Mountains, 1600 feet above the sea—Moravia, Austria—

two-and-a-half miles from the Hradisch station (about one and three-quarter hours from Vienna) of the North Austrian Railway.

Cold gaseous springs containing chloride of sodium and carbonate of soda, combined in proportions especially suitable for the treatment of congestion of the liver and portal system, chronic gastric catarrh, and for gouty exudations and gravel.

There are four springs which contain, in sixteen ounces, from

33 to 61 grains of carbonate of soda,

4 to 5 grains of carbonate of lime,

23 to 33 grains of chloride of sodium,

14 to 50 cubic inches of carbonic-acid,

together with small quantities of chloride of potassium bromide and iodide of sodium and protoxide of iron.

Neuhaus, an indifferent thermal bath in Styria, about ten miles (two hours by diligence) from Cilli, on the railway between Gratz and Trieste. The temperature of the springs is 95 degrees Fahrenheit. They are employed in the same cases as those of Romerbad, and resemble also the baths of Ragatz. It is 1220 feet above the sea, and is very beautifully situated on the spurs of the Sulzbach Alps, with charming walks and excursions in all directions. There is a good Kurhaus and also an establishment for the whey cure.

Nieder-langeman :

Nieder-langenau, near Glatz, in Silesia, has a pure iron spring, containing 0.28 grains of bicarbonate of iron in 16 ounces, with 2½ grains of carbonate of lime and 35 cubic inches of carbonic-acid. It is situated in a beautiful, well-protected valley, more than 1100 feet above the sea, and is well suited to those who require a pure iron course in a quiet retreat. There are mud baths here and a milk cure.

Pyrmont, in Waldeck-Pyrmont, near Hanover, is well-known for its chalybeate springs. It is one of the oldest and most celebrated in Germany. Besides its iron spring, it possesses a common salt spring, which is fortified in the bath with "mutter lye" obtained from the salt works. It also has another weaker salt spring, with much carbonic-acid, which is used for drinking; it resembles the Kissingen springs in containing a small quantity of sulphate of magnesia, but it has less carbonic-acid.

The iron spring is a decidedly strong one—stronger than those drunk at Schwalbach and St. Moritz—but it is not so agreeable to drink, as it contains a small quantity of the bitter sulphate of magnesia. Pyrmont is situated in a deep valley 400 feet above the sea. It has a healthy, mild, and agreeable climate. These waters are applicable to cases of anæmia and chlorosis, and to certain female maladies; also to cases of atonic dyspepsia, and others for which the Kissingen course is prescribed.

park for the same purpose, with a salt-water fountain, 40 feet high.

In the inhalatorium it has been calculated that a cubic metre of air contains from six to forty grammes of salt, according to the distance from the apparatus. The inhalation of this brine-spray is found very useful in promoting expectoration in cases of chronic bronchial catarrh. It is also a resort for cases of chronic stationary phthisis, on account of its moderate sub-Alpine climate. It is, however, sometimes very hot in summer. Like other salt baths, it is suitable to chronic, lymphatic, and scrofulous affections, and cases of asthma and emphysema, as well as catarrhal cases are treated here in the pneumatic chamber.

The *sool* is very strong, containing as much as 23 per cent. of chlorides, and it has to be diluted for the baths, and still more for drinking. A little too much, perhaps, is made of the fact that it contains a small quantity of bromide of magnesium. Altogether, Reichenhall is a very important curative station.

Rippoldsan :

Rippoldsan, in the Black Forest, nearly 2000 feet above the sea, is in repute both for its climate and its mineral springs. The latter contain iron in varying amount ; some are strong, some mild. They also contain considerable amounts of sulphate of soda and carbonic-acid. It is also a station for the whey cure and for hydropathic treatment. It is situated in a beautiful wooded district at

the foot of the Kniebis, and has a mild, but invigorating sub-Alpine climate. The promenades and excursions into the surrounding forest and mountains are numerous and delightful. Pine-leaf, sool, and other baths are given there. Cases of phthisis are sent there to pass the summer months in the open air amidst the pine-woods.

Römerbad :

Römerbad, in Styria, near Cilli, on the railway between Gratz and Trieste. A hot bath and a favourite and picturesque resort, charmingly situated, with a good Kurhaus and fine pleasure-grounds, in a wide valley, 755 feet above the sea. It was known to the Romans.

Its summer climate is pleasant and equable. The baths vary in temperature from 93 to 99 degrees Fahrenheit. It is considered to be suited to the same kind of cases as Schlangenbad—viz., hysteria and chronic uterine maladies, and also, to some extent, to the same cases as Gastein— viz., chronic gouty exudations, rheumatism, and certain forms of paralysis. It is an *indifferent thermal* spring.

Rosenheim :

Rosenheim, 1466 feet above the sea, situated in the broad valley of the Inn, at the junction of Munich-Salzburg and Innsbruck railways; is a pleasant little town in the neighbourhood of fine Alpine scenery, and possesses sool baths (receiving the concentrated brine from the salt works at Reichenhall) a sulphur spring, and a pure gaseous iron spring.

Salzbrunn :

Salzbrunn, in Silesia, about one-and-a-half hour from Frei-burg station, has been known since the beginning of the seventeenth century. It has been called, from the composi-tion of its spring, the *cold Ems;* but it differs from Ems in having scarcely any chloride of sodium and more carbonate of soda and carbonic-acid. The following gives the con-stituents in 16 ounces:

		Grains.
Bicarbonate of soda	..	18·0
„	lime ..	3·6
„	magnesia	3·8
„	protoxide of iron	0·002
Chloride of sodium	..	1·3
Sulphate of soda	..	4·7
Carbonic-acid	cubic ins.	38

The accommodation is good, and it is much visited, espe-cially by cases of bronchial catarrh and some forms of phthisis, as well as some nervous cases. It is agree-ably situated in a wooded valley 1220 feet above the sea. It is a well-known station for the whey cure; it also has moor baths.

Salzungen, in the Duchy of Meiningen, near Eisenade, on the Werra railway, a salt bath in a very picturesque situa-tion, 778 feet above the sea. It affords excellent accommo-dation and pleasant walks in mountain and forest air.

Its salt springs are abun-dant, and contain from 28 to 30 per cent. of chlorides. With these the *mother-lye*, contain-ing 30 per cent., is mixed in the baths, which are usually given *strong*. The spring for drinking contains 91 grains of chloride of sodium and 10 grains of chloride of calcium in 16 ounces, and is abundantly charged with carbonic-acid. It possesses a graduation building, which is roofed in and surrounded by glass walls for the inhalation of air im-pregnated with saline vapour.

Cases of scrofula and early phthisis are sent here.

Schandau :

Schandau possesses a pure but weak chalybeate spring. It is finely situated in Sazon, Switzerland, and is a favourite resort in summer on account of its fresh and agreeable climate. It is on the railway between Dresden and Prague.

Schingnach :

Schingnach, a well-known Swiss sulphur-bath, close to Brugg in Canton Aarau, about 1000 feet above the sea. The temperature of the spring is 95 degrees Fahrenheit, and it contains a considerable amount of free sulphuretted hydrogen. The temperature of the water is usually raised for the baths, and they are generally ordered of long duration—one to two hours. These springs also contain lime salts and a small amount of chloride of sodium. The "iodine water" of Wildegg, three miles distant, is often taken here with benefit.

Spa :

Spa, the well-known Belgian resort, is close to the German frontier, and is agreeably situated in a valley of the Ardennes at an elevation of more than 1000 feet above the sea. It has, like Schwalbach, very pure iron springs, of which the Pouhon is the

PAGE

best known, but they contain less iron and much less carbonic - acid than those of Schwalbach. There are good arrangements for all kinds of baths. Spa is adapted to the treatment of all those cases usually sent to iron springs, such as anæmia, chlorosis, general loss of tone, and certain female maladies. It has the advantage of affording many pleasant walks and drives in a mild and agreeable climate.

Teinach :

Teinach, a small village in a sheltered position in a beautiful valley in the Black Forest, near Caleo (railway station), Wurtemberg. It is only a few miles from Wildbad. It is situated 1225 feet above the sea, and has a weak acidulated alkaline spring and chalybeate one ; there is

PAG

also an establishment there for hydropathy and for electrical and other treatment. There is comfortable accommodation in the Kurhaus, which is crowded in July and August. This is a quiet, cool retreat in summer, and its mild mountain climate renders it an agreeable and useful summer residence for feeble, nervous patients and for cases of chronic phthisis.

Traunstein :

Traunstein, in Bavaria, 1929 feet above the sea, is in the Salzburg Alps, on the railway between Munich and Salzburg, and about seventy miles from the former. It has salt works and good arrangements for sool baths. The brine is brought in pipes from Reichenhall, twenty-five miles distant. The town contains 4500 inhabitants ; it is situated on a slope above the Traun, and affords agreeable summer quarters.

CHARLES DICKENS AND EVANS, CRYSTAL PALACE PRESS.